Handbook Digital Farming

Jörg Dörr · Matthias Nachtmann
Editors

Handbook Digital Farming

Digital Transformation for Sustainable Agriculture

Editors
Jörg Dörr ⓘ
Technical University of Kaiserslautern
Chair of "Digital Farming" and Fraunhofer
Institute for Experimental Software
Engineering (IESE), Extended Institute
Management
Kaiserslautern, Germany

Matthias Nachtmann
BASF SE, Agricultural Solutions
Sustainable Business Model Developer,
Limburgerhof, Germany and Chairman
Förderverein Digital Farming e.V.
(Friends of Digital Farming)
Kaiserslautern, Germany

ISBN 978-3-662-64377-8   ISBN 978-3-662-64378-5 (eBook)
https://doi.org/10.1007/978-3-662-64378-5

© The Editor(s) (if applicable) and The Author(s), under exclusive license to Springer-Verlag GmbH, DE, part of Springer Nature 2022
This work is subject to copyright. All rights are solely and exclusively licensed by the Publisher, whether the whole or part of the material is concerned, specifically the rights of translation, reprinting, reuse of illustrations, recitation, broadcasting, reproduction on microfilms or in any other physical way, and transmission or information storage and retrieval, electronic adaptation, computer software, or by similar or dissimilar methodology now known or hereafter developed.
The use of general descriptive names, registered names, trademarks, service marks, etc. in this publication does not imply, even in the absence of a specific statement, that such names are exempt from the relevant protective laws and regulations and therefore free for general use.
The publisher, the authors and the editors are safe to assume that the advice and information in this book are believed to be true and accurate at the date of publication. Neither the publisher nor the authors or the editors give a warranty, expressed or implied, with respect to the material contained herein or for any errors or omissions that may have been made. The publisher remains neutral with regard to jurisdictional claims in published maps and institutional affiliations.

Responsible Editor: Markus Braun
This Springer imprint is published by the registered company Springer-Verlag GmbH, DE part of Springer Nature.
The registered company address is: Heidelberger Platz 3, 14197 Berlin, Germany
(www.springernature.com/mycopy)

Preface

The digital transformation of the economy and society has gained pace over the last years. Digitalisation is regarded as "enabler" to facilitate our life, for example, by making processes more efficient and/or more customised. However, digitalisation does not only bring benefits, but also bears risks, such as digital divides between regions or population groups. In 2019, the European Commission has set "A Europe Fit for the Digital Age" as a headline ambition for its political agenda and adopted a Communication on the "Digital Decade" in 2021 proposing a compass for a way forward.[1] Proactive steering and capacity building are essential to use the potential of digitalisation effectively and manage the transformative processes responsibly.

The agricultural sector is no exception. Digital technologies more and more shape the day-to day-work of farmers. There have been waves in the digitalisation in agriculture, driven by e.g. the increasing availability of data, noteworthy by GPS[2] data and freely available satellite imagery, as well as by certain technological advancement, for instance in the development of apps. While 20 years ago, the main focus was on precision farming, the scope of digitalisation in the sector has quickly extended to smart farming and the digitalisation of supply chains. While in the early stages, digital technologies were predominantly associated with key decision-making support, nowadays, as the portfolio of applications has increased rapidly, farmers often need decision aid to take the right choice of smart farming tools.

In agriculture, digital technologies can support all types of farming—e.g. conventional, organic, large- and small-scale farming. Digitalisation allows to efficiently and effectively link competitiveness and sustainability ambitions, which are often regarded as trade-offs. Digital technologies can serve as "enabler" to maintain productivity, and food security, while increasing environmental benefits.

Against this background, digital technologies in agriculture have received much political attention globally. Their development and uptake have been supported through numerous commitments and programmes. For instance, the Communique of the Global Forum for Food and Agriculture (GFFA) in 2019[3] highlights the importance of digital innovation in supporting farmers, while acknowledging the projected increase in the world's population; the agricultural ministers of 74 nations called for smart solutions in increasing agricultural production in full commitment to the goals of the 2030 Agenda for Sustainable Development and the Paris Agreement on Climate Change. Subsequently, the Food and Agriculture Organisation of the United Nations (FAO) has launched the process to implement an International Digital Council for Food and Agriculture[4]. In the European

---

1 Communication from the Commission to the European Parliament, the Council, the European Economic and Social Committee and the Committee of the Regions "2030 Digital Compass: the European way for the Digital Decade", see https://ec.europa.eu/info/sites/info/files/communication-digital-compass-2030_en.pdf.
2 "GPS" stands for "Global Positioning System".
3 https://www.gffa-berlin.de/wp-content/uploads/2015/10/GFFA-2019-Communique.pdf.
4 http://www.fao.org/3/ca7485en/ca7485en.pdf.

Union digitalisation will form an inherent part of the so-called cross-cutting objective of the Common Agricultural Policy (CAP) post-2022, complementing the CAP general goals and specific objectives and acknowledging its enabling potential.[5]

Fascinating and performing innovative digital solutions for agriculture are available on the market. However, to achieve wide outreach among farmers, barriers to the uptake of digital technologies have to be addressed. Therefore, for the digital transformation of the agricultural sector, more is needed than innovative digital tools. While digital technologies may serve as "enabler", an environment enabling farmers to take up and effectively deploy them is indispensable. A prerequisite is broadband; beyond that, for instance digital skills, and/or advice to farmers are essential. Capacities to invest and to cooperate may determine the level of advancements in using digital technologies and generating data.

Data and data technologies are important determinants for the effectiveness of digital technologies. For instance, freely available satellite or meteorological data can enable many farmers to benefit from digitalisation. More cost-intensive data collected with sensors allows higher levels of precision in digital applications and thus more tailored production. The data pool considered in the analytics to enhance production can be increased with reference data from other farms, and other regions. Hence, data sharing is a vehicle to boost the digital transformation. Data sharing is also required for most farmers to get their data analysed and interpreted. Agricultural data is of value not only for famers, but also for actors in the up- and down-streaming sectors and for public good purposes, such as policy monitoring. At the same time, agricultural data may form business data, including secrets, such as the keys to successful production. Next to trust, clarity about the rights of the individual actors involved in data sharing, about data ownership and data interoperability is essential for the digital transformation of the sector.

Data sharing in agriculture and interoperability are only two fields linked to the digital transformation of the sector, where systemic action appears to be advantageous. Here, the question for policy-makers arises, whether to intervene or not, and whether to take horizontal or sector-specific action, and at which level, e.g. national or EU, considering that agri-food supply chains have often a global dimension. Next to possible (self-)coordinating or regulatory actions, other types of interventions can be instrumental to boost digitalisation in the sector. For instance, support through the provision of data, data infrastructure and digital infrastructure, as well as furthering innovation, the provision of advisory services and training in digital skills, and tailored investment measures are potentially relevant fields of interventions. To be able to take such policy decisions, it is important to foster a strategic approach and to take stock of the state of play of digital

---

[5] The "cross-cutting objective of modernising the sector by fostering and sharing of knowledge, innovation and digitalisation in agriculture and rural areas, and encouraging their uptake" is set out in Article 5 of the Regulation (EU) 2021/2115. The general objectives set for the CAP post-2022 are (a) to foster a smart, competitive, resilient and diversified agricultural sector ensuring long-term food security; (b) to support and strengthen environmental protection, including biodiversity, and climate action and to contribute to achieving the environmental and climate-related objectives of the Union, including its commitments under the Paris Agreement; (c) to strengthen the socio-economic fabric of rural areas. ▶ https://eur-lex.europa.eu/legal-content/EN/TXT/?uri=CELEX:32021R2115

technologies in agriculture in a certain region or country, seeing it in an international context.

The Handbook

Why a handbook? One may raise the question, whether a book is the right format to discuss digital technologies in agriculture in the digital age, a time characterised by the trend towards shorter and more frequent and numerous pieces of information. A time, where "innovation" in technologies may be regarded "common practice" after a very short period of time, and might better be captured in fast-track scientific papers. Without doubt, in one year from now, there will be technological advancements in the field of digital agriculture.

Yet, the primary objective of a handbook is not to present the latest novelties, but to take stock of the state of play as well as of past and future trends to provide the reader a comprehensive overview of a complex subject at the intersection of multiple disciplines. This handbook sheds light on the situation of digitalisation in agriculture in Central, Eastern and Western Europe. It will guide the digital transformation of the sector in the decade to come. A period, in which, innovation ecosystems, business models, or data sharing regimes evolve and benefit from documented and shared experiences.

This book is a valuable help for public authorities and policy-makers in Europe, when developing digitalisation strategies and/or designing interventions for the agricultural sector; and also actors from other parts of the world will be able to transfer the lessons learnt presented in this book.

This handbook is important also for farmers and their advisors to take informed decisions on the digital future of farms. This book will facilitate capacity building for taking decisions, which digital tools to deploy and for performing human oversight of digital technologies. Moreover, farmers have to be in a position to judge how far digital technologies take over tasks on the farm, and in which way they best continue to follow their passion as a farmer. Such a personal decision may also be influenced by the fact that agriculture is not only a business, but also a culture, and is linked to deep-rooted traditions in many regions across Europe going along with expectations at community and societal levels.

At the same time, this book may provide insights for the interested public, and reveal that an increased use of technologies in farming does not necessarily go along with intensification and increased impact on the environment, explaining the opportunities and challenges to achieve sustainability ambitions through digital technologies.

Scientists will gain from this handbook to set their work in the field of digitalisation in agriculture into context, and get inspired in the development of comprehensive study designs, as this handbook closes an important gap in the literature, complementing a portfolio of scientific papers and reports.

I am grateful to the authors for joining forces and bringing this handbook to life. All of them are widely acknowledged experts in their fields, including e.g. law, agricultural economics, as well as digital and machinery technologies. The tailored compilation and interplay of their contributions in this handbook adds value to authors' works, and results in outstanding nonfiction.

Personally, I may opt for the printed, rather than for the e-edition of this handbook: this book is thought-provoking, and invites to scribble, and I am convinced that I will revisit it several times over the next decade to reflect on the progress in the digital transformation of the agricultural sector.

Doris Marquardt
Directorate-General for Agriculture and Rural
Development of the European Commission
(The information and views set out in this preface are those of the author and do not necessarily reflect the official opinion of the European Commission.)

Acknowledgements

We warmly thank the friends of Digital Farming (Förderverein Digital Farming e. V.) and all experts who helped to shape and produce this book. Starting at Agritechnica 2019, this was an exciting journey and we thank the authors for their commitment and openness. It was a pleasure working with you on this book. We especially thank Felix Möhrle, Julian Schill, Sonnhild Namingha, and our lecturer Mr. Markus Braun from Springer for their support throughout the whole process. Finally, we thank our families, who missed us on several Saturday and Sunday mornings, and also some late evenings. We thank them for their patience and support so we could follow our passion for Digital Farming.

Contents

1	**Introduction**	1

Jörg Dörr, Matthias Nachtmann, Christian Linke, John Crawford, Knut Ehlers, Frederike Balzer, Markus Gandorfer, Andreas Gabriel, Johanna Pfeiffer, Olivia Spykman, Beat Vinzent, Mathias Olbrisch and Ines Härtel

1.1	**Motivation and Overview**	3
1.1.1	Motivation	3
1.1.2	Rationale for This Handbook and Scope	5
1.1.3	Overview on Contents	6
1.1.4	Information for Reading This Book	8
1.2	**Today's Farming Practice—Challenges and Options**	9
1.2.1	Challenges	9
1.2.2	Options	12
1.2.3	Conclusions and Outlook	16
1.3	**Sustainability Systems Perspective**	17
1.3.1	UN Sustainability Goals and Dimensions	18
1.4	**Agriculture and the Environment: Where Are We Headed? A German Case Study**	22
1.4.1	Introduction	23
1.4.2	Agriculturally Relevant Environmental Targets	24
1.4.3	Conclusion and Outlook	29
1.5	**Adoption and Acceptance of Digital Farming Technologies in Germany**	30
1.5.1	Introduction	30
1.5.2	Adoption and Acceptance	30
1.5.3	Discussion and Conclusions	33
1.6	**Agricultural Digital Policy**	34
1.6.1	Enabler and Discursive Reference Point for the Digital Transformation	34
1.6.2	Global Policy Level: Transnational Multi-Stakeholder Governance	35
1.6.3	EU Policy Level: Supranational Impetus for the Digital Transformation of Agriculture	36
1.6.4	German Policy Level: Digitally Transformed Agriculture as a Cross-Cutting Objective Across Ministries	38
1.6.5	The Future: Federal Diversity and Multi-Level Integration	39
1.7	**Agricultural Digital and Data Law**	40
1.7.1	Contexts of Agricultural Digital Law	40
1.7.2	The Normative Framework: Right to Food and SDGs	40
1.7.3	Agricultural Data Sovereignty and Agricultural Data Space	42
1.7.4	Self-Regulation—Code of Conduct as Private Soft Law	43
1.7.5	Regulated Self-Regulation: Non-Personal Data	45
1.7.6	Privacy According to the EU General Data Protection Regulation	46
1.7.7	Artificial Intelligence: Security Law and Liability Law	48
1.7.8	Further Fields of Law	50
1.7.9	Outlook	51
	References	51

2	**Framework for the Digital Transformation of the Agricultural Ecosystem**...	59
	Carsten Gerhardt, Stefanie Bröring, Otto Strecker,	
	Michael Wustmans, Débora Moretti, Peter Breunig, Leo Pichon,	
	Gordon Müller-Seitz and Borris Förster	
2.1	**From Farm to Fork and Back: History and Roadmap of Digital Farming**............	61
2.1.1	Introduction..	61
2.1.2	View on the Agriculture Industry Overall................................	62
2.1.3	The Roots of Digital Farming...	64
2.1.4	Implications for the Industry: From Product to Service..................	67
2.1.5	Digital will Change the Face of Farming................................	68
2.1.6	Outlook: From Farm to Fork and Back....................................	70
2.2	**Beyond Digitalization: Major Trends Impacting the AgFood System of the Future**...	71
2.2.1	Introduction..	71
2.2.2	Innovation is Multi-Systemic—Main Disruptions from Farm to Fork.........	72
2.2.3	Focus: Digital Disruption and Its Implications for Involved Agribusiness Companies...	75
2.2.4	Concluding Questions..	78
2.3	**Economic Benefit Quantification**......................................	79
2.3.1	Introduction..	79
2.3.2	Fundamentals of Economic Value Creation................................	79
2.3.3	Cost Structure Fundamentals of Digital Solutions and Economic Benefit...	82
2.3.4	Limitations of Economic Benefit Quantification for Decision Making......	84
2.3.5	Example for Economic Benefit Quantification............................	85
2.3.6	Summary and Outlook...	86
2.4	**Successfully Disseminating Digital Tools for Farmers: A French Perspective**.......	86
2.4.1	Introduction..	86
2.4.2	Material and Method...	87
2.4.3	Results...	88
2.4.4	Discussion..	90
2.4.5	Conclusion..	91
2.5	**Business Model Innovation and Business Model Canvas**.................	91
2.5.1	Statement of the Problem..	92
2.5.2	Business Model Innovation as a Distinct Form of Innovation.............	93
2.5.3	Business Model Innovation in Practice—The Business Model Canvas........	93
2.5.4	Reflections on the (Mis-)Use of the Business Model Canvas and Conclusions........	97
2.6	**Accelerators & Partnerships: Anticipating the Unknown is Hard: An Experience Report**...	97
2.6.1	Introduction..	98
2.6.2	Understanding and Managing Direct and Indirect Impact on Success—Borrowing from Physics and Finance....................................	99
2.6.3	The Current State of Technology in the Food Value Chain................	102
2.6.4	Looking Ahead: From Stand-Alone Programs to Multi-Corporate Innovation Platforms...	104
	References..	106

3	**Technology Perspective**	109
	Thomas Herlitzius, Patrick Noack, Jan Späth, Roland Barth,	
	Sjaak Wolfert, Ansgar Bernardi, Ralph Traphöner, Daniel Martini,	
	Martin Kunisch, Matthias Trapp, Roland Kubiak, Djamal Guerniche,	
	Daniel Eberz-Eder, Julius Weimper and Katrin Jakob	
3.1	**Efficient Systems Engineering for Automation and Autonomous Machines**	111
3.1.1	Introduction	111
3.1.2	Characteristics of Agricultural Machinery Development	112
3.1.3	Mechanization and Automation as Drivers of Productivity on the Farm	115
3.1.4	Paradigm Shift from "Bigger, Faster, Wider" to Sustainability, Robotics, and Autonomy	116
3.1.5	Challenges of Autonomous Systems	118
3.1.6	Summary and Outlook	120
3.2	**Precision Farming**	120
3.2.1	Introduction and History	121
3.2.2	Technology	122
3.2.3	Applications	125
3.3	**Safe Object Detection**	127
3.3.1	Introduction	127
3.3.2	V-Model	129
3.3.3	Challenges in Safe Surround Sensing	129
3.3.4	Solutions for Safe Object Detection	132
3.3.5	Conclusion and Outlook	137
3.4	**Interoperability and Ecosystems**	137
3.4.1	Introduction	138
3.4.2	A Reference Architecture for Integrated Open Platforms and Key Components for Interoperability	139
3.4.3	A Lean, Multi-Actor Approach for Ecosystem Development	142
3.4.4	Conclusions and Future Development	144
3.5	**Artificial Intelligence**	145
3.5.1	AI in the Agriculture Context	145
3.5.2	Capturing the Environment	146
3.5.3	Data Exchange and Shared Understanding	146
3.5.4	Interpretation, Analysis, and Decision Support	147
3.5.5	Getting Smarter: Machine Learning	148
3.5.6	Artificially Intelligent Robots	151
3.5.7	Economics of AI	152
3.5.8	Outlook: Individualized Optimization	153
3.6	**Agricultural Data and Terminologies**	153
3.6.1	The Data Landscape in the Agricultural Sector	154
3.6.2	A Global Data Space—Achievable or Wishful Thinking?	157
3.6.3	Controlled Vocabularies, Thesauri, and Ontologies	159
3.6.4	Conclusion and Outlook	162

3.7	**The Role of Geo-Based Data and Farm-Specific Integration: Usage of a Resilient Infrastructure in Rhineland-Palatinate, Germany**	162
3.7.1	Introduction	162
3.7.2	Overview	163
3.7.3	Site-Specific Resilient and Climatic Smart Farming	164
3.7.4	The GBI as an Example for Infrastructural Resilience	166
3.8	**Technology Outlook**	170
3.8.1	Introduction	170
3.8.2	Key Enablers for Digital Agriculture	170
3.8.3	Autonomous Machinery for Tomorrow's Agriculture	175
3.8.4	Digital Twin: The Farmer as a Factory Manager	178
3.8.5	Democratization of Agriculture Through Digitalization	179
	References	180
4	**Agronomy Perspective**	**191**
	Jörg Migende, Johannes Sonnen, Sebastian Schauff, Julian Schill, Alexa Mayer-Bosse, Theo Leeb, Josef Stangl, Volker Stöcklin, Stefan Kiefer, Gottfried Pessl, Sebastian Blank, Ignatz Wendling, Sebastian Terlunen, Heike Zeller, Martin Herchenbach, Fabio Ziemßen and Wolf C. Goertz	
4.1	**The Development of Agricultural Distributors into Solution Providers: Who is Helping Farms to Successfully Apply Smart Farming?**	195
4.1.1	Status of the Use of Digital Technologies	195
4.1.2	Reasons for the Limited Use of Smart Farming Technologies by Farmers	195
4.1.3	Common Features of these Obstacles	196
4.1.4	The Term "Solution"	196
4.1.5	Overcoming the Obstacles	197
4.1.6	Who Can Help to Overcome the Obstacles?	198
4.1.7	From Product Seller to Solution Provider	199
4.1.8	Summary and Outlook	200
4.2	**Cross-Manufacturer Data Exchange Interoperability as a Basis for Efficient Data Management in Agriculture**	200
4.2.1	Introduction	200
4.2.2	Technology Development—History, Important Actors and Current Projects	202
4.2.3	Market Development—Current Status	207
4.2.4	Conclusion and Outlook	209
4.3	**E-Commerce and Logistics**	210
4.3.1	Types of Digital Distribution Channels	210
4.3.2	Differentiation Strategy of ag.supply	212
4.3.3	Requirements for the Online Trade of Agricultural Input Goods	212
4.3.4	Effects of Digital Distribution Channels on Agricultural Contribution Margin Accounting	213
4.3.5	Sustainability Effects of Digital Distribution Channels	214
4.3.6	Outlook	214
4.4	**The Digital Eco-System of Sustainable Farming: Agricultural Insurance as a Glue**	214
4.4.1	Introduction	214
4.4.2	When Agricultural Insurance is Parameterized, Their Digital Loss Assessments Result in Immediate Pay-Outs	215

4.4.3	Typical Insurance Covers Along the Agricultural Supply Chain	216
4.4.4	Insurance as a Glue in a Digital Ecosystem of Sustainable Farming	217
4.4.5	Digitally Enabled Risk Analysis and Mitigation—Farming Risk Profiles	218
4.4.6	Enabling More Sustainable Production: Risk Profiles as Ally	218
4.4.7	Conclusion	219
4.5	**Soil and Seed Management**	220
4.5.1	Introduction	220
4.5.2	Methods of Soil Management	220
4.5.3	Increasing the Performance of Soil	221
4.5.4	The Role of Crop Rotation	222
4.5.5	The Impact of Climate Change	222
4.5.6	Seed Management	223
4.5.7	Outlook	225
4.6	**Nutrient Supply: From Whole Fields to Individual Plants**	226
4.6.1	Introduction to Metering and Spreading	226
4.6.2	Determination of Machine Settings	227
4.6.3	GPS-Based Automation Systems	228
4.6.4	Field-Zone-Specific Nutrient Identification and Supply	229
4.6.5	Pneumatic Fertilizer Spreaders	230
4.6.6	Organic Fertilization	231
4.6.7	Conclusion and Outlook	232
4.7	**Crop Protection: Diverse Solutions Ensure Maximum Efficiency**	233
4.7.1	More Than 50 Years of Modern Plant Protection Technology	233
4.7.2	How Did the Plant Protection Technology Commonly Used Today Come About?	233
4.7.3	Which Progress can be Expected in the Next 5 Years?	235
4.7.4	What are the Long-Term Prospects 2025–2030?	239
4.7.5	Conclusion: Future Crop Protection will be More Specific and Diverse	240
4.8	**Weather and Irrigation**	241
4.8.1	Introduction	241
4.8.2	Current Status and Tool Kits Availability for Mitigation of Risks for Farmers	242
4.8.3	Key Components and Technologies Presently Available	243
4.8.4	Future Development	245
4.9	**Harvest Sensing and Sensor Data Management**	246
4.9.1	Introduction	246
4.9.2	Vehicle-Based Sensing Systems	247
4.9.3	Remote Sensing Support	250
4.9.4	Data Quality Management (Post Correction)	250
4.9.5	Automation	251
4.9.6	Outlook: Ubiquitous Sensing and the Autonomy Challenge	252
4.10	**Direct Agricultural Marketing and the Importance of Software: It's not Possible Without Digitalization! Which Software Solutions Help to Digitalize the Administrative Work Steps of Direct Agricultural Marketing?**	253
4.10.1	Direct Agricultural Marketing—Blessing or Curse?	253
4.10.2	The Evolution of Direct Agricultural Marketing	254
4.10.3	Current Software Solutions for the Digitalization of Administrative Work Steps in Direct Agricultural Marketing	256
4.10.4	Future Developments of Direct Agricultural Marketing	257

4.11	**Challenges and Success Factors on the Way to Digital Agricultural Direct Marketing: "We Do Not Need a Homepage"**	258
4.11.1	Introduction	259
4.11.2	Status Quo of Direct Marketing	259
4.11.3	Challenges of Digital Direct Marketing	260
4.11.4	Success Factors of Digital Direct Marketing	260
4.11.5	COVID-19 and the Digitalization	262
4.11.6	Conclusion	263
4.12	**Digitalization in the Food Industry**	263
4.12.1	How Digitalization is Changing Food Sales	263
4.12.2	More Transparency: Close to the Customer and Yet Far Away	264
4.12.3	"Direct to Consumer": Not New, but Different	264
4.12.4	E-Food Startups: New Models, Processes, and Infrastructure	265
4.12.5	New Distribution Channels: "Tiny Stores, Dark Stores, Ghost Stores"	265
4.12.6	Agile Approach: Learning from the Start-ups and Joining in	266
4.12.7	The Digital Path to "B2B2C"	266
4.13	**Artificial Intelligence and Sustainable Crop Planning: Better Planning and Less Waste Through Digital Optimization**	267
4.13.1	Artificial Intelligence—Mystery or Helpful Tool	267
4.13.2	Why Don't You just Go Sustainable?	268
4.13.3	Thinking Vegetable Production and AI Together	268
4.13.4	Conclusion	271
	References	271
5	**Farming System Perspective**	277
	Tom Green, Emmanuelle Gourdain, Géraldine Hirschy, Mehdi Sine, Martin Geyer, Norbert Laun, Manuela Zude-Sasse, Dominik Durner, Christian Koch, Noura Rhemouga, Julian Schill, Christian Bitter and Jan Reinier de Jong	
5.1	Arrival of Digital Ag at Scale: The Farming Perspective	280
5.1.1	Background and Context	280
5.1.2	What's New?	281
5.1.3	How Come?	281
5.1.4	So What? The Prize	282
5.1.5	Who Cares?	284
5.1.6	Where's the Catch?	285
5.1.7	Outlook	286
5.2	**The Digital Revolution, a Performance Accelerator from a French Perspective: The Issues and a Panorama of Possibilities for French Cereal Crops**	286
5.2.1	Introduction	286
5.2.2	Observe	288
5.2.3	Record	289
5.2.4	Analyze and Decide	291
5.2.5	Act	292
5.2.6	Impact of Digital Agriculture on the Multi-Performance of Farms	293
5.2.7	Conclusion	294
5.3	**Digital Transformation of Vegetable Production**	295
5.3.1	Seedling Cultivation and Planting	296
5.3.2	Fertilization	297

5.3.3	Plant Protection and Irrigation	297
5.3.4	Climate	298
5.3.5	Weed Control	298
5.3.6	Harvest	298
5.3.7	Processing	300
5.3.8	Outlook	302
5.4	**Digital Transformation of Fruit Production**	302
5.4.1	Challenges in the Supply Chain of Fresh Fruit	302
5.4.2	Irrigation	302
5.4.3	Crop Load Management	304
5.4.4	Fruit Quality Post-Harvest	306
5.4.5	Conclusions	306
5.5	**Digital Transformation in the Wine Business**	307
5.5.1	Innovation Versus Tradition	307
5.5.2	The Wine Value Chain	308
5.5.3	The Potential of the Vineyard	308
5.5.4	Wine as the Role Model for an Authentic and Sustainable Agricultural Product	309
5.5.5	From the Wine Value Chain to an Operational Network	310
5.5.6	Challenges in Communication	310
5.5.7	Digital Transformation in the Wine Business	311
5.5.8	Conclusion	315
5.6	**Transparency of Animal Welfare Through Digitalization: A Dairy Farming Example of "Hofgut Neumühle"**	315
5.6.1	Dairy Farming in Germany	316
5.6.2	Legal Basis	318
5.6.3	Automation in Dairy Farming	318
5.6.4	Outlook	323
5.7	**German Farmers Perspective on Digitalization**	323
5.7.1	Cross Farm Comparison	323
5.7.2	Field-Specific Open Data	327
5.7.3	Conclusion	329
5.8	**A Farm Case Study from the Netherlands**	329
5.8.1	The Arable Farm of the Family De Jong	330
5.8.2	Precision Agriculture	332
5.8.3	Future	333
	References	334
6	**Sustainability Perspective**	341
	Jörg Dörr, Keith A. Wheeler, Markus Frank, Friedhelm Taube, Klaus Erdle and Isabel Roth	
6.1	**Role of Multi-Stakeholder Organizations in Digital Sustainable Agriculture Transformation**	343
6.1.1	Introduction	343
6.1.2	Discussion	345
6.1.3	Conclusion	348
6.2	**Digitalization Towards a More Sustainable AgFood Value Chain**	350
6.2.1	Setting the Stage: Sustainable and Resilient Agriculture	350
6.2.2	Sustainability Assessment and Management Systems	351

6.2.3	Case Studies for Effective Interoperability of FMIS and Sustainability Assessment	352
6.2.4	Value Creation Through New Business Models	355
6.2.5	Bringing Sustainability to Life—The Role of Digitalization	357
6.3	**From Ecological Intensification to Hybrid Agriculture—The Future Domain of Digital Farming**	**357**
6.3.1	Introduction	357
6.3.2	Why Should the Production Narrative be Questioned?	358
6.3.3	Why is Ecological Intensification the New Paradigm?	359
6.3.4	Is an Ecological Intensification in Western Europe Justifiable in View of the Global Hunger Problem?	360
6.3.5	Animal Sourced Foods—Less and Better?	361
6.3.6	Back to the Roots: Integrated Crop—Livestock Systems Including Grass-Clover Leys?	362
6.3.7	Implementation of Ecological Intensification Through Hybrid Agriculture and Public Goods Boni?	362
6.3.8	Conclusion	364
6.4	**Digitization as Co-Designer for Cropping Systems: Technology Shapes Cropping Systems, Now and in Future**	**365**
6.4.1	Mechanization and Its Excesses on Farming	365
6.4.2	Precision and Efficiency, the Evil Twins	365
6.4.3	Back to Reality	367
6.4.4	Digitization for Initiating Promising Cropping Concepts	367
6.4.5	Conclusion	370
6.5	**Fighting Climate Change Through "Carbon Farming": A Future Business Opportunity for Digital Farming?**	**371**
6.5.1	The Carbon Cycle and Its Interaction with the Nitrogen Cycle	371
6.5.2	Carbon Sequestration in Agricultural Soils—A Synopsis	372
6.5.3	Digitization as Key Enabler of Carbon Trading in Agriculture	377
6.5.4	Conclusion	377
	References	**378**
7	**Summary**	**385**
	Jörg Dörr and Matthias Nachtmann	
7.1	**Key Insights from the Experts Views**	386
7.2	**Remaining Challenges and Vision for the Future of the Digital Agricultural Ecosystem**	391
	Supplementary Information	
	Glossary	398

Editors & Contributors

About the Editors

Jörg Dörr Technical University of Kaiserslautern, Chair of "Digital Farming" and Fraunhofer Institute for Experimental Software Engineering (IESE), Extended Institute Management, Kaiserslautern, Germany

Matthias Nachtmann BASF SE, Agricultural Solutions, Sustainable Business Model Developer, Limburgerhof, Germany and Chairman Förderverein Digital Farming e.V. (Friends of Digital Farming), Kaiserslautern, Germany

Contributors

Frederike Balzer German Federal Environment Agency, Dessau-Roßlau, Germany

Roland Barth ITK Engineering GmbH, Rülzheim, Germany

Ansgar Bernardi Deutsches Forschungszentrum für Künstliche Intelligenz GmbH (DFKI), Kaiserslautern, Germany

Christian Bitter BASF Digital Farming GmbH, Münster, Germany

Sebastian Blank European Technology Innovation Center (ETIC), John Deere GmbH & Co. KG, Kaiserslautern, Germany

Peter Breunig Weihenstephan-Triesdorf University of Applied Sciences, Freising, Germany

Stefanie Bröring Institute for Food and Resource Economics, Chair for Technology, Innovation Management and Entrepreneurship, University of Bonn, Bonn, Germany

John Crawford University of Glasgow, Glasgow, UK

Jan Reinier de Jong Farm Owner in the Netherlands, Odoorn, Netherlands

Jörg Dörr Technical University of Kaiserslautern, Chair of "Digital Farming" and Fraunhofer Institute for Experimental Software Engineering (IESE), Extended Institute Management, Kaiserslautern, Germany

Dominik Durner Weincampus Neustadt, Neustadt an der Weinstraße, Germany

Daniel Eberz-Eder DLR Rheinhessen- Nahe- Hunsrück, Bad Kreuznach, Germany

Knut Ehlers German Federal Environment Agency, Dessau-Roßlau, Germany

Klaus Erdle Office for Rural Areas, Hochtaunuskreis, The District Committee, Bad Homburg, Germany

Markus Frank Nürtingen-Geislingen University, Institute of Applied Agricultural Research, Nürtingen, Germany

Borris Förster Amathaon Capital GmbH, Munich, Germany

Andreas Gabriel Institute for Agricultural Engineering and Animal Husbandry, Bavarian State Research Center for Agriculture, Freising, Germany

Markus Gandorfer Institute for Agricultural Engineering and Animal Husbandry, Bavarian State Research Center for Agriculture, Freising, Germany

Carsten Gerhardt Kearney and Circular Valley Foundation, Düsseldorf, Germany

Martin Geyer Leibniz-Institut für Agrartechnik und Bioökonomie (ATB) Potsdam, Potsdam, Germany

Wolf C. Goertz Netrocks GmbH, Osnabrück, Germany

Emmanuelle Gourdain ARVALIS - Institut du végétal, Boigneville, France

Tom Green Spearhead International, Cambridge, UK

Djamal Guerniche RLP AgroScience, Neustadt an der Weinstraße, Germany

Martin Herchenbach Industrieverband Agrar e.V., Frankfurt am Main, Germany

Thomas Herlitzius Institute of Natural Materials Technology (INT), Chair of Agricultural Systems and Technology, Technische Universität Dresden, Dresden, Germany

Géraldine Hirschy Naïo Technologies, Escalquens, France

Ines Härtel Justice of the Federal Constitutional Court, European University Viadrina Frankfurt (Oder), Research Center for Digital Law, Frankfurt (Oder), Germany

Katrin Jakob California Business Associates, Alameda, USA

Editors & Contributors

Stefan Kiefer AMAZONEN-WERKE H. DREYER SE & Co. KG, Hasbergen, Germany

Christian Koch Hofgut Neumühle, Neumühle, Germany

Roland Kubiak RLP AgroScience, Neustadt an der Weinstraße, Germany

Martin Kunisch Kuratorium für Technik und Bauwesen in der Landwirtschaft e. V. (KTBL), Darmstadt, Germany

Norbert Laun Dienstleistungszentrum Ländlicher Raum Rheinpfalz (DLR), Neustadt, Germany

Theo Leeb HORSCH Maschinen GmbH, Schwandorf, Germany

Christian Linke Ingenieurbüro Dr. Linke, Preetz, Germany

Daniel Martini Kuratorium für Technik und Bauwesen in der Landwirtschaft e. V. (KTBL), Darmstadt, Germany

Alexa Mayer-Bosse Munich Re AG, Munich, Germany

Jörg Migende BayWa AG, Munich, Germany

Débora Moretti Institute for Food and Resource Economics, Chair for Technology, Innovation Management and Entrepreneurship, University of Bonn, Bonn, Germany

Gordon Müller-Seitz Technische Universität Kaiserslautern, Kaiserslautern, Germany

Matthias Nachtmann BASF SE, Agricultural Solutions, Sustainable Business Model Developer, Limburgerhof, Germany and Chairman Förderverein Digital Farming e.V. (Friends of Digital Farming), Kaiserslautern, Germany

Patrick Noack Chair of Agricultural System Technology, Hochschule Weihenstephan-Triesdorf, Freising, Germany

Mathias Olbrisch Research Center for Digital Law, European University Viadrina Frankfurt (Oder), Frankfurt (Oder), Germany

Gottfried Pessl Pessl Instruments GmbH, Weiz, Austria

Johanna Pfeiffer Institute for Agricultural Engineering and Animal Husbandry, Bavarian State Research Center for Agriculture, Freising, Germany

Leo Pichon Institut Agro, Montpellier, France

Noura Rhemouga Hochwald Foods GmbH, Thalfang, Germany

Isabel Roth Nürtingen-Geislingen University, Institute of Applied Agricultural Research, Nürtingen, Germany

Sebastian Schauff ag.supply GmbH, Münster, Germany

Julian Schill MHP Management- und IT-Beratung GmbH, Ludwigsburg, Munich, Germany

Mehdi Sine ACTA, Région de Paris, France

Johannes Sonnen DKE-Data GmbH & Co. KG, Osnabrück, Germany

Olivia Spykman Institute for Agricultural Engineering and Animal Husbandry, Bavarian State Research Center for Agriculture, Freising, Germany

Jan Späth ITK Engineering GmbH, Rülzheim, Germany

Josef Stangl HORSCH Maschinen GmbH, Schwandorf, Germany

Otto Strecker AFC Consulting Group, Bonn, Germany

Volker Stöcklin RAUCH Landmaschinenfabrik GmbH, Sinzheim, Germany

Friedhelm Taube University of Kiel, Grass and Forage Science / Organic Agriculture (GFO), Institute of Crop Science and Plant Breeding, Kiel, Germany

Sebastian Terlunen FlexFleet Solutions GmbH, Münster, Germany

Ralph Traphöner Empolis Information Management GmbH, Kaiserslautern, Germany

Matthias Trapp RLP AgroScience, Neustadt an der Weinstraße, Germany

Beat Vinzent Institute for Agricultural Engineering and Animal Husbandry, Bavarian State Research Center for Agriculture, Freising, Germany

Julius Weimper Universität Trier, Trier, Germany

Ignatz Wendling Meisenheim Am Glan, Germany

Keith A. Wheeler Wheeler Consultancy, State College, PA, USA

Sjaak Wolfert Wageningen Economic Research, Wageningen University and Research, Wageningen, The Netherlands

Michael Wustmans Institute for Food and Resource Economics, Chair for Technology, Innovation Management and Entrepreneurship, University of Bonn, Bonn, Germany

Heike Zeller aHEU – regionale Vermarktungsstrategien, Munich, Germany

Fabio Ziemßen Zintinus GmbH, Düsseldorf, Germany

Manuela Zude-Sasse Leibniz-Institut für Agrartechnik und Bioökonomie (ATB), Potsdam, Germany

Introduction

Jörg Dörr, Matthias Nachtmann, Christian Linke, John Crawford, Knut Ehlers, Frederike Balzer, Markus Gandorfer, Andreas Gabriel, Johanna Pfeiffer, Olivia Spykman, Beat Vinzent, Mathias Olbrisch and Ines Härtel

Contents

1.1	**Motivation and Overview – 3**	
1.1.1	Motivation – 3	
1.1.2	Rationale for This Handbook and Scope – 5	
1.1.3	Overview on Contents – 6	
1.1.4	Information for Reading This Book – 8	
1.2	**Today's Farming Practice—Challenges and Options – 9**	
1.2.1	Challenges – 9	
1.2.2	Options – 12	
1.2.3	Conclusions and Outlook – 16	
1.3	**Sustainability Systems Perspective – 17**	
1.3.1	UN Sustainability Goals and Dimensions – 18	
1.4	**Agriculture and the Environment: Where Are We Headed? A German Case Study – 22**	
1.4.1	Introduction – 23	
1.4.2	Agriculturally Relevant Environmental Targets – 24	
1.4.3	Conclusion and Outlook – 29	
1.5	**Adoption and Acceptance of Digital Farming Technologies in Germany – 30**	
1.5.1	Introduction – 30	
1.5.2	Adoption and Acceptance – 30	
1.5.3	Discussion and Conclusions – 33

© The Author(s), under exclusive license to Springer-Verlag GmbH, DE, part of Springer Nature 2022
J. Dörr and M. Nachtmann (eds.), *Handbook Digital Farming*,
https://doi.org/10.1007/978-3-662-64378-5_1

1.6	**Agricultural Digital Policy – 34**
1.6.1	Enabler and Discursive Reference Point for the Digital Transformation – 34
1.6.2	Global Policy Level: Transnational Multi-Stakeholder Governance – 35
1.6.3	EU Policy Level: Supranational Impetus for the Digital Transformation of Agriculture – 36
1.6.4	German Policy Level: Digitally Transformed Agriculture as a Cross-Cutting Objective Across Ministries – 38
1.6.5	The Future: Federal Diversity and Multi-Level Integration – 39
1.7	**Agricultural Digital and Data Law – 40**
1.7.1	Contexts of Agricultural Digital Law – 40
1.7.2	The Normative Framework: Right to Food and SDGs – 40
1.7.3	Agricultural Data Sovereignty and Agricultural Data Space – 42
1.7.4	Self-Regulation—Code of Conduct as Private Soft Law – 43
1.7.5	Regulated Self-Regulation: Non-Personal Data – 45
1.7.6	Privacy According to the EU General Data Protection Regulation – 46
1.7.7	Artificial Intelligence: Security Law and Liability Law – 48
1.7.8	Further Fields of Law – 50
1.7.9	Outlook – 51
	References – 51

Introduction

1.1 Motivation and Overview

1.1.1 Motivation

The starting point of Digital Farming as we see it today was Monsanto's acquisition of Precision Planting in 2012 followed by acquisition of The Climate Corporation in 2013. Michael Porters article on "How smart, connected products are transforming competition," published in 2014, lifted data-based strategies to the center of attention of leading ag and tech industries, start-ups, scientists and strategic farmers. Also, policy makers on state, member state and EU level consider Digital Farming solutions as key enabler for a more profitable, sustainable, diverse and resilient farming.

Despite the general understanding that Digital Farming is the way to go, today's situation is mixed. On the one hand Digital Farming solutions are available on farm level. On the other hand, studies confirm Digital Farming implementation is behind expectations, from an on-farm efficiency and a market adoption point of view. There is a gap between the Digital Farming ambition to transform farming toward more sustainable agriculture and today's farming practice.

In December 2019 the European Union published the Green Deal, later translated to farming by the Farm to Fork Strategy as part of the new Common Agricultural Policy (CAP). Digital Farming was named key to translate these political objectives to farm operations big scale. If this is the tipping point for large-scale adoption of Digital Farming, the question is how to support this development and ensure that Digital Farming is applied at a large scale.

The following convictions shaped the content of the book. In principle, end consumers' acceptance and willingness to invest are the best drivers for transformation. The digital transformation toward more sustainable agriculture requires in addition:

- Clear targets to transform farming towards a defined more sustainable future
- Sound methodologies to scale from single successes to broad-scale adoption
- Holistic technology overview to identify and address key usability issues
- Agronomy process thinking to assure products interoperating seamlessly
- Farming product and territorial requirements to specify value adding solutions
- Sustainability associations, methods and directions to lower entry hurdle

The content of the book comes from two different perspectives. The top-down perspective understands farming as part of an ecosystem incl. nature and environment, society, politics and law, but also technologies and standards (see ◘ Fig. 1.1). Scaling up Digital Farming requires policies, funding programs and product portfolios targeting key environmental and farming needs.

The bottom-up perspective understands farming as an efficient production process with key production factors measurable, but not completely adjustable like soil, precipitation, disease pressure, etc. This production process includes the upstream value chain incl. technology, investment goods and input providers, the actual farming production and the downstream value chain including commodity traders, primary processors, food brands and retails and finally the consumer (see center of ◘ Fig. 1.1). Scaling new approaches and methods requires a better performance in terms of sustainable yield level, production costs, and process efficiency (incl. set up times) as with the current state of the art.

Increasing the adoption of Digital Farming in farm practice for the benefit of farmers, environment, market players and society in this environment is a challenge. It requires a joined understanding of the farmers' profitability and sustainability needs, societal and political targets, as well as agronomic processes and technology

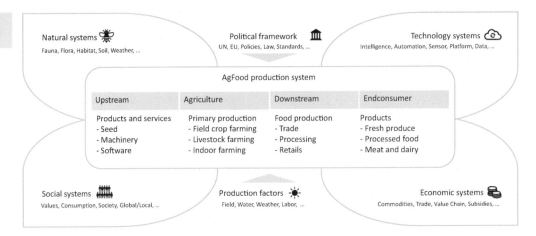

● Fig. 1.1 Overview on agricultural ecosystem elements and influencing factors

challenges. The editors experienced in multiple project discussions, even when we look at and talk about the same things, that the novelty of the subject, the missing standards and evaluation methods make it hard to meet the expectations.

On the business side, companies are proud to deliver a minimum viable product. The customer underestimates the training, setup and maintenance efforts required to use the first products. The data manager is happy about the first data sets, but actually requires much more data with higher time and local resolution. The budget owner needs to match the total sales potential with the final product, not with the initial MVP feature. And finally, the methods are premature to proof to farmers, why a Digital Farming tool finally helped to increase yield by, e.g., 3% compared to last season.

On the technology side, more and more Digital Farming systems are emerging. As the term Digital Farming is assigned with different meanings, we define it for this book as "the software-supported optimization and automation of agricultural work and business processes as well as the enablement of innovative business models." The terms Agriculture 4.0 and Smart Farming are seen as synonyms. In the realm of Digital Farming, various system classes of Digital Farming solutions emerged in the last decades (see ● Fig. 1.2): various information systems such as GIS emerged that each offered specialized functionalities to the farmers. The information systems developed into full-fledged connected Farm Management Information Systems (FMISs) that are nowadays often platform-based and offer a multitude of functionalities ranging from GIS functionality via field diary functionalities to sophisticated decision support. In the embedded world a multitude of sensors and actuators were developed, e.g., the widely adopted GPS sensors. Those sensors and actuators can be standalone. Nowadays, they work together in highly connected cyber-physical and autonomous systems to provide highly automated or even autonomous functionalities like in a tractor, its implements, harvesters, or autonomous robots in the field. In addition, with the development of smartphones and tablets, various apps emerged for these mobile systems, e.g., for detecting weed or for workforce management. All these systems together, even though they are often not highly interoperable, shape a sophisticated digital agricultural ecosystem.

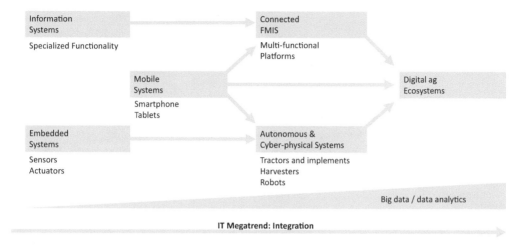

Fig. 1.2 Overview on system classes of Digital Farming solutions shaping the digital agricultural ecosystem

1.1.2 Rationale for This Handbook and Scope

As described in the motivation, there are fast advances in the state of the art and state of the practice of Digital Farming. The future of digitally enabled, sustainable farming will be co-created as farming is a remote process, which is supported by multiple stakeholders, industries, systems and products. This handbook summarizes the baseline about what was achieved during the "rise of Digital Farming opportunities" decade (2010–2020) and provides structure, terminology, technology and examples for the "scale-up Digital Farming" decade (post 2020).

Digital Farming is a multifaceted topic. Therefore, this handbook aims to connect statements of key experts on their view from a multitude of different perspectives. For readers with an interest in methodological and technological know-how, especially ▶ Chaps. 2 and 3 will be of interest. There, the authors present the elements of a framework for the digital transformation of the agricultural ecosystem and describe key technological areas. Exemplary questions answered in these chapters are: "What methodologies and technologies are available today?", "What is their cost–benefit ratio?", and "How will this technology area develop in future, e.g., with regard to performance or cost?". For readers with an interest in agronomy, ▶ Chaps. 4 and 5 will provide insights in the current use of Digital Farming solutions in various agronomic areas and in different farming systems in Europe. Example questions answered in these chapters are "How to create and capture value with Digital Farming technologies?"; "What digital products are currently used?", "Which Digital Farming tools are used for specific farming systems?", and "What are the still unsolved needs, where can Digital Farming improve?". For readers with an interest in sustainability, ▶ Chap. 6 will explain how Digital Farming can be used to contribute to sustainability, either directly or indirectly by supporting other, non-digital measures. Example questions answered are "Which sustainability goals affect Digital Farming?", and "How can Digital Farming contribute to sustainable farming approaches?".

This handbook is intended for farmers, managers and developers in businesses, scientists, students and policy makers. It

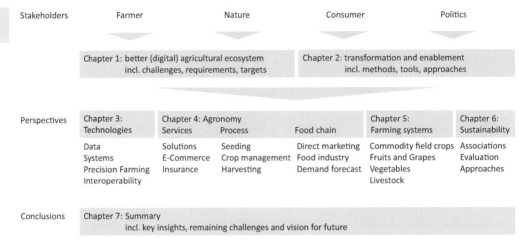

Fig. 1.3 Overview on various stakeholders, process steps, and cross-cutting technologies in the agricultural value chain

describes the state of the art and state of the practice with regard to Digital Farming. It provides existing challenges in various dimensions, such as technology, business and sustainability. It sketches an outlook about current developments and future extensions of Digital Farming solutions. It is also meant as an invitation for discussion: the digital transformation of the agricultural food chain will be an interactive, interdisciplinary process with a lot of testing and discussion. This book provides the baseline for all friends of Digital Farming to these discussions. It provides a foundation of necessary terminology and views, fostering the discussion about new farming practices and new cross value chain partnership opportunities.

This handbook has an intended, limited scope. Technology-wise, all relevant Digital Farming technologies are in scope without territorial or farming system restrictions, but with a focus on technology for crop farming. From an agronomy perspective, the handbook focusses on the European market with humid, maritime climates with yield levels larger than 5t/ha including central European countries like France, Germany, Netherlands, UK, as well as Eastern European Markets. We believe that many views are transferrable to many other areas in the world as well. From a farming system perspective, the main focus is on field crops, but also specialty crops and livestock farming examples complete the picture. The perspective beyond field cash crop farming is especially important, as more sustainable farming in the future will require closing the nitrogen and carbon circle, thus closer interaction between the different farm production systems.

1.1.3 Overview on Contents

In this section we provide an overview on the contents of this book, which is also visualized in ◘ Fig. 1.3.

Chap. 1 describes the current challenges, requirements and targets in the agricultural ecosystem when it comes to digital transformation. It begins with a farmer's perspective and moves through sustainability and environmental perspectives to a description of the state of the practice in the adoption and acceptance of Digital Farming technologies. It provides key requirements and open issues to be addressed when developing or using Digital Farming technologies. Finally, it provides insights how policy makers see the topic of Digital Farming and how the current status of agricultural digital and data law gives a frame for Digital Farming technologies.

▶ Chap. 2 presents methods, tools and approaches on the transformation process itself and its enablement. It starts with a view on the history of Digital Farming technology and a first view on a potential future roadmap. Then it presents major (macroeconomic) trends that are impacting the current digital agricultural ecosystem. From this macroeconomic perspective we switch the view toward possibilities to make an economic benefit quantification on a farm level, knowing that questionable economic viability is among the top causes for neglecting Digital Farming technologies. This is followed by a French perspective how Digital Farming tools can be successfully disseminated. We provide the key lessons learned on how the stakeholders in our agricultural ecosystem should behave in order to support a better adoption of Digital Farming technologies. Besides the dissemination of these technologies, many companies (farmers as well as companies in the upstream and downstream of agriculture) consider how to establish new, successful business models. Therefore, a model and approach for business model innovation with a concrete example application in agriculture are outlined. This chapter ends with an experience report on accelerators and partnerships in the agricultural domain. It describes why establishing successful business models is hard and states how to adopt innovative Digital Farming programs to better prepare for the new challenges ahead.

▶ Chap. 3 starts with a perspective on how the efficient system engineering for automation and autonomous machines will change in the future into the direction of highly automated, flexible equipment systems. This is followed by an overview of current Precision Farming technologies ranging from positioning systems, via sensors systems to Farm Management Systems. Typical Precision Farming application scenarios are outlined as well. Further, one article highlights the current state of the art and specific challenges of safe object detection and how such systems can be engineered. A large challenge in the digital agricultural ecosystems is the interoperability of the different Digital Farming solutions. Therefore, a reference architecture on how to achieve better interoperability is presented. As more and more Digital Farming solutions rely on artificial intelligence, the use of AI in agriculture is outlined by sketching technology and key application scenarios. In order to successfully develop innovative Digital Farming solutions, agricultural data must be available and accessible. Therefore, a view on data sharing in agriculture and relevant technologies such as ontologies and terminologies for describing the data is given. As one key area of agricultural data is geo-related data, we highlight the role of geo-based data and its farm-specific integration. As closing of ▶ Chap. 3, a comprehensive outlook on technology developments in various technological areas is provided.

After these technology-oriented perspectives in ▶ Chaps. 2 and 3, ▶ Chap. 4 provides an agronomy perspective along the agricultural production process and value chain (see also ◘ Fig. 1.3). We start with a view on the change in sales in the agricultural upstream sector from products to solutions, why this is a trend and how the sales will change. Then we switch to cross-cutting and enabling topics that support the primary production processes (see "Services" in ◘ Fig. 1.3) presenting solutions in the area of interoperability of Digital Farming solutions, E-commerce and logistics solutions for the upstream and downstream sector, as well as new insurance models related to Digital Farming solutions. After this, we switch to the primary agricultural production process and describe how Digital Farming solutions support soil and seed management, the nutrient supply of plants (i.e., fertilizing), the crop protection, capturing information about weather and usage of equipment for irrigation, and finally for the harvesting of the crops. Then we switch perspective to the use of digital tools for the sales of the produced goods of farmers. We start with describing the relevance and possibilities of software tools for direct marketing, followed by highlighting the economics point of view of direct marketing. Further, we present how advanced start-ups are challenging the established food retailers and describe which new infrastructures, processes and business models are emerging in this respect. We close this chapter with an outlook on how Artificial Intelligence could be used for demand forecasting and sustainable crop planning.

▶ Chap. 5 takes the farming systems perspective. At the beginning we acknowledge the key challenge is not only to show the yield, cost, and sustainability benefits achieved with Digital Farming solutions, but make the solutions available beyond early mover, strategic thinking, big farms, and contract farmers. Another challenge is supporting the low and average performing farms striving toward the biological optimum of crops and location. This is followed by a perspective on cereal-based production systems, from a crop management process (e.g., observe, analyze, decide, act and record), a country and cross farm point of view (e.g., France, Germany and the Netherlands). To complete the requirements from additional farming perspectives, the chapter further focusses on farming systems, which are more closely connected with the food value chain, as they a) grow and market fresh produce like vegetables and fruits, b) processed products like wine or c) livestock farming products like meat, milk or other dairy products. These sections focus on the specific requirements like high manual harvest efforts, such as labor costs, the particularly high harvesting efforts with fruits and vegetables and specific requirements linked with permanent plantations like fruits and grapes or livestock requirements like animal welfare. In addition to the production-specific topics the section on wine puts more attention to downstream opportunities and the question, how Digital Farming can support differentiation in the end-consumer market. We close with an experience report from a family farm in the Netherlands taking a look back into a long history and into the future.

The topic of sustainability gets more and more attention when it comes to the digital transformation of the agricultural sector. Therefore, ▶ Chap. 6 deals with this important topic. It starts with a view on sustainable farming associations and the frameworks they developed. It continuous with a perspective on how digital technology can contribute to sustainability including some case studies. We then switch to the perspective of hybrid agriculture as an emerging concept to deal with sustainability and continue this chapter with a view on how digitalization can support addressing sustainability especially for cropping systems. We close the chapter with a perspective on carbon farming and how digital technologies enable this new approach.

▶ Chap. 7 summarizes this handbook. The key lessons learnt about the current state of the practice are presented in a chapter-by-chapter fashion. Next, a selection of the remaining challenges for future developments of Digital Farming is summarized.

1.1.4 Information for Reading This Book

▶ Section 1.1 and ▶ Chap. 7 are written by the co-editors of this book. The remainder of the book (i.e., ▶ Sect. 1.2 to ▶ Chap. 6), contains the views and perspectives on Digital Farming as written by corresponding experts in the field. Some articles are written as experience reports, which is indicated for the reader in the subtitle. All articles are mostly self-contained, with the exception of back references to related topics when additional relevant information is provided there. Forward references are not used in this book. The individual chapters align papers with regard to the perspectives outlined in ▶ Sect. 1.1.2:

- ▶ Chap. 2 presents elements of a framework for the digital transformation of the agricultural ecosystem.
- ▶ Chap. 3 describes key technological areas.
- ▶ Chap. 4 provides insights in the current use of Digital Farming solutions in various agronomic areas according to the value chain and production process steps.
- ▶ Chap. 5 provides insights in the current use of Digital Farming solutions according to different farming systems in Europe.
- ▶ Chap. 6 explains how Digital Farming can be used to contribute to sustainability.
- ▶ Chap. 7 provides a summary of the insights we gained in the previous chapters and outlines remaining challenges from the viewpoint of the editors.

A glossary of key terms is included at the end of the book.

1.2 Today's Farming Practice— Challenges and Options

Christian Linke

Abstract

This section describes the external factors and challenges impacting farm operations, farming for income in industrialized areas, mainly in the EU, and the consequential development options farmers have.

Environmental issues, limited resources, climate change, negative public perception of agriculture in many industrialized countries, agricultural policy, limited labor availability, new technologies and a growing interest of investors in agriculture lead to increasing costs for land, labor and farm inputs, a rapidly growing number of regulations and booming bureaucracy. At the same time yield increases are slowing down and prices of most agricultural products are volatile but stagnating and similar worldwide due to globalization and worldwide commodity trade. Hence, farm producers in countries with lower land prices, lower labor costs and fewer regulations are at a significant competitive advantage.

This situation forces farmers to either massively grow their operation in size to leverage economy of scale or to differentiate, e.g., to shift to premium or niche products, direct marketing, renewable energy, contract production or tourism. However, these markets are limited and the majority of farmer producers will remain in the commodity business. Hence, many farmers will move to part-time farming and finally sell or lease out their operations. Additionally, some farm production will shift to countries with less regulations, lower labor costs, lower land prices and a less critical public.

In summary the structural change and consolidation in farming will continue and further accelerate. Farming structure and production will become much more diverse and new opportunities for people with entrepreneurial mindset will appear. Digitalization intensifies these developments. Politics will remain very influential and not predictable but won't be able to stop or reverse these trends.

1.2.1 Challenges

1.2.1.1 Environmental Issues

Agricultural crop production impacts the environment since natural vegetation is removed, landscape elements are modified (e.g., by soil tillage) and foreign materials such as seeds, crop protection products or fertilizers are added [Bra32], [Ell96]. This leads to environmental issues such as a reduction of biodiversity, soil damage, an increased discharge of nutrients and crop protection products or additional greenhouse gas emission [Can17]. Since these effects usually do not directly impact farm production and profitability ("external cost"), governments and authorities enact laws and regulations on farm production to reduce the negative environmental impact. On farm level this leads to higher cost and more bureaucracy.

1.2.1.2 Limited Resources

The demand of farm products (food, bio-energy, bio-renewables) is growing worldwide due to population growth and increasing wealth [Fao+19]. Hence, more and more resources, mainly fossil fuel [Fao11], water [Fao17], phosphate [CW15] and agricultural land, are required.

In Europe the volume of farm production has been about stagnant in the last decades [Fao20a]. The availability of fossil fuel, water (except of some areas) and phosphate has not yet been limited in Europe. The utilized agricultural area in Europe has remained about constant in the last decade [Eur20a], but land rents and land prices have risen in all European countries [Eur20b], [Eur20c], [Sav20]. In Europe land prices vary considerably between regions and countries, and prices in Denmark, Germany, Ireland, Italy, Netherlands and UK belong to the highest worldwide.

1.2.1.3 Climate Change

Climate change impacts agricultural production significantly. Increasing temperatures and weather extremes (droughts, storms, heavy rain fall, etc.) lead to higher production risks and force farmers to adjust crop selection and cropping systems and to invest in risk management measures, e.g., insurances [Elb15]. This will lead to shifts of production areas, production intensity and impact farm profitability, both negatively and positively [LMH+09], [OTK+11], [VMM17].

1.2.1.4 Public Perception of Agriculture

The acceptance of modern agriculture by the public is declining. Environmental issues, extensive use of resources and animal welfare in agriculture are widely discussed in the public and in the media, often with a negative view on modern agriculture [BPL06], [WR00], [Wir20], [ZIB+13]. The increasing awareness of environmental issues and climate change, a growing focus on health and nutrition, a changing perception of nature and animals, food in abundance, decreasing experience with agriculture in daily life and activities of pressure groups are important drivers of this development.

As a consequence, politics implements a growing number of laws and regulations on agriculture which usually lead to higher cost and more bureaucracy on farm level. Additionally, farmers face increasing resistance to further develop and expand their operations and are increasingly demotivated [Cop20].

1.2.1.5 Agricultural Policy

Sufficient food supply is a prerequisite for political and social stability. Hence, agricultural policy is of high importance in most countries and agriculture is heavily regulated [OEC19c]. Additionally, environmental and sustainability issues and the critical public perception of modern agriculture forces politics to manage public expectations and conflicting targets, mainly sufficient and secure supply of affordable safe food, environmental und resource protection, animal welfare, sufficient farm income and rural development.

In the European Union, agricultural policy is conducted within the framework of the Common Agricultural Policy (CAP) and adjusted and executed on national, regional and local level. This leads to a rapidly growing number of laws, directives, regulations and decisions and a booming bureaucracy [Eur20d]. Due to the complexity, unexpected side-effects and contradicting regulations are inevitable. Furthermore, political decision making and execution are getting more and more complex and time-consuming and important decisions increasingly delayed. On farm level this causes growing cost and efforts for bureaucracy and complying with regulations. Additionally, the investment risk increases due to unforeseeable changes.

1.2.1.6 Labor Availability

Agriculture competes on labor with other industries regarding wages and social standards such as daily working hours, free weekends, vacation or working conditions, career opportunities and job satisfaction. The aging of societies intensifies this situation. Hired labor can easily find an employment on other farms or in other industries, while family farms lack successors as the younger generation increasingly decides for other opportunities. However, studies indicate that job satisfaction is often high in agriculture [HG15], [JWH18].

As a consequence, farmers must provide attractive and competitive working conditions not only regarding payment but also a cooperative management style, regular days off, summer vacations, modern and comfortable machines, appealing facilities and other benefits. Automation and robots can relieve of hard work, improve the quality of work, increase labor productivity and save labor, but may also lead to higher total cost.

1.2.1.7 New Technologies

Technologies are production factors which improve the competitiveness of an operation, provided the benefits, e.g., cost reduction, comfort gains or payments for environmental benefits, are larger than the total cost of ownership. In a competitive market environment it is indispensable to utilize new beneficial technologies to remain competitive. In farming a multitude of new technologies has been adopted until today [PB13].

It is often argued that the implementation of new technologies, especially digitalization, can reduce or even offset cost disadvantages of European farmers. Most technologies are available worldwide today due to globalization, especially digital technologies. Hence, only a fast development and adoption of new technologies can create a short-term advantage. In the EU, limited internet access in many rural areas, the strong restrictions on plant biotechnologies and related farm products, and a very cumbersome and restricted homologation of crop protection products are not advantageous for farmers.

Software systems are a special case. A software-system is always developed for specific tasks or use-case, e.g., text processing, herd management or decision support, and requires the user to follow a defined process. While the development and maintenance of software is usually very costly, the costs for scaling, i.e., adding additional users, are typically very low. That is why, a software-system specifically developed for one operation is very costly—often several hundred thousand Euros—and hence viable for very large operations only.

Today's software-solutions are usually cloud-based apps and can serve millions of users. This allows service offers for very low cost provided enough users with the same use case can be acquired. However, the limited number of farms in Europe and the highly diverse farm structure make it difficult to acquire a sufficient number of users for non-generic farm software solutions at a price point accepted by farmers. Hence, many software providers try to create additional revenue streams from user data, e.g., using customer data for marketing or selling data to 3rd parties. This can weaken the market position of the farmers, especially if suppliers or purchasers provide the software [Wes18].

1.2.1.8 Investors

A growing number of agricultural and non-agricultural investors invest in agricultural land and farming operations. Ongoing population growth, growing demand for high value food, the shift toward bio-energy and renewable resources and limited natural resources, especially agricultural land, makes this field increasingly attractive [Eur20c], [Sav20]. Additionally, some states buy foreign land and farms to secure domestic food supply, while companies buy land and farms to ensure raw material supply. Furthermore, land titles are a very secure investment in most countries and large commercial farm operations can be very profitable. Moreover, even very large farms are small companies compared to other industries and hence the investments and risks are limited. Finally, interest rates are at historical low level and investors with high liquidity look for investments, increasingly under sustainability aspects [SS17].

As a consequence, land prices and valuations of farm operations continue to increase, and exiting of low performing farming operations becomes more attractive.

1.2.1.9 Globalization

Most farm products are commodities, i.e., basic goods defined by specific standards are interchangeable with other goods of the same type. Today agricultural products are traded worldwide and national protection and tariffs of agricultural products have been significantly reduced in the last decades [BGJ19]. Since transport cost of most agricultural products are low today, and markets are highly competitive (trade boards), prices for many agricultural products are similar worldwide. That is why, farmers cannot set the price of their products, i.e., farmers are price-takers, and farm profitability depends mainly on cost of production and subsidy payments.

In summary, farmers in Europe face high and further growing cost for land, labor and compliance with regulations, while product prices tend to be stagnant but increasingly volatile [Fao20b]. The European community provides significant subsidy payments to farmers to support domestic production and compensate for regulations. However, the payment level is around OECD average and decreasing [OEC20a], while cost for land, labor and compliance is rapidly increasing. Furthermore, the payments are intensively discussed and under strong political and public pressure.

This leads to a strong and further increasing economic pressure on farm operations in Europe [NS20]. Farm producers in countries with lower land prices, lower wages and fewer regulations are at a significant competitive advantage.

1.2.2 Options

The challenges described above lead to massive and ongoing changes in European agriculture. This part discusses options to deal with these challenges.

1.2.2.1 Massive Growth

As mentioned earlier, the majority of farm products are commodities and will remain so. Commodity producing farmers are price-takers, i.e., they have no influence on the prices of their products. Price-takers must reduce the unit cost and increase production volume to maintain profitability in a competitive environment.

Compared to other industries, today's farms are very small businesses. 1,000 ha intensive grain production means significantly less than 2 Mio Euro annual turnover, which is a micro-business according to definitions of the European Commission [EUC20].

In some regions and countries such as Ukraine, Russia, Kazakhstan, China or South America extremely large farm operations exist [IAM20]. However, many very large farm holdings and cooperatives are not visible in the statistics since usually only individual farms are recorded.

Economies of Scale

Very large farm operations, i.e., operations with € 50 Mio total revenue or more [EUC20], have a number of significant competitive advantages. Large input purchasing volumes and sales volumes lead to a stronger market position and allow to skip intermediate trade. Additionally, large product volumes can make processing and packaging of own products profitable, which creates additional revenue and improves the position in the value chain.

The size requires division of labor and hiring specialized experts such as crop care agronomists, controllers or lawyers, which leads to higher productivity. Career opportunities, higher wages and social benefits of a larger operation attract talents. Furthermore, professional management and implementation of corporate standards such as ISO 9000ff. certifications or international accounting standards improves the access to markets, capital and investors.

The implementation and use of technology on farms require capital and skilled managers and workers. Technologies require a minimum production volume or operational size to be profitable, e.g., a minimum annual area to harvest for combine harvesters or a minimum turnover for an Enterprise-Resource-Planning System (ERP), and unit costs decrease further with growing volume or size. Furthermore, automation, modern supervision, monitoring and communication technologies and IT-supported corporate style organizational architecture (ERP) enable further growth and make management of very large, complex cross-regional farm operation easier and much more efficient. Finally, very large farm operations can work across regions to compensate regional issues such as local weather events or political issues and to further increase utilization of machines.

Challenges in Scaling

However, simple linear upscaling of a farm operation is not successful. Growing workload and management issues will more than offset the advantages of size. Adjustments of structures and processes similar to mid-size companies in other industries are indispensable. Large operations require significantly different management skills than average family farms. Managing and leading people, delegation of work, process orientation and standardization and business acumen are more important for success than driving a tractor properly. Many owners of rapidly growing farm operations struggle with this. Additionally, high labor cost and internal bureaucracy and administration can lead to significant overheads and significantly impact profitability.

Moreover, public and politics believe that smaller family farm operations are superior and worth pursuing compared to large farm operations. That is why a growing number of regulations impact directly or indirectly farm size, e.g., digression and caps on area payments or significantly higher regulatory requirements on large livestock operations.

Organization Structures

Very large farm operations are rarely structured as a single operational entity, but usually consist of many operational entities.

Holding structures are very common for large farms, i.e., a mother company owns several smaller farm operations and additional entities such as pack houses or farm services. A holding often starts when a farmer takes over a neighboring farm. The concrete legal and ownership structures depend on location, tax aspects, decision making, risk mitigation, and more. Very large farm holdings are usually integrated both horizontally, i.e., several farm production entities, and vertically, i.e., along the value chain and often active across countries.

Another approach to create large production structures are contract farming or franchise systems. A large farm operation or a processor/trader contracts with local farmers the production of a specific product with fixed quality traits und defined production rules. The products are taken for an agreed price scheme from the farmers. In many cases, inputs and genetics are provided to the supplying farmers and consulting and monitoring schemes are implemented. These systems enable a controlled production of large volumes of products with defined traits and standards and require less investment than a complete takeover of farms. The risk and profitability of production remain with the local farmers while the market risk for the farmer is reduced. Furthermore, the flexibility regarding income and labor standards and privileges of family farms, such as higher livestock density, can be utilized and the family farm structure is maintained for the public. These systems are standard in poultry production and become common in pig production, while the dairy sector, vegetable production and organic farming move toward these systems.

Another opportunity to utilize economies of scale is the formation of cooperatives. The success of cooperatives depends on many factors [Cen88]. Engagement and support of the members are indispensable, while issues of decision making and transactional cost can be challenging [Rey97]. In addition, large successful cooperatives often tend to adopt corporate structures and behavior and pay little attention to the needs of their members. With special but efficient structures, farm cooperatives can be economically very successful, while fully considering the needs of the members [Lim20].

1.2.2.2 Differentiation

Differentiation of products and services creates income and growth opportunities beyond commodity production. However, differentiation requires a stronger focus on marketing and customer needs, openness for change, creativity and innovation and an entrepreneurial mindset. Moreover, these markets are usually limited in size. This paragraph discusses opportunities for farm producers to differentiate.

Premium Products

Premium products have a higher value according to customer perception compared to standard products. That is why customers are willing to pay a higher price. The value-add can be based on product properties and quality, production schemes, delivery process, branding and image, comfort, and more. Examples are organic food, "free-of-products" (gluten, lactose, vegan, nuts, sugar, etc.), regional products, functional food or convenience food.

The production of premium products is usually more costly, e.g., due to higher labor demand, more expensive inputs, specific production, certification or higher marketing efforts, while yields can be significantly lower, e.g., in organic farming or

Introduction

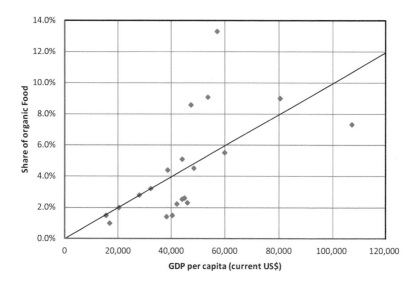

Fig. 1.4 Share of organic food purchased on total food purchase by country (diamonds)

with old breeds [SRF12]. Sometimes, the production of premium products is supported by additional governmental payments, e.g., organic farming.

Premium product markets are limited in volume. Besides other factors, such as mind set, the volume is mainly determined by the price premium compared to the standard product and per capita income. Figure 1.4 shows the share of organic food purchased on total food purchase [WL19] depending on per capita gross domestic income by country [Wor20] in 2017.

As long as the consumer has the choice, the impact of price premium and income on market volume of organic products will remain. Hence, organic farming associations try to restrict market access for additional producers to prevent a price decline [Top20], [Sch20]. Additionally, premium products, e.g., organic grain or organic milk, can be interchangeable and traded internationally, too. This can lead to a similar market dynamics as for commodity products, i.e., larger organic operations tend to have a competitive advantage.

Direct Marketing

In most cases, the value chain from farming to consumers consists of many steps and stakeholders, i.e., only a small share of the price a consumer pays reaches the farmer and the farmers are rather disconnected from the consumer. Hence, an increasing number of farmers sell products directly to consumers. This creates not only additional income opportunities, but also establishes a closer relationship between farmers and consumers and stimulates a better understanding of farming in public.

Direct marketing requires open, communicative and entrepreneurial people and causes significant additional cost, e.g., for salesrooms, additional staff, processing and marketing. Moreover, the location, e.g., proximity to cities or tourist areas, is important for success and seasonality of products, and incomplete assortment can limit sales. Collaboration of farmers, sales platforms and new innovative concepts, e.g., [Mar20], help to overcome these limitations and create new opportunities for farmers [WSM18].

Niche Products

Production of niche products, such as rare fruits and vegetables, old breeds, herbs, medical plants, fiber plant or hay for pets provide interesting opportunities for entrepreneurial and innovative farmers. In spite of the fact that specific expertise for production and marketing is required and the development of new products is usually risky, time-consuming and costly, a multitude of farmers and start-ups test new ideas and concepts and create new products and markets [Fff20].

Renewable Energy

Renewable energy, such as biogas, bio-fuels, photovoltaic or wind energy, can be attractive business branches for farmers. It usually requires significant long-term investments, while implementation and profitability is highly determined by the regulatory framework. The investment risk through politics is declining, while curtailment and price risks are becoming more important [Egl20].

Services

Farmers do not only produce products, but also provide services in rural areas. Farm contractor services such as combine harvesting or maize seeding and road maintenance services have been offered for decades. Touristic offers provided by farmers are well-established and further growing and developing [MVN20]. Natural conservation programs can be attractive for farmers but are often bureaucratic. Moreover, creative and innovative farmers, entrepreneurs and start-ups are testing and developing new services, e.g., care services for dementia patients on farms [EPT19], creating new perspectives in rural areas.

1.2.2.3 Exiting

The number of farms in Europe is significantly decreasing [Eur20e], i.e., a large number of farmers has given up farming in the last decades. In the past, exiting was usually a slow process of dissolution over many years characterized by reduced investments, some land sales and giving up livestock, and finally, quitting due to age and lack of a successor. Today, an increasing number of complete farm operations are sold or rented out to other farmers or investors to preserve assets and generate income for retirement.

1.2.2.4 Dislocation

According to economic theory, it can be expected that in a globalized sector with limited trade barriers, production will shift from countries with high cost and strong regulation to countries with less regulations, lower labor costs, lower land prices and a less critical public. Moreover, entrepreneurial farmers and investors will prefer countries with more competitive conditions. Though statistical data do not yet indicate a general major shift of production in the last decade in Europe [Eur20f], comprehensive studies on dislocation effects on agricultural production due to cost and regulations in Europe are not available.

1.2.3 Conclusions and Outlook

The demand of agricultural products will further increase worldwide while environmental issues and climate change will become more challenging, i.e., a sustainable intensification of agriculture is urgently needed.

In Europe the largest part of agricultural production will remain commodity production and cost leadership will remain crucial for competitiveness of these farm operations. Since organic farming is not a niche anymore, it will be increasingly exposed to the same driver as conventional commodity farming.

The structural change in agriculture is expected to accelerate further and the number of traditional family farms will decline in an accelerated pace. New technologies and digitalization accelerate this development. Knowledge and technologies are available today to run extremely large farm operations very efficiently. Politics may slightly slow this trend in some areas, but will definitely not be able to stop or revert it.

Farming structure and production will become much more diverse in the future. Agriculture provides very attractive opportunities for farmers, new entrants with an entrepreneurial mindset and for financial and strategic investors.

Politics will remain a wildcard. The trend to populism, nationalism and trade wars, more regulations and increasing intervention by the public sector strongly impacts agriculture today and in the future. Furthermore, it is unclear how politics will move agriculture toward sustainability, while securing sufficient and save supply of food and raw material.

Digitalization impacts agriculture already today significantly and is expected to do even more in the future. Digitalization contributes to sustainability of agricultural production through process improvements and efficiency increases and improved transparency of farm operations. However, massive adjustments of operations and significant investments are necessary to fully leverage potential benefits. Therefore, larger and financially strong organizations will benefit more.

As mentioned before, most technologies are available worldwide, especially digital technologies. Hence, only a fast development and adoption of new digital technologies and systems can create a short-term advantage. Due to rapidly increasing regulations, high cost of land and labor and limited internet infrastructure in many rural areas in the EU, it cannot be expected that digitalization will create a competitive advantage for European farmers compared to farmers in other regions in the next years. However, digital technologies help farmers to better deal with growing requirements on efficiency, precision, quality, sustainability and bureaucracy and by this, contribute to remain in business. Furthermore, digitalization supports the development of new business opportunities in farming, especially new service offers. Digital technologies in farming are expected to grow in the following areas:

- Low cost or free software and apps financed by farm input providers, farm data utilization of 3rd parties and data sales
- Further integration of digital technologies in machines and machine systems (automation and autonomous systems)
- Trading platforms for farm inputs and farm products
- Comprehensive, integrated farm management software solutions (ERP-Systems) for very large operations, holdings and cooperatives.

1.3 Sustainability Systems Perspective

John Crawford

Abstract

The Sustainable Development Goals (SDG) are the blueprint of the United Nations to achieve a better and more sustainable future. This section describes the role of agriculture in these goals, the importance of the understanding of agriculture as a system of systems, and why data and technology is important to reach the SDGs until 2030.

» *"When we try to pick out anything by itself we find that it is bound fast, by a thousand invisible cords that cannot be broken, to everything else in the Universe."*
John Muir, 1869 (Naturalist and Co-founder of the Sierra Club).

1.3.1 UN Sustainability Goals and Dimensions

The Sustainable Development Goals (SDGs) lie at the heart of the 2030 Agenda for Sustainable Development that all of the UN Member States signed up to in 2015. The 17 SDGs form an interconnected set of targets that aim to end global poverty and deprivation recognizing that this goes hand-in-hand with improving education and health, ending inequalities and injustices, supporting economic growth, tackling climate change and sustaining global biodiversity.

1.3.1.1 Role of Agriculture in the SDGs

Increasing the productivity of global agriculture goes hand-in-hand with enabling access in ensuring that we end hunger as described in the second Sustainable Development Goal (SDG). However, agriculture has a far larger role in meeting the SDGs than ending hunger. For the poorest people in the world, subsistence farming provides their nutrition, supports their health and is a vital source of income. It enables access to healthcare and education and has a crucial role to play in reducing gender inequality. In short, agriculture is one of the few human enterprises that has an essential role to play in meeting all of the sustainable development goals (see ◘ Fig. 1.5).

On the other hand, agriculture is by far the largest factor in the destruction of natural habitats, contributes about a quarter of the emissions of greenhouse gas and is the most significant contributor to biodiversity loss globally. The accelerating demand for food is resulting in rapid conversion of natural habitats to farmland and an increase in direct contact between humans and animals. Deforestation is a particularly severe threat to both biodiversity and human health. Around two-thirds of the Earth's biodiversity resides in tropical forests that

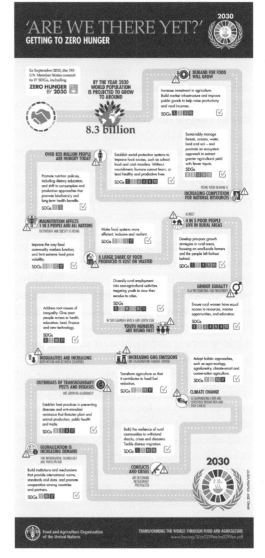

◘ Fig. 1.5 The impact of agriculture on all of the SDGs. Reproduced with permission from the FAO

constitute only one-tenth of the land area. [Gia17] has calculated that if rates of deforestation remain unabated, the resulting magnitude in the loss of biodiversity will be similar to previous mass extinction events [Alr17]. In the past 20 years, deforestation for agriculture has directly caused a third of new and emerging diseases because of the increase in the rate of zoonotic cross-infections resulting from increased proximity of animals and humans [DPH+20]. The number of countries experiencing significant disease outbreaks during the previous ten years has doubled [DPH+20]. Zoonotic diseases already infect over a billion people each year and result in millions of deaths [Who14]. HIV and COVID-19 are recent examples of pandemic zoonoses that are still ongoing and have had a devastating impact on lives and the global economy.

It is clear that agriculture interconnects with all of the SDGs and that the SDGs themselves are interconnected—agriculture and the SDGs form a complex system of interacting parts. In aiming to deliver a more sustainable agriculture, we must embrace this complexity by accounting for the whole system of interconnections to ensure positive outcomes for all of the SDGs. To date, we lack a theoretical and practical framework for doing this and we will explain why this is critically important. Indeed, COVID-19 and the current environment and climate emergency are all symptoms of historic failures to take a whole-system perspective on how our food system interacts with nature. To understand the importance of taking a whole-system approach, we need to first understand why systems are so much more than the sum of their parts.

1.3.1.2 Systems: Why More is Different

In common parlance, a system is simply a collection of a large number of interacting parts. The scientific method together with reductionism (the study of the parts in isolation) have been spectacularly successful in driving innovation and will continue to be so. However, emphasis on reductionism has resulted in us paying far less attention to how, when or even if, knowledge of the parts leads to a better understanding of the whole. While in-depth knowledge of parts has yielded tools to combat pests, diseases and optimize nutrition in agriculture, this has not resulted in an adequate understanding of agricultural systems to avoid systemic collapse. Indeed, many of the global challenges we face are the unintended consequences of a failure to connect up our increasingly fragmented (reductionist) knowledge about agricultural systems.

We can understand linear systems entirely from knowledge of their parts. The behavior of the whole is equal to the sum of the behavior of the parts. For almost 300 years after the scientific renaissance, this was the way we understood the world. The intuitive or predictable, aspects of life are generally linear systems, and science explained them. Because the equations describing linear systems can be solved exactly, the behavior of linear systems is entirely predictable, and we can understand any effect in terms of a cause. Indeed, the publication of Newtons Principia and the exact solution to the orbital motion of one body about another under the force of gravity (e.g., the moon about the earth or a planet about the sun) provided evidence for a predictable 'clockwork' universe. However, nature is overwhelmingly nonlinear–linear systems are a tiny subset—and we cannot obtain a complete understanding of natural systems by only studying their parts. For example, we cannot find a general solution of Newton's equations describing the motion of more than two bodies because the behavior of the system can be unpredictable (e.g., chaotic). In general, non-linear systems are less predictable, and consequently it is harder (or impossible) to determine an effect in terms of a cause. We had to wait for the advent of accessible computing power to provide specific solutions to the equations that describe non-linear systems, and this allowed us to extend our intuition beyond the easily predictable.

The behavior of a system that is qualitatively different from the behavior of its parts, is known as "emergent" behavior—it 'emerges' not from the parts but from their interaction. The ability of soil to undertake the myriad of complex biochemical reactions that convert organic matter into nutrients, to store water and to convey nutrients and water to plants is an emergent consequence of the self-organizing interactions between microbes and soil particles. Often the emergent behavior of the system can affect the behavior of the parts. The effect of soil carbon dynamics on climate is an example—carbon released by soil degradation changes climate, and a changed climate can affect soil in a way that accelerates degradation and release of carbon. Emergent behavior is usually hard to predict and there is no simple relation between cause and effect (e.g., does microbial activity in soil change the climate or vice versa). Natural systems are less predictable than the emergent behavior in other kinds of systems because they tend to be "critical" i.e., they are poised on the edge of a stable state, close to a highly unstable state. The reason they are poised in this way is because evolution acts to 'weed out' systems that cannot adapt to change, and systems that are in a critical state can react more rapidly to a perturbation than those that are not. Therefore, we should not assume that a natural system will behave in a predictable way, and this has fundamental consequences for the way we manage these systems. [Kau19] has introduced the concept of the adjacent possible: the idea that natural systems are fundamentally different from most physical and engineered systems in that they have an unknowable future beyond their immediate trajectory (they are non-ergodic). If this holds generally for natural systems, then it will only be possible to manage in real time and it will not make sense to aim for a long-term outcome. Optima can only make sense in terms of concepts that are local in time. The only guiding principle for long-term management is to aim to eliminate changes that limit the system's future options.

When Stephen Hawking stated in 2000 that this would be the "century of complexity" he was pointing to the fact that the scientific methods developed to date are unsuited to the study of systems, and yet our future wellbeing depends on it. As our capacity to measure everything increases, one of the biggest outstanding scientific challenges (particularly in life sciences) is to determine from all the almost infinite things we could measure in a system (cell, organism, community, earth system): "What do we really need to measure?" "What is the 'healthy' state?", and "How can we restore it once it is lost"?

1.3.1.3 Agriculture as a System and Why It Matters

Agriculture can be thought of as ecology in action. It is our attempt to manage a system of interacting organisms and the physical environment to favor a particular species above all others. As Humboldt illustrated so clearly in his Naturgemälde (also known at the Chimborazo Map) we have known for many centuries that the interactions between the parts of the earth system are so strong that ecosystems share many properties of a single organism. Humans themselves may be more appropriately thought of as Holobionts—an assemblage of organisms building niches for each other and sharing an economy of nutrients and signals that impact on the functioning of the system as a whole. The concept of the gut-brain is a particularly clear instantiation of this. Crop plants are the same—they comprise a system of above and below ground organisms that exist in an information economy that affects and is affected by the physical environment. Indeed, domesticated agricultural species can also be thought as part of the human holobiont—neither can now thrive without the other.

In many ways, the Green Revolution was one of greatest achievements of science, saving more lives and alleviating misery to a greater extent than almost any other field of knowledge. However, the Green Revolution had a single goal—producing more. The unintended consequences of achieving that goal have inflicted substantial societal, ecological and climate damage precisely because we failed to properly account for the system of interactions that connects that goal to the wider earth system. As explained above, if we had acknowledged these interactions, it would have been much harder to identify and manage the desired outcomes, and probably beyond what was possible at the time. Nevertheless, we are living with the consequences now.

In order to take a more sustainable approach, we need to understand what it is we are trying to optimize—what is a "healthy" state of the agri-environment? The concept of planetary boundaries is an attempt to explore the idea that there is a limit to the forcing that human activity can exert on the earth system before there is systemic collapse. It has helped us to think in terms of the critical variables that affect the functioning of ecosystems and the climate. Sceptics point to the fact that the boundaries interact and that there is no evidence of tipping points in ecological systems subjected to extreme degrees of stress. Therefore, the idea that there is a "safe operating space" in which agriculture can live may be simplistic, but it may help us begin to frame the challenge.

We have good evidence that living systems persist in a critical state and that the human-agriculture-environment system is a highly interconnected living system. For these living systems, and those involving humans in particular, the emergent behavior of systems can be unpredictable in principle and future states may be un-presta-teable. This means that we need to think very differently about how we monitor and manage agricultural systems.

There are two significant consequences of this view of agricultural systems and their connection to the SDGs. The first is simple: we must abandon the notion of predictability and embrace profound uncertainty. This goes beyond the usual statistical concepts of precision and accuracy, to consider the system as subject to un-presta-teable, discontinuous and systemic transformation when subject to smoothly changing conditions. As well as known unknowns, there will always be unknown unknowns.

The second is more subtle. The number of variables in any agriculture system is overwhelming and includes soil parameters; climate; cropping/rotational system; details of interactions with the embedding ecosystem; agronomy; financial constraints, local and global markets, policies and farmer risk appetite. We cannot use traditional field trials if we take a systems view of agriculture—the combinatorics mean it takes so long to conduct a representative diversity of treatments that the present challenges will have long since played out. Therefore, we need to use existing farms as living laboratories and so the billions of global farmers become partners in the scientific enterprise. This opens up legitimate concerns about the use of farmer data, informed consent to participation and appropriate incentives. It is critical to get this right from the outset, because without farmer trust and participation, a systems approach to sustainable agriculture will not deliver on its potential.

1.3.1.4 The Way Forward and the Role of Data and Technology

As [FM20] has pointed out, the challenges we face have come at a fortuitous time in our evolution. For the first time in the past 200,000 years, humans are now capable of coordinating at a global scale. We have the potential to record all of the fine-scale data we need, share knowledge and improve understanding of what to measure, how to integrate and interrogate that data, and how to govern the resulting complex socio-technical system.

In light of the topics discussed above, there are six observations that relate to building a roadmap to sustainable agricultural systems.

- To properly define and deliver a sustainable agriculture with maximal impact across the SDGs requires a systems approach.
- Complexity arising from a systems approach means that uncertainty and unpredictability are an inevitable consequence that must be properly integrated into strategies.
- There will be no push-button solution. Intrinsic unpredictability means that solutions have to be actively managed on an ongoing basis and combine hard and soft systems methodologies. Pervasive monitoring and evaluation are an essential consequence.
- Pervasive data and related technologies should be regarded as essential tools in transforming agriculture, and their use as a means to refine the status quo should be seen as a distraction. Transformation will not happen by playing the wrong game better.
- Agriculture and agricultural science are fundamentally linked in a systems approach through the need to observe and monitor the full range of realizations of agricultural systems—farming and science form a new partnership.
- Farmers will be major agents of change, delivering a wide range of public and private goods and services that must be appropriately resourced and incentivized.

Knowing what we now know about systems and the geopolitical and environmental context of agriculture and the food system there are three important questions:

- What should sustainable agriculture optimize?
- What do the trade-offs between food production and the SDGs look like? What is the appropriate balance? How do we engage with all elements of the system (ecology, society, economy)? Searching and staying within planetary boundaries may not be enough or even relevant.
- What do we need to measure to identify interventions and monitor impacts in order to optimize sustainable agriculture and SDG impacts over time? There is no protocol to identify the components and interactions that must be included in a description of a system to answer a specific question. This is one of the outstanding challenges of life, social and economic sciences. In light of that fact, how do we prioritize data and monitoring campaigns?
- How do we deliver impact at the pace and scale needed to address the food, climate and environment crisis?
- We cannot eliminate uncertainty and we do not have time to wait for perfect solutions. How do we establish the right governance structure to ensure the coordination and transparency that will establish a learning platform?

1.4 Agriculture and the Environment: Where Are We Headed? A German Case Study

Knut Ehlers and Frederike Balzer

Abstract

Agriculture is an open system. It works in and with the environment. Agriculture is therefore always relevant to the environment. At the same time, it shapes our cultural landscape and has created habitats worth protecting in the first place. However, increasing intensification is associated with negative environmental impacts that affect all environmental elements. Soil, water, air, climate and biodiversity are influenced significantly by agricultural use. Germany has developed a large number of quantitatively measurable environmental targets, the majority of which are to be achieved by 2030. The impact of agriculture on the achievement of these environmental targets are outlined in this section, using Germany as an example. At present, there is a considerable gap between the actual situation and the agreed target values. In the coming years, environmental protection and resource conservation will therefore increasingly be defining criteria that require changes in agriculture. Policymakers and society will modify agriculture's scope for action accordingly. The digital transformation can support these changes.

1.4.1 Introduction

Agriculture and the environment are interdependent—inevitably so. Environmental conditions—better known in the agricultural sector as site conditions—have a decisive influence on the way agriculture can take place in a natural area. Conversely, agriculture shapes and influences the environment. This, too, is unavoidable because, unlike in the industrial sector, where classical end-of-the-pipe cleaning mechanisms can effectively reduce negative environmental impacts, this is only possible to a limited extent in agriculture. A few exceptions, such as exhaust air purification systems for stables, confirm this rule. As a general rule, agriculture forms an open system, working in and with the environment. Agriculture is therefore always environmentally relevant.

The special thing is: Historically, it was agriculture that first created many types of biotopes that we now consider particularly worthy of protection (such as alpine meadows, heaths, nutrient-poor grasslands).

These diverse habitats have boosted biodiversity in Germany. However, the increasing intensification of agriculture is associated with negative environmental impacts, which have increasingly become the focus of society in recent decades (see ▶ Sect. 1.2). The range of issues affects all environmental elements: soil, water, air, climate and biodiversity. Landmark studies on planetary boundaries, such as [RSN+09] or [SRR+15], have made it clear: In all areas where planetary boundaries are exceeded or endangered and where we have gone beyond a safe space of action, agriculture plays a major role. This applies in particular to the areas of genetic diversity, nitrogen and phosphorus cycles, changes in land usage and climate change.

The United Nations Global Sustainable Development Goals therefore speak of the need for sustainable agriculture (see ▶ Chap. 6). But how can sustainable agriculture be measured in ecological terms? Germany has developed a large number of quantitatively measurable environmental goals, the majority of which are to be achieved by 2030. Some of them are declarations of intent and thus at most politically binding, while others are legally binding—with the threat of legal proceedings and sanctions in the event of failure to comply. Germany's conviction by the European Court of Justice in 2018 for violation of the EU Nitrates Directive is a prime example of the importance of these legally binding obligations [EC19a]: Faced with the threat of substantial fines, Germany slid into the role of the party being driven and once again had to toughen up its fertilizer legislation, which had just been amended in 2017, within a very short time. This example shows that it would be advisable in the future to keep an early eye on the progress made in achieving the environmental targets. First and foremost, an agri-environmental policy is called for. But agriculture and its upstream sectors should also keep an eye on these targets in order to be able to act with foresight and anticipate developments.

1.4.2 Agriculturally Relevant Environmental Targets

The impact of intensive agriculture on the protected elements climate, biodiversity, air, soil and water will be considered in more detail in the following subsections. The structure and the content of these subsections are strongly based on an article published by Ehlers and Messner in 2020 for the updated data situation [EM20].

1.4.2.1 Greenhouse Gas Emissions from Agriculture

Agriculture is both a cause and a victim of climate change. More than almost any other sector, it will have to adapt to the realities of climate change. In climate policy, the role of agriculture is quite significant. It directly accounts for up to 7.4 percent of total greenhouse gas emissions in Germany [Umw20]. This might not appear to be much at first glance and is partly due to the fact that large quantities of fossil fuels continue to be burned in Germany, which overshadows the relative importance of other sources of emissions. On the other hand, it is due to the limitations of the analysis. If indirect greenhouse gas emissions from agriculture were to be factored in, e.g., greenhouse gas emissions from the agricultural use of former moorland sites or from the production of nitrogen fertilizer, the emissions would be roughly twice as high [DWH+09]. Globally, the sector causes 21 to 37 percent of the global greenhouse gas emissions when the upstream and downstream sectors of the global food system are also considered [EM20], [IPC19].

However, let us focus on direct greenhouse gas emissions from agriculture: Here, nitrous oxide and methane emissions from agriculture are of central importance. Nitrous oxide emissions are mainly a result of nitrogen fertilization (mineral and organic), while methane emissions are mainly emitted by cattle and occur during the storage and spreading of manure.

Greenhouse gases from agriculture declined rapidly immediately after German reunification (see Fig. 1.6). This can be largely attributed to the decline in livestock numbers in the eastern German states in the early 1990s. In addition, there was a simultaneous increase in the efficiency of nitrogen fertilization, also in the new federal states. After that, emissions stagnated. Since 2016, they have been decreasing slightly again. This current decline is essentially based on drought-related declines in crop yield and falling livestock numbers. Whether this trend will continue remains to be seen [EM20] (see Fig. 1.6).

In the 2019 Climate Protection Act, Germany stipulated that greenhouse gas emissions from agriculture, which include direct greenhouse gas emissions from agriculture plus those from agricultural traffic, must be reduced from about 68 million tons of CO_2 equivalents today to 58 million tons[1] by 2030 [Bun19]. This corresponds to a required decrease of around 15 percent compared to 2019.

1.4.2.2 Biodiversity

The loss of biodiversity has developed an alarming dynamic globally. The driving force behind this is humans—including our agriculture [IPB19]. On the national level, the loss of biodiversity is especially a problem in the agricultural landscape [EM20].

As an indicator of biodiversity, the population of representative bird species in Germany is used in the German Sustainability Strategy. This data clearly indicates that the situation in Germany has massively deteriorated since the 1970s. The indicator has been stagnating at a low level for around 20 years (see Fig. 1.7). Biodiversity in the agricultural landscape is particularly affected by this negative development. Causes include the loss of habitats and the disruption of

---

[1] The German Climate Protection Act was amended in 2021 and the sector targets tightened. The new sector target for agriculture is now 56 million metric tons of CO2 eq in 2030.

Introduction

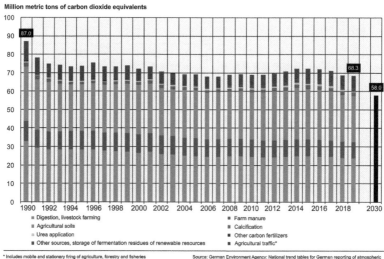

Fig. 1.6 Greenhouse gas emissions of agriculture by sector of the Climate Protection Plan 2050, incl. target value for 2030 according to the Climate Protection Act

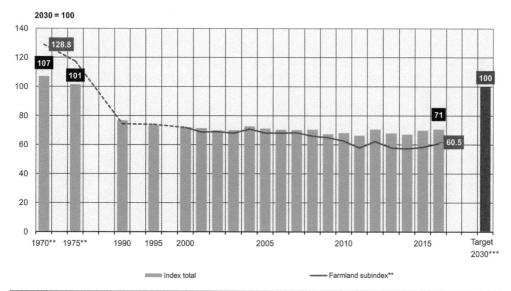

Fig. 1.7 Population of representative bird species in different types of landscapes and habitats; Sub-indicator agricultural land

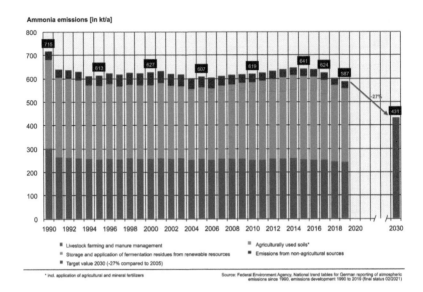

Fig. 1.8 Ammonia emissions in Germany; Focus on agricultural emissions and target achievement

food chains. These are made worse by the clearing of the landscape, the standardization of agricultural cultivation systems, as well as by the constant enrichment with nutrients and the intensive use of pesticides. As a result, the index value for biodiversity in the agricultural landscape fell to 60.5—the target for 2030 is a value of 100. This target is currently a long way off and will not be achievable without comprehensive changes in agriculture [Bfn17], [EM20], [Sta18].

1.4.2.3 Air

Ammonia emissions, in particular, play an important role in agriculturally influenced air quality. Ammonia is a gaseous nitrogen compound, which, among other things, leads to the formation of Particulate Matter and contributes to nutrient enrichment in ecosystems via atmospheric deposition. This affects the functionality and the composition of the species in an ecosystem. Ecosystems that are adapted to low nutrient levels are affected particularly severely. Of Germany's ammonia emissions, 95 percent come from agriculture, and about 60 percent are directly linked to livestock farming. Correspondingly, there was a reduction in emissions immediately after 1990, following German reunification. Since then, emissions have stagnated at a largely constant, high level [EM20] (see Fig. 1.8).

However, a reduction in ammonia emissions is urgently needed, as, under the European Directive on National Emission Ceilings for Certain Air Pollutants (NEC Directive), Germany has committed to reducing ammonia emissions by 29 percent by 2030, compared to 2005 [EU16]. Since ammonia emissions actually increased slightly between 2005 and 2015, emissions must continue to be reduced by 29 percent in the coming decade compared to 2018.

1.4.2.4 Soil

At the global level, the United Nations Sustainable Development Goals (SDGs) specify the target of Land Degradation Neutrality to achieve soil protection [UN15]. This target is to be achieved by 2030 and means that the loss of soil—for example in the form of soil degradation—must be minimized and unavoidable soil degradation must be compensated for by restoration measures. On balance, the soil situation must not deteriorate further until 2030.

Since around half of the land in Germany is used for agriculture, the agricultural sector will also be affected by this goal. This illustrates the responsibility that agriculture bears for soils. After all, agriculture can also endanger soils. The result is soil degradation, for example in the form of erosion, soil compaction or loss of humus. When it comes to soil protection, it can be seen that short-term and long-term interest are sometimes in conflict in the agricultural sector. While soil-degrading practices may well reduce costs in the short term, they undermine one of the most important production factors in agriculture—fertile soils—in the long term [EM20].

It is planned to transfer the target of Land Degradation Neutrality into the German Sustainability Strategy as well. So far, this has not been done. Work is still underway to develop a meaningful and easy-to-measure soil indicator. Based on this indicator, the target is to be made measurable. Making the problem measurable would also be a first important step toward strengthening soil protection in agriculture [Bun17], [EM20].

1.4.2.5 Water

In the water sector, there are several environmental goals that are of importance for agriculture. One of the main problems caused by agriculture is still the pollution of water bodies with nitrogen and phosphorus from agricultural fertilization. This leads to Eutrophication and thus contributes to species extinction in water bodies and algae growth, facilitates algal blooms and reduces the possibility to use water bodies for the production of drinking water [EM20].

Nitrate pollution in groundwater remains a major problem area. The EU Nitrates Directive defines the target that all monitoring stations should comply with the limit of 50 mg N/l maximum [Bun17]. This target remains a distant prospect. Although a trend can be observed that the number of monitoring sites and the average concentration at the particularly affected monitoring sites that do not comply with the limit are improving, more than 27 percent of the monitoring stations affected by agriculture still exceed the limit. Among the monitoring sites not affected by agriculture (such as those located under forests or settled areas), this figure is only five percent. The nitrate problem in groundwater in Germany is therefore largely caused by agriculture [BB20].

The pressure to achieve an improvement here will remain high. Particularly affected are the finishing regions with high animal population density in northwestern Germany, the central German drylands—where low seepage rates lead to high concentrations of nitrates—and the Rhine-Main area—where the problem is mostly caused by the high proportion of vegetable cultivation and specialty crops (see ◘ Fig. 1.9). Ultimately, however, the nitrate content in groundwater always also represents the past—a short-term improvement in nitrate leaching on the surface may take years and decades to be reflected to a significant extent in the nitrate content of groundwater bodies [BB20].

Surface and coastal waters are also affected by agricultural nitrogen and phosphorus pollution. In the case of nitrogen, agriculture is responsible for around 75 percent of the inputs, while for phosphorus, the figure is around 50 percent [BB20].

For surface waters, the EU Water Framework Directive defines the target that all waters should be in good ecological status by 2027. This good ecological status is also defined by the phosphorus levels in flowing waters. Even though the long-term trend is clearly positive, only 44 percent of the flowing waters in Germany meet the target value in terms of phosphorus load. The rest, and thus the majority, exceed it [BB20].

Fig. 1.9 Mean nitrate levels at the monitoring sites of the EEA monitoring network for the period 2016–2018

What is in flowing waters usually ends up in the seas at some point—i.e., in the North Sea and in the Baltic Sea. In the North Sea, only six percent of the water achieves good status in terms of eutrophication. 55 percent are too heavily polluted and 39 percent cannot be assessed. The situation in the Baltic Sea is even worse—here 100 percent of the Baltic Sea waters are too heavily eutrophicated [BB20].

1.4.3 Conclusion and Outlook

The list reveals that wherever there are agreed, quantitative targets for the environmental impacts of agriculture, there is currently a considerable gap between the actual situation and the target values. It is likely that the pressure for change outlined above will rather tend to increase even more, also against the backdrop of climate change and its impacts on German agriculture. In the EU Commission's "Farm to Fork" strategy, more environmentally friendly agriculture is included in the European list of priorities and also sets quantitative environmental targets for agriculture. For example, the use of pesticides and antibiotics in livestock farming is to be reduced by 50 percent by 2030. In the same period, the amount of fertilizer used is to be reduced by 20 percent and nutrient losses by 50 percent [EC20e]. In order for these and existing environmental targets to be achieved in the next decade, agriculture will have to undergo significant changes. The need for agriculture to adapt to climate change is another challenge that agriculture will face in the process.

We are facing a break in trend: In recent decades, specialization and intensification of agriculture have, over time, led to the agriculture that dominates today. It will not be possible to continue this trend unchanged. Instead, in the next ten years, environmental protection and resource conservation will increasingly be defining criteria that require change. Policymakers and society will modify agriculture's scope for action accordingly. Instruments for shaping agriculture, such as regulatory law, taxes and levies and agricultural subsidies, will change. For agriculture to adapt successfully, it is to be hoped that the transition will be approached in a planned and long-term manner—also to prevent situations like the ad-hoc amendment of the fertilizer legislation in 2020. Instead of maximum yield, optimum yield—in terms of environmental impact—will establish itself in the future as a new benchmark for agricultural action.

Focusing on optimal yield from an environmental perspective will essentially happen in two ways: On the one hand, there will be a stronger focus on environmental performance per unit of land farmed. Organic farming is particularly efficient in this regard. Therefore, the goal of the German Sustainability Strategy—expansion of organic farming to 20 percent of the agricultural land by 2030—as well as the corresponding, but somewhat more ambitious goal of the "Farm to Fork" strategy (expansion of organic farming to 25 percent of the agricultural land by 2030), will be pursued more intensely. On the other hand, there will be a much stronger focus on eco-efficiency. The goal will therefore be to optimize the environmental impact per kilogram of grain or per liter of milk. This path has great potential in organic farming, but also in conventional agriculture.

The digital transformation of agriculture is an instrument that can contribute significantly to both areas—improvement of the environmental impact per unit of land and improvement of the environmental impact per unit of product (eco-efficiency). In order to exploit the potential of this development and to ensure that the technological developments can still have an effect in good time—after all, most of the environmental targets mentioned above are to be achieved by 2030—it will be necessary for agriculture

1.5 Adoption and Acceptance of Digital Farming Technologies in Germany

Markus Gandorfer, Andreas Gabriel, Johanna Pfeiffer, Olivia Spykman and Beat Vinzent

Abstract

Although Digital Farming technologies have been available on the market for many years, their adoption in agricultural practice is currently still limited, with the exception of a few specific technologies. There are multiple reasons for this. In general, the fundamental attitude of society toward digitalization in agriculture can be assessed as positive.

1.5.1 Introduction

The issue of the actual adoption of digital technologies in agricultural practice is of central importance for different stakeholders and has therefore been the subject of intensive research for many years. For example, some addressed the use of computers in farm management [RW04], and others studied the adoption of precision farming technologies in Germany [RJ09]. Internationally, the topic has also received considerable attention for years. One prominent example is the Precision Agriculture Dealership Survey in the USA, which in 2020 was conducted for the 19th time already and thus provides one of the longest time series in this context [EL20].

The issue of the adoption of digital technologies in agriculture is often followed directly by the issue of possible barriers to acceptance. It is not uncommon to observe that the practical use of digital technologies associated with desired positive environmental effects (e.g., approaches to variable rate nitrogen fertilization) falls short of expectations [Gan19]. In the past, the issues of acceptance outlined below predominantly focused on the perspective of agricultural practice.

Most recently, the issue of acceptance of digital technologies in agriculture by society has also been raised very frequently [PGG20]. The reason for this is that the use of these technologies can address some central global challenges (climate change, biodiversity, animal welfare, soil and water protection, see ▶ Sect. 1.4). The use of digital technologies is thus also associated with the hope of improving social acceptance of agricultural production.

Against the background of the issues raised above regarding the adoption and acceptance of digital farming technologies, empirical data on these areas will be presented and discussed in the following. Even though the focus is on Germany, the presented insights can largely be transferred to other countries, provided the underlying agricultural structures and sociopolitical conditions are comparable.

1.5.2 Adoption and Acceptance

1.5.2.1 Adoption in Agricultural Practice

In order to gain an insight into the current adoption of digital technologies in agriculture, the authors conducted an online survey among farmers in Germany from November 2019 to January 2020. The farmers were approached via various channels (e.g., online agriculture trade journal portals), so this survey cannot claim to be representative. However, due to the relatively large sample of 550 questionnaires that could be evaluated, from farms with a

Introduction

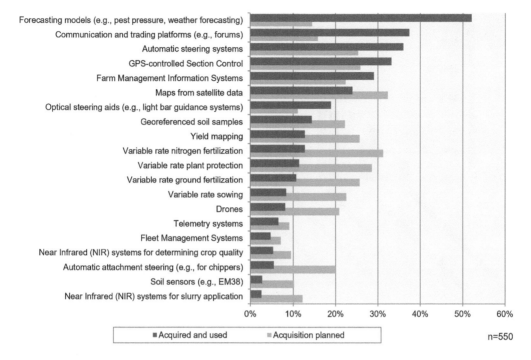

Fig. 1.10 Use of digital technologies in Germany [GGS21] (translated)

production focus on crop farming and forage and on the specialty crops sector (67% full-time farms; 90% of them doing conventional farming) as well as a realistic regional distribution of the farms throughout Germany, the results of the survey provide good insights into the current situation of agricultural practice (see Fig. 1.10). The results reveal that some technologies and applications already enjoy rather widespread adoption. These include, in particular, app-based offerings (e.g., monitoring of pests, weather forecast) or digital communication and trading platforms, which can often be used free of charge (see Fig. 1.10). Furthermore, it can be observed that technologies that are particularly beneficial in terms of reducing the workload and improving work quality (e.g., automatic steering systems, section control) are comparatively widespread. However, if we look at classical precision farming applications such as variable rate fertilization, seed-

ing and crop protection, adoption rates still fall short of expectations. This is a key finding, as it is exactly with these technologies that various positive environmental effects could be achieved. In addition to the question regarding the current use of specific technologies, the survey also asked about planned acquisitions in the next five years (see data series "Acquisition planned" in Fig. 1.10). This shows that there is a great interest overall in investing in digital technologies, particularly in the area of variable rate technologies.

1.5.2.2 Barriers to Acceptance in Agricultural Practice

When interpreting empirical data regarding the adoption of digital technologies in practice, the question inevitably arises as to existing barriers that hold back this adoption. This question can be pursued with different methodological approaches, which include, in particular, quantitative surveys

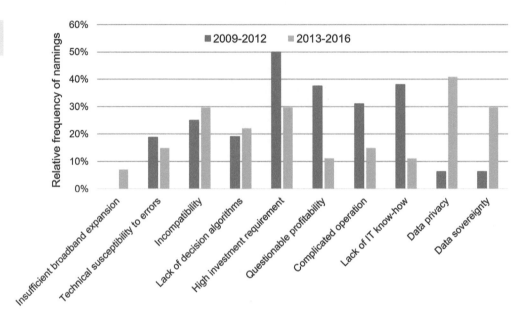

◘ **Fig. 1.11** Results of a media analysis on barriers to the acceptance of digital technologies in agriculture (source: [SG18], translated)

(e.g., [RJ09]) and qualitative approaches such as focus group discussions or workshops (e.g., [PGE18]). In this context, [SG18] conducted a media analysis and analyzed agricultural journals over a longer period of time, comparing two consecutive time periods (see ◘ Fig. 1.11). It was found that the high investment requirements, and in recent years also increasing concerns about data protection and data sovereignty, as well as interfaces and compatibility problems constitute key barriers to acceptance. Furthermore, a lack of decision algorithms could be identified as a barrier to acceptance. This means that many technologies generate large amounts of data, but there is a lack of corresponding decision support and the collected data is unable to facilitate optimized decisions. [PGE18] arrived at similar conclusions in the context of a stakeholder workshop and additionally highlighted the aspect of usability as another important barrier to investment in digital technologies.

1.5.2.3 Acceptance by Society

As described above, many digital technologies in agriculture offer the potential to reduce negative external effects of agricultural production (see ▶ Sect. 1.4). The question therefore arises whether the use of these technologies can have a positive influence on the acceptance of agriculture by society. An extensive online survey of 2,012 residents of Germany (aged 18 and over, sample representative in terms of age, gender, size of town of residence and educational attainment) revealed a fundamentally positive attitude toward the use of digital technologies in agriculture (see ◘ Fig. 1.12). A majority of the respondents agreed with the statements that digitalization has positive effects in terms of environmental protection and animal welfare and that the use of digital technologies can have a very positive impact on the working conditions of farm families. However, the results of this survey also showed that the proportion of people who are undecided is

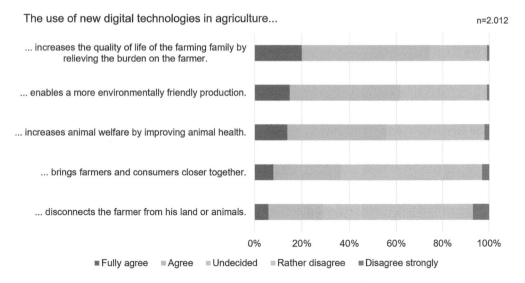

Fig. 1.12 Acceptance of digital technologies in agriculture by society [PSG+19] (translated)

still very high. Thus, there is a distinct possibility that the attitude of this proportion of the respondents, as well as those with a positive attitude, may change. This would require a fundamental reassessment of the situation.

1.5.3 Discussion and Conclusions

The results and interrelationships presented above clearly show that the adoption of digital technologies in agricultural practice heavily depends on the specific application. It is true that agriculture is very advanced in terms of digitalization and can certainly compete with other sectors of the economy. However, it must also be taken into account that the degree of digitalization of farms, which is sometimes communicated as being very high, is driven by individual applications. For example, the results presented here show that 84% of the farmers surveyed use at least one of the digital technologies offered for selection. This result is consistent with a survey conducted by [RKR20], which arrived at the following conclusion: "8 out of 10 farmers use digital technologies". However, if we exclude the two categories "Communication and trading platforms" and "Forecast models" from the survey presented here, the degree of digitalization decreases by more than 10 percentage points. Among the remaining technologies, the current focus is clearly on user-friendly automation solutions that reduce the workload.

There are many reasons why a wide range of digital technologies have only been adopted very tentatively in agricultural practice. They range from economic aspects, compatibility problems and usability aspects that should be improved to concerns about data protection and data sovereignty. At this point, it is primarily the manufacturers who are called upon to act. In addition, applied agricultural research is needed when it comes to the development of decision algorithms to generate benefits from the collected data. Moreover, publicly funded research can help to reduce uncertainties regarding the cost-effectiveness of individual technologies through scientifically sound technology assessment.

The survey results we have presented on the acceptance of digital technologies in agriculture by society give cause for hope that, on the one hand, the increased use of these technologies can improve the attitude of society toward agricultural production. On the other hand, the large proportion of undecided people shows that this finding may change at any time as a result of current developments. This makes it necessary to conduct studies on the acceptance of digital technologies in agriculture by society at regular intervals in order to be able to evaluate the impact of new developments (e.g., technological innovations, social discourse, media reporting). As a matter of principle, all players are called upon at this point. Digital technologies must be presented objectively to the general public, with their advantages shown in a wide variety of formats. However, in doing so, potential problem areas and conflicting goals always need to be pointed out as well.

1.6 Agricultural Digital Policy

Mathias Olbrisch

Abstract

The digital transformation is to be understood as a technological-social development process whose course depends to a large extent on its political framing. In this respect, the political landscape is characterized by a complex coexistence of very different concepts. The aim of this section is to provide an overview of the related political concepts and agendas of the various actors at the global, European and German levels. The main focus is on the description of conceptual convergences and disparities in order to make the range of the agricultural digital policy instrument mix tangible as a discursive reference point for future legal design.

1.6.1 Enabler and Discursive Reference Point for the Digital Transformation

The digital transformation of the agricultural sector toward a comprehensive integration of field robots, drones and precision farming into AI-supported smart farming offers enormous agronomic potential for increasing yields for farmers and new business models for agribusiness companies (see ▶ Sect. 1.1). Furthermore, it is also in sight as the first realistic milestone for the digitalization of the economy as a whole, as the agricultural sector—unlike the automotive sector for example—enables the possibility of an arguably relatively risk-free testing and application of innovative technologies. In this respect, digital-transformative processes are suitable for developing the potential of a multifunctional agriculture that can make a decisive contribution to achieving global sustainability and climate protection goals beyond food security (see ▶ Sect. 1.4).

An agricultural policy promotes the actual implementation of the digital transformation if it succeeds in striking an appropriate balance between the affected stakeholder interests in the collection, exchange and use of interoperable and high-quality data as a resource, on the one hand, and legitimate interests in the protection of personal and business-related data, on the other, by establishing a reliable governance framework orchestrated by the rule of law [JCW+20]. In addition, a modern agricultural digital policy is capable of setting effective incentives and synergistically links digitalization with other socio-economic and ecological objectives.

The concept of agricultural digital policy integrates in a vertical perspective the political agendas directed toward the digital transformation of agriculture within the European-German multi-level system, including its references to the global policy level. With agricultural policy on the one hand, and digital policy on the other, it merges the digitalization-related implications of both policy fields in a horizontal perspective. The concept of agricultural digital policy forms a discursive reference point that can serve to make scientific discussions regarding policy-making in this sector more tangible and to functionalize it in a transdisciplinary perspective as the basis for a consistent "Agricultural Digital Law" [Här19].

1.6.2 Global Policy Level: Transnational Multi-Stakeholder Governance

The global policy level is characterized by approaches of transnational multistakeholder governance with a convergent orientation toward the digital transformation of agriculture.

In 2019, the Global Forum for Food and Agriculture (GFFA) was able to reach an intergovernmental consensus between more than 70 national ministries of agriculture that considers the digital transformation of agriculture as a driver for increasing food production, improving sustainability and improving living conditions in rural areas [GFF19]. In continuation of this postulate, follow-up conferences in 2020 and 2021 pushed for integrity and traceability along the value chain, including the use of blockchain technology, as well as the creation of an agricultural market information system for statistical recording of price volatilities by the FAO and the launch of an "e-agriculture" platform for the digitalization of agriculture and food, which pools information on scientific and political activities on digital transformation and maintains a directory of good practices for the implementation of digitally driven technologies [GFF20], [GFF21]. With the creation of the International Digital Council for Food and Agriculture, which also goes back to the GFFA, an impetus for the digital transformation is also to be institutionalized at international level, which will provide policy advisory and promote conceptual exchange at international level [IDC20].

As an international organization, the OECD also forms a multilateral forum for coordinated and dialog-based decision-making at the international level and, as a political actor, advocates the usability of data as a globally tradable economic good and driver of economic development. In this regard, the Directorates for Science, Technology and Innovation as well as Trade and Agriculture have presented a joint document for a G20 Digital Economy Task Force, which is to enable a trustworthy cross-border data flow [OEC20b]. This is relevant for multinational companies whose businesses are based on processing data in a jurisdiction other than the one in which the data was obtained. A mix of instruments, including plurilateral arrangements (such as the OECD Guidelines for the Protection of Privacy and Transborder Flows of Personal Data), unilateral measures (e.g., in the form of recognition of foreign data law regimes) and private standardization initiatives, should enable a global offering of novel data-driven business models [OEC20b]. At the global trade law level, the WTO adopted a joint statement on e-commerce in 2019 in this context [WTO19]. Beyond its function as a resource that can be used by the private sector, the OECD suggests using data assets to improve agricultural policy. For example, the case-by-case appropriateness and site suitability of compliance-oriented administrative decisions could be promoted on the basis of artificial intelligence [Här20] and a market-based remuneration of ecosystem services could be made possible [OEC19a]. In this context, the Farm Level Analysis Network operated by the OECD aims to make micro-level geodata usable for agricultural governance purposes.

To promote open data in agriculture (such as satellite data), the multistakeholder initiative Global Open Data for Agriculture and Nutrition (GODAN) brings together actors along the value chain, provides policy advice and advocates for common data standards. It is supported, among others, by the FAO, the US Department of Agriculture, the Federal Republic of Germany, the governments of the Netherlands, and India, as well as by research institutions such as the Kuratorium für Technik und Bauwesen in der Landwirtschaft (KTBL) or the Leibniz Center for Agricultural Landscape Research (ZALF), but also by private actors such as the Kellogg Company or the Bill and Melinda Gates Foundation.

Founded by the UN Conference on Sustainable Development and directed toward the goal of transdisciplinary sustainability research, the Future Earth Association promotes the establishment of a worldwide community of scientists from relevant disciplines. Sponsors include NASA, ESA, UNESCO, the DFG and the UN Environment Programme [FE21]. Aiming into a similar direction, the Consultative Group on International Agricultural Research (CGIAR) acts as a broker to bring together institutions and scientists to initiate projects on the digital transformation of agriculture. Its sponsors include the EU Commission, the German Federal Ministry for Development and Economic Cooperation, the French Ministry of Research, the World Bank, and the Bill and Melinda Gates Foundation [CG21].

The global policy level is thus characterized by multi-layered multi-stakeholder governance with regard to the digital transformation of the agricultural economy. This has a potentially harmonizing effect on enabling cross-border digital agribusiness.

1.6.3 EU Policy Level: Supranational Impetus for the Digital Transformation of Agriculture

The supranational impulses for the digital transformation of agriculture at the level of the European Union can be divided into a core area, at the center of which is research funding, and a peripheral area.

The core area is formed by political programs directly aimed at digitalization. These include first and foremost the European Union's digital strategy, in which the Commission formulates an overarching matrix of goals for the Union to launch the digital transformation, consisting of a fair, competitive digital single market based on the rule of law that ensures sustainability and human dignity [EC20a]. To this end, the use of quantum computing, artificial intelligence [EC20b] and blockchain technology is to be promoted, a high-capacity 5G or 6G network is to be built [EC16a] and people's digital literacy [EC21a] is to be supported while protecting critical infrastructure and establishing data interoperability [EC20c]. Within the reliable framework formed in this way, the private sector should be able to develop its potential as an implementer of the digital transformation [EC20a]. As part of the private sector, agriculture is also included in this legal policy framework. The digital strategy is flanked by the Union strategy for a digital single market, in which the use of artificial intelligence is seen as a key technology for reducing pesticide use [EC17]. In addition, the creation of a European data space is planned, in which a data rights respecting exchange of data sets from European big data inventories for digitally driven applications—also for the benefit of agriculture— is to be made possible [EC20d]. The core area of digital policy for the agricultural sector is reinforced by the Green Deal as the Union's overarching sustainability concept, which defines innovative, digitally supported precision agriculture as an explicit goal [EC19b].

As a first step toward implementing the clear Union policy choices in favor of the digital transformation of agriculture, there is a need not only for the elaboration of a reliable policy and legal framework, but also for technological research and development. Since investments in research and development of innovative technologies at the basic research stage are hardly profitable, especially for financially weaker start-ups, there is a need for publicly funded research. In this respect, sustainable primary production, characterized by resource savings, plant health and biodiversity are also mentioned as a priority area of the specific inter- and transdisciplinary research strategy of the European Union for the agricultural sector, which is to be achieved, among other things, through IT-supported smart and precision farming [EC16b]. Research funding is mainly implemented through the EU's "Horizon 2020" program, whose financial framework amounted to €80 billion for the period from 2014 to 2020. Research projects with an insightful focus on the digital transformation of agriculture include, for example, the INSYLO project for the development of an IoT-based solution to optimize the value chain in the livestock sector, or the SMARTFISH H2020 project, which aims to optimize the fisheries sector on the basis of machine learning, sensor technology and artificial intelligence. The successor program "Horizon Europe" provides a budget of 95.5 billion euros for the period from 2021 to 2027 for the funding of research projects [EC18a]. The eligible clusters continue to include the digital transformation of agriculture. The cooperation of member states in the implementation of research projects is to be stimulated via so-called ERANETs (European Research Area Networks). The research consortia should also include actors along the value chain in accordance with a multi-stakeholder approach [EC16b].

For this purpose and in order to bundle different funding opportunities for the implementation of research projects in a coordinative manner, the Agricultural European Investment Partnership (EIP-AGRI) offers a one-stop shop for agricultural innovations [EC21b]. As a counterpart at the national level, this form of innovation brokerage is complemented by the German Networking Agency for Rural Areas, which—based at the Federal Agency for Agriculture and Food—establishes federal networking between the federal government and the federal states. In order to bring disruptive innovations to the market, the European Innovation Council, which is to be distinguished from this, supports the acquisition of venture capital, which is particularly relevant for start-ups in agribusiness [EIC21].

The periphery of the agricultural digital policy is enriched at the Union level by the Farm-to-Fork strategy, which primarily aims at modernizing the value chain, but which sees a digitally transformed agriculture as an enabler for sustainability improvements, in particular by making sustainability indicators recordable in databases [EC20e]. For this purpose, the Farm Accountancy Network is to be transformed into a Farm Sustainability Network. The Common Agricultural Policy, as a central EU legal steering instrument for the European agricultural economy, is now also pushing the digital transformation of agriculture as a cross-cutting objective and is concretizing this as a specific objective with regard to the development of future national strategy plans by the member states. Following on from this, the member states will in future be accountable for the measures with which digitalization is promoted in the agricultural sector [EC21c]. In Germany, for example, concrete intervention descriptions for the expansion of the data network infrastructure are being developed on the basis of a SWOT analysis. Within the strategy plans, the member states will also have to implement so-called eco-schemes in the future. These are an element of the first pillar of the Common Agricultural Policy, which is aimed at promoting agriculture, and which in turn must be distinguished de lege ferenda from the second pillar, which is aimed at the development of rural areas. Eco-schemes provide financial incentives based on hectares, which are granted for non-compulsory measures directed at climate protection, biodiversity or improvements in ecosystem services, and sustainability [LR19]. In view of the above-mentioned digitally transformative objectives of Union policy, such measures may also include the use of digitally based technologies aimed at this end [EC20f].

Finally, as an actor of European civil society, the industry association "Digital Europe" is active at the Union level as a policy advisor. It also supports the expansion of 5G infrastructure in rural areas and the use of open data in agriculture [DE20].

1.6.4 German Policy Level: Digitally Transformed Agriculture as a Cross-Cutting Objective Across Ministries

At the national level, the goal of a digital transformation of the agricultural sector is being pursued by various ministries. In its arable farming strategy, the Federal Ministry of Food and Agriculture paradigmatically presupposes a digitally transformed agriculture as a driver for sustainability and efficiency [FMF19a]. To this end, it intends to establish a center of expertise for geoinformation and a research center for agricultural remote sensing with the "Future Programme for Digital Policy in Agriculture," which is to bundle the stocks of geodata already available in Germany [FMF18]. On behalf of the Federal Ministry of Food and Agriculture, the Fraunhofer Institute for Experimental Software Engineering has analyzed the political, legal and technological framework conditions for the conception of a state agricultural data platform within the framework of a transdisciplinary feasibility study and, on this basis, has developed concrete implementation proposals for the integration of public and private data pools as the basis for IT-supported smart farming services for the agricultural sector [FSE20]. During a conference on the digital transformation of the agricultural value chain held in December 2020, the possibility of a uniform agricultural data space guaranteeing data sovereignty was also explored as part of a governance framework for agricultural data stocks [FMF20] on the basis of a legal opinion [Här21].

The digital agenda of the Federal Ministry for the Environment, Nature Conservation and Nuclear Safety, in which nature conservation, agriculture and water management are combined as a transformation field, aims in a similar direction. According to this, public and operational data are to be bundled as a basis for decision support systems and made usable for biodiversity monitoring [FME20]. According to the Federal Biodiversity Strategy, the spatial data infrastructure should provide a basis for legal decisions in this area [FME07]. With an interdisciplinary joint research project, the Federal Ministry of Education and Research is also promoting the development of an agricultural management system that aims to use the potentials of digitalization to improve the provision of ecosystem services [DAK21]. In order to test innovative smart farming applications, the Federal Agency for Food and Agriculture is also providing administrative support for "experimental fields" worth 50 million euros in total. An interdisciplinary competence network with experts from science and industry evaluates the results found there [FMF19b].

The Federal Ministry of Transport and Digital Infrastructure is pushing the expansion of the technical data network infrastructure in rural areas [FMT17]. For such improvements to the agricultural structure, the Basic Law for the Federal Republic of Germany (Grundgesetz) also provides for cooperation between the federal government and the federal states. In future, agriculture, which is networked in this way and increasingly digitally transformed, is to be grouped together as an agricultural domain and integrated into the GAIA-X data infrastructure planned by the Federal Ministry for Economic Affairs and Energy. This is intended to network decentrally stored data for AI-supported applications in an interoperable manner on the basis of European data protection standards, thus enabling Europe's digital sovereignty [FMEA20]. This overall strategic concept can be linked to the delivery model of the Union's digital strategy, with which the EU Commission intends to make its data resources usable and integrable in an interoperable manner [EC18b].

If the different political approaches of the individual ministerial departments are coordinated at national level in the sense of a whole-of-government approach, there is great synergy potential for an agricultural digital policy that promotes the digital transformation [OEC19b]. The party-political consensus, which is articulated across parliamentary groups, is likely to have a supportive effect in that, for example, the CDU [CDU19], SPD [SPD19], Greens [GR19] and FDP [FDP21] are in principle in favor of a digital transformation of agriculture. Likewise, the German Farmers' Association as the civil society representative of agriculture and the Bitkom Association as the representative of the interests of the digital economy are in favor of this objective [GFA21]. With regard to the future design of the agricultural digital policy agenda, the German Farmers' Association calls for data sovereignty of farmers, while the Bitkom Association argues in principle for open data access and standardized data formats [Bit19].

1.6.5 The Future: Federal Diversity and Multi-Level Integration

An overall view of the individual political levels shows a fundamental cross-level consensus on the great potential of a digitally transformed agriculture, which is universally supported as such. In order to realize this objective, the ministries involved are producing a wide range of different conceptual approaches.

In the future, it will be necessary to link these concepts in order to generate synergy potentials. Remaining political diversities in the implementation of transformation concepts are based on the federal division of competences and can be seen as a fruitful basis for democratic policy-making in the sense of competitive federalism, which in this respect form a creative pool of innovative political concepts [Här12]. Such a pool enables an agricultural digital policy that is capable of reacting adaptively to disruptive leaps in innovation on the one hand and, on the other, of shaping the framework in an agile manner in which the stakeholders involved can turn the vision of a digitally transformed agriculture into reality.

- **Acknowledgements**

This section was written in connection with the joint research project DAKIS (Digital Agricultural Knowledge and Information System), funded by the Federal Ministry of Education and Research, funding code 031B0729J. The author is responsible for the content of this publication.

1.7 Agricultural Digital and Data Law

Ines Härtel

Abstract

A well-ordered agricultural digital and data law, with its differentiated forms and contents of regulation, underpins a sustainable Agriculture 4.0. Data sovereignty, regulated self-regulation, codes of conduct and the GDPR protection regime on privacy form the cornerstones for this. These are specified by special law, such as for security and liability in AI.

1.7.1 Contexts of Agricultural Digital Law

A well-ordered agricultural digital and data law [Här19], with its differentiated forms and contents of regulation, underpins a sustainable agriculture 4.0. European values and principles, as they have been shaped since the Enlightenment, modernity and today's late modern age, form the basis for this. The development of digitalization—in agriculture as well as in other areas—with its commitment to human dignity, autonomy, human rights/basic rights and liberal democracy based on the rule of law, is paradigmatic for a "third way." The liberalist strategies of the USA with rather few regulations on one hand, and China's path with comprehensive centralized state guidelines and quasi-dictatorial controls on the other, are seen as antipodes. Because of the wide range of areas of application, bringing the interests of the different actors into an adequate balance, is particularly important when establishing a legal framework for agriculture 4.0. For farmers, the legal institution of agricultural data sovereignty, which also corresponds to (the more comprehensive) digital sovereignty, and its incorporation into the legal framework, is of fundamental importance. In this context, digital-transformative agriculture offers considerable opportunities to better meet both social and ecological requirements. The establishment of an adequate legal and technical framework for Digital Farming, that takes into account the specific needs of the agricultural sector, remains an important task for the EU. A task which is highly dynamic considering the rapid digital and technological developments. Such a legal framework must cover an extensive spectrum of agri-digital technologies, in particular satellite technology, sensor technologies, algorithm and big data analysis, cloud computing, artificial intelligence, robotics, blockchain, drones and assisted/partially autonomous driving. Areas of application for products and services in the field of agriculture 4.0 include: digital farm management systems for arable farming and livestock farming, agricultural apps, intelligent tractors and equipment that is structurally combined in a network, agricultural drones / agricultural copters, robotic systems for animal feeding and milking, environmentally friendly optimization of cultivation, as well as agronomic and harvesting processes. These applications are increasingly cloud-based.

1.7.2 The Normative Framework: Right to Food and SDGs

1.7.2.1 Right to Food and Sustainable Development Goals

Digitalization as a technology is not an end in itself but should serve to achieve normative goals. Thus, the digital transformation in the AgFood sector along its value chains and value networks contributes to the effectiveness of the right to food as well as to sustainable development. The human right to food as defined in Article 25 (1) of the 1948 Universal Declaration of Human Rights (UDHR) and Article 11 of the 1966 International Covenant on Economic, Social and Cultural Rights (ICESCR) encompasses a right to food security. This right is fulfilled "when all people, at all times, have physical and economic access to sufficient, safe and nutritious food to meet their dietary needs and food preferences for an active and healthy life" [FAO96]. Goal 2 of the Sustainable Development Goals formulated by the United Nations in its "2030 Agenda for Sustainable Development" is to "end hunger, achieve food security, and improve nutrition and promote sustainable agriculture." This is further substantiated in several sub-goals.

1.7.2.2 GFFA Communiqué

The 2019 communiqué on "Agriculture goes digital – smart solutions for future farming" adopted by the Global Forum for Food and Agriculture (GFFA)[2] is particularly worth mentioning. Agriculture ministers of 74 nations have signed this communiqué. The aim is to use digitalization in order to make agriculture more efficient and sustainable and to improve life in the countryside. To this end, the development of appropriate, site-adapted and scalable digital solutions in agriculture is to be promoted. Globally, the objective is to create the necessary "digital infrastructure" for farmers and accelerate its expansion. Cooperatives and cooperative models in implementing digitalization in agriculture are going to be given support. Research and education in the field of agriculture 4.0 will be promoted in order to make the value chain more efficient and sustainable. Digital solutions will also be used to provide farmers with appropriate information and improved market access (including e-markets for food and agriculture). In addition, nine important goals with regard to improving data use, ensuring data security and data sovereignty in the interest of farmers are enunciated. Those goals are formulated in terms of a sustainable digital agriculture:

— working on international solutions with agricultural stakeholders in order to develop standards and reduce global differences in regulations on data collection, data security and data use;
— enabling the effective use of digitally recorded data;
— improving the interoperability of digital systems in order to enhance the possibilities for data exchange, data use and data analysis by farmers, academia, industry and policy makers;
— ensuring that farmers are not dependent on individual digital systems and that the intellectual property rights and privacy rights of users with regard to digital innovation and information are protected;
— increasing trust and transparency in relation to data governance principles, including rules on authorization and oversight in data collection and use and promoting data use models where farmers can, in compliance with national rules, decide themselves on whether to pass on their operating, machine and business data;
— making public data available through appropriate platforms;
— promoting digital solutions in order to strengthen the transparency, efficiency and integrity of supply chains and taking effective steps against counterfeits, fraud and smuggling;
— promoting international digital data infrastructures in order to strengthen the cross-border fight against animal and plant pests and diseases and

---

2   GFFA is an international conference regarding central questions of global agriculture and food economics, taking place annually since 2009 as part of International Green Week in Berlin, status as of 15 July 2019, available at ▶ https://www.gffa-berlin.de/.

- developing digital methods at the World Organisation for Animal Health (OIE) as part of the modernization of the OIE-WAHIS notification system (World Animal Health Information System).

The FAO is being asked to develop a concept for establishing an international Digital Council for Food and Agriculture that will advise governments, promote the exchange of ideas and experience and help everyone to harness the opportunities presented by digitalization. Although these declarations are merely soft law,[3] in political, legal and practical terms they are ground-breaking for the design and implementation of the agricultural policies of the states and the EU—especially since soft law can later on lead to binding regulations. In this sense, the sovereign authorities have a corresponding responsibility within the framework of their domestic and foreign agricultural policy.

1.7.2.3 The Common Agricultural Policy of the European Union After 2020

Future secondary legislation on the Common Agricultural Policy (CAP) will contain an explicit normative reference to digitalization. This results from the European Commission's legislative proposals on the CAP beyond 2020, which set a new cross-cutting goal for the digitalization of agriculture. The agricultural sector is to be modernized through financial support for digitalization in agriculture and in rural areas.[4] This cross-cutting goal affects all agricultural target areas (food security, protection of ecological resource protection, strengthening rural areas). The application system for agricultural subsidies will also be further digitalized, e.g., through a geodata-based application by the farmer and a geodata-based application system by the agricultural administrations.[5] Financial support for digital AgTech can accelerate the progress of a sustainable agriculture and food industry but requires that farmers use AgTech in a legally compliant manner. At the same time, digital data sovereignty, and the corresponding (more comprehensive) digital sovereignty is of fundamental importance for the farmer.

1.7.3 Agricultural Data Sovereignty and Agricultural Data Space

Regarding data and also agricultural data, there is no exclusive property right in the traditional sense. The (constitutional) right to property (Article 17 Charter of Fundamental Rights of the European Union, Article 14 of the Basic Law for the Federal Republic of Germany (Grundgesetz)) protects trade and business secrets. But most of the data generated on farms is not covered by this. However, the need for protection

---

3 For the legal and practical significance of Soft Law see [Mon14].

4 Article 5 lit. b Regulation of the European Parliament and of the Council of 2 December 2021 establishing rules on support for strategic plans to be drawn up by Member States under the Common Agriculture Policy (CAP Strategic Plans) and financed by the European Agricultural Guarantee Fund (EAGF) and by the European Agricultural Fund for Rural Development (EAFRD) and repealing Regulations (EU) No. 1305/2013 and (EU) No. 1307/2013.

5 Cf. Article 66 para. 1 lit. b Regulation of the European Parliament and of the Council of 2 December 2021 on the financing, management and monitoring of the Common Agricultural Policy and repealing Regulation (EU) No. 1306/2013 ('Horizontal CAP Regulation')

in the new digital agricultural world requires a new legal institution. The EU legislator should therefore in future regulate digital agricultural data sovereignty as an expression of part of the freedom to conduct a(n agricultural) business [Här21].

Farmers should have a right to access their own agricultural data (including the metadata, and if applicable consent to deletion), a right to confidentiality in the sense of freedom to decide on the transfer of data to the self-selected provider and a right to authenticity as the knowledge of the provider and his business conduct.[6] Data sovereignty and overall digital sovereignty are ultimately an expression of fundamental rights. The lack of protection for farmers under the applicable law, contractual practices as well as the high importance of agriculture for the European food sovereignty, are why a binding legal provision of agricultural data sovereignty is needed. In particular the fact that farmers of small and medium-sized agricultural enterprises lack the necessary contractual capacity, leads to problematic contractual practices. Information and negotiation asymmetries between farmers and suppliers are the reasons for a lack of balance of interests in private data rights. Furthermore, dependencies on a single digital provider can be observed, making it difficult or impossible for the farmers to switch to other providers (vendor lock-in effect).

The legal institution of digital agricultural data sovereignty would be a fundamental component of the common European agricultural data space, which is to be established according to the European Commission's Communication "A European Strategy for Data" published on February 19, 2020.[7] In the Commission's view, the common European agricultural data space should serve to improve the sustainability and competitiveness of the agricultural sector.

Agricultural data law is a vital area of legislation for the European agricultural data space and must be further developed. This refers mainly to the business-to-business (B2B) use of data—for example in the context of agricultural data-driven business models and agricultural data ecosystems. However, it also extends to Government-to-Business (G2B) and data sharing between public authorities (G2G), for example in the context of e-government and geodata use.

1.7.4 Self-Regulation—Code of Conduct as Private Soft Law

For farmers and suppliers, the important questions in the context of B2B contractual relationships related to agricultural digitalization are: Who has what kind of rights to what data or which data sets? And what kind of obligations correspond to these rights? The "EU Code of conduct on agricultural data sharing by contractual agreement," which was signed by nine organizations and associations[8] from the agricultural sector in 2018, provides orientation for the design of contractual provisions under agricultural data law. This code was developed in the course of private self-regulation and represents private soft law due to

---

6   See DLG, Digitale Landwirtschaft, 2018, p. 5; Friends of the Earth Europe, Digital Farming, February 2020, p. 16; Fraunhofer IESE, Kalmar/Rauch, Wie schafft man Datensouveränität in der Landwirtschaft, May 2020.
7   COM(2020) 66 final, p. 37.
8   The signatories include Copa and Cogeca (The united voice of farmers and their cooperatives in the European Union), CEMA (European Agricultural Machinery Association), fertilizers Europe (The European Fertilizer Manufacturers Association), CEJA (European Council of Young Farmers), ECPA (European Crop Protection Association), EFFAB (European Forum of Farm Animal Breeders), FEFAC (European Compound Feed Manufacturers' Federation) und ESA (European Seed Association).

its voluntary nature. Nevertheless, the signatories appeal to all players in the Ag-Food value chain to base contracts on this code of conduct. The code of conduct is divided into the following sections: introduction; definitions; attribution of the underlying rights to derive data; data access, control and portability; data protection and transparency; privacy and security; liability and intellectual property rights; annexes (definitions of different types of data in the agri-food sector; four case studies; EU regulatory framework for the sharing of agricultural data; main legal principles to have a balanced contract—contract check list for agricultural data).

Good approaches for achieving an adequate balance of interests between market participants can certainly be found in the code of conduct. However, a number of guidelines remain vague in their wording and thus do not constitute concrete recommendations for potential conflicts of interest. This applies in particular to the question of data authorship/data ownership. It is not clear exactly which rights farmers are entitled to in the event of multiple data authorship (several actors are data authors) within the digital data value chain. Nor do the case studies in the appendix (especially examples 2 and 4), which hint at this problem, lead to any concrete legal solution. Furthermore, the possible right to data access is insufficiently addressed. Data security/IT security/cyber security have only been included to a rudimentary extent and require further specification. Liability is dealt with very incompletely. Data portability has been formulated on the basis of the General Data Protection Regulation (GDPR). But the code of conduct does not clarify the question regarding a possible guarantee of data portability given by the provider to the farmer. There is also a lack of sufficient information on the interoperability offered to the farmer. Another deficit is that there is no indication of possible legal consequences or sanctions in the event of breaches of contractual provisions.

A further development of the EU code of conduct as soft law should take these points into account. A look at foreign codes of conduct is helpful in this respect—and can to a certain point also be an inspiration regarding binding Agricultural Data Law.

In 2014 the American Farm Bureau Federation, the largest farmers' association in the USA, established the "Privacy and Security Principles for Farm Data." On May 5, 2015, a number of agricultural organizations and companies signed this code of conduct. With its three pages, the document is more concise than the EU code of conduct. Thirteen points are listed and referred to as core principles. The first principle is "education" of farmers. The industry should work to develop programs that create educated customers who understand their rights and responsibilities. In addition, contracts should be written in simple, easy and understandable language. Under the heading "ownership," it is stated that farmers have ownership of data generated during their farm operations. In contrast to the EU code of conduct, certification is envisaged in the US as a tool for enforcing the guidelines/principles for agricultural data contracts. Since 2016, the "Ag Data Transparent Seal of Approval"[9] has been awarded.

The "New Zealand Farm Data Code of Practice (NZ Farm Data Code)" was introduced in 2014 to provide guidelines for effective data sharing in the New Zealand agricultural industry. The code explicitly refers to the data of primary producers (farmers). A certification system is used to implement the code, and a seal is awarded. In the certification process, the AgTech company must demonstrate in a self-disclosure that it complies with the Farm Data Code. It must fill out a questionnaire

---

9 Regarding the method of operation and the following remarks see ▶ https://www.aglaw.us/agdatatransparent.

(checklist) and provide evidence of the respective answer. This compliance checklist is much more comprehensive than that used in the US certification system.

The "Australian Farm Data Code" of February 1, 2020 has been drawn up by the National Farmers Federation and the agricultural industry. In contrast to the codes in the USA and New Zealand, no certification system is envisaged. After an introduction, the principles of agricultural data include transparency, comprehensibility, honesty, fair and equitable use of data, as well as data access rights of the farmer and portability.

In Germany there is a joint industry recommendation (Branchenempfehlung) of seven associations[10] on the "data sovereignty of the farmer" of February 28, 2018, which refers to the "collection, use and exchange of digital farm data in agriculture and forestry." One positive aspect is that the recommendation briefly and comprehensively sets out important aspects in a concise manner. A basic principle is the "ownership of data."

From the farmers' point of view, the problem with the existing codes of conduct is that they do not effectively guarantee the sovereignty of agricultural data. Certification systems seem to be able to provide some remedy. The certification system in New Zealand is more differentiated than in the USA.

---

10 Deutscher Bauernverband/German Farmers' Federation (DBV), Bundesverband der Maschinenringe/ Federal Association of Machinery Pools (MR), Bundesverband der Lohnunternehmen/ Federal Association of German Contractors (BLU), Deutsche Landwirtschaftsgesellschaft/ German Agricultural Society (DLG), Deutscher Raiffeisenverband/ German Raiffeisen Association (drv), LandBau-Technik-Bundesverband/Agricultural Technology Association, Verband Deutscher Maschinen- und Anlagenbau/ German Engineering Federation (VDMA).

1.7.5 Regulated Self-Regulation: Non-Personal Data

The EU legislator has so far used the concept of regulated self-regulation for the area of B2B data rights. According to Article 6 (1) of the Free Flow of Data Regulation (EU) 2018/1807, codes of conduct should cover the following aspects:
- procedures to facilitate the change of service provider and the transfer of data in a structured, common, machine-readable format or open standard format;
- rules on minimum information in order to ensure that the user receives accurate, clear and transparent information, prior to the conclusion of the data processing contract, on the processes, technical requirements, time limits and charges applicable to a user wishing to change to another service provider or to transfer data back to his own IT systems;
- approaches to certification schemes;
- communication plans to disseminate the rules of conduct to relevant stakeholders.

These aspects relate to some key self-regulation requirements in general terms but should be specified for the agricultural sector. Furthermore, the concepts of self-regulation and regulated self-regulation are not sufficient to effectively guarantee agricultural data sovereignty. Therefore, regulation through binding specifications is necessary. Regulations specific to agriculture should serve to solve practical problems (such as the vendor lock-in effect in particular) and aim to remedy digital-related structural hazards in agriculture. For non-personal data in the B2B sector in particular, the data principles of transparency, fairness, data portability, interoperability, data security and data quality must be legally anchored, as well as agricultural-specific regulations on general terms and conditions and contract law and regulations on agricultural-specific data access right.

1.7.6 Privacy According to the EU General Data Protection Regulation

In the European agricultural data area, the privacy protection regime under the General Data Protection Regulation (GDPR) makes only a limited contribution to the practical guarantee of agricultural data sovereignty for farmers. One reason for this is the limited scope of application of the regulation for personal data of natural persons, which at the same time causes legal uncertainties for market participants. For example, there are legal grey areas with regard to the question of which of the types of data generated by Digital Farming are personal and which are not.

The term "personal data" is widely understood. According to Article 4 No. 1 GDPR, this includes "any information relating to an identified or identifiable natural person." The broad term thus covers directly personal data, i.e., data which can be directly collected, but also indirectly personal data which can only be obtained by consulting further information and/or by further activities. Depending on whether the processed information has a (direct or indirect) personal reference, the obligations under data protection law are either completely or not at all applicable. It is questionable when it is sufficient that a personal reference can be established. Recital 26 of the GDPR provides guidance on this (objective factors such as costs, time required and available technology).

Behavioral patterns usually have a personal reference. This concerns, for example, deductions from soil properties to the farmer's ability to do farming. Factual information such as financial and ownership relationships can also constitute personal data. Which agricultural data generated by agricultural machines, agricultural robots and drones are to be qualified as personal data remains in need of clarification.

Agricultural data includes in particular all basic operational data (e.g., location, size), production-related data (e.g., yield planning), machine-related data (e.g., machine equipment), key business data (e.g., financing) and other machine-/system-related data (e.g., quality characteristics of the harvested goods, type/quantity—use of fertilizers, pesticides). Due to the combination with other information, the respective factual information mentioned could allow conclusions to be drawn about an individual farmer. However, it may also be qualified as non-personal data.

If personal and non-personal data are "inseparably linked," the GDPR applies in full to the entire mixed data set (cf. Article 2 (2) Regulation 2018/1807). Inseparability should be present if the "separation is impossible or is considered by the person responsible to be economically inefficient or technically not feasible."[11]

Furthermore, the personal data must refer to "natural persons." If agricultural holdings are organized as legal entities, data records relating to the holding are generally to be classified as non-personal data. However, using the case law of the CJEU in the Schecke case,[12] information about legal persons can also constitute personal data if the name of the legal person is identical to that of a natural person who owns the legal person or if the information relates to a specific or identifiable natural person. This distinction is also associated with legal uncertainties.

Another legal problem regarding the assignment of information to personal data concerns the area of anonymization and aggregation of data. In principle, anonymized data and the processing thereof do not fall

---

11 COM(2019) 250 final, p. 9.
12 CJEU, Judgment of 9 November 2010, joined cases Volker and Markus Schecke GbR (C-92/0) and Hartmut Eifert (C-93/09)/Land Hessen, ECLI:EU:C:2010:662, para. 52.

within the scope of the GDPR (see Recital 26). However, due to the new technical possibilities of far-reaching big data analyses, there is an increased risk of re-identification. Recital 9 of Regulation (EU) 2018/1807 contains the following note in this regard: "If new technological developments make it possible to convert anonymized data back into personal data, these data must be treated as personal data (…)."

The contribution of the GDPR to agricultural data sovereignty must also be assessed differently with regard to data protection principles and the rights of data subjects. For example, there are considerable doubts about the mass suitability of consent as a legal instrument effective for data protection. The ideal of fully informed and voluntary consent encounters barriers in practice. In the absence of parity between the contractual partners, the premise "take it or leave it" applies. With regard to the rights of the data subject, it should be noted that there is no data access right of the data subject that extends to real time. The right to data portability (data transferability) in accordance with Article 20 GDPR gives the data subject (farmer) the right "to receive the personal data concerning him/her that he/she has made available to a responsible party in a structured common and machine-readable format (…)." However, the person responsible is not obliged to provide and maintain a permanent interface.

Notwithstanding the aforementioned limitations of the General Data Protection Regulation, this regulation also contains a far-reaching data protection mechanism within its scope. The far-reaching and effective data protection is demonstrated, for example, not least by the provisions for the transfer of personal data to third countries pursuant to Article 44ff. GDPR. The jurisprudence of the European Court of Justice in the European Data Protection Regulation also sets a strict and consistent standard for the interpretation of the legal provisions of Article 44ff. GDPR, which as a result serves an effective basic data protection right in the sense of Article 8 Charter of Fundamental Rights of the European Union. For example, in its judgment of 16 July 2020 in the Schrems II case,[13] the CJEU rightly declared the European Commission's adequacy finding on the EU-US Privacy Shield Agreement[14] to be invalid. The CJEU concludes that the law of the USA does not guarantee the level of protection required under Article 45 GDPR in the light of Articles 7, 8 and Article 47 Charter of Fundamental Rights of the European Union. The reason for this lies in the extensive monitoring powers of the US security authorities for foreign intelligence. This case illustrates in a special way the globally existing tension between security and freedom, which extends into the B2B data rights relations.

Increased data protection requirements in terms of transparency and data quality could in future apply to AI systems in agriculture. The special standard Article 22 GDPR, which applies to automated decisions, has hardly gained any significance in practice so far. In the agricultural sector, however, automated decisions with legal effect are conceivable in the future, e.g., in the context of predictive maintenance for AgMachinery, where, in the case of necessary maintenance identified by AI, this is initiated directly without the farmer placing a separate order; in this case, the farmer has only agreed to the system in advance by means of a framework agreement. Such

---

13 CJEU, Judgment of 16 July 2020, C-311/18, Facebook Ireland and Schrems, ECLI:EU:C:2020:559, para. 199; Opinion of AG, 19 December 2019, C-311/18, ECLI:EU:C-2019:1145.
14 Commission Implementing Decision (EU) 2016/1250 of 12 July 2016 correspondent to Directive 95/46/EG of the European Parliament and the Council regarding the adequacy of protection offered by the EU-US privacy shield, OJEU, No. L 207/1, 1 October 2016.

fully automated decisions are subject to increased transparency requirements (see Article 22 (2) lit. a and c in conjunction with Article 22 (3) GDPR). The responsible person would have to inform the farmer in a meaningful way about the existence of automated decision making, the logic involved and the scope and effects of the processing operations.[15] The program code or algorithm code does not have to be disclosed, because it is covered by trade secrecy and intellectual property rights. However, AI applications in the agricultural sector that are not covered by Article 22 GDPR could also be subject to high data protection requirements in the future. This is shown in particular by the position paper of the Conference of the Independent Data Protection Supervisory Authorities of the Federal Government and the Federal States (Konferenz der unabhängigen Datenschutzaufsichtsbehörden—DSK) in Germany of November 6th, 2019. Here the necessary technical and organizational measures (TOM) for different phases of the life-cycle of AI systems are differentiated.

1.7.7 Artificial Intelligence: Security Law and Liability Law

In order for AI systems with their advantages for sustainable agriculture to be more widely used in the future, the farmer needs specific digital technology acceptance in this area in addition to agricultural data sovereignty. This in turn requires that the law provides appropriate solutions for AI-specific liability risks in agriculture. The existing product safety law and product liability law do not yet provide an adequate legal framework with regard to AI specifics.

New impulses for the discussion on the development of a legal framework for AI at the EU level were given in particular by the White Paper of the European Commission "On Artificial Intelligence—A European Concept for Excellence and Trust" of February 19, 2020,[16] the report of the Expert Group on Liability and New Technologies "Liability for Artificial Intelligence," also published on February 19, 2020, and the "Report on the Impact of Artificial Intelligence, the Internet of Things and Robotics on Security and Liability,"[17] published by the Commission. It follows from this that the following special features of AI (as well as IoT) must be taken into account in the further development of product safety and liability law[18]:

— the complexity (diversity of actors in digital ecosystems, multitude of digital components—hardware, software, services),
— Opacity (opacity of processes, black box through self-learning, makes it difficult to predict the behavior of an AI-based product),
— Openness (updates/updates or improvements/upgrades, for this purpose interaction with other systems or data sources),
— Autonomy (tasks with less human control, changing algorithms),
— Data dependency (dependency on external information, which is not pre-installed, but generated by built-in sensors or communicated from outside)
— Vulnerability (particular vulnerability to cyber security breaches due to the openness and complexity of Digital Ecosystems).

---

15 Cf. Martini, in: Paal/Pauly, DSGVO, 2nd ed. 2018, Art. 22 para. 4.

16 COM (2020) 65 final.
17 COM (2020) 64 final.
18 See COM(2020) 64 final, p. 20; Expert Group on Liability and New Technologies, Report, European Union, 2019, p, 32–34.

The AI specifics mentioned above are also immanent AI-specific liability risks in agriculture. If the use of AI systems causes personal injury, property damage or environmental damage, the question arises for farmers but also for other AI actors involved or third parties as to who is liable for what. AI damage cases are e.g., conceivable[19] if an AI-controlled harvesting robot picks strawberries that are still unripe and not yet marketable, if farm animals such as chickens or pigs are injured during automated feeding, if animal diseases are not detected by an AI-controlled animal health management system or if the wrong medication is administered, if a robot applies fertilizers or pesticides in such a way that the water's edge is not maintained or the application limits are exceeded, or if a cyber attack causes the entire (future) farm 4.0 to collapse. Violations of agricultural specialized right (among other things fertilizer, plant protection agent right, animal protection right) bear the risk that agricultural subsidies on EU or national level are shortened or even completely cancelled.

Conventional product safety law should develop into AI safety law that incorporates the AI specifics listed above and also extends to AI services.[20] With a view to effective risk management and risk control, it is necessary to develop new behavioral obligations for the responsible actors in autonomous systems. In addition, the use of the product during its entire life-cycle would have to be included in the risk assessment, for example. With regard to the vulnerability of AI systems (due to their openness and complexity), legally binding basic requirements for cyber security must be anchored in the future. The current legal act on cyber security (Regulation 2019/881) is inadequate, as it only provides a voluntary framework for the cyber security certification of products.[21] In order to ensure the proportionality of regulatory intervention, a "risk-based approach" is to be chosen with regard to the security requirements for AI applications. The higher the risk posed by an AI application, the higher the legal requirements for it. While the Data Ethics Commission in Germany [Zwe19] has already divided algorithmic decisions into five risk levels, the EU Commission's AI White Paper initially only distinguishes between high-risk and non-high-risk AI applications with regard to the legal framework. A strict regulatory framework is proposed for high-risk AI applications. For other applications, the introduction of a voluntary labelling system is considered as an option. A differentiated regulatory regime for AI is intended by the drafted Artificial Intelligence Act of the EU.[22]

With regard to AI liability cases, the proposals focus on non-contractual liability law. It is proposed to create a new legal regime on strict liability for operators of AI systems. With regard to the subject of liability, a flexible concept of the "operator" is proposed. Liability is assumed by the person who controls the risk associated with the operation of the new digital technologies and benefits from their operation (the so-called operator).[23]

In addition, the existing strict product liability is to be adapted to the specifics of AI, especially with regard to the terms product, defect and manufacturer. The national fault-based tort liability is also to be modified with regard to AI specifics by Union law, especially with regard to the rules on the burden of proof [Här20]. In contrast to contractual liability, tort liability covers damage

---

19 See *Wilde-Detmering*, InTeR 2019, 174, 179.
20 For details see the Commissions demands, COM (2020) 64 final, p. 6–14.
21 COM (2020) 64 final, p. 7.
22 Proposal for a Regulation of the European Parliament and of the Council laying down harmonised rules on artificial intelligence (Artificial Intelligence Act) and amending certain Union legislative acts, COM (2021) 206 final.
23 Expert Group on Liability and New Technologies, Report, European Union, 2019, p. 40.

to absolutely protected legal interests, such as life, health, property or the general right of personality, and aims at their appropriate restitution. The strict liability for increased risks of damage is to be accompanied by a mandatory liability insurance by law, which covers all damages caused by AI.[24]

1.7.8 Further Fields of Law

A separate area of law concerns the mobility of assisted/partially autonomous/autonomous driving up to unmanned flight systems (agricultural drones) in agriculture. With regard to semi-autonomous driving tractors in Germany, road traffic law on the one hand and product safety law on the other hand apply. Associated with this are the applicable regulations on hazardous and fault-based liability as well as the data processing regulations in the German Road Traffic Act. However, special regulations for autonomous driving in the field are still required. Other EU member states have their own legislation. The legal framework for the use of agricultural drones is determined by the EU Civil Aviation Regulation 2018/1139, which takes a risk-based approach to the certification and operation of drones. Further regulations can be found in German aviation law (drone driving license, operating license, etc.). With regard to liability risks that may arise from the use of agricultural drones (as unmanned aircraft), the legal framework e.g., in Germany for the implementation of EU law[25] is already satisfactorily designed. There is a strict liability of the operator when operating an (agricultural) drone according to Sect. 33 (1) sentence 1 German Aviation Act (Luftverkehrsgesetz) and the obligatory liability insurance according to Sect. 43 (2) sentence 1 German Aviation Act.

With regard to cyber security, reference should be made to Regulation (EU) 2019/881 on the European Union Agency for Cyber Security (ENISA) and on the certification of the cyber security of information and communication technology, as well as to Directive (EU) 2016/1148. The legal framework for cyber security in Germany relates in particular to critical infrastructures and thus also to the food sector.

With regard to cloud computing, a coherent set of cloud regulations is to be created in the EU that will enable cloud providers and users to access competitive, secure, and fair cloud services.[26] This will secure important storage options for agricultural data. GAIA-X, which combines the integration of cloud-to-cloud systems with Edge Systems and secure blockchain encryption, is also a further development from a legal perspective.

In agriculture, blockchains with smart contracts enable the traceability of food/agricultural products in the value chain, fast access to land registry entries, and the securing of contracts (e.g., commodity futures exchanges), payments and leases. A specific legal framework is currently being developed.

For the use of open data, in particular geodata, in agriculture (e.g., on land use, weather conditions, water networks), the Federal Spatial Data Access Act and the Spatial Data Infrastructure Acts of the federal states are in place to implement EU secondary legislation. There are restrictions on open data access due to the protection of personal data, intellectual property rights and company and business secrets.

In the meantime, a large number of agricultural data platforms/agricultural data ecosystems exist. From a legal point of view, a distinction must be made between private sector and sovereign platforms. Of particular importance are questions of interoperability, data integrity/security,

---

24 Expert Group on Liability and New Technologies, 2019, Nr. 33, p. 61.
25 Regulation (EU) 2018/1139 – civil aviation regulation.

---

26 COM (2020) 66 final, p. 21 f.

responsibilities of the operator and, in the future, facilitation of data exchange platforms under agricultural competition law (in the form of meta-cloud systems).

1.7.9 Outlook

In view of the diversity of the areas to be regulated, the digital transformation requires well-ordered agricultural digital law, which includes in particular data law, the effects of specific digital technologies, and an orientation toward liberal, constitutional democracy with its fundamental values. A discursive and creative approach in the political-legal multi-level system of the European Union and its member states is indispensable, as is, as far as possible, cooperation in the global system of states. The digital-technological transformation of agriculture serves primarily to secure food supplies on the basis of efficient, resource-light and socially secure agriculture. The new approaches to reconciling the agricultural economy and ecology that have been made possible by digital technologies (especially in Digital Farming) can also contribute to compliance with the planetary guard rails in the sense of a "digital sustainability society"[27] in the anthropocene. However, the requirements of a well-ordered digital law related to this—from algorithmic rules to AI and robotics rules—require considerable efforts not only in the current digital technological challenges, but also in connection with other developments that will shape the agricultural and food sector in the future. On the one hand, this includes all advanced and new (digital) technologies, including questions of ethical and cultural digitality, the design of human–computer interaction and anthropotechnologies, assistive digital technologies and post-human enhancement. On the other hand, new breedings based on the findings of genetics and epigenetics, developments in the fields of biodesign/synthetic biology and proteins, and biological-technological convergences, bioeconomy and biointelligent value creation in significant advancements of existing cultivation methods and "Farm to Fork" strategies are to be included. The importance of the sphere of law with its normative reference to order therefore lies not only in the current formation of an agricultural digital and data law, but in the future also in the enabling and taming of novel (agrarian) technologies in the perspective of a basic value-based humanism that transcends the relationship between ends and means.

References

[Alr17]   Alroy, J. 2017. Effects of habitat disturbance on tropical forest biodiversity. *Proceedings of the National Academy of Sciences of the United States of America* 114(23):6056–6061.

[BB20]   Bundesministerium für Umwelt, Naturschutz und nukleare Sicherheit [German Federal Ministry for the Environment, Nature Conservation and Nuclear Safety], and Bundesministerium für Ernährung und Landwirtschaft [German Federal Ministry of Food and Agriculture], Ed. 2020. *Nitratbericht 2020*. ▶ https://www.bmu.de/fileadmin/Daten_BMU/Download_PDF/Binnengewaesser/nitratbericht_2020_bf.pdf.

[Bfn17]   BfN—Bundesamt für Naturschutz [German Federal Agency for Nature Conservation], Ed. 2017. *Agrar-Report 2017. Biologische Vielfalt in der Agrarlandschaft*. Bonn. ▶ https://www.bfn.de/fileadmin/BfN/landwirtschaft/Dokumente/BfN-Agrar-Report_2017.pdf.

[BGJ19]   Bureau, Jean-Christophe, Houssein Guimbard, and Sebastien Jean. 2019. Agricultural trade liberalisation in the 21st century: has it done the business? *Journal of Agricultural Economics* 70(1):3–25. ▶ https://doi.org/10.1111/1477-9552.12281.

[Bit19]   Bitkom Bundesverband Informationswirtschaft, Chancen der Digitalisierung nutzen – Offener Zugang und standardisierte Datenformate für eine zukunftsfähige Landwirtschaft 4.0. 2019.

---

27 WBGU, Unsere gemeinsame Zukunft – Zusammenfassung des Gutachtens, 2019, p. 1.

[BPL06] Buijs, Arjen. E., Bas Pedroli, and Yves Luginbühl. 2006. From hiking through farmland to farming in a leisure landscape: changing social perceptions of the European landscape. *Landscape Ecology* 21(3):375–389. ▶ https://doi.org/10.1007/s10980-005-5223-2.

[Bra32] Braun-Blanquet, Josia. 1928. *Pflanzensoziologie*. Berlin: Julius Springer. English Edition: Fuller, G.D., and Conrad, H.S. 1932. *Plant Sociology—the study of plant communities* (trans.: Fuller, G.D., and Conrad, H.S.). London: McGraw-Hill.

[Bun17] German Federal Government. 2017. *Deutsche Nachhaltigkeitsstrategie – Neuauflage 2016*. ▶ https://www.bundesregierung.de/resource/blob/975292/730844/3d30c-6c2875a9a08d364620ab7916af6/deutsche-nachhaltigkeitsstrategie-neuauflage-2016-download-bpa-data.pdf.

[Bun19] German Federal Government. 2019. *Bundes-Klimaschutzgesetz vom 12. Dezember 2019*. Federal Law Gazette I p. 2513, Annex 2 to Sec 4.

[Bun21] Bundesministerium für Ernährung und Landwirtschaft: GEFA-Kommuniqué. 2021. ▶ https://www.gffa-berlin.de/deckblatt-communuque/.

[Can17] Canter, Larry W. 2017. *Environmental Impact of Agricultural Production Activities*. CRC Press. ISBN 978–1–315–89269–6.

[CDU19] Christian Democratic Union of Germany, Für eine Zukunft mit Landwirtschaft – für eine Landwirtschaft mit Zukunft, Beschluss des 32. *Parteitages der CDU Deutschlands*. 2019.

[Cen88] Centner, Terence J. 1988. The role of cooperatives in agriculture: historic remnant or viable membership organization? *Journal of Agricultural Cooperation* 03:94–106. ▶ https://doi.org/10.22004/ag.econ.46213.

[CG21] Consultative Group on International Agricultural ResearchAction Areas. Retrieved 13.03.2022. ▶ https://www.cgiar.org/research/action-areas/.

[Cop20] Copa-Cogeca, European Farmers European Agri-Cooperatives. *Over 30% of French and German farmers are reported as feeling demotivated because of 'agri-bashing'*. Press Release. Accessed 25 Feb 2021.

[CW15] Cordell, Dana, and Stuart White. 2015. Tracking phosphorus security: Indicators of phosphorus vulnerability in the global food system. *Food Security* 7(2):337–350. ▶ https://doi.org/10.1007/s12571-015-0442-0.

[DAK21] Digital Agricultural Knowledge and Information System (DAKIS). *Ökosystemleistungen einen Wert geben*. ▶ https://adz-dakis.com. Accessed 4 Mar 2021.

[DE20] Digital Europe, How to spend it: A digital investment plan for Europe 2020.

[DPH+20] Dobson, A. P., et al. 2020. Ecology and economics for pandemic prevention Investments to prevent tropical deforestation and to limit wildlife trade will protect against future zoonosis outbreaks. *Science* 369(6502):379 ff.

[DWH+09] Döhler, H., S. Wulf, H.-D. Haenel, B. Eurich-Menden, C. Rösemann, and A. Freibauer. 2009. *Nationale Klimaschutzziele – Potenziale und Grenzen der Minderungsmaßnahmen*. KTBL – Kuratorium für Technik und Bauwesen in der Landwirtschaft e.V. (Schrift 485).

[EC16a] European Commission, 5G for Europe: An Action Plan, COM(2016) 588 final.

[EC16b] European Commission. 2016. A Strategic Approach to EU Agricultural Research and Innovation, final paper.

[EC17] European Commission, Communication from the Commission to the European Parliament, the Council, the European Economic and Social Committee and the Committee of the Regions on the Mid-Term Review on the implementation of the Digital Single Market Strategy, A Connected Digital Single Market for All, COM(2017) 228 final.

[EC18a] European Commission, Communication form the Commission to the European Parliament, the European Council, the Council, the European Economic and Social Committee and the Committee of the Regions, A renewed European Agenda for Research and Innovation – Europe's Chance to shape its Future, COM(2018) 306 final.

[EC18b] European Commission, Communication to the Commission, European Commission Digital Strategy, A digitally transformed, user-focsed and data driven Commission, C(2018) 7118 final.

[EC19a] European Commission, Communication from the Commission to the European Parliament, the European Council, the Council, the European Economic and Social Committee and the Committee of the Regions, The European Green Deal, COM(2019) 640 final.

[EC19b] European Commission. 2019. *Nitrat im Grundwasser: Kommission mahnt Deutschland zur Umsetzung des EuGH-Urteils*. ▶ https://ec.europa.eu/germany/news/20190725-nitrat_de. Accessed 17 Feb 2021.

[EC20a] European Commission, Communication from the Commission to the European Parliament, the Council, the European Economic and Social Committee and the Committee of the Regions, A European Strategy for Data, COM(2020) 66 final.

[EC20b] European Commission, White Paper on Artificial Intelligence – A European Approach to Excellence and Trust, COM(2020) 65 final.

[EC20c] European Commission, Joint Communication to the Parliament and the Council, The EU's Cybersecurity Strategy for the Digital Decade, JOIN(2020) 18 final.

[EC20d] European Commission, Proposal for a Regulation of the European Parliament and of the Council on European Data Governance, COM(2020) 767 final.

[EC20e] European Commission, Farm to Fork Strategy, For a fair healthy and environmentally-friendly food system. 2020. ▶ https://ec.europa.eu/food/sites/food/files/safety/docs/f2f_action-plan_2020_strategy-info_en.pdf.

[EC20f] European Commission, Working with Parliament and Council to make the CAP reform fit for the European Green Deal. 2020.

[EC21a] European Commission, Digital Education Action Plan. 2021.

[EC21b] European Commission, EIP-Agri, Agriculture and Innovation, Innovation and Support Services, brochure. 2021.

[EC21c] Regulation of the European Parliament and of the Council of 2 December 2021 establishing rules on support for strategic plans to be drawn up by Member States under the Common Agriculture Policy (CAP Strategic Plans) and financed by the European Agricultural Guarantee Fund (EAGF) and by the European Agricultural Fund for Rural Development (EAFRD) and repealing Regulations (EU) No. 1305/2013 and (EU) No. 1307/2013.

[Egl20] Egli, Florian. 2020. Renewable energy investment risk: An investigation of changes over time and the underlying drivers. *Energy Policy* 140. ▶ https://doi.org/10.1016/j.enpol.2020.111428

[EIC21] European Innovation Council, About the EIC Fund. ▶ https://ec.europa.eu/research/eic/index.cfm?pg=investing. Accessed 3 Mar 2021.

[EL20] Erickson, Bruce, and James Lowenberg-DeBoer. 2020. *2019 Precision Agriculture Dealership Survey*. ▶ https://ag.purdue.edu/digital-ag-resources/wp-content/uploads/2020/03/2019-CropLife-Purdue-Precision-Survey-Report-4-Mar-2020-1.pdf. Accessed 20 Jul 2020.

[Elb15] Elbehri, Aziz. 2015. *CLIMATE CHANGE AND FOOD SYSTEMS—Global assessments and implications for food security and trade*. FAO. ISBN 978–92–5–108699–5.

[Ell96] Ellenberg, Heinz. 1996. *Vegetation Mitteleuropas mit den Alpen in ökologischer, dynamischer und historischer Sicht*. Ulmer. ISBN 3–8001–3430–6.

[EM20] Ehlers, K., and D. Messner. 2020. Gegen Dürre und Überdüngung: Landwirtschaft neu denken. *Blätter* 6:93–102. ▶ https://www.blaetter.de/ausgabe/2020/juni/gegen-duerre-und-ueberduengung-landwirtschaft-neu-denken.

[EPT19] Eriksen, Siren, et al. 2019. Farm-based day care services—a prospective study protocol on health benefits for people with dementia and next of kin. *Journal of Multidisciplinary Healthcare* 12:643–653. ▶ https://doi.org/10.2147/JMDH.S212671.

[EU16] European Union. 2016. Directive (EU) 2016/2284 of the European Parliament and of the Council of 14 December 2016 on the reduction of national emissions of certain atmospheric pollutants, amending Directive 2003/35/EC and repealing Directive 2001/81/EC. Annex II, Table B.

[EU91] European Union. 1991. *Council Directive 91/676/EEC of 12 December 1991 concerning the protection of waters against pollution caused by nitrates from agricultural sources*.

[EUC20] ▶ https://ec.europa.eu/growth/smes/business-friendly-environment/sme-definition_en, *"What is an SME?—Internal Market, Industry, Entrepreneurship and SMEs"*, European Commission. Retrieved 13.03.2022.

[Eur20a] ▶ https://ec.europa.eu/eurostat. *"Utilised agricultural area by categories"*, Eurostat. Retrieved 13.03.2022.

[Eur20b] ▶ https://ec.europa.eu/eurostat. *"Agricultural land renting prices for one year by region"*, Eurostat. Retrieved 13.03.2022.

[Eur20c] ▶ https://ec.europa.eu/eurostat. *"Agricultural land prices by region"*, Eurostat. Retrieved 13.03.2022.

[Eur20d] ▶ https://eur-lex.europa.eu/statistics/legal-acts/2007/legislative-acts-statistics-by-type-of-act.html. *"Legal acts—statistics"*, EUR-Lex. Retrieved 13.03.2022.

[Eur20e] ▶ https://ec.europa.eu/eurostat. *"Farm indicators by agricultural area, type of farm, standard output, legal form and NUTS 2 regions"*, Eurostat. Retrieved 13.03.2022.

[Eur20f] ▶ https://ec.europa.eu/eurostat. *"Data Navigation Tree—Tables by Themes—Agriculture, forestry and fisheries—Agriculture (t-agr)—Agricultural production (t_apro),"* Eurostat. Retrieved 13.03.2022.

[FAO96] Food and Agriculture Organisation of the United Nations: World Food Summit. 1996. *Rome Declaration on World Food Security*. ▶ http://www.fao.org/WFS/.

[Fao11] FAO. 2011. *Energy-smart food for people and climate*. FAO Issue Paper.

[Fao17] FAO. 2017. *Water for sustainable food and agriculture—a report produced for the G20 presidency of Germany*.

[Fao20a] ▶ www.fao.org/faostat. *FAOStat, "Production Indices"*. Retrieved 13.03.2022.

[Fao20b] ▶ www.fao.org/worldfoodsituation/foodpricesindex/en. *"FAO Food Price Index", FAO World Food Situation*. Retrieved 13.03.2022.

[Fao+19] FAO, IFAD, UNICEF, WFP and WHO. 2019. *The state of food security and nutrition in* the world 2019—safeguarding against economic slowdowns and downturns. FAO. ISBN 978–92–5–131570–5.

[FDP21] Free Democratic Party, *Digitale Landwirtschaft 4.0*. ▶ https://www.fdp.de/forderung/digitale-landwirtschaft-40. Accessed 02 Mar 2021.

[FE21] Future Earth, A Systems Based Approach to Transformations in Global Sustainability. ▶ https://futureearth.org/about/. Accessed 02 Mar 2021.

[Fff20] f3 Farm.Food.Future. 2020. *Nische, öffne dich! Neue, landwirtschaftliche Standbeine als unternehmerische Chance*", Edition Summer

[FM20] yFlack, Jessica, and Melanie Mitchell. 2020. *Uncertain times, The pandemic is an unprecedented opportunity – seeing human society as a complex system opens a better future for us all*. ▶ https://aeon.co/essays/complex-systems-science-allows-us-to-see-new-paths-forward. Accessed 20 July 2021.

[FME07] Federal Ministry for the Environment, *National Biodiversity Strategy*. 2007.

[FME20] Federal Ministry for the Environment, *Nature Conservation and Nuclear Safety, Digital Policy Agenda for the Environment*. 2020.

[FMEA20] Federal Ministry for Economic Affairs and Energy, GAIA-X: Technical Architecture. 2020.

[FMF18] Federal Ministry for Food and Agriculture, Digitalisierung in der Landwirtschaft, Chancen nutzen – Risiken minimieren. 2018.

[FMF19a] Federal Ministry for Food and Agriculture, 2035 Arable Farming Strategy, Prosepcts for Productive and Diverse Crop Farming. 2019.

[FMF19b] Federal Ministry for Food and Agriculture, Leuchttürme der Digitalisierung in der Landwirtschaft, Pressemitteilung. ▶ https://www.bmel.de/SharedDocs/Pressemitteilungen/DE/2019/209-experimentierfelder.htm. Accessed 26 Feb 2021.

[FMF20] Federal Ministry for Food and Agriculture, Digital Transformation of the Agricultural Value Chain – Opportunities, Challenges and the Role of Science. ▶ https://www.eu2020.de/eu2020-de/veranstaltungen/-/2365104. Accessed 23 Feb 2021.

[FMT17] Federal Ministry of Transport and Digital Infrastructure, Gigabit Initiative for Germany. 2017.

[FSE20] Fraunhofer Insitute for Experimental Software Engineering, Machbarkeitsstudie zu staatlichen digitalen Datenplattformen für die Landwirtschaft. 2020.

[Gan19] Gandorfer, Markus. 2019. *Digitale teilflächenspezifische Stickstoffdüngung – eine ökonomisch-ökologische Perspektive*. In *Ökologie und Bioökonomie – Neue Konzepte zur umweltverträglichen Nutzung natürlicher Ressourcen, 48. Rundgespräche Forum Ökologie, Bayerische Akademie der Wissenschaften*, ed. Forum Ökologie, pp. 105–112. München: Verlag Dr. Friedrich Pfeil.

[GFA21] German Farmers' Association, Beispiele für Nutzen und Nutzenpotentiale der Digitalisierung. ▶ https://www.bauernverband.de/themendossiers/digitalisierung/themendossier/digitalisierung-nutzen-und-nutzenpotentiale. Accessed 3 Mar 2021.

[GFF19] Global Forum for Food and Agriculture Communiqué, *Agriculture Goes Digital – Smart Solutions for Future Farming*. 2019.

[GFF20] Global Forum for Food and Agriculture Communiqué 2020, *Food for All! Trade for Secure, Diverse and Sustainable Nutrition*. 2020.

[GFF21] Global Forum for Food and Agriculture Communiqué 2021, *How to Feed the World in Times of Pandemics and Climate Change?* 2021.

[Gia17] Giam, X. L. 2017. Global biodiversity loss from tropical deforestation. *Proceedings of the National Academy of Sciences of the United States of America* 114(23):5775-5777.

[GGS21] Gabriel, Andreas, Markus Gandorfer, and Olivia Spykman. Nutzung und Hemmnisse digitaler Technologien in der Landwirtschaft. *Berichte über Landwirtschaft* 99(1). ▶ https://doi.org/10.12767/buel.v99i1.

[GR19] Alliance 90/The Greens North Rhine-Westphalia, Zukunft gestalten – Digitale Transformation als Chance für NRW in einer globalisierten Welt. 2019. ▶ https://gruene-nrw.de/2019/06/zukunft-gestalten-digitale-transformation-als-chance-fuer-nrw-in-einer-globalisierten-welt/. Accesses 4 Mar 2021.

[Här12] Härtel, Ines. 2012. *Handbuch Föderalismus, vol. II, Probleme, Reformen, Perspektiven des deutschen Föderalismus*. Berlin: Springer.

[Här19] Härtel, Ines. 2019. *Agrar-Digitalrecht für eine nachhaltige Landwirtschaft 4.0*, NuR, pp. 577–586.

[Här20] Härtel, Ines. 2020. *Künstliche Intelligenz in der nachhatligen Landwirtschaft—Datenrechte und Haftungsregime*, NuR, pp. 439–452.

[Här21] Härtel, Ines. Gutachten zum Thema "Europäische Leitlinien bzw. Regeln für Agrardaten" (European Agricultural Data Governance). ▶ https://www.bmel.de/SharedDocs/Downloads/DE/_Digitalisierung/agrardaten-gutachten-haertel.html. Accessed 5 Mar 2021.

[HG15] Harrison, Jill Lindsey, and Christy Getz. 2015. Farm size and job quality: mixed-methods studies of hired farm work in California and Wisconsin. *Agriculture and Human Values* 32(4):617–634. ▶ https://doi.org/10.1007/s10460-014-9575-6

[IAM20] ▶ www.largescaleagriculture.com/data, *"Large Scale Agriculture", IAMO Leibniz Institute of Agricultural Development in Transition Countries*. Retrieved 13.03.2022.

[IDC20] International Digital Council for Food and Agriculture, Platform "e-agiculture". ▶ https://www.fao.org/e-agriculture/international-digital-council-food-and-agriculture. Accessed 5 Mar 2021.

[IPB19] IPBES—Intergovernmental Science-Policy Platform on Biodiversity and Ecosystem Services. 2019. *The global assessment report on Biodiversity and ecosystem services. Summary for policymakers*. Bonn. ▶ https://ipbes.net/sites/default/files/inline/files/ipbes_global_assessment_report_summary_for_policymakers.pdf.

[IPC19] IPCC—Intergovernmental Panel on Climate Change. 2019. *Climate change and Land. SRCCL. Summary for Policymakers*. ▶ https://www.ipcc.ch/srccl/chapter/summary-for-policymakers/

[JCW+20] Jouanjean, Marie-Agnes, Francesca Casalini, Leanne Wiseman, and Emily Gray. 2020. *Issues Around Data Governance in the Digital Transformation of Agriculture: The Farmers' Perspective*.

[JWH18] Jantsch, Antje, Tobias Weirowski, and Norbert Hirschauer. 2018. *Arbeits- und Lebenszufriedenheit abhängig Beschäftigter in der ostdeutschen Landwirtschaft (Pflanzenproduktion)— Eine explorative Auswertung der Jahre 2000–2015*. Conference Paper, 58. Jahrestagung der GEWISOLA "Visionen für eine Agrar- und Ernährungspolitik nach 2020". Kiel.

[Kau19] Kauffman, Stuart A. 2019. *A world beyond physics—the emergence and evolution of life*. Oxford University Press.

[Lim20] ▶ www.limagrain.com/en/an-efficient-decision-making-system. *An efficient decision-making system—What is the decision making process with Limagrain?* Limagrain. Retrieved 13.03.2022.

[LMH+09] Lavalle, Carlo et al. 2009. Climate change in Europe. 3. Impact on agriculture and forestry. A review. *Agronomy for Sustainable Development* 29(3):433–446. ▶ https://doi.org/10.1051/agro/2008068.

[LR19] Latacz-Lohmann, Uwe, and Norbert Röder. 2019. *Eco Schemes: Was kommt auf die Bauern zu? top agrar* 7:34–36.

[Mar20] ▶ https://marktschwaermer.de. Marktschwärmer. Retrieved 13.03.2022.

[Mon14] i.a. Monien in: Härtel (ed.), Nachhaltigkeit, Energiewende, Klimawandel, Welternährung, 2014, p. 790–792.

[MVN20] Maroto-Martos, Juan Carlos, Andreas Voth, and Aida Pinos-Navarrete. 2020. The importance of tourism in rural development in Spain and Germany, Chapter 9 (pp. 181–205). In *Neoendogenous Development in European Rural Areas: Results and Lessons*. Springer Geography. ISBN 978–3–030–33462–8.

[NS20] Niegsch, Claus, and Michael Stappel. 2020. *Branchenanalyse—Deutsche Landwirtschaft unter Druck*. DZ Bank AG: Research-Publikation der DZ BANK AG.

[OEC19a] Organisation for Economic Co-operation and Development. 2019. *Digital Opportunities for Better Agriculture Policies, Paris*. OECD Publishing.

[OEC19b] Organisation for Economic Co-operation and Development. 2019. *Regulatory effectiveness in the era of digitalization*.

[OEC19c] OECD. 2019. *Agricultural Policy Monitoring and Evaluation 2019*. OECD. ISBN: 9789264632622.

[OEC20a] ▶ https://data.oecd.org/agrpolicy/agricultural-support.htm. *Agricultural support. OECD*. Retrieved 13.03.2022.

[OEC20b] Organisation for Economic Co-operation and Development. 2020. *Mapping Approaches to Data and Data Flows, Report for the G20 Digital Economy Task Force*.

[OTK+11] Olesen, J. E. et al. 2011. Impacts and adaptation of European crop production systems to climate change. *European Journal of Agronomy* 34(2):96–112. ▶ https://doi.org/10.1016/j.eja.2010.11.003.

[PB13] Pardey, Philip G., and Jason M. Beddow. 2013. *Agricultural Innovation: The United States in a Changing Global Reality*. The Chicago Council on Global Affairs.

[PGE18] Pfeiffer, Johanna, Markus Gandorfer, and Klaus Erdle. 2018. Smarte Technik, große Wirkung? *Schule und Beratung* 5–6:55–66.

[PGG20] Pfeiffer, Johanna, Andreas Gabriel, and Markus Gandorfer. 2020. Understanding the public attitudinal acceptance of digital farming technologies: A nationwide survey in Germany. *Agriculture and Human Values*. ▶ https://doi.org/10.1007/s10460-020-10145-2.

[PSG+19] Pfeiffer, Johanna, Sebastian Schleicher, Andreas Gabriel, and Markus Gandorfer. 2019. *Gesellschaftliche Akzeptanz von Digitalisierung in der Landwirtschaft*. Referate der 39. GIL-Jahrestagung in Wien, Meyer-Aurich et al. (Ed.), pp. 151–154.

[Rey97] Reynolds, Bruce J. 1997. *Decision-Making in Cooperatives With Diverse Member Interests*. USDA, RBS Research Report 155. ▶ https://doi.org/10.22004/ag.econ.280001.

[RKR20] Rohleder, Bernhard, Bernhard Krüsken, and Horst Reinhardt. 2020. *Digitalisierung in der Landwirtschaft 2020*. ▶ https://www.bitkom-research.de/system/files/document/200427_PK_

Digitalisierung_der_Landwirtschaft.pdf. Retrieved 13.03.2022.

[RJ09] Reichardt, Maike, and Carsten Jürgens. 2009. Adoption and future perspective of precision farming in Germany: Results of several surveys among different agricultural target groups. *Precision Agriculture* 10:73–94.

[RSN+09] Rockström, Johan, Will Steffen, Kevin Noone, Åsa Persson, F. Stuart Chapin III, Eric F. Lambin, Timothy M. Lenton, Marten Scheffer, Carl Folke, Hans Joachim Schellnhuber, Björn Nykvist, Cynthia A. de Wit, Terry Hughes, Sander van der Leeuw, Henning Rodhe, Sverker Sörlin, Peter K. Snyder, Robert Costanza, Uno Svedin, Malin Falkenmark, Louise Karlberg, Robert W. Corell, Victoria J. Fabry, James Hansen, Brian Walker, Diana Liverman, Katherine Richardson, Paul Crutzen, and Jonathan A. Foley. 2009. Planetary boundaries: exploring the safe operating space for humanity. *Ecology and society* 14(2).

[RW04] Rosskopf, Karin, and Peter Wagner. 2004. *Der digitale Landwirt: Die Nutzung des Computers im Betriebsmanagement.* Referate der 25. GIL Jahrestagung in Bonn. Schiefer et al. (Ed.), pp. 121–124.

[Sav20] Global Farmland Index, Savills Research, UK, 2021. ▶ https://pdf.euro.savills.co.uk/uk/rural---other/spotlight-global-farmland-index---2021.pdf. Retrieved 13.03.2022.

[Sch20] ▶ www.schweizerbauer.ch/politik-wirtschaft/milchmarkt/umfrage-biomilch-markt-wie-weiter. *"Umfrage: Biomilch, wie weiter?,"* Schweizer Bauer, 22.01.2020. Retrieved 13.03.2022.

[SG18] Schleicher, Sebastian, and Markus Gandorfer. 2018. *Digitalisierung in der Landwirtschaft: Eine Analyse der Akzeptanzhemmnisse.* Referate der 38. GIL-Jahrestagung in Kiel. Ruckelshausen et al. (Ed.), pp. 203–206.

[SPD19] Social Democratic Party of Germany, Positionspapier *"Unser Vorschlag für eine Gute Gemeinsame Agrarpolitik".* 2019.

[SRF12] Seufert, Verena, Navin Ramankutty, and Jonathan A. Foley. 2012. *Comparing the yields of organic and conventional agriculture. Nature* 485:229. ▶ https://doi.org/10.1038/nature11069

[SRR+15] Steffen, W., K. Richardson, J. Rockström, S.E. Cornell, I. Fetzer, E.M. Bennett, R. Biggs, S.R. Carpenter, W. de Vries, C.A. de Wit, C. Folke, D. Gerten, J. Heinke, G.M. Mace, L.M. Persson, V. Ramanathan, B. Reyers, and S. Sörlin. 2015. Planetary boundaries: Guiding human development on a changing planet. *Science* 347(6223).

[SS17] Spiess-Knafl, Wolfgang, and Barbara Scheck. 2017. *Impact investing—Instruments, Mechanisms and Actors.* Springer International Publishing. ISBN: 978–3–319–66555–9.

[Sta18] Statistisches Bundesamt [Federal Statistical Office of Germany]. 2018. Nachhaltige Entwicklung in Deutschland—Indikatorenbericht 2018. ▶ https://www.destatis.de/DE/Themen/Gesellschaft-Umwelt/Nachhaltigkeitsindikatoren/Publikationen/Downloads-Nachhaltigkeit/indikatoren-0230001189004.pdf?__blob=publicationFile.

[Top20] ▶ www.topagrar.com/oekolandbau/news/biofach-forum-bei-der-umstellung-auf-oeko-auf-die-bremse-treten-11981278.html. Forum: 'Bei der Umstellung auf Ökoauf die Bremse treten. topagrar online. Retrieved 13.03.2022.

[Umw20] Umweltbundesamt [German Federal Environmental Agency]. 2020. *Nationale Trendtabellen für die deutsche Berichterstattung atmosphärischer Emissionen seit 1990.* Emissionsentwicklung 1990 bis 2018. ▶ https://www.umweltbundesamt.de/themen/luft/emissionen-von-luftschadstoffen. Accessed 17 Feb 2021.

[UN15] United Nations. 2015. *Sustainable Development Goals.* ▶ https://www.un.org/sustainabledevelopment/sustainable-development-goals/. Accessed 17 Feb 2021.

[VMM17] Van Passel, Steven, Emanuelle Massetti, and Robert Mendelsohn. 2017. *A Ricardian analysis of the impact of climate change on European agriculture. Environmental and Resource Economics* 67(4):725–760. ▶ https://doi.org/10.1007/s10640-016-0001-y.

[Wes18] Wessling, Ewald. 2018. Behalten Sie die Datenhoheit! topagrar, No. 05, 2018

[WHO14] WHO. 2014. *EMRO Zoonotic disease: emerging public health threats in the Region.* Who Regional Committee For The Eastern Mediterranean, Tunisia 19–22 October 2014. ▶ http://www.emro.who.int/about-who/rc61/zoonotic-diseases.html. Accessed 31 Aug 2020.

[Wir20] ▶ www.wir-haben-es-satt.de. Wir haben Agrarindustrie satt! Retrieved 13.03.2022.

[WL19] Willer, Helga, and Julia Lernoud. 2019. *The World of Organic Agriculture—Statistics & Emerging Trends 2019.* Research Institute of Organic Agriculture FiBL & IFOAM. ISBN 978–3–03736–118–4.

[Wor20] ▶ https://data.worldbank.org/indicator/NY.GDP.PCAP.CD. GDP per capita (current US$). The World Bank—Data. Retrieved 13.03.2022.

[WR00] Wachenheim, Cheryl, and Richard Rathge. 2000. *Societal Perceptions of Agriculture. Agribusiness and Applied Economics Report*, No. 449.

[WSM18] Wille, Stefan Clemens, Achim Spiller, and Marie von Meyer-Höfer. 2018. *Lage, Lage, Lage? Welche Rolle spielt der Standort für die landwirtschaftliche Direktvermarktung?* Georg-August-Universität Göttingen, Department für Agrarökonomie und Rurale Entwicklung (DARE), Göttingen, Diskussionsbeitrag 1808.

[WTO19] Word Trade Organisation. 2019. *Joint Statement on Electronic Commerce*, January 25th 2019, WT/L1056.

[ZIB+13] Zander, Katrin et al. 2013. *Erwartungen der Gesellschaft an die Landwirtschaft*. Thünen-Institut, Abschlussbericht: Stiftung Westfälische Landschaft.

[Zwe19] Datenethikkommission, Gutachten. 2019. p. 177 regarding the pyramid of criticality (Kritikalitätspyramide) of algorithmic systems. the risk matrix was developed by Katharina Zweig, s. in: *Zweig, Ein Algorithmus hat kein Taktgefühl*, 2019, p. 234 ff.

Framework for the Digital Transformation of the Agricultural Ecosystem

Carsten Gerhardt, Stefanie Bröring, Otto Strecker, Michael Wustmans, Débora Moretti, Peter Breunig, Leo Pichon, Gordon Müller-Seitz and Borris Förster

Contents

2.1	From Farm to Fork and Back: History and Roadmap of Digital Farming – 61	
2.1.1	Introduction – 61	
2.1.2	View on the Agriculture Industry Overall – 62	
2.1.3	The Roots of Digital Farming – 64	
2.1.4	Implications for the Industry: From Product to Service – 67	
2.1.5	Digital will Change the Face of Farming – 68	
2.1.6	Outlook: From Farm to Fork and Back – 70	
2.2	Beyond Digitalization: Major Trends Impacting the AgFood System of the Future – 71	
2.2.1	Introduction – 71	
2.2.2	Innovation is Multi-Systemic—Main Disruptions from Farm to Fork – 72	
2.2.3	Focus: Digital Disruption and Its Implications for Involved Agribusiness Companies – 75	
2.2.4	Concluding Questions – 78	
2.3	Economic Benefit Quantification – 79	
2.3.1	Introduction – 79	
2.3.2	Fundamentals of Economic Value Creation – 79

© The Author(s), under exclusive license to Springer-Verlag GmbH, DE, part of Springer Nature 2022
J. Dörr and M. Nachtmann (eds.), *Handbook Digital Farming*,
https://doi.org/10.1007/978-3-662-64378-5_2

2.3.3	Cost Structure Fundamentals of Digital Solutions and Economic Benefit – 82	
2.3.4	Limitations of Economic Benefit Quantification for Decision Making – 84	
2.3.5	Example for Economic Benefit Quantification – 85	
2.3.6	Summary and Outlook – 86	

2.4 Successfully Disseminating Digital Tools for Farmers: A French Perspective – 86

- 2.4.1 Introduction – 86
- 2.4.2 Material and Method – 87
- 2.4.3 Results – 88
- 2.4.4 Discussion – 90
- 2.4.5 Conclusion – 91

2.5 Business Model Innovation and Business Model Canvas – 91

- 2.5.1 Statement of the Problem – 92
- 2.5.2 Business Model Innovation as a Distinct Form of Innovation – 93
- 2.5.3 Business Model Innovation in Practice—The Business Model Canvas – 93
- 2.5.4 Reflections on the (Mis-)Use of the Business Model Canvas and Conclusions – 97

2.6 Accelerators & Partnerships: Anticipating the Unknown is Hard: An Experience Report – 97

- 2.6.1 Introduction – 98
- 2.6.2 Understanding and Managing Direct and Indirect Impact on Success—Borrowing from Physics and Finance – 99
- 2.6.3 The Current State of Technology in the Food Value Chain – 102
- 2.6.4 Looking Ahead: From Stand-Alone Programs to Multi-Corporate Innovation Platforms – 104

References – 106

2.1 From Farm to Fork and Back: History and Roadmap of Digital Farming

Carsten Gerhardt

Abstract

Global supply of agricultural products surpasses demand. This puts the industry under permanent price pressure. Digital Farming as a mean to improve yields and become more cost-effective has entered the market around 2010. It will continue to be applied to a steadily increasing fraction of the global farmland. This will heavily impact the agricultural input industry, which will transition from a product to a service provider. Plus, digital will change the face of farming, as it allows to move away from ever bigger machinery to small, autonomous swarm robots. Ultimately, digital in agriculture is a key enabler for the transition to a bio-economy where farmland will provide inputs to a variety of industries, well beyond today's food and feed.

2.1.1 Introduction

Agriculture globally has been characterized by large production increases of the last decades to feed a growing population. Yield increases and additional farming land have been driving this overall production increase. The former could be achieved due to a very professional Ag input industry, providing high-yielding seed, fertilizer and AgChemicals. The latter often has come at the expense of turning natural habitats like rainforest into farmland (see ▶ Sect. 1.4). Both ultimately have led to environmental degradation and are not sustainable. Monoculture, loss of biodiversity and soil degradation characterize agriculture in large parts of the world in 2020. A further production increase cannot be achieved by the means of the past. Great hopes and expectations put into novel technologies like second- and third-generation genetically modified seeds or biologicals for crop protection have not materialized.

At the same time, the industry now is under price pressure. The farm-level prices of agricultural commodities corrected for inflation mostly have declined over the last centuries.

The highly consolidated farm input providers had nicely profitable businesses and benefited from an overall market growth for the last decades—it was a tide that lifted all boats. However, this trend began to slow down around 2015.

Digital will drastically change agriculture moving forward. Digital Farming has been emerging after 2010 as a possible solution to improve cost structures, increase yield, and at the same time lower the environmental footprint of agriculture. Digital means can increase the efficiency and effectiveness of the existing processes before, on and after the farm. This increases production at improved cost positions and optimizes quality. Furthermore, digital may provide a means to better communicate quality features of the agricultural products through tracking and tracing technologies, ultimately realizing a better product price at the retail level. Digitally operated machinery in the field will totally change the face of farming in the long term.

Finally, in an economy that is turning more and more circular (Circular Economy), agriculture will play an important role: on the one hand as provider of bio-based feedstock, and on the other hand as off-taker of post-consumer materials such as sludge, compostable organic waste and biodegradable plastics. For the management of these circles, digital support will also be of vital importance.

2.1.2 View on the Agriculture Industry Overall

The agriculture sector from farm input providers over distributors, farmers, processors down to retail as a whole has been growing steadily and in parts profitably for decades, driven by two fundamental demand drivers (see ◘ Fig. 2.1). Population increase, on the one hand, is going from 5 bn in 1990 to almost 8 bn in 2020 and is forecasted to near 10 bn in 2050. Along with the increase in number of people went an increase in calorie consumption on the other hand very much triggered by a heightened meat consumption. Given the limitation of land, increased production could only be achieved by the help of a professional seed and Ag-Chemical industry. The crop protection market alone has almost tripled from 1990 to 2020, from slightly over 20 bn USD to almost 60 bn USD. A similar development could be observed in the seed industry, both in genetically modified and conventional seeds. Higher yielding and seeds better adapted to regional specifics were the main productivity driver besides improved agronomic practices.

But there are imperfections in the "conventional story" of lasting industry growth. As early as 2015, the OECD long-term agriculture outlooks started to paint a bleaker picture of production growth, with annual growth rates for key commodities like wheat, soybean, corn or poultry meat decreasing from historical values between 2–4% to almost half of that (see ◘ Fig. 2.2). Nominal commodity prices were forecast to stay stable or even decrease. This anticipated price development is very much in line with an overall rather weak price development of agricultural commodities in the past. Except for variations that were most likely caused by speculation, the overall price increase, for example, for wheat has barely increased by 30% from 1990 to 2020. If adjusted for inflation, it has even decreased, resembling a trend that could be observed during the complete last century. Other, non-agricultural commodities have shown a substantial price increase in the same period. The copper price from 1990 to 2020, for example, has risen almost 300%.

Besides the challenging growth and price outlook, there are additional challenges for the agriculture industry. Multiple

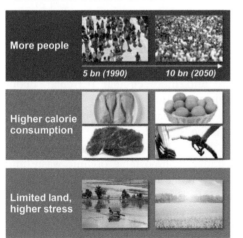

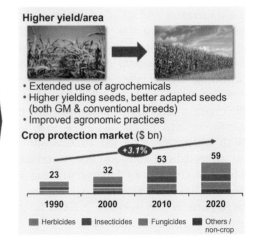

◘ Fig. 2.1 Fundamental demand drivers provided a growth story to the Ag input providers

Framework for the Digital Transformation ...

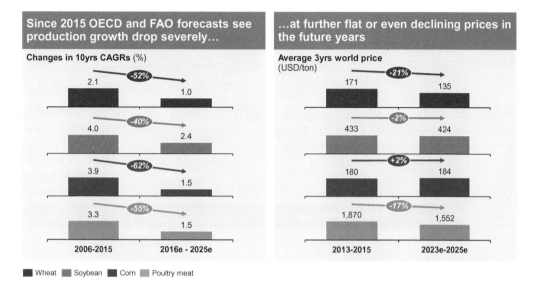

Fig. 2.2 OECD/FAO forecasts since 2015 forecast declining growth rates and flat prices

trends are superimposing and affecting the industry's development.

Many technologies are reaching their limits. This can be seen from the increase of resistances in crop protection or chemical substances backfiring on crop yield. Also, soil fertility is being reduced in many global geographies due to biodegradation.

Especially in mature markets social scrutiny is increasingly turning against modern intensive farming. A significant part of consumers is against GMO or the conventional high-input agriculture. Percentages differ strongly between regions, from low double-digit percentage in the USA to almost two-thirds of consumers in France. Ultimately this consumer skepticism will result in further tightened secondary standards that drive down the use of AgChemicals and regulatory approvals and registrations that will be much harder to obtain going forward.

Then, there are many innovations with a disruption potential: They can broadly be structured by whether they impact demand or supply and their degree of certainty (see ◘ Fig. 2.3). The latter is ranging from already existing trends like novel traits, farm consolidation, extension of cropland or no-till farming over already known-trends that are currently materializing like vertical farming to "unknowns" like space farming. Specifically, in the area of crop protection, robotics and automation are likely to further replace today's crop protection technologies. One major trend that will heavily impact the whole agriculture value chain is alternative proteins/artificial meat. Having been known for several years now, 2019 was the year of successful market entry. Next to plant-based and insect-based meat alternatives, cultured meat can clearly be seen to be evolving. All these new products have the potential to not only disrupt the multi-billion dollar global meat industry but also the whole value chain due to the impact on feed demand, especially in corn and soy. Our research shows that in 20 years from 2020 onwards less than half of global meat consumption will still come from conventional meat sources. The rest will be novel vegan meat replacements and cultured meat [GSZ+19].

It is against the above-described background of a challenged, rapidly changing industry that we now take a look at the development of "Digital" in agriculture.

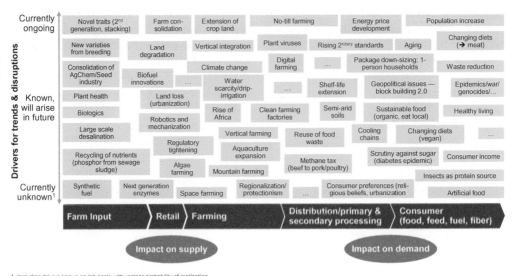

■ **Fig. 2.3** Disruptive trends in the industry

2.1.3 The Roots of Digital Farming

For the purpose of this section, we define Digital Farming as all farming methods that use the means of digital to optimize agriculture, which is in line with the definition in this book. We consider optimized agronomical advice, based on big data insights generated from a multitude of sources as the most important building block.

Hence, it is much more than just digitizing individual parts of the value chain, like e.g., digital sales channels or e-commerce. Precision farming as the ability to very precisely plant, fertilize, spray, and harvest is also only one component of Digital Farming. It constitutes an important enabler for Digital Farming, though. Terms like Smart Farming or Farming 4.0 in our view can be used as synonyms for Digital Farming.

Farming always has been at the forefront of innovation, from a variety of mechanization methods in the eighteenth century to the introduction of the steam engine. Tractors with steam engines were in use as early as from the 1870s on soils that could bear the weight. The adoption and early introduction of novel technologies not only has tremendously helped increase yields per hectare but also brought down labor needs in the field. One farmer today can harvest in excess of 100 hectares. Work productivity in the field has increased by a factor of well over 100 in the last century.

Not surprisingly, farming was an area of our economy that embraced "digital" very early. Already at the beginning of the century increasingly more equipment parts got digital features (combines, tractors, etc.). For many years, this was primarily to better capture information on performance indicators like product use or yield by the farmers. But the basic necessary building blocks to arrive at digital farming solutions were present (see ■ Fig. 2.4).

Digital Farming soon has been identified as an attractive market where all prerequisites were given to realize a substantial value potential. This starts with the technical feasibility. Data collection devices on the equipment have been in place since the start of the century in the form of cameras, sensors as standard equipment for many new combines and tractors with more than 200 horse power.

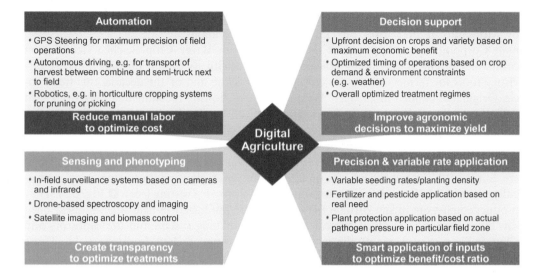

Fig. 2.4 Technological elements of Digital Farming—need to be brought together

The potential to increase yield could be taken as a given. In project work with leading agricultural chemistry and seed companies globally broad acre crops like cereals, corn and soy were identified to have a yield increase potential of at least 15–25%, with the biggest levers for yield increase being seed variety, fertilizer and crop protection. With the generational change already in 2015 some 20–30% of growers were identified to be willing to apply Digital Farming in Europe, a number expected to double by 2024. The estimated total value creation potential is estimated to reach up to 20 bn p.a. in the broad acre crops in the decade of 2020. This translates into a value capture potential of up to 7bn USD p.a. for the service providers at a 30/70 profit split between industry and the farmers. In the long term, i.e., beyond 2040, the added value of a crop production globally increased by 25% and more is in the order of magnitude of 200 bn €.

Many players from a variety of industries went after this value. Seed and crop protection companies, distributors and equipment manufactures saw the potential first. Digital farming start-ups sensed the opportunity to disrupt an established market as well as IT & data analytics companies. They could be well differentiated by the range of their offering and the level they were willing to put skin into the game. The offering ranged from single famer job steps over multiple job steps in crop management to a comprehensive crop farm management offering. Level of engagement spread from simply providing information or providing specific assessments over tailored advice including guidance to implementation/application to assurance of the targeted benefit and sharing risk with the famer. So far, no provider could win with a comprehensive crop farm management offering with assurance of the benefit. Most solutions are confined to several job steps and giving tailored advice.

At first, a plethora of Farm Management and Information Systems (FMIS) emerged, seeking to support the farmer in managing his plots and internal processes. Other systems were developed to better manage singular dimensions relevant to the farmer, e.g., weather events.

The first landmark in a true development towards Digital Farming that provides a comprehensive recommendation scheme was set by the acquisition of Climate Corp by Monsanto in 2013 for 1 bn USD. Climate Corporation had been one of the first companies aiming at developing holistic advisory tools for farmers along the crop cycle. Originally founded in 2006 as "Weather Bill" the company had initially focused on providing weather insurance to farmers but also other industries dependent on weather effects, like ski resorts or large event providers. Since 2010, the pure focus was on agriculture with the Total Weather Insurance product coming out in late 2010 on the large row crops corn and soy. Over the next years, Climate Corp moved out of the insurance business and around the time of its acquisition by Monsanto targeted digitally supported decision making for the farmer with Climate Basic and Climate Pro. Those developed into integrated service offerings with a focus on nitrogen management and field health on a per field level, later accordingly re-branded as Climate FieldView. Striving to gain more focus on digital advisory Climate Corp in 2015 intended to sell its hardware activities in Precision Planting LLC to John Deere. The deal was ultimately closed with AGCO in 2017 due to anti-trust considerations.

With Bayer Crop Science stepping on stage with its digital farming offering xarvio at the mid of the decade, this trend towards holistic recommendations from seed selection over fertilization/nutrition to crop protection was expedited.

The holistic digital farming offerings focus on the complete crop cycle from planning over planting, nurturing, crop protection to harvesting (see ◘ Fig. 2.5). They cover optimized in-field operations, a decision-making system and the capturing and processing of data from the field. Those range from historical field data over soil analytics to exact plant nutrition information, pathogen occurrence to weather information. Source can both be proprietary field data or publicly provided data of the respective plots.

These input data—historical and current—will be fed into a decision-making system that combines agronomic understanding with artificial intelligence and algorithms and modeling tools to provide recommendations for all steps of the crop

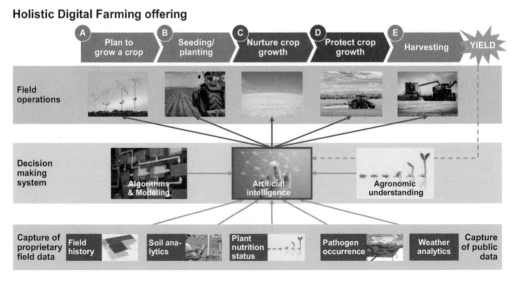

◘ Fig. 2.5 Holistic digital farming offering

cycle. It is important to note that the power of the system results from the breadth of data it is supplied with. Data from local fields, combined with data from other fields in comparable soil and climate conditions, plus research and development data from the farm input providers and distributors. Several other players from a variety of industries (equipment producers, distributors, tech companies, etc.) also began to increasingly invest in the area and consolidated smaller companies, e.g., satellite imagery providers into their offerings—those companies to a large extent had their origins as consulting and software providers already in the 1980s.

Hence after starting broadly, the years 2015–2017 showed a clear focusing: Agriculture input providers moved towards agronomic advisory. Agricultural equipment providers focused on completing their portfolio with more precision application solutions.

Then, overall, after 2017 the development stalled. The financial potent players in the market were focusing on the consolidation of their "classical" business with a series of mergers and acquisitions and other players like distributors did not step in, due to the lack of financial strength and not the same degree of R&D experience.

Kearney expects that in the coming years the trend towards fully integrated farming solutions as a true differentiator in the market will gain speed again and we will see providers with a sophisticated offering, targeting the professional farming sector, in particular in Eastern Europe, North America and the large farms in Brazil, with a complete offering:

– Digital Solution Platforms - cloud-based, machinery integrated SaaS solutions for farm management along the whole crop cycle
– Advanced Satellite Image Analytics: agronomical and crop yield analytics, prognostics and monitoring services
– Farm Management Information and Operations Systems: end-to-end farm business and regulation management software
– Precision Farming Machinery and Services: software integrated precision farming equipment (trading model) and services.

This will need to be accompanied with the change of business models from product sales to offering solutions as we are seeing it in the market already.

2.1.4 Implications for the Industry: From Product to Service

The traditional farm input model is changing significantly. Still, seed, fertilizer and agricultural chemistry are brought to the grower primarily through farm retailers/distributors. The grower is receiving separate agronomic advice. The future model will likely rather have a digital recommendation and application platform for holistic in-field crop management (see ◘ Fig. 2.6), as long as legally possible. This can substantially alter the balance of the power in the market, similar to other industries. This is similar to retail, where today also buying information and recommendations are given together with the opportunity to order.

As the most substantial change for the Ag input industry, however, we foresee the transition from product to service. Until now, in almost all parts of the agriculture input industry the focus was on selling a product—seed, herbicides, a tractor, etc. However, the main intention of a grower is not to buy a specific amount of herbicides, but rather to have a weed-free field with ideally minimal long-term detrimental impact on the soil. The solution of the agriculture industry moving forward,

Traditional versus Digital Farming value chain model

Traditional farm input model

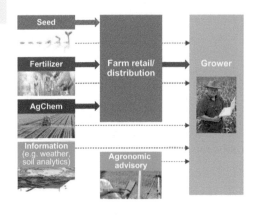

Digital recommendation and application platform-based farm input model

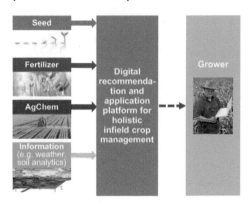

➡ Input flow ┄┄▶ Information flow

Fig. 2.6 Changing business models

hence, should be to meet exactly that demand of its customer. Provide a weed-free field with minimal environmental impact. Thus far, the thinking of the big input providers is product-oriented—revenue and profit are more or less directly proportional to the amount of product sold. And incentive schemes are linked to that. A facility manager tasked with cleaning an airport is not paid by the amount of cleaning products used, but father by the area cleaned. Big Ag will need to adopt a similar model. When quantity of product sold is no longer the key performance indicator, the industry can quickly pick up service models. That comes with big transformations, though. For example, industries so far were used to highly centralized production and then distributing via a variety of sales channels to even the remotest parts of this planet. They would make profits in excess of 20% EBITDA with a comparatively small workforce and limited equipment investments except for the central production facilities. A service model in the future will bind more capital and manpower for servicing machinery, drive down relative margins and need substantial capital investments/financing.

But future-proof business models even go beyond products and services in the long run. In terms of technology applied, crop protection and seed need to be complemented by the beforementioned data science and equipment, e.g., robots.

Products will be added services and holistic solutions up to an integrated business. There are three main additional offerings we see: automated services provision, running agricultural service contractor business, and perhaps in the long-term running owned farming business (see ◻ Fig. 2.7).

2.1.5 Digital will Change the Face of Farming

The most visible impact of digital will be regarding the type of machinery used in the field. Ever since field labor came up it was the ambition in industrialized agriculture to minimize the costly la-

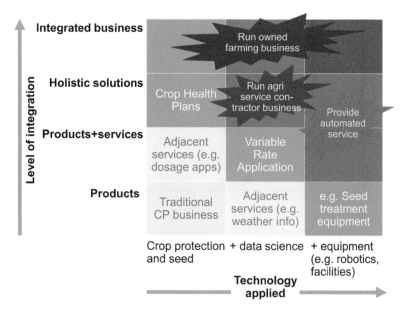

Fig. 2.7 Future-proof business models for the Ag input industry beyond products and services

bor part. This led to machinery becoming bigger and bigger, with only a single operator and combines with a width of 14 m or sprayers 50 m wide, which are not only able to cover large amounts of land fast but also could be operated by a single person. This drove the labor cost down. With the advent of autonomous robots taking over more and more jobs in the field, ultimately from planting to harvesting, size will no longer matter and we will come back to small, independently operated swarm robots in the fields (see Fig. 2.8). They will likely be powered by renewable energy (photovoltaics, e-batteries, fuel cells or synthetic fuels). The main challenge today is that they do not have the power for intense physical work like plowing, harrowing or harvesting. However, we understand that this is no principal problem but only needs some more development.

For agricultural robots to take over broader market shares, we see four criteria that need to be met.

1. Proper sensing technology, e.g., spectrometry of leave color to detect plant health status and camera technology to identify pathogens based on shape recognition. Weeds need to be identified among crops. Fungi or volatile organic compounds need to be identified, e.g., through high-speed gas chromatography.

2. Artificial intelligence/algorithms: pattern recognition to identify shapes of weeds, insects, fungus induced decomposition, etc. It is important to check against thresholds, e.g., characteristic patters for economically relevant pathogen pressure. Decision making will need to be based on pattern and threshold comparison.

3. Actuation and application: in-field movement needs to be automated, application technology needs to be developed for spraying, spreading, etc. Mechanical weeding technologies like pulling out, stamping down, cutting off, etc., need to be further developed. Advanced technologies like laser-based weeding and insect control need to be implemented.

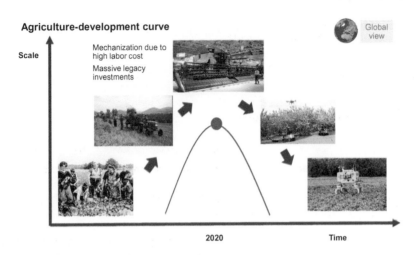
Fig. 2.8 Agriculture development curve back to smaller machinery

4. In-field infrastructure/logistics: energy supply, e.g., via on-board photovoltaic panels, supplementing charging stations at field borders to re-charge, ideally based on renewable sources. Inputs like fertilizer, AgChemicals and other consumables likewise need to be supplied.

2.1.6 Outlook: From Farm to Fork and Back

The sections before have mainly described how the existing agriculture value chain—mainly up to the farm—will likely change due to digital.

On their way from the farm gate to the retail shelf most agricultural products get substantial price mark-ups, often in the order of magnitude of a factor of 5–10 or more. This is especially the case with processed foods. The value of corn, sugar and fat in 1 kg of cornflakes is below 50 Cents; the retail price hits 5 € and more, though. For sustainable products, this retail price is likely to double or triple, far beyond what most consumers are willing or able to pay. Mark-ups at the retail level of up to 10% are acceptable to roughly two thirds of consumers, if—and only if—they can be sure that the products meet the stated specifications [GPD20]. This provides another opportunity for digital in agriculture. Tracking and tracing production methods in the field and onwards to the consumer—from farm to fork. Digital may provide a means to better communicate quality features of the agricultural products through tracking and tracing technologies, ultimately realizing a better product price at the retail level.

Finally, in an economy that is turning more and more circular, agriculture will play an important role: on the one hand as provider of bio-based feedstock and on the other hand as off-taker of post-consumer material. For the management of these circles, digital support will also be of vital importance.

The "bioeconomy" will use biotechnology for the production of bio-based goods from biomass as the main feedstock. In several geographies, governments and regulators are crafting bioeconomy strategies accordingly. The OECD started as early as 2006 and the EU followed in 2012.

The potential is huge, with over 15 bn tons of biomass being produced annually. The more of this biomass is taken from the fields, though, the greater is the need to feed post-consumer material back into the circle and onto the fields. This poses huge challenges. The agriculture industry on a massive scale ships proteins and nutrients around the globe and latest at the consumer loses track of composition of its products. To take the nutrients from post-consumer products back into the fields will require additional digital support in sophisticated reverse supply chains with tracking, tracing and testing.

With circular economy becoming the most relevant future topic, agriculture can redefine its role in the environment. The focus as of now is primarily on closing the carbon cycle and reducing the emission of CO_2 from fossil carbon sources. This has been widely understood in the wake of the Paris agreement and finds its way into the company reality with 2020 being a key year in that regard. Activist activity like Fridays for Future and regulators alike (EU Green Deal) in combination with the financial markets drive fossil carbon reduction into implementation, opening up room for the next circles to be closed.

2.2 Beyond Digitalization: Major Trends Impacting the AgFood System of the Future

Stefanie Bröring, Otto Strecker, Michael Wustmans and Débora Moretti

Abstract

Interrelated disruptions on agriculture are not only broadening the horizon of change but are also clarifying opportunities and challenges guided by continuous innovation in the marketplace. Data from one of the richest European databases for trend analysis, Trendexplorer, reveal that currently, 16 different mega-trends are affecting the AgFood system. From these mega-trends, we identify three major disruptions that may change the rules in food and agriculture, namely digital-driven disruption, sustainable-driven disruption and societal-driven disruption. By drawing on selected case studies, we also discuss how the different trends and resulting disruptions relate to each other. We thus further explore specifically the impact of the digital disruption on the AgFood system, providing an analysis of different scenarios, in which the blurring of the boundaries between the different sectors and technologies affecting current industry structures is illustrated. Based on this discussion, firms of different industry origins may better understand the opportunities that are emerging, the necessary resources and capabilities needed to conduct strategic renewal, and how this affects both their positioning and the fit of their strategy in this game.

2.2.1 Introduction

Following a worldwide trend, the AgFood system has been subject to a profound transformation driven by the application of new technologies previously used elsewhere and fostered by the increasingly demand for efficiency, food security and sustainability (see ▶ Sect. 1.3 and 2.1). Such transformation opens new opportunities for innovation and induces new behavior patterns [BLW20]. Accordingly, the digitalization of the AgFood system—although inevitable, one could argue—comprises only partly the renewal process. Smart sensing, but also biotechnology play a big role, for instance, in the reduction of pesticide use. This suggests that the combination of different knowledge areas and technologies is necessary to reach a major goal, composing a System of Systems (SoSs) [PH94].

Moreover, one may not forget that technological disruptions do not occur in the vacuum and may hinder or reinforce other trends. Technical change will not only provoke the evolution of the economic system but also shape new societal rules [Per02]. In order to fully exploit this potential, the actors involved should try to think systemically, towards the entire innovation ecosystem, spanning industry borders [AK10]. Ecosystems are very dynamic and often emerge from the convergence of different, hitherto separately functioning business sectors, such as IT and agriculture. The possible convergence of such industries [Bro10], triggering new ecosystems, increases the complexity of knowledge and innovation management mechanisms involved in intra- and inter-organization interactions. However, only an expanded view will allow firms to successfully identify the potential for new business models and opportunities.

We, therefore, use the following questions as a guide to our reasoning: Which are the main disruptions affecting the AgFood system and how do they influence each other? Focusing on digitalization, what are the main challenges, resulting in strategic options, and needed capabilities that firms must develop, first to survive in the marketplace and second, to exploit new opportunities?

By shading some light to those issues, we highlight three main contributions of this section. First, we expand the scope of what is usually understood of agriculture 4.0 and bring other perspectives (for example social) to the table, balancing the technocratic bias of this (r)evolution. Second, we turn our attention specifically to the digitalization of agriculture and provide an analysis of strategic management practices, which companies may use to deal with such transformation. We finalize the section by merging both macro- and micro-perspectives and end with questions that may influence decision-makers in their strategies.

2.2.2 Innovation is Multi-Systemic—Main Disruptions from Farm to Fork

To navigate in times of change and uncertainty is naturally challenging although to some extent predictable. To understand which paths society is following, we use trend data from one of the richest European databases for trend analysis, the Trendexplorer from TRENDONE. Trend data include textual information about emerging technologies, research developments, and product launches and are a common source used by practitioners in foresight activities to identify innovation fields [DU08]. The TRENDONE approach subdivides trends into three categories, namely micro-, macro- and mega-trends. **Micro-trends** consist of short descriptions with the above-mentioned content. They are allocated to **macro-trends** that describe change occurring within a medium timeframe. Macro-trends frame jointly a **mega-trend**, such as globalization, demographic development or digitization that describe long-term change. From the database, we take a broad approach looking for trends related to agriculture and extract 16 mega-trends, from which we identify three major disruptions that may fundamentally change the rules in the AgFood system: digital-driven, sustainability-driven and societal-driven (see ◘ Fig. 2.9).

Each of the 16 mega-trends exhibits 3 to 9 macro-trends, which are composed of 330 micro-trends in total. For instance, the mega-trend food culture encompasses the following four macro-trends: Newtrition, Food Fashion, Slow Food, and Performance Food. Due to topics such as alternative protein, the mega-trend food culture appears more frequently than artificial intelligence. This analysis allows us, on the one hand, to trace the multiple influences shaping innovation in agriculture and, on

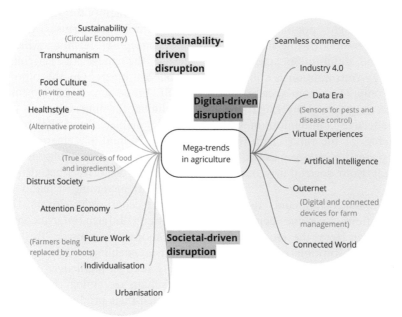

Fig. 2.9 16 Mega-trends related to agriculture. Practical examples are highlighted inside some of the mega-trends of most importance to the AgFood system

the other, to recognize the plurality of impacts that transformation in agriculture may provide.

The **digital-driven disruption** originates from the advances of artificial intelligence, Big Data, and IoT, which must be adapted to the agriculture sector both in terms of functionality and compatibility, considering the several systems that are mandatory to the user, mainly the farmer. Interestingly, the disruption encompasses expected trends—Industry 4.0 and Data Era, for instance—and emerging concepts, as Outernet, which represents the level of digital integration of previous pure physical things. The French start-up MyFood (myfood.eu) represents an example, as it developed a small greenhouse to be installed in houses and restaurants in the city. Such cases are rarely developed without sensors that can be controlled by online platforms and apps. Thus, the separation of digital and physical, offline and online, is becoming blurred. Apart but not necessarily detached from digitalization, the **sustainability-driven** disruption encompasses four mega-trends: Sustainability (which includes concepts such as Circular Economy and Zero Waste, see ▶ Sect. 1.3 and 2.1), Healthstyle, Food Culture, and Transhumanism. Of interest, Healthstyle points to personalization, to which the macro-trend Data Era plays an important role. In its turn, Food Culture regards to new fashions and new alternative sources of nourishment, leading to further exploration of biodiversity and the recombination of existing resources. Case in point, bioengineering, including CRISPR-Cas, represents the mega-trend Transhumanism, which relates to the ability to modify organisms with biotechnology tools. Last but not least, **societal-driven disruption** will relate to mega-trends that are both cause and effect of innovations. For instance, Urbanization is a growing trend, which calls for solutions that allow the accommodation of the majority of the population in urban spaces. To tackle this challenge,

Vertical Farms and rooftop farms are becoming increasingly popular, providing fresh and healthy food, while saving transport costs and diminishing land use, although energy consumption is still a challenge. Not only, trust from society should not be taken for granted, implying transparency and effective communication among different actors of the value chain (farmer to end-consumer, for instance) as two of the major trends from societal-driven disruption.

As mentioned beforehand, these different disruptions to some degree reinforce each other. Here, the cross-influence among the three disruptions opens up room for new business models that design value propositions matching the different mega-trends (see ◘ Fig. 2.10).

For instance, the US company Aspire (aspirefg.com) draws on robotics and automated data collection to grow insect protein on digitized farms. Alternative protein sources are a sustainability-driven trend (first-order driver) and have been supported by modern technologies (second-order driver). The company connects several farms via Internet of Things, allowing high predictability for yields and reproducibility among the farms. It creates a whole new concept of farming, which might redesign the image of agriculture understood by society. On the other hand, the BASF brand xarvio (xarvio.com) focusses on digital farming solutions such as the "field manager" and, thereby, takes advantage of multiple emerging digital technology systems enabling precision farming (first-order driver) but also provides an answer to increasing demand for sustainability (second-order driver). Other examples are (1) Infarm (infarm.com), an urban farm model that provides fresh food grown in cities, enabling increasing urbanization; (2) AgriLedger (agriledger.io), which uses blockchain technology to help farmers in Haiti sell their produce at better prices; (3) Nourished (get-nourished.com), a business fostered by individualization trends, that supply personalized 3D-printed high-impact vitamins; and Vital farms (vitalfarms.com) an initiative to approximate buyers and farmers, by labeling every egg carton with the names of the farms where they came from and then providing 360° view on the respective farm.

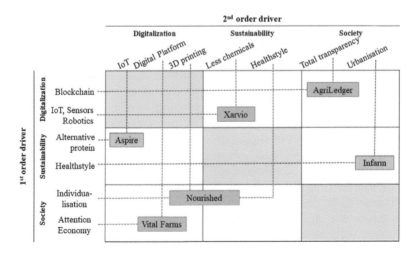

◘ **Fig. 2.10** Start-up and company responses to mega-trends. Cross-influence among the three disruptions: digital-driven, sustainability-driven, and societal-driven. The disruptions can be both the main goal of a new value proposition of the start-up examples, acting as a first-order driver, or the enabler of another disruption, serving as a second-order driver

2.2.3 Focus: Digital Disruption and Its Implications for Involved Agribusiness Companies

Digitalization as a mega-trend has the potential to disruptively change AgFood technologies as well as existing business processes and business models. In agriculture, for example, digitalization is an essential lever to use resources more efficiently, to facilitate work processes, to be more animal-friendly, and to produce and sell sustainable, high-quality food. The players in the AgFood system, which includes traditional companies, global players, and numerous AgTech start-ups, have recently recognized the **potential** of digitalization for themselves and the entire sector [HWB19]. Still, this leads to the following three major challenges for agricultural players.

2.2.3.1 First Challenge: Dealing with an Increasingly Complex Knowledge-base

The blurring of boundaries between the AgFood system and information technology (IT) as well as the trends allocated to digital-driven disruption indicates that the knowledge-base for all players along the value chain is expanding. So, what are the key capabilities and knowledge areas for a digitalized AgFood system? While looking at the knowledge base of different digital technologies, one can observe that next to rather obvious knowledge areas such as data science, new knowledge areas emerge, i.e., bioinformatics, synthetic biology, geoinformatics or nutrigenomics to name a few (see ◘ Fig. 2.11). Some grounding technologies and innovations were highlighted as examples connected to the disruptions mentioned in the previous section, which are spanning over different steps of the value chain. A clear change in the value chain relates to its circular potential, as technologies also allow the reabsorption of outputs and waste again into the chain. Moreover, the technology systems are increasingly connected with each other. This connection and the emergence of novel technological systems is not only driven by the technology push (i.e., emergence of new functionalities and applications of enabling technologies) but also increasingly by societal pull triggering novel regulations (i.e., increasing ban of using certain pesticides). Here, the EU Green Deal will certainly foster the diffusion of smart farming technology systems such as smart spraying systems allowing to reduce the usage of pesticides, as, e.g., the smart sprayer project of Amazone, Bosch and xarvio nicely demonstrates [Ama21].

2.2.3.2 Second Challenge: Dealing with New Players from Outside the Industry due to Convergence

A look at the impact of digitalization on the AgFood system shows that also the value creation structure in the AgFood system is becoming increasingly complex, as not only new fields of science and technology become relevant (see ◘ Fig. 2.11) but also new players from outside the industry are entering the market, and industry boundaries are dissolving [Bro05], [HWB19]. More precisely, the blurring of boundaries between the AgFood system and the digital economy can be described in more detail using four different scenarios (see ◘ Fig. 2.12). These four different scenarios are not mutually exclusive but run in parallel, with individual players even participating in different scenarios at the same time.

In **Scenario 1**, the AgFood system is the driver of converging technologies and responsible for the increasing blurring of industry boundaries by developing and integrating digital skills. This scenario occurs when agricultural companies train the

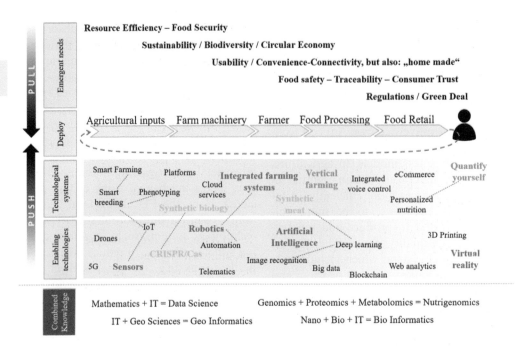

■ Fig. 2.11 The AgFood system between Tech PUSH (the combination of distant knowledge fields to the emergence of new Technological systems) and societal PULL

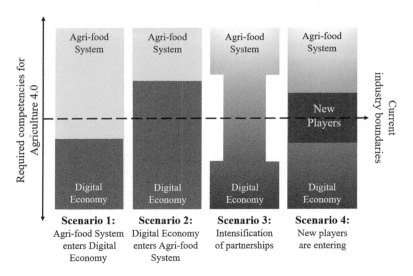

■ Fig. 2.12 Scenarios to depict the blurring of boundaries between the AgFood system and the digital economy. Source: Authors, based on [BPS+15] and [HWB19]

existing staff, hire new staff that is already trained, or buy and integrate IT-driven companies. For example, in 2017 John Deere acquired the start-up Blue River Technology as it focused on computer vision, robotics and machine learning applied to smart machines [Dee17]. Another example is the fertilizer supplier Yara, who purchased the Berlin-based AgTech start-up Trecker.com in 2018 to extend its recently established business unit "Digital Farming" [Yar18].

In **Scenario 2**, on the other hand, IT-driven companies, such as Amazon, Google, IBM or Microsoft, penetrate the AgFood system. Platforms such as IBM Watson and Microsoft FarmBeats aim to help the farmer to make decisions [MR18]. In **Scenario 3**, AgFood and IT companies enter into cooperation. For instance, the pig farming corporation Dekon Group and pig feed supplier Tequ Group cooperate with Alibaba Cloud, aiming the use of sensors to analyze the behavior of pigs to digitally record pig pregnancies or diseases at an early stage, and provide appropriate feedback so that respective measures can be initiated [Pen18]. **Scenario 4** shows typical investments in start-ups that create new players on the borders between the AgFood system and the digital economy. Some of those start-ups are able to join multiple worlds: The Israeli Phytech was invested by Syngenta, Tencent Holdings, and Mitsui & Co; corporations, respectively, coming from the agriculture, IT and trading sectors.

2.2.3.3 Third Challenge: Exploiting the Right Strategy and Identifying the Necessary Capabilities to Thrive in such a Complex Ecosystem

Up to this point, we explored the complexity of the current AgFood system [SEW+10] and provided an overview of players and knowledge fields that are contributing to it. This brings us to the third challenge: the plethora of actors and systems must be well aligned towards unique value propositions. For companies, this means that it is not anymore about producing and selling products or offering specific services, but engaging with the whole ecosystem. Therefore, we ask: which strategic options do companies have and what does that imply in terms of capabilities? We start our answer following [HE18], who describe three strategies in ecosystems: the system, the component and the bottleneck strategies (Table 2.1).

The **system strategy** is characterized by a higher level of control, where one company chooses to simultaneously enter multiple components, reducing its dependency on complementors. If different components are to be produced and commercialized by one company, integrative capabilities are essential, i.e., the organization must be highly capable of combining products, resources and knowledge to secure in-house development [HR18]. The John Deere Com-

Table 2.1 Strategies and capabilities to navigate the ecosystems. Source: Authors, based on [HE18], [NS11] and [HR18]

Strategies	Capabilities
Bottleneck—enter the bottleneck component at the founding, and new ones as they emerge	**Innovation capabilities**—opportunities through product sequencing
Component—enter one or a few components and cooperate for the rest	**Scanning and sensing capabilities**—towards core products as well as complementary asset providers
System—enter multiple components and minimize cooperation	**Integrative capabilities**—introduction and modification of products, resources, and business models

pany seems to apply this reasoning. Taking advantage of its large resource base, they developed several digital platforms in-house, namely MyJohnDeere, Field Connect™, AgLogic™ & DigiConnect, targeting the different ecosystem actors. The **component strategy** relates to parts of systems that may take innovation to the next step. Such a strategy is less resource-intensive in terms of development, but rather requires scanning and sensing capabilities not only to keep innovating the developed component but also to identify complementors that will increase value creation. The case of the Israeli start-up Prospera illustrates this strategy, as they received investment from Cisco and Qualcomm, two hardware leaders. Both corporations enter the agribusiness with a component, and secure value creation through complementors [Pee17]. Finally, the **bottleneck strategy** is as complex as potentially successful. It can be regarded as a specialized type of component strategy where the component is a bottleneck for the whole ecosystem to grow, due to poor quality, poor performance or short supply [HE18]. Microsoft FarmBeats came into place not only as a management platform for farms but also to solve a bottleneck issue: farm connectivity. It is expected that Microsoft continuously innovates in this bottleneck, bringing connectivity to the most remote areas, whereas the company must maintain its eyes open to possible shifts in this bottleneck. Another example comes from Agrirouter (DKE-Data) that allows data integration in a single system independent from the technology suppliers.

2.2.4 Concluding Questions

We started this section by calling attention to the substantial change that the AgFood system is currently facing. This change is not restricted to the use of new technologies, but may indeed transform society, as well as technology development is influenced by societal trends. To shed some light on the different disruptive forces, we draw upon a trend databank to depict the three major disruptions potentially transforming the AgFood system, namely digital-driven, sustainability-driven and societal-driven disruptions. It seems pivotal to be alert to these dynamic developments since all disruptions are connected and not only influence but also potentially reinforce each other. It is, however, not clear which vision do the incumbent corporations have regarding the future and especially if such vision differs among companies previously coming from the AgFood system or the digital economy. Moreover, it is unclear how incumbents should best partner with start-ups who are perhaps more agile to design and test their value propositions.

However, the definition of such a vision is not only important to the firms themselves, as it affects their strategy and triggers their renewal. It seems also of utmost importance to governments and other stakeholders that are willing to influence sustainable development. For instance, digitalization is only a concern regarding unemployment, if the work force is not reallocated (and accordingly educated) to the new knowledge and application fields that are emerging. Therefore, what is the role of universities and their faculty structures—perhaps these need to cooperate even more to account for the needed knowledge combination as innovation in the AgFood system happens at the interface of different knowledge fields?

We further highlight three main challenges for companies that are embedded in this context or that are focusing it. The digitalization of agriculture brings new players to the game, different knowledge fields, and therefore different strategies. As a consequence, new threats or new opportunities for collaborations are at place. In order to allow for a timely response if not a proactive action, one should scan its own resources to understand how prepared the company is to absorb new knowledge from related ecosystem partners, or implement innovations out of its core competencies. Are the companies entrepreneurial enough to risk out of their comfort zone? Are they aggressive enough to aim for system strategies? How does the business model need to be adapted? Will they follow or orchestrate emerging (digital) platforms and eco-systems? What industry will be more successful and act as orchestrators in the AgFood system of the future: Big IT- or AgTech? What will be the role of the farmer in the future—just owning land, or even less if the farm-free food movement diffuses?

Moreover, one could ask: do all those changes impact and change value chains? If robots and autonomous vehicles substitute farmers, artificial intelligence platforms substitute advisors, and marketplaces connect farmers and consumers directly, changes are to be seen. Up to this point, it is rather clear that value chains are not enough to encompass all the relevant actors and the larger ecosystem perspective should guide as further.

2.3 Economic Benefit Quantification

Peter Breunig

Abstract

This section provides an overview of how economic value is created through digital solutions, which cost is involved in using these technologies, and how the economic benefit is calculated. In addition, the limitations of the economic benefit model are described followed by an example.

2.3.1 Introduction

Although economics are not the only driver for the adoption of technologies (see ▶ Chap. 1), understanding the economic benefits of digital solutions from a farm-level perspective is relevant for farms themselves as well as the companies offering these technologies. Farmers need to understand the economics behind digital solutions to do the right investment decisions for their operation. Agribusiness companies on the other hand need to understand the farm-level economics of their offering for value-based pricing and to convey the value to customers during market introduction. Even early in product development, when final costs of a solution are not yet foreseeable, quantifying the economic value of potential solutions might help to prioritize development projects.

The following content focusses on arable farming only, although the underlying logic can be transferred to other types of farming as well.

2.3.2 Fundamentals of Economic Value Creation

To understand the economic benefits of digital solutions, we must first understand how digital solutions create value for arable farming in a way that can be quantified. There are five main ways of how value is created:

1. **Improve job execution**
   Digital solutions help to execute a job like planting, spraying or fertilizer application better. These improvements are based on two dimensions: higher precision and increased output.
   1. **a) Higher precision**

Higher precision is achieved in three ways:
- Less variability of defined job quality parameters: Technologies like advanced planter monitors allow to adjust tractor speed to ensure a defined singulation quality and placement of seed. NIRS-based nutrient sensing in organic fertilizer application ensures a more precise application of actual nutrients. In harvesting grain, cameras enable automated settings adjustments which ensure a consistent grain sample in varying crop conditions.
- Reduced overlaps through technologies like autosteer and section/nozzle/row control: GNSS autosteer, especially using RTK-correction signals, reduces overlaps between machine swaths in the field. GNSS-controlled shut-off of sections, single nozzles, and planter rows minimizes overlaps on headlands and irregularly shaped fields.
- Adopting input application and machine settings to sub-field variability based on soil, slope, weed distribution, and other factors affecting crop growth: Technologies like Variable Rate Application and spot spraying allow applying inputs on a sub-field level optimized to a specific zone or even a single plant. Variable tillage and seeding depth allow machine settings to vary on a sub-field level based on the requirements.

1. **b) Increase output**

Digital solutions can improve job execution by increasing output per operator hour, i.e., hectares worked per operator hour.
- Reduced overlaps between machine swaths in the field enabled through GNSS autosteer increase the output per operator hour: This is especially relevant for large equipment with wide working widths.
- Higher speed: Digital solutions that analyze machine performance or automatically adjust machine settings based on conditions allow machines to always run at the maximum speed possible. Examples for these technologies are monitors or speed-automation systems for planters (TIM) and combine harvesters. These systems outperform in most cases speed adjustments by the operator, especially in longer shifts, and lead to higher average working speeds.
- Autonomous operation: Output per operator hour can be drastically increased when machines operate autonomously. This means that one operator can manage several machines at the same time.
- Less downtime: Digital technologies enable remote diagnostics of machines, remote support, and other solutions to reduce machine downtime and increase output.

2. **Improve management processes**

Digital solutions can help to speed up management processes and reduce errors.

2. **a) Simplify job planning, controlling and documentation**

Digital farm management information systems (FMIS) increase speed and reduce errors in job planning and execution. In combination with telemetry systems, job plans can be sent remotely to machines and the execution can be controlled from the office. Documentation can also be simplified and even automated using digital FMIS.

2. **b) Improve purchasing and selling**

Digital marketplaces allow farmers to get quotes from input suppliers faster and simplify selling of commodities including logistics. Furthermore, these trading systems often provide access to more potential sellers and buyers of inputs and commodities compared to current practices. In addition, disintermediation (i.e., the reduction of intermediaries like commodity traders or input dealers within the value chain) as well as the aggregation of demand (e.g., several farms aggregate their demand and purchase inputs together) enable better selling and purchasing conditions.

3. **Improve decision making**
Besides improving the execution of jobs and processes ("doing things right"), digital solutions also create value by enabling better decisions ("doing the right things").

3. **a) Agronomy**
Agronomy decisions include questions on which operations should be done when and how as well as which inputs should be applied at what rate and point in time. When applying inputs on a sub-field level, i.e., based on defined zones or even single plants, the number of required decisions increases significantly. Digital systems using crop/disease models, expert systems, or machine learning can lead to better decisions (e.g., higher yield and less inputs) and/or faster decisions. Currently, available solutions are moving from tools that support decisions to prescriptions that almost fully automate decision making.

3. **b) Equipment related**
Farms need to take equipment-related decisions about machine logistics (which machines should do what, when and where?), machine settings, and repairs/maintenance (when to change which parts?). Telemetry systems using machine sensors in combination with smart analytical tools allow farms to improve decision making and can enable better logistics, improved machine settings, and optimized repair timing (predictive maintenance).

3. **c) Business related**
There are numerous business decisions farmers need to take throughout the year like input purchase or crop selling. On a longer-term perspective, farmers need to also take business decisions on renting or purchasing land and how to use their fields. Digital platforms can support regular business decisions by providing price and market information. Also, strategic decisions around land expansion and land use can benefit through tools like profit zone field-maps and digital platforms that provide land value and land productivity information.

4. **Enable new production systems**
Besides improving single jobs, digital technologies can also enable new production systems, i.e., the sequence of jobs like tillage, planting, crop care, etc. to establish a crop. For example, controlled traffic farming is enabled through RTK autosteer and leads to fewer tillage passes. Strip Tillage is another production system enabled by RTK autosteer technology which reduces tillage to a small zone around the crop rows. Potentially, smaller autonomous machines could make new diverse copping patterns with various crops within one field possible.

5. **Provide data for partners along the value chain**
Digital farm data can provide value to up- and downstream partners along the value chain. Examples are machine or agronomic data that helps machine manufacturers or input companies to optimize their offering. Although there are only very few cases so far in which farms are paid directly for their data (e.g., Farmobile LLC), this would be possible and would create additional revenue for farms.

Farm data can also provide value downstream the value chain. In this case, certain production methods or environmental benefits can be traced through digital data leading to possibilities for farms to differentiate their commodities and directly react to customer needs and wants. Although these systems are still in development, there seems to be a significant potential for higher prices and additional revenue (e.g., carbon market) for farms.

To quantify the economic value of digital solutions, we need to connect the ways value is created with revenue and cost on the farm level.

Revenue is yield multiplied by price plus additional revenue streams. The relevant cost groups are direct cost (cost for seed, fertilizer, plant protection plus variable irrigation cost, crop insurance and drying energy cost) as well as operating cost (variable machinery cost like repairs, fuel cost, depreciation, finance for machinery, labor cost, and contractor cost). Overhead cost (building depreciation and interest, land, property taxes, building insurance, and miscellaneous items) are usually not influenced by digital solutions.

The overall economic value created by a digital solution on the farm level is equal to the changes in revenue, direct cost and operating cost of the farm's production system in comparison with the situation without this solution.

Economic Value to the Farm = RvC + DCC + OCC = (Y*PC + YC*P + YC*PC + AR) + DCC + OCC

RvC = Revenue Change, Y = Yield, YC = Yield Change, P = Crop Price, PC = Crop Price Change, AR = Additional Revenue, DCC = Direct Cost Change, OCC = Operating Cost Change

It is important to note that one solution could offer an advantage on one revenue or cost item but have a disadvantage on another one. E.g., spot spraying decreases herbicide costs (DCC) but increases operating costs due to lower speed (OCC).

▪ Table 2.2 shows how the ways digital solutions create value relate to relevant revenue and cost items on the farm level.

2.3.3 Cost Structure Fundamentals of Digital Solutions and Economic Benefit

Besides the economic value that digital solutions provide, there is also cost involved in using these technologies. These costs can be divided in variable cost (depended on the utilization) and fixed cost (independent of the utilization).

Variable cost of digital solutions consists of the following three cost items:
1. Repair and maintenance cost: Hardware components of digitals solutions may require repairs or maintenance. One example would be the protection glass of NIRS sensors used in self-propelled forage harvesters which needs to be replaced regularly based on usage.
2. Variable labor costs: E.g., if application maps are created on the farm, there is a certain amount of labor required to do these operations including operational inefficiencies, partly due to interoperability issues. This amount is dependent of the number of fields for which application maps are created.
3. Variable licensing and data cost: Sometimes licensing fees, software subscriptions, data and data transmission cost, and data transformation or adaption cost are based on the usage and therefore variable costs.

Fixed cost are made up of these four cost items:
1. Depreciation: To allocate the usage cost of a tangible asset over its useful life, depreciation is used as part of fixed costs. Usually depreciation is calculated as follows: (Purchase price—salvage value) / usage.
2. Fixed labor cost: Labor cost involved to start-up a technology (once or several

Table 2.2 Relationship between value created by digital solutions and relevant revenue and cost items on the farm level

	1. Improve job execution								2. Improve management processes	3. Improve decision making			4. Enable new production systems	5. Provide data for partners along the value chain
	a) Higher precision			b) Increase output						a) Agronomic	b) Equipment related	c) Business related		
	Less variability of defined job quality parameters	Reduced overlaps	Adopting input application and machine settings to sub-field variability	Reduced overlaps	Higher speed	Autonomous operation	Less downtime							
Examples	Harvester automation, organic nutrient sensing, planter automation	Autosteer, section control	Variable-rate application, spot spraying	Autosteer	Harvester automation, planter automation	Autonomous weeding robots	Remote diagnostics, remote support	Farm management information systems, digital trading platforms	Disease models, crop models, Seeding rate prescriptions	Predictive maintenance	Profit zone field-maps	Controlled traffic farming, strip tillage	Certification and tracing systems
Yield increase	X		X						X			X	
Price increase	X							X			X		X
Additional revenue													X
Direct cost decrease	X		X					X	X			X	
Operating cost decrease		X		X	X	X	X	X		X		X	

times during usage) independent of the total usage amount. For example, this could be the installation of a crop sensor on a machine.
3. Learning cost: To be able to realize the value of technologies, users must learn how to utilize them. These learning costs might include labor costs, costs for seminars, travel costs, etc. Learning costs are part of the fixed cost because they are independent of the utilization of the technology. As with depreciation, learning costs must be allocated over the useful life of a technology.
4. Fixed licensing and data cost: In addition to depreciation for hardware fixed licensing, data and data transmission cost might occur.
5. Interest: Instead of using financial assets to purchase technology solutions or for learning they could also create value through interest on, e.g., a bank account. This opportunity cost has to be considered in the cost calculation and is calculated as follows: (purchase price + fixed licensing cost + learning cost — salvage value) / 2 * interest rate.

The **total cost** of utilizing digital solutions can be summed up as follows:

Total Cost of Technology Usage = VC + FC = (RC + VLC + VLiC) + (D + FLC + LC + FLiC + I)
  VC = variable cost, FC = fixed cost, RC = repair and maintenance cost, VLC = variable labor cost, VLiC = variable licensing cost, D = depreciation, FLC = fixed labor cost, LC = learning cost, FLiC = fixed licensing cost, I = interest

To understand if and how profitable the investment in a certain digital solution on the farm level is, we can now calculate the economic benefit:

Economic benefit = economic value to the farm — total cost of technology usage

2.3.4 Limitations of Economic Benefit Quantification for Decision Making

There are several aspects that are important to consider when using economic benefit quantification to understand decision making on the farm level.
1. In arable farming, the economic benefit of a technology can vary substantially between years and between farms. Due to changing weather and market conditions, the economic value of a digital solution can change drastically from one year to another. Also, farms can be hugely different in regard to soil types, crops grown, production systems, existing machinery fleet, labor availability, skill level, etc. Whenever economic benefit quantification is used to understand decision making on a multi-year and market-level instead of a single-year and single farm-level, these variations need to be considered.
2. Several aspects that drive decision making are hard to quantify in economic terms. Some examples are: Increased comfort because of automation features due to a lower stress and activity level of the operator; peace of mind caused by sensing and monitoring systems, e.g., on a planter; increasing social status of farmers due to the technology leadership image that is supported by digital solutions, which could also help to attract workers; complexity costs that occur if technologies create operational complexity for farms.
3. Some aspects of the economic benefit are not fully visible for most farms. Especially when it comes to yield effects of digital solutions, most farms are not able or willing to do precise trials to measure these effects. So, although there is an economic value that could be quantified, it is not visible for the farmer.

To give an example for these limitations, let us consider a farm that has to decide between two digital solutions that offer in this case the same economic benefit for the farm: sprayer section control and variable rate application (VRA) for nitrogen. Most likely the farm will decide to invest in sprayer section control. Why? In comparison with VRA, section control delivers value independent of yields and crop prices, it improves comfort and does not increase complexity. In addition, the economic value of VRA is not directly visible and can only be quantified with yield trials on the farm.

2.3.5 Example for Economic Benefit Quantification

In the following, the economic benefit of spot spraying herbicide in sugar beets and corn will be quantified. This digital solution detects crops and weeds and applies herbicide only on the weeds.

The assumptions are as follows:

- Farm size: 600 ha
- Crops: 300 ha wheat, 150 ha sugar beet, 150 ha corn
- Spot spraying only works for herbicide application and only in sugar beets and corn. It will require a slower speed than during broad application
- The spot spraying system is an option built on top a trailed sprayer. The purchasing price for the option is 150.000 €, its usage life is 10 years and there is no salvage value
- Herbicide savings through spot spraying (DCC):

| Sugar beet: | -20% -> 60 €/ha | =9000 €/year |
| Corn: | -30% -> 30 €/ha | =4500 €/year |

- Revenue increase due to less herbicide damage (RvC):

| Sugar beet: | +3% yield | =12.000 €/year |

- Higher operating cost due to slower speed in spot spraying (OCC):

| Sugar beets: | | =-800 €/year |
| Corn: | | =-200 €/year |

- Repair and maintenance cost (RC), variable labor cost (VLC), variable licensing cost (VLiC) and fixed licensing cost (FLiC) are all zero for the spot spraying system

Depreciation (D) for the spot spraying system (150.000 € purchase price and 10 years usage)		=15.000 €/year
Learning cost (LC)		=100 €/year
Interest (I) for the spot spraying system		=1.500 €/year

Based on these assumptions above the results for the economic value, the total costs and the economic benefit are as follows:

Economic value = RvC + (DCC + OCC)

| =12.000 €/year + (13.500 €/year—1000 €/year) | =24.500 €/year |

Total cost = D + LC + I

| =15.000 €/year + 100 €/year + 1500 €/year | =16.600 €/year |

Economic benefit = economic value—total cost

| =24.500 €/year— 16.600 €/year | **=7900 €/year** (26 €/ha) |

It is important to mention that this example only provides a positive economic benefit because of the yield increase in sugar beet. As mentioned above, this yield increase is usually not visible for farmers which might decrease the adoption of this technology. One solution to this challenge could be an outcome-based pricing model, where suppliers of a technology help to measure yield effects (e.g., through remote sensing) and

2.3.6 Summary and Outlook

Quantifying the economic benefit of digital solutions is essential for farms as well as digital agriculture companies. These solutions provide value to customers through improving job execution, management processes, and decision making as well as through enabling new production systems and providing value to partners along the value chain. There are also variable and fixed cost involved in using digital solutions, which need to be considered to understand the overall economic benefit. But there are also limitations to economic benefit quantification: variability between farms and years, hard to quantify factors like peace of mind and limitations of farms to quantify benefits themselves limit this approach.

2.4 Successfully Disseminating Digital Tools for Farmers: A French Perspective

Leo Pichon

Abstract

In France, agriculture is currently undergoing many changes and society's expectations of it are evolving. The so-called Agro-Ecological transition is tending to rethink agricultural models by relying on less chemistry but using more knowledge. Digital technology offers tools for acquiring and sharing this knowledge to support agriculture in its transition. Many digital tools have now reached a high level of technological maturity and their lower costs make them accessible to a large majority of farmers. Despite this, adoption levels remain relatively low and the use of these tools is struggling to become more widespread (see ▶ Sect. 1.5).

The objective of this section is to understand the factors that would allow a better diffusion of these digital tools to farmers in the French context. To tackle this issue, a collective approach has been set up between companies in the field of agriculture and digital technology and teaching and research institutes grouped within a consortium called the AgroTIC chair.

Multidisciplinary working groups analyzed real cases of successes and failures in the diffusion of digital tools to farmers. The conclusions were then shared and discussed with some 30 stakeholders of the sector. This work showed that the distributor plays a central role in the dissemination of these tools. In order for them to play their role, it is essential that these actors clearly identify the value they can find in the distribution of digital tools. This value is not necessarily financial or direct. It may, for example, be found in the improvement of his image or the quality of his relationship with his customers. This study also showed that to ensure the proper diffusion of digital tools, it is important that the distributor is involved at a very early stage in the design process.

2.4.1 Introduction

French agriculture is currently undergoing many changes driven by changes in its environment (adaptation to climate change), its relationship to society and biodiversity (agro-ecological transition) or even the organization of its sectors (separation of sales and consulting). Digital technology, because it enables observations, information or advice to be collected, stored, enhanced or shared more easily and more quickly, from the within field level to the regional scale, offers tools to support agriculture in its transitions.

In the past twenty years, some studies have started to investigate the adoption of these tools by farmers [DM03], [PCP13]. Most of them provide objective evidence on the number of farmers equipped and their level of use [MGB+17]. These studies often focus on the technical or socioeconomic factors that influence the dissemination of these tools to farmers [PT17]. However, they often focus on the obstacles that exist at the farmer level without taking into account all the actors in the value chain and their role in the dissemination of digital tools to farmers. In particular, these studies rarely focus on the value that each actor involved in the dissemination of digital tools to farmers could perceive.

The objective of this section is to (i) make an inventory of the actors of the value chain influencing the diffusion of digital tools to farmers, (ii) to identify the value they perceive or could perceive and (iii) to propose good practices to be implemented by the actors to promote this diffusion.

2.4.2 Material and Method

2.4.2.1 The AgroTIC Chair

The AgroTIC chair is a structure grouping together 3 teaching and research institutes in digital agriculture and 28 companies among the main editors and distributors of digital tools for farmers in France. Its objective is to lead collective reflections on digital technologies in agriculture, their dissemination and adoption by farmers. Its composition and the work carried out there are conducive to exchanges between all the actors involved in the dissemination of digital tools. The people who participate in the AgroTIC chair's activities are all experts in the field and generally occupy strategic positions within their companies. It is these people who have contributed as experts to this study.

2.4.2.2 The Focus Groups

The study was based on the expert analysis of use cases [Mit83] of digital tools for farmers. The use cases were selected from tools in which the experts were involved in the conception or the commercialization and thus had access to detailed data on the dissemination of these tools.

The study was carried out in the form of a focus group [KC00] in order to promote exchanges and discussions among the experts. The workshops were repeated several times with the same group of experts according to the "repeated focus group" methodology [MFG08] in order to allow the experts to formulate complex reasoning and to offer them the opportunity to mature their thinking between 2 workshops.

Three initial focus groups were conducted in January, June and October 2019 with six experts representing editors and distributors. These workshops enabled an initial analysis of the obstacles and best practices to emerge, which were then submitted to all the experts of the AgroTIC chair (around 50 people) in order to develop a collective and shared vision. This vision was then disseminated to the general public in the form of a professional conference and in a document intended for stakeholders in the sector (◘ Fig. 2.13).

2.4.2.3 Results' Analysis

During the focus groups, experts first identified the actors who played a role in the dissemination of digital tools to farmers. They then identified on the basis of use cases the value that each actor could derive from this dissemination. This value perceived by the actors was classified into four main categories. The direct financial value corresponds to the direct sale of a tool that brings money to the actor who sells it. The indirect financial value corresponds to the sale of other tools or services that is allowed by the diffusion of the digital tool by the actor concerned. This is for example the case of a decision support tool that allows the sale of a global service including the digital tool, advice and a product. The human value corresponds to the fact that an actor will be able to improve its relationship with its customers or suppliers

□ Fig. 2.13 People participating to the focus group

□ Table 2.3 Experts identify the value perceived by each actor at different levels in the digital tool's distribution chain

Value	Actors			
	Editors	Distributors	Farmers	Influencers
Financial direct				
Financial indirect				
Human				
Environmental				

by distributing a digital tool. This is, for example, the case of an actor who will be able to better know his customers and their expectations and thus provide them with personalized advice. Finally, the environmental value corresponds to a better protected environment or a better control of pollution thanks to the dissemination of digital tools. This is, for example, the case of a farmer who will be able to better control his inputs thanks to decision support tools.

2.4.3 Results

2.4.3.1 Actors of the Value Chain and the Value They Perceive

The results of the focus groups identified four main types of actors in the dissemination of digital tools to farmers (□ Table 2.3):
- **Editors** who design the tools: In France, the companies identified by the experts are often mid-sized companies that have been established for several years

or start-ups and also some more traditional input suppliers. According to the experts, the editors mainly find direct financial value in the selling of their tools. According to them, this turnover can also be indirect by allowing for example the development of customer loyalty or improving the way their client sees their company.
- **Distributors** who sell the tool to the user and provide support: In France, the distributors identified by the experts are mainly cooperatives, traders or accounting centers. According to these experts, the value perceived by the distributors can be financial by being either direct through the simple resale of the tools or indirect by increasing the value of a product (e.g., decision tools optimizing the use of phytosanitary products), by allowing the distributor to better value its technical expertise, or by allowing the distributor to gain in productivity in its advice. Finally, the value perceived by the distributors can be human, modifying the relationship with their customer. The fact that the farmer uses a digital tool allows the distributors to better understand their needs, to accompany the evolution of their practices, or to increase the quality of their advice. These changes then tend to differentiate the distributors from their competitors, to build customer loyalty, and to enable them to obtain new customers.
- **Farmers** are the users of these digital tools: According to the experts, the value they perceived can be i) economic, by bringing a margin gain per hectare that is easily understood by the farmer, ii) human, by allowing the farmers to optimize their interactions with their advice and mutual aid circles or to improve their working comfort, or iii) environmental, by reducing the impact of his practices and promoting their sustainability.
- **Influencers** are people or structures that modify the behavior of other actors through the knowledge they disseminate (agricultural education, higher education, research), the advice they provide (technical institutes, independent advisers), the opinions they express (farmers' unions, politicians, media) or the funding they offer (public financiers, AgFood industries): According to the experts, these actors derive value only from the recognition they receive from other actors who trust them.

2.4.3.2 Recommendations for a Better Diffusion of Digital Tools

The recommendations below are not injunctions but a contribution of the group of experts, based on their experience, to a reflection that seems to be necessarily collective. These good practices are addressed to the actors of the value chain.

Editors: let's put ourselves in the place of others!
- Let's think about our end users, the farmers: Each farmer has his or her own way of working and the tools we design must be able to fit their specificities. Reconciling technological or agronomic added value on the one hand, and ergonomics and ease of use on the other, is a real challenge. Let's not neglect either of these two aspects.
- Distributors represent us: They are the ones who, in the field, convince farmers that our tool has value. Let's make it easy for them. Let's describe and document the benefits that our tools bring to the farmer. Our literature often devotes too much space to describing the features of the tools and too little to describing the benefits and the concrete evidence they provide.
- Distributors are also our customers: They must perceive a value in having their farmers use our tools. Let's identify this use value and integrate it into the design of our tools from the very beginning.

- Our tools must be able to fit into an existing technological ecosystem: Farmers want their new tool to integrate easily with those they already have, without re-entering existing information or becoming familiar with a new interface. Each tool is at the center of the system, but interoperability is a major objective that can only be achieved collectively. Let's promote the interoperability of our tools!

2.4.3.3 Distributors: let's be ambitious!

- Let's quantify the complementary value to define an ambitious strategy: Why are we distributing digital tools to farmers? It makes sense because of the complementary value that these tools bring to us the editors and our farmers. Let's quantify this value, let's make it tangible. We can then integrate it into our objectives, build an ambitious strategy and convince our teams to implement it.
- Let's train our teams. Our teams need support to familiarize themselves with tools that are sometimes new to them. Above all, they need to be supported in the change of profession and mentality that the transition from selling products to selling services implies.
- Let's co-construct the offer with the editors. We are the ones who will present the tools to farmers. Let's invest in their design. Let's test the tools and feedback our opinion and that of the farmers to the editors!

2.4.3.4 Influencers: let's get involved!

- Let's get out of the futuristic vision of digital tools for agriculture: Today, tools already exist and they are valuable to farmers. A number of these tools are mature and accessible to all. Some are available for a few hundred euros and are accessible even for small structures.

- Let's communicate their benefits for farmers: We are independent and we are recognized in our field of expertise. Farmers need our point of view to be reassured and get started. Let's share with them the evidence we have identified for the benefits they will find in the use of these tools. Let's encourage them!

2.4.3.5 Farmers: get started!

- Test existing tools: Make up your own mind by trying out the solutions on the market. Today, there are many solutions to test the tools yourself (test platforms, equipment loans, etc.).
- Share your experiences: Have you used and adopted a digital tool, even a simple one, and found value in it? Talk about it to your neighbors or fellow farmers. Don't hesitate to share your more mixed experiences as well. Share your customer feedback with other farmers and your suppliers.
- Trust your advisors: The people who surround you and advise you can help you in the use of digital tools. They can help you to see more clearly in the offer of services and to make the right choice according to your expectations and your context. Ask them!
- Depending on the tools, the costs can be relatively affordable and the risk limited: In any case, in a decision-support tool, it is always you who decides what actions to take in the field. Try them out!

2.4.4 Discussion

This study collectively produced recommendations to support the dissemination of digital tools to farmers. It is possible that some recommendations may be specific to the French agricultural context. For example, distributors play a particularly important role in France. It is likely that this role will be different in other European countries depending on the way the agricultural sectors are structured. Nevertheless, the method that was used to get the actors of the value chain to discuss with each other can be applied in every context. One of the major obstacles to the dissemination of these digital tools is the lack of mutual understanding of the issues and working methods for each actor. Setting up places and times for sharing point of views between these actors is a way to bridge this gap. It is important that these discussions occur in neutral places and that they are led by trusted institutions. Actors involved in this type of discussions must not expect immediate economic benefits. They must adopt a posture of building a common culture that will promote a better understanding of the reciprocal issues at stake. The digital tools will therefore be better adapted to the farmers' needs, be better distributed, and, therefore, be more easily disseminated. These collective approaches make even more sense in unstable climatic and agricultural contexts. All the actors in the value chain will have to adapt strongly and quickly. They will do so all the better if they share their point of view and know the issues and concerns of the other actors.

2.4.5 Conclusion

French agriculture is currently undergoing many changes, and society's expectations are evolving. The agro-ecological transition is tending to rethink agricultural models by relying on less chemistry but using more knowledge. In agriculture, as in many fields, knowledge has always been based on observation. Observations that are then discussed, shared, put into perspective to build general or, on the contrary, very local and specific knowledge.

Numerical tools allow for more and more objective collection of observations. They allow these observations to be stored and shared more easily and quickly. These tools also make it possible to build new knowledge on the basis of these observations. Digital tools can therefore help agriculture and its actors in the transitions they are experiencing today.

In recent years, these tools have reached a certain level of technological maturity that allows them to be accessible to farmers at reasonable costs. The multiplication of uses also makes it possible to have a perspective on the direct or indirect value that they can bring to farmers and other actors in agricultural sectors.

For these tools to have a real impact in supporting the changes that French agriculture is undergoing, a significant number of farmers must use them. This significant number will only be reached if all members of the chain are committed. First of all, farmers who have an interest in adopting these tools to improve the economic and environmental sustainability of their farms. Second, editors, of course, by continuing to invest in research and development for ever greater value, and distributors, who will create a new source of direct and indirect value for their organizations. Finally, the influencers who can make a decisive contribution to the transformation of French agriculture.

2.5 Business Model Innovation and Business Model Canvas

Gordon Müller-Seitz

Abstract

The prevalent need for innovations in today's world is commonly accepted and virtually pervades all segments and organizations (see ▶ Chap. 1). Hence, it is not surprising that farming is also impacted and the observation holds particularly true as business environments are changing ever faster fueled by the digital transformation. In this connection, product and service innovations pervade the discourse and managerial thinking about innovations. However, compared to product or service innovations, business model innovations remain rather unexplored, but offer valid chances to develop new agricultural ecosystems (see ▶ Sect. 2.1). Against this background, the present section presents business model innovations as a distinct innovation form that is of utmost importance in light of digitalization. Building upon the conceptual introduction, the so-called business model canvas as a managerial tool is introduced, applied to the Digital Farming setting, and critically reflected upon.

2.5.1 Statement of the Problem

To remain competitive, organizations across industries and countries pursue innovations [Bau02], [CGM+16], [TB18]. Staying ahead of competition due to being innovative is deemed beneficial for several reasons, e.g., to gain Schumpeterian pioneer advantages [Sch06], a positive image for marketing purposes, or heightened attractiveness as an employer [JMD12].

In addition, the digitalization changes the business landscape dramatically [BM16] and results in new opportunities to foster innovations with unprecedented pace [LMR17], e.g., building up industry-wide platforms (for an overview [Mül22]). Common outcomes of this trend are process innovations relating to organization internal operations (e.g., improving auction operations [TB18]), digital services surrounding existing products (e.g., predictive maintenance services; see in general [BDP+15]), or collaborations in the form of interorganizational service networks (e.g., the way Bosch acts as a supplier and service (network) provider for different industries [Mül15]).

Bearing these observations in mind, it is striking that the managerial debates in practice and theory primarily revolve around product and service innovations [TB18]. Though of increasing interest across industries [AHR+15], business model innovations are rather seldom discussed from a science perspective. For the purposes of this section, we relate to business model innovations as changes in the underlying logic and operations of an organization and its environment [OP10], [ZA10]. A frequently cited example for business model innovation is the streaming platform Netflix. The organization started as a service for renting DVDs in the USA by mail. Soon after its inception, Netflix did not charge its customers anymore on a pay-per-use-basis, but started to charge them on a subscription basis. About ten years later, Netflix changed its course again, offering streaming services that are by now globally available. Though having a large customer base and being a prime example, in terms of profitability challenges remain ahead for Netflix. This also applies to other frequently hailed business models, such as those of Airbnb or Uber. Against this background, it becomes obvious that future research and managerial practice still need to elucidate the multifaceted phenomenon of business model innovations in light of digitalization—not only, but also in the field of digital farming (for practical perspectives in this regard see [TA17]).

The present section's objective is to address the aforementioned managerial and theoretical void by offering insights into business model innovations as a distinct innovation form that is of utmost importance in light of digitalization with special emphasis on Digital Farming.

The section is structured as follows: First, we introduce business model innovations as a distinct form of innovation and elaborate upon the business model canvas as a managerial tool and apply it to the digital farming setting, illustrating our idea against the backdrop of [Xar20]. Thereafter, we discuss the managerial implications and conclude with a summary, critical reflection and outlook for future research as well as managerial practice.

2.5.2 Business Model Innovation as a Distinct Form of Innovation

Despite being a comparatively young discourse, within the managerial debate about what makes up a business model, definitions are abound. One prominent definition by [AHR+15] suggests a value-based approach, focusing upon the value created for the customers, how the value is delivered to the customer, and finally how much value is captured (i.e., the amount of revenue that an organization receives for its part in providing an offer). Along similar lines, Gassmann and colleagues [GFC20] deliberate who the target customers are, what is actually offered, how value is generated, and how the offering is produced. In contrast, [Tee10] suggests that it is decisive to analyze how a firm delivers value to customers and converts it into profits and [ZA10] (see also [ZA07], [ZA08]) lay emphasis on the fact that a business model transcends organizational boundaries.

In sum, one can identify no generally accepted definition. However, it is commonly agreed upon that business models emerge from interaction of components. Moreover, a business model offers a foundation for dynamic strategies for companies to achieve and contain competitive advantage ([AHR+15]; see also [LMR17], [MZB+18]).

Building upon these observations, business model innovations represent a novel form of innovation. Maybe the most prominent innovation forms are product innovations (e.g., the iPhone in the case of Apple; see [CGM+16] for an overview) or service innovations (e.g., online banking; see [BDP+15] for an overview). Business model innovations have gained increasing attention, though they are still less debated when being compared to product or service innovations.

As for business model innovation definitions, some authors stress that they relate to "two or more elements of a business model are reinvented to deliver value in a new way" [LRS+09], while others stress that "A business model improvement is any successful change in any business model element [...] that delivers substantially enhanced ongoing sales, earnings and cash flow advantages versus competitors and what customers can supply for themselves" [MC04].

2.5.3 Business Model Innovation in Practice—The Business Model Canvas

Within the debate about business model innovations, there exists a broad range of tools for managerial practice. One of the most prominent examples is the so-called business model canvas, being introduced by [OP10]. In this connection, the authors define a business model innovation as follows: "Business model innovation is not about looking back, because the past indicates little about what is possible in terms of future business models. Business model innovation is not about looking to competitors, since business model innovation is not about copying or benchmarking, but about creating new mechanisms to create value and derive revenues. Rather, business model innovation is about challenging orthodoxies to design original models that meet unsatisfied, new, or hidden customer needs" [OP10].

The authors conceptually build the business model canvas on ideas of how a value chain can be sketched within an organization as put forward already by [Por85] and the idea of the resource-based view [Bar91]. Porter's basic idea was to illustrate the flow of input, throughput, and output within an organization and how this is related to overall performance. Barney suggests that an internal organizational resource can be the source of a sustainable competitive advantage, if it is valuable, rare, inimitable, and non-substitutable. These ideas coalesce in the business model canvas that offers an overview of organizational activities and can help to identify how an organization can generate value and capitalize thereupon; that is, setting the basis for a critical reflection of the status of the business model and offering a springboard for business model innovation.

The business model canvas consists of a reflection of the following elements (see Fig. 2.14):

- **Key partners:** This element relates to the most important collaboration partners of an organization. It stretches usually across the value chain, including suppliers and key customers of an organization. As is the case with any interorganizational collaboration, benefits (e.g., deriving economies of scale and scope) and risks (e.g., losing one's core competences) need to be carefully reflected upon [SSM16].
- **Key activities:** With regard to this element, a reflection is needed in terms of what operations add most value with regard to the existing or future business model. Towards this end, not only internal organizational operations, but also the input and output operations need to be considered. Also managing patents or trademark rights are viable other options to scale one's business model innovations (see also [CGM+16]).
- **Value proposition:** In terms of the value proposition, an organization needs to be able to define clearly, what value is being offered to customers. The value can be manifold, be it related to an additional service and data being offered, the reduction of a flaw, the prediction of errors, a gain in efficiency, or a gain in reputation to mention only a few options.
- **Customer relationships:** In this connection, key questions revolve around the way an organization deals with its customer base. Here, a key concern relates to the way customers are being integrated into the business model (e.g., think of the extreme integration of customers in the way IKEA customers usually assemble their furniture on their own in contrast to conventional furni-

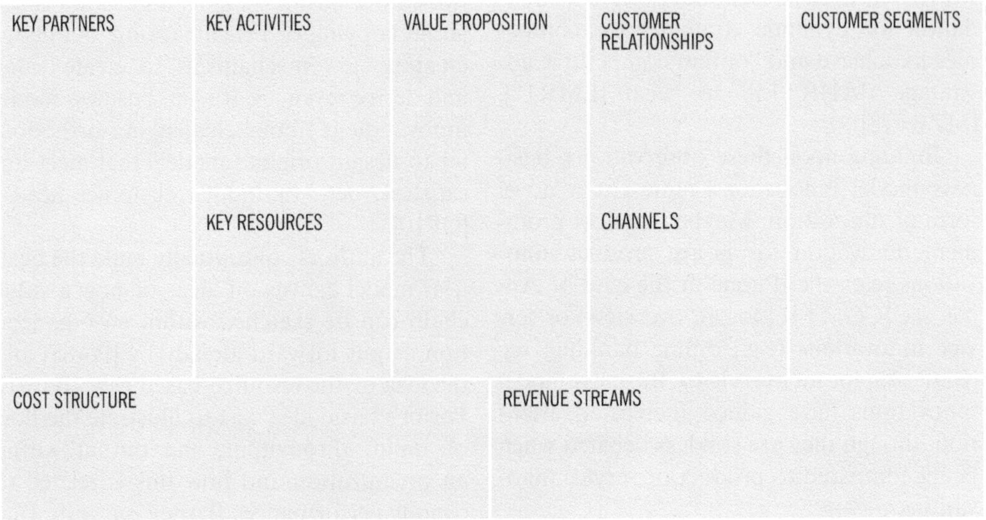

KEY PARTNERS	KEY ACTIVITIES	VALUE PROPOSITION	CUSTOMER RELATIONSHIPS	CUSTOMER SEGMENTS
	KEY RESOURCES		CHANNELS	
COST STRUCTURE			REVENUE STREAMS	

Fig. 2.14 Business Model Canvas, source adapted from [Str20]

ture stores). The relationship can vary. For instance, it might be the case that you have close interpersonal ties as in the case of private banking where relationships can last for a lifetime. In contrast, merely transactional relations are also possible as in the case of buying an item at an online platform such as Amazon or Alibaba.

- **Customer segments:** Analyzing an organization's customer base along different segments offers the possibility to prioritize customers. A consequence can be to (re-)align the key account management activities. Moreover, it is also possible to ignore customer segments that are not beneficial for the organization (e.g., because they harm the reputation of an organization or the value captured is insufficient).
- **Key resources:** The key resources of an organization describe the most valuable material as well as immaterial resources, such as personnel, machinery, data, fields, or factories. As pointed out above, here the ideas of the resource-based view are most evident [Bar91].
- **Channels:** With regard to channels, [OP10] target the relevant media by means of which the focal organization interacts with the customers. These types of channels can be differentiated along various dimensions. One dimension might be the customer life-cycle, e.g., making a distinction between the first customer touch points to after-sales services. Alternatively, one might make an analytical distinction between the exchange format, ranging from face-to-face communications to digital exchange situations supported by digital devices such as mobile phones or laptops and personal computers.
- **Cost structure:** The cost structure directs attention to the costs occurring for the different parts of an organization's operations. Towards this end, [Str20] suggests reflecting upon, among others, the resources or the activities that cause the highest costs to come up with suggestions on how to curb down costs.
- **Revenue streams:** Reflecting upon the revenue streams is informative in many ways. For instance, valuable insights might be generated from differentiating between one-off (e.g., in the course of a single transaction) and continuous revenue streams (e.g., in the course of a subscription model as in the case of streaming services, such as Netflix). In addition, checking customer preferences might be beneficial, e.g., in terms of the customer's willingness to pay, the (digitally supported) payment options, or how the different value streams contribute to the overarching financial results.

Bearing these elements of a business model in mind, we would like to apply it to the digital farming setting based upon an example of a farming-related app, such as that of [Xar20] being offered by BASF:

- **Key partners:** BASF needs to consider the farmers, which represent the final customers or at least final beneficiaries of xarvio. Depending upon the specific setting, BASF also cooperates with a corporation for AgMachinery and original equipment manufacturer corporations (e.g., Bosch) and retailing corporations. Finally, public institutions might also be key partners in the foreseeable future as they oftentimes provide freely available data (e.g., with regard to weather forecasting the Joint Research Centre of the European Commission).
- **Key activities:** Visualizing and analyzing data are the essence of such products. They are capable of analyzing data generated from different sources and systems to come up with novel business model opportunities. That implies that farmers adopt the role of data providers (i.e., suppliers) while at the same time being customers who shall buy the

products. Additionally, activities need to be aligned (especially technological interfaces) with collaboration partners of AgMachinery. Finally, providing a cloud or making use of cloud services of another company might be another critical key activity. As a result, interfaces need to be organized and can be further developed (see along similar lines with regard to cloud services in a farming setting: [TA17]).

- **Value proposition:** Farmers will be supported in the management of their fields, in effect improving their crop management, reducing costs and risks (see ▶ Sect. 2.3). For instance, farmers are able to make use of the app to check a specific disease of their plants by taking a photo of a plant being affected and immediately request advice what crop protection product is necessary. Another service that comes handy might be fertilizing the fields as precisely as possible based upon a close data exchange between the aforementioned parties involved, e.g., AgMachinery companies. What is more, positioning sustainability as a further value to be promoted is another instance. Take for instance the so called "Lerchenbrot" (lark bread), a joint activity by BASF and other partners along the value chain up until the final consumer including the bakeries. In this case, the companies collaborate to jointly safe breeding grounds for larks (provided by xarvio/BASF) and nonetheless benefit also financially from the activity while being able to derive higher profit margins for the final product, the larks bread at the bakery.
- **Customer relationships:** The relationships to be managed are likely to be rather transactional insofar as the app might be offered for farmers around the globe. However, to a certain extent a community might be built up in terms of a Q&A forum integrated into the respective app. However, market access can also take place via other stages of the value chain, e.g., food producers or traders.
- **Customer segments:** As pointed out, the primary customers are farmers and contract farmers. They might be divided into different segments, e.g., depending upon their production philosophy, machinery brands used, or crop rotation.
- **Key resources:** Key resources comprise software and technology development teams as well as the necessary related financial resources. Moreover, having data available for analytical ('big data') purposes represents another key resource. To put it differently, this relates to the ability to connect with multiple market partners and establish sustainable business models.
- **Channels:** The digital products represent the primary medium for interacting with the farmers. Hence, digital channels dominate. In this case, a multi-channel approach is deemed suitable, so that customers can be approached through different channels depending on the country.
- **Cost structure:** Data center investments, app developers, and the customer support services might account for the largest cost chunks. Over time, customer service might be outsourced to a chatbot and others solutions in parts where possible. That is, if the gathered data is substantial enough so as to offer adequate big data generated insights from farming operations.
- **Revenue streams:** In terms of the revenue streams, farmers pay for the app on a subscription basis. However, a freemium represents the starting point with regard to some functionalities so as to entice customers.

2.5.4 Reflections on the (Mis-) Use of the Business Model Canvas and Conclusions

To sum up, reflecting upon business models and business model innovations seems to be a key managerial concern, not only, but also in particular due to the digital transformation [LMR17]. Against this backdrop, the business model canvas turns out being an easy-to-use managerial tool that helps managers to reflect the different elements of a business model and—if changing its elements—business model innovation.

The popularity of the business model canvas can be traced back to several advantages: The canvas is easy to visualize and, thus, easy to use and convey information ensuing from the canvas. Moreover, it covers the key issues making up a business model and, depending upon the specific needs, data to be analyzed and the ensuing discussion regarding potential implications can vary in mass-customized fashion. This is also particularly attractive insofar as the agricultural sector appears being in many sectors a commodity market, where business model innovations targeting higher product segments and alternative market access is a valid—albeit difficult to achieve—strategic option for farmers and Ag market segments (e.g., focusing on old varieties, specialty products, co-production systems like agricultural-energy-farmers, see ▶ Sect. 1.2). An externality effect such as the European Green Deal (see ▶ Sect. 1.6) might result in further threats and opportunities for actors in industry as it is most likely to entail novel dynamics and altering value chains, which in turn result in new business model opportunities awaiting further exploration and exploitation (e.g., biodiversity, carbon market).

Nonetheless, the business model canvas is not without limitations. Apart from its static nature—owing to the observation that it heavily relies upon the eclectic consolidation of static conceptions, such as that of [Por85] and [Bar91]—it does not offer any advice on how extensive data collection and analysis shall be. Moreover, the elements are partially overlapping, making the decomposition oftentimes problematic. Finally, due to its intra-organizational focus, the organizational environment is by large neglected: For instance, the competitive landscape, societal or cultural factors or in light of digitalization the collaboration across organizations. Thus, future managerial practice and business model research might consider taking a look at business model generation and innovation on the whole network level of analysis [PFS07], that is, how business model innovation might unfold on the level of interorganizational networks [MZB+18], [Thi19]. This might comprise cloud services (for a practical discussion see [TA17]) or other platform operations [GC14]. Given the networked nature of (digital) farming, this topic awaits further exploration, incl. requirement definitions in initial business model versions or minimum viable business models.

2.6 Accelerators & Partnerships: Anticipating the Unknown is Hard: An Experience Report

Borris Förster

Abstract

The sustainable production of food is one of the most complex and pressing challenges of our time (see ▶ Chap. 1). The traditional food value chain across all actors from input companies through to retail is facing unprecedented pressure through environmental, demographic, societal and regulators trends. These trends will lead to significant structural changes over the forthcoming decade on the products consumed and how they are produced and moved through the supply chain. The food supply chain will change substantially, and this in turn will affect the way business is done and value is created. Some segments and its constituent intermediaries in the value chain will cease to exist for obvious reasons and others will change in the way they operate and what they produce today. Incumbents are scrambling to deploy a variety of innovation measures to confront the change with varying degrees of success.

In this section, I will give an insight into the conflicts of interest and associated challenges within much hailed acceleration and corporate partnership programs. I endeavor to provide my view and personal experience on the current landscape and where we need to head in terms of foundations and backbone infrastructure to move faster and more efficiently towards the future of farming and food production. Partnerships, co-development and new industry players will have to play a significant role to leap forward in the industry. We need to re-think collaboration.

2.6.1 Introduction

Predicting the future is a tough task, being right even tougher. This is essentially what incumbents are trying to do, when launching accelerators, corporate venture capital vehicles, start-up partnership programmes and the like. The real challenge, however, appears frequently misunderstood. To stop extrapolating the current core portfolio and instead start thinking in terms of future operational environments and to target actions accordingly in combination with available assets and likely future customer needs (see ► Sect. 2.5).

◾ Figure 2.15 depicts a French farmer in the year 2000, painted by Jean-Marc Côté. It stems from a series of artworks issued in France between 1899 and 1910 showing the view of the future of farming in the year 2000. As is often the case, the predictions fell some way off the mark, failing to go far enough in thinking outside the confines of their current technological milieu. We tend to under- and over-estimate the changes in the future as we extrapolate our personal experience into the future, instead of putting our focus of thought on predicting probable future environments to derive solutions that will be necessary to operate under those conditions. So, when embarking on new paths to build solutions and new products in partnerships or advancing early start-up companies through accelerator programmes, incumbents must ensure they focus on a vision of the future industry operating environment instead of a specific outcome and extrapolating todays "ways of doing things". This proves especially hard for established businesses as the majority of their operation is geared toward preservation, prolonging product lifetime, and incremental innovation.

To cope with this uncertainty, incumbents require a clear vision to build upon. In order to do so, first, we must understand agriculture and food as one single supply chain that it is not linear, but looping back at several points. Secondly, we must understand the basic requirements to support and grow innovation in different contexts.

The digital transformation of agriculture, food production and supply is perhaps the most complex challenge of our time and thus requires joint efforts across multiple disciplines and actors. The route to successful new solutions is full of pitfalls in terms of problem-solution and solution-market fit in the first place. Today's incumbent actors, while mostly intending to foster and support new solutions through accelerators, corporate venture capital and partnerships fall short on delivering on their own expectations. In fact, experience

Fig. 2.15 "a very busy farmer" by Jean Marc Côte [Cot10] (cropped)

has shown established companies frequently compound the challenges, making it harder and overestimating their ability to accelerate at the current point on their transformational roadmap. To make significant leaps forward collaboration and exchange across and within value chain segments must increase drastically.

To get there, current thinking requires a change of direction: away from owning 100% of a small pie towards enlarging the pie for everyone. We need to understand why and how to create equal partnerships between unequal partners. We need to kill pet start-up projects and accelerator programs that are not delivering or become unlikely to deliver real value to the goals of the organization and the supply chain at large, sooner rather than later. We need to start treating start-up-us like companies and entrust them with paid projects for grown-ups, instead of grants and special programmes treating founders as hobbyists and enthusiasts. This, however, necessitates that the goals and vision of the organization are clear and commonly understood by all actors—which often it is not.

2.6.2 Understanding and Managing Direct and Indirect Impact on Success—Borrowing from Physics and Finance

To understand the workings behind success and failure of accelerators and fruitful partnerships, it helps to understand what you are up against. Three at first sight unlikely concepts lay the groundwork to give direction and prepare the field before sowing:
1. Newton's Laws of Motion [Bri20a],
2. Newton's Law of Universal Gravitation [Bri20b],
3. "Modern Portfolio Theory" or in other words the risk-averse behavior of investors [Mar52].

2.6.2.1 Newton's Laws: What Physics can Teach us about Accelerators & Partnerships

Let us understand the basics of Newton's Laws of Motion and Universal Gravitation to make better decisions. Starting with motion:

Newton's First Law of Motion "A body at rest will remain at rest, and a body in motion will remain in motion unless it is acted upon by an external force." [Bri20a, Bri20b].

This simply means that things cannot start, stop or change direction all by themselves. It requires some external force to cause such a change.

Newton's Second Law of Motion "The acceleration of an object increases with increased force and decreases with increased mass". It also states that the direction in which an object accelerates is the same as the direction of the force." [Bri20a], [Bri20b].

Assuming constant mass means in order to accelerate an object a force, it needs to be exerted upon the object. Moreover, the direction of force exertion is equal to the direction of acceleration.

Newton's Third Law of Motion "For every action, there is an equal and opposite reaction". [Bri20a], [Bri20b].

In other words, whenever two objects interact with one another, they exert forces on each other.

In summary, objects in the physical world do not move of their own accord. Nor do they in economics and business. Within the context of innovation and building new businesses, this notion provides a fundamental understanding and scaffolding for success. No business starts without the deliberate application of a force, i.e., action. This, by the way, is also the reason why good ideas by themselves are worthless, if they are not acted upon. It needs concerted action and execution. The entrepreneur or intrapreneur provides this initial action to identify a need, derive a solution and business model, test and build it. In its absence there would be no new businesses. In fact, there would be no economy as we know it today. Moreover, only the continued effort of the entrepreneur, employees and other stakeholders will keep things moving. This is important, because in business, operations rarely maintain their level; much less grow by themselves. In fact, the opposite is true. There is always friction in the business environment: interest rates and terms of contracts change; competitors enter the market; economic volatility affects demand and so on. Consequently, things tend to decay over time if we stop acting. Revenues drop off and costs increase. The bottom line erodes.

The larger the business, the more entrenched its bureaucracy, the greater the mass you are pushing against. It requires the application of extraordinary energy to get it to move even a small amount. Often, the easiest course of action is simply to do what you are doing. Everyone knows the routine. Generally speaking, no additional expenditure of time, effort, or money is required. It feels familiar and comfortable. It is only when someone seizes the initiative and takes an action that things change.

Translating the laws of physics to the world of corporate start-up partnerships and accelerator programs brings some fundamental problems to light. In order to accelerate effectively, i.e., wasting as little energy as possible, the common direction of acceleration must be known to all parties and aligned by all parties exerting force. In the context of corporate accelerators and partnerships this leads to a dilemma if the vision and strategy of the corporate is not 100% clear and aligned with the start-ups and partners (remember: a clear corporate strategy is important). Furthermore, the direction and size of the force exerted is heavily influenced by corporate politics and the agendas of individuals, which in turn is influenced by corporate culture.

Once direction is aligned between parties, the focus shifts to the force to be exerted in order to accelerate or at least keep it moving at the same speed, not slowing business down. This is where corporate acceleration and partnership programs tend to exaggerate their abilities. Experience shows, because actors heavily underestimate the effect of their own size, direction of movement and misalignment on strategy and vision within their organization. At the same time, actors tend to overestimate transferability of know-how and the required motivation and competence within their own organization. This brings us to the last key learning from Newton in regards to transformation through accelerators and partnerships, pulling it all together.

Newton's law of universal gravitation "every particle attracts every other particle in the universe with a force that is directly proportional to the product of their masses and inversely proportional to the square of the distance between their centers" [Bri20a+b].

In other words, attraction between objects is subject to two variables: mass and distance. Taking Newton's laws of motion into account this means that the object with the larger mass will be a greater influencer on the direction of the smaller object than vice versa.

The tendency of an object to resist changes in its current state of motion varies with mass. Hence, the more massive an object, the greater its tendency to resist changes in its state of motion. The same is true for businesses. It indirectly asserts that the force to change direction must be much larger on a big organization than on smaller start-up company.

This effect, unaccounted for, can be detrimental for the relationship and result in losing the optimal trajectory (i.e., roadmap and strategy) desired by the start-up management and the corporate supporter. Imagine SpaceX miscalculated the mass of the earth and thus its gravitational force when shooting a rocket into space. An underestimation would pull the rocket back down toward earth and never reach its actual destination.

Lastly, Newton's Laws offer one potential explanation of why we do not find much quantitative evidence today that the general speed boat doctrine of digital and innovation gurus globally is panning out as prophesied. Visualizing the concept, it would either need dozens of speed boats (business pilots and start-ups) moving in exactly the same direction to eventually move the ship (a sizeable organization) or a speed boat that has already matured into the size of a tug boat, once you bring it back closer to the parent company.

2.6.2.2 Modern Portfolio Theory: How Investors Think and Act

Modern Portfolio Theory shows that an investor can construct a portfolio of multiple assets that will maximize returns for a given level of risk. Likewise, given a desired level of expected return, an investor can construct a portfolio with the lowest possible risk. Based on statistical measures such as variance and correlation, an individual investment's return is less important than how the investment behaves in the context of the entire portfolio. [Mar52].

Continuing the speed boat example, this means that dozens of speed boats moving into exactly the same direction to move an organization would leave investment risk skyrocketing. It would substantially defeat our understanding of modern portfolio theory guiding large organizations today.

In an ideal world, business would be run like a group of independent boats and ships including speed boats (new opportunities), fast ferries (growth stage commercialization) and tankers (the cash cows of the core business). Each boat is given an equal chance to grow into a tanker and sufficient independence. If successful crews are moved around, added and withdrawn as necessary and we slingshot smaller businesses through the fleet in order to grow faster utilizing existing assets.

Modern Portfolio Theory assumes that investors are risk-averse, meaning they prefer a less risky portfolio to a riskier one for a given level of return. This implies that an investor will take on more risk only if he is expecting more reward. The expected return of the portfolio is then calculated as a weighted sum of the individual assets' returns. This in turn assumes that the desired level of risk and return is known and well understood by all. Experience and data on corporate accelerators and innovation returns has shown that this is not the case in many organizations. More often, there seems to be a distinct discrepancy between the accepted risk and the desired returns. The relatively low risk appetite among companies in Germany, compared with our anglo-saxon counterparts, has to do with our past and industrial systems. Germany is a world leader at selling units of physical products with exceptional quality. Once a unit is sold that same unit cannot be sold again. This is different with digital products, i.e., the potential return and marginal cost function is different and often seems not well understood by major stakeholders.

In conclusion, theory and evidence suggests that long-established businesses are losing the competencies and drive is required to build and scale business due to their long-standing position of relative strength in a given market. They are organized to keep the business running, mending, maintaining, and incrementally growing or securing the existing business. They understand their playing field exceptionally well and the effects that sizable competitors have on their core KPIs. But they underestimate the effort, organizational setup, transparency, resources, strategic clarity, and vision necessary to gain positive returns on invest from partnerships and accelerator programs.

The first two (Newton's laws of motion and gravitation), translated to business, will impact your accelerator portfolio and partnerships in terms of development, commercialization speed, and success toward significant returns. The third (Modern Portfolio Theory), will, to a large extend, affect your freedom and resources to operate as a practitioner, leading innovation and transformation programs in an organization. Though I have rarely experienced any arguments against the applicability and importance of these forthcoming ground rules as a framework for operational execution, understanding the negative implications, if ignored has frequently been met with surprise and disbelief in executive boards—until they shut initiatives down.

For clarity, employees, business owners, partners, or anyone else spending time on the business in partnerships or within the accelerator and other innovation initiatives are in fact investors. And they need good and sustainable incentives to move towards the right and the same goals.

2.6.3 The Current State of Technology in the Food Value Chain

Technological innovation, ever stricter regulation and global challenges and trends are putting significant pressure on the dynamics of the traditional food supply chain (see ▶ Sect. 1.2). New entrants are challenging the market share and growth prospects of settled incumbents with advanced biology, technology and new foods and challenging the status quo of the supply chain. Traditional farm production is squeezed in the middle between global oligopolies trying to protect their markets and novel high-tech food company entrants are skipping the traditional chain from grower to food company and in some cases even retailers altogether.

The AgFood sector has seen a significant rise and venture capital investment (corporate and private) and the number of corporate accelerator programmes has been mushrooming over the past years. In fact, venture capital invested in AgFoodTech including accelerator programs has grown by 32% CAGR between 2012 and 2019 to a total of EUR 17.8bn according to Agfunder [Agf20]. Currently, the largest part of the investment is circumventing Europe. This is partly due to the fact that two of the main tech trends are driven by developments in genetics and AI. Two things, to say the least, that face severe obstacles in the EU from an international perspective. However, in the long run, technology will change farm operations and the entire supply chain radically, also in the EU. This will have significant effects on the supplier industry. The question is, will incumbents in the EU be a driver and benefactor of this change or fall victim to progress in other parts of the world. The EU Green Deal may pose a regulatory opportunity to set European companies on a path of future competitive advantage, if they see and cease the opportunity (see ▶ Sect. 1.6).

Already today the fight for vertical control, integration and circumvention of previously established segments in the food supply chain is on. The pressure from upstream incumbents like retail giants is mounting as they push to gain deeper control of Upstream production to reduce costs, increase efficiencies, and deliver to consumer demands at low prices. At the same time, input companies are developing "X-as-a-service" and guarantee-based business models to build new future-proof business. The up and downstream pressure is concentrating at the farm production level.

Traditional food supply takes place in a highly complex linear supply chain (see ◨ Fig. 2.16) with mostly product-driven players, comparatively low R&D intensity and low efficiency (hard to optimize holistically, leading to high wastage) in which each segments' performance is dependent— at least in part—on the level of technological capability, transparency and efficiency of the previous segment. Upstream agricultural production is perhaps the most complex segment. It is the least transparent, least digitally enabled, least technologically integrated chain-link [Mck20] and is highly exposed to unforeseeable circumstances. This vulnerability circles back on the rest of the value chain creating a butterfly effect, with ever-increasing uncertainty and unpredictability.

The status quo is preventing the scalable deployment of digital & physical services and applications, because such are dependent on information and feedback from other participants. This problem has been one of the drivers for the increasing trend to vertically integrate. This makes acceleration of digital solutions difficult, especially the one relying on information flows from different parts of the supply chain.

To counteract the described changes in the supply chain and future operating environment, incumbents have ploughed significant resources into start-up innovation initiatives like accelerators with limited success. According to estimates from Doblin innovation consulting [Dob19], 96% of all innovation initiatives—across a selection of major industries—fail to make a return on investment.

As described earlier, large organizations are generally slow-moving, process-driven and risk-averse. They are optimized for stability and preventing failure by tackling incremental projects. But to innovate and "accelerate", you need to move fast, often ignore process and take substantial calculated risks. This requires an organizational setup geared towards the testing of ideas and accepting failure as part of the game. The cultural shift that is necessary is often the biggest impediment to achieving success. Acceleration programs setup as a side show in fancy office buildings with bean bags and tabletop soccers were supposed to get around these obstacles. However, this will not work by itself. The organization has to provide a compelling value proposition to start-ups and partners, be able to deliver on it and ensure strategic alignment.

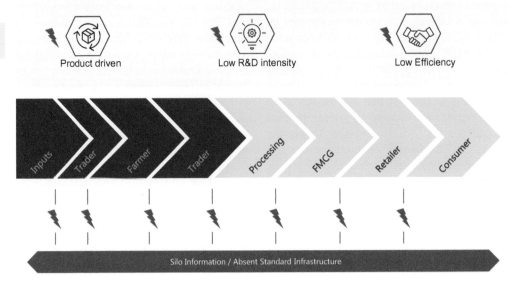

Fig. 2.16 Inefficiencies in today's AgFood value chain

A frequently used indicator of success for accelerator programmes and Corporate Venture Capital (CVC) units is the number of partnerships between the big companies and the start-ups post-investment or after the corporate accelerator program. According to research from CB Insights [Cru19] on corporate innovation across all major industries, only 10% announced a formal partnership after CVCs invested (see Fig. 2.17). Accelerators only announced partnerships with participating companies 1% of the time. That is only 1 in every 100 start-ups. With the usual two annual batches of 12 start-ups it would take more than four years and several million Euros to form one single partnership.

Organizations are acting and quietly shutting down programs or cutting budgets as their success stays behind expectations. In fact, 60% of corporate accelerators shut down within two years and one-third of new corporate venture capital activities stopped investing after five years [Cru19].

Corporate accelerators, start-up partner programmes, and venture investment today need a combined vision and the determination to move the needle. Experience has shown that frequently agriculture dedicated innovation vehicles think too small, are too internal-faced, ill equipped, led by people with expertise in areas that do not match the task at hand, and are too dependent on and involved in day-to-day corporate politics. Incentive and budget structures are not geared toward supporting risky and time intensive start-up acceleration and partnership projects with uncertain outcomes.

2.6.4 Looking Ahead: From Stand-Alone Programs to Multi-Corporate Innovation Platforms

In order to solve the challenges ahead, we need to move toward interconnected food value chains and accept that the roles of incumbents within the segments will change over time. Innovation initiatives like accelerators and partnership programs can be part of the foundation, if the influencing

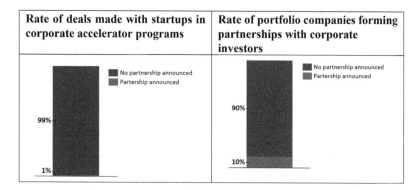

Fig. 2.17 Start-up deals and partnerships with corporate ventures [Cru19]

factors are well understood. We have to re-calibrate collaboration across the entire supply chain.

Solutions developed by start-ups today can, to a large extent, only unfold their true potential if the right data is supplied to and by them to the adjacent segment at the right time to make actual decisions and later guide and activate physical actions on fields, handle farm logistics, and improve the efficiency of the wider food supply chain (see **Fig. 2.18**). It is about enlarging the pie for everyone.

Business models are changing from product driven to service and results driven offerings. The required industry infrastructure to support the business model evolution, however, is not yet in place. This is slowing transformation, cost-effective development, adoption and monetization. To enable the orchestrated deployment of connected services and applications in concert this must change to-

Characteristics change:

 Results driven (Focus on outcomes)

 Increased R&D (Data availability bringing down costs)

Increased Efficiency (Better communication & orchestration)

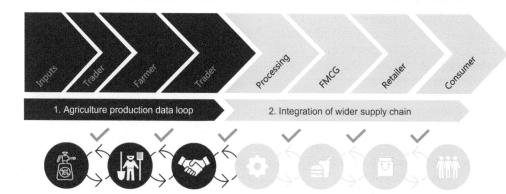

Fig. 2.18 Necessary changes to the AgFood value chain

wards an open collaboration platform for multiple players, start-ups, and corporates alike.

To be successful, incumbents should provide their cohorts and partners with an Unfair Advantage to build and deploy their services and solutions across the Ag-Food supply chain. Help to connect the upstream and downstream value chain allows direct interactions from retail to farm inputs. Enabling all participants to build, test, deploy, and scale business models with their cohorts, in start-up partnerships and with their corporate venture portfolio. So future and current programs would do well to join forces with adjacent incumbents along the value chain to create innovation ecosystems together acting as a foundation for multi-corporate acceleration programs, innovation labs, start-up partnerships and company builders. This would be a true game changer.

Lastly, the way organizations are setting up their initiatives takes into strong consideration where they currently are on their digital and cultural transformation journey. It has to be clear what can truly be provided to start-ups and partners to create an Unfair Advantage. Initiatives need to follow a clear vision, be joined closely with corporate strategy, if early on in the transformation. In a constantly changing AgFood value chain (see ▶ Chap. 1), it requires a softer approach on execution and roadmaps to successfully correct assumptions and successfully position in the AgFood transformation.

References

[AK10] Adner, R., and R. Kapoor. 2018. Value Creation in Innovation Ecosystems: How the Structure of Technological Interdependence Affects Firm Performance in New Technology Generations. *Strategic Management Journal* 31(3):306–333.

[Agf20] Agfunder. 2020. *"Agri-FoodTech Investing Report"*. Accessed 15 Juni 2020.

[AHR+15] Aversa, P., S. Haefliger, A. Rossi, and C. Baden-Fuller. 2015. From Business Model to Business Modelling: Modularity and Manipulation. *Advances in Strategic Management* 33:151–185. ▶ https://www.researchgate.net/publication/283357082_From_Business_Model_to_Business_Modelling_Modularity_and_Manipulation.

[Ama21] Amazone. 2021. *SmartSprayer joint project. The smarter sprayer! Crop protection technology meets camera technology and expert knowledge*. ▶ https://go.amazone.de/go2020/agritechnica/2019/neuheiten-en-us/pflanzenschutztechnik-en-us/smartsprayer-gemeinschaftsprojekt-en-us/. Accessed 15 Feb 2021.

[Bar91] Barney, J. 1991. Firm Resources and Sustained Competitive Advantage. *Journal of Management* 17(1):99–120. ▶ https://josephmahoney.web.illinois.edu/BA545_Fall%202019/Barney%20(1991).pdf.

[Bau02] Baumol, W. 2002. *The Free-Market Innovation Machine: Analyzing the Growth Miracle of Capitalism*. Princeton: Princeton University Press.

[BDP+15] Barrett, M., E. Davidson, J. Prabhu, and S.L. Vargo. 2015. Service Innovation in the Digital Age: Key Contributions and Future Directions. *MIS Quarterly* 39(1):135–154. ▶ https://www.researchgate.net/publication/283825550_Service_Innovation_in_the_Digital_Age_Key_Contributions_and_Future_Directions.

[BLW20] Bröring, Stefanie, Natalie Laibach, and Michael Wustmans. 2020. Innovation types in the bioeconomy. *Journal of Cleaner Production* 266:121939.

[BM16] Brynjolfsson, E., and A. McAfee. 2016. *The Second Machine Age*. New York: Norton. ISBN-13 : 978-0393350647.

[BPS+15] Bauernhansl, Thomas, Dominik Paulus-Rohmer, Anja Schatz, Markus Weskamp, Volkhard Emmrich, and Mathias Döbele. 2015. *"Geschäftsmodell-Innovation durch Industrie 4.0: Chancen und Risiken für den Maschinen- und Anlagenbau"*. Fraunhofer IPA.

[Bri20a] Encyclopædia Britannica (UK) Ltd. ▶ www.britannica.com/science/Newtons-law-of-gravitation. Accessed 13 Mar 2022.

[Bri20b] Encyclopædia Britannica (UK) Ltd. ▶ www.britannica.com/science/Newtons-laws-of-motion. Accessed 13 Mar 2022.

[Bro05] Bröring, Stefanie. 2005. *"The front end of innovation in converging in-dustries: The case of nutraceuticals and functional foods"*. Zugl.: Münster, Univ., Diss., 2005 (1. Aufl.). Gabler Edition Wissenschaft. Betriebswirtschaftliche Studien in forschungsintensiven Industrien. Wiesbaden: Deutscher Universitäts-Verlag.

[Bro10] Bröring, Stefanie. 2010. "Developing innovation strategies for convergence – is "open innovation" imperative?". *International Journal Technology Management* 4(1/2/3):272–294.

[CGM+16] Corsten, H., R. Gössinger, G. Müller-Seitz, and H. Schneider. 2016. *Grundlagen des Technologie- und Innovationsmanagement*, 2. Aufl. Munich: Vahlen, ISBN-13 : 978–3800651320.

[Cot10] Jean-Marc Côté, "a very busy farmer". 1910. ▸ https://commons.wikimedia.org/wiki/File:France_in_XXI_Century._Farmer.jpg Accessed 13 Mar 2022.

[Cru19] CB Insights. 2019. ▸ www.cbinsights.com/research/corporate-accelerator-failure/. Accessed 13 Mar 2022.

[Dee17] Deere John. 2017. *"Deere to Advance Machine Learning Capabilities in Acquisition of Blue River Technology"* ▸ https://www.deere.ca/en/our-company/news-and-announcements/news-releases/2017/corporate/2017sep06-blue-river-technology/. Accessed 3 Feb 2021.

[DM03] Daberkow, S.G., and W.D. McBride. 2003. Farm and Operator Characteristics Affecting the Awareness and Adoption of Precision Agriculture Technologies in the US. *Precision Agriculture* 4:163–177.

[Dob19] Doblin Innovation Consulting. ▸ www.doblin.com. Accessed 13 Mar 2022.

[DU08] Daheim, Cornelia, and Gereon Uerz. 2008. Corporate foresight in Europe: From trend based logics to open foresight. *Technology Analysis & Strategic Management* 20(3):321–336.

[GC14] Gawer, A., and M.A. Cusumano. 2014. Industry Platforms and Ecosystem Innovation. *Journal of Product Innovation Management* 31 (3): 417–433. ▸ https://www.researchgate.net/publication/261330796_Industry_Platforms_and_Ecosystem_Innovation.

[GFC20] Gassmann, O., K. Frankenberger, and M. Choudury. 2020. *Business Model Navigator: The strategies behind the most successful companies.* London: Financial Times Publishing.

[GPD20] Gerhardt, Carsten, Markus Plack, and Natalia Drost. 2020. „*Why today's pricing is sabotaging sustainability*".

[GSZ+19] Gerhardt, Carsten, Gerrit Suhlmann, Fabio Ziemßen, Dave Donnan, Mirko Warschun, and Hans-Jochen Kühnle. 2019. *"How will cultured meat and meat alternatives disrupt the agriculture and food industry?"*.

[HE18] Hannah, Douglas P., and Kathleen M. Eisenhardt. 2018. How firms navigate coopera-tion and competition in nascent ecosystems. *Strategic Management Journal* 39(12):3163–3192.

[HWB19] Hartjes, Katrin, Michael Wustmans, and Stefanie Bröring. 2019. „*Digitale Transformation.: Wie etablierte Unternehmen und Start-ups aus dem Agribusiness digitale Kompetenzen aufbauen*". In 15. Symposium für Vorausschau und Technologieplanung (Ed.), Vorausschau und Technologieplanung (Vol. 2019). Paderborn.

[HR18] Helfat, Constance E., and Ruth S. Raubitschek. 2018. Dynamic and integrative capabilities for profiting from innovation in digital platform-based ecosystems. *Research Policy* 47(8):1391–1399.

[JMD12] Jones, T., D. McCormick, and C. Dewing. 2012. *Growth Champions: The Battle for Sustained Innovation Leadership.* Chichester: Wiley.

[KC00] Krueger, R.A., and M.A. Casey. 2000. "Focus Groups. A Practical Guide for Applied Research (3rd Ed.)", *Forum Qualitative Sozialforschung / Forum: Qualitative Social Research*, 3(4):28.

[LRS+09] Lindgardt, Z., M. Reeves, G. Stalk, and M. Deimler. 2009. *Business model innovation: when the game gets tough change the game.* Boston: The Boston Consulting Group, ISBN-13: 978–1118591703.

[LMR17] Lingnau, V., G. Müller-Seitz, and S. Roth, eds. 2017. *Management der digitalen Transformation: Interdisziplinäre theoretische Perspektiven und praktische Ansätze.* Munich: Vahlen, ISBN-13: 978–3800655403.

[Mar52] Markowitz, H., "Portfolio Selection". 1952. *Journal of Finance.* ▸ www.math.ust.hk/~maykwok/courses/ma362/07F/markowitz_JF.pdf. Accessed 13 Mar 2022.

[MC04] Mitchell, D., and C. Coles. 2004. Business model innovation breakthrough moves. *Journal of Business Strategy* 25(1):16–26. ▸ https://www.emerald.com/insight/con-tent/doi/10.1108/02756660410515976/full/html.

[Mck20] McKinsey Research. 2019. *"Strategy in the face of disruption: A way forward for the North American building-products industry"* ▸ www.mckinsey.com. Accessed 15 June 2021.

[MFG08] Morgan, D., C. Fellows, and H. Guevara. 2008. *"Emergent approaches to focus group research."*, *Handbook of Emergent Methods*, 189–205. USA.

[MGB+17] Miller, N.J., T.W. Griffin, J. Bergtold, A. CiampittiI, and A. Sharda. 2017 "Farmers' Adoption Path of Precision Agriculture Technology". *Advances in Animal Biosciences: Precision Agriculture (ECPA)* 8(2):708–712.

[Mit83] Mitchell, J.C. 1983. Case and Situation Analysis. *The Sociological Review* 31(2):187–211.

[MR18] Mello Ulisses, and Sriram Raghavan. 2018. *Bringing the power of Watson to farmers.* ▸ https://www.ibm.com/blogs/research/2018/09/smarter-farms-agriculture/. Accessed 3 Feb 2021.

[Mül15] Müller-Seitz, G. 2015. Strategische Führung in Industrial Service Networks: Leitgedanken zu Chancen und Grenzen aus Sicht von KMU. *zfbf* 69(15):17–34. ▸ https://link.springer.com/article/10.1007/BF03372932.

[Mül22] Müller-Seitz, G. forthcoming. Plattform-Management. In *Handbuch Digitalisierung*, Eds. H. Corsten and S. Roth. Munich: Vahlen.

[MZB+18] Müller-Seitz, G., D. Zühlke, T. Braun, D. Gorecky, and T. Thielen. 2018. Netzwerkbasierte

Geschäftsmodellinnovationen – Das Beispiel der Industrie 4.0-Anlage SmartFactoryKL. *Die Unternehmung* 72(2):146–168. ▶ https://www.nomos-eli-brary.de/10.5771/0042-059X-2018-2-146/netzwerkbasierte-geschaeftsmodellinnovationen-das-beispiel-der-industrie-4-0-anlage-smartfactorykl-jahrgang-72-2018-heft-2.

[NS11]   Nambisan, Satish, and Mohanbir Sahwney. 2011. Orchestration Processes in Net-work-Centric Innovation: Evidence From the Field. *Academy of Management Perspectives* 25(3):40–57.

[OP10]   Osterwalder, A., and Y. Pigneur. 2010. *Business model generation*. Hoboken: Wiley.

[PCP13]   Pierpaoli, E., C. Carli, E. Pignatti, and M. Canavari. 2013. Drivers of Precision Agriculture Technologies Adoption: A Literature Review. *Procedia Technology* 8:61–69.

[Peer17]   Peer Boaz. 2017. *Prospera Series B Shows Connectivity is Driving Growth in AgTech*. ▶ https://insights.qualcommventures.com/prospera-series-b-shows-connectivity-is-driving-growth-in-agtech-92471b11729b. Accessed 3 Feb 2021.

[Pen18]   Peng Tony. 2018. *Alibaba AI Detects Pig Pregnancies*. ▶ https://syncedreview.com/2018/12/18/alibaba-ai-detects-pig-pregnancies/. Accessed 3 Feb 2021.

[Per02]   Perez Carlota. 2002. *"Technological revolutions and financial capital: The dynamics of bubbles and golden ages" (Repr)*. Cheltenham: Edward Elgar Pub.

[PFS07]   Provan, K.G., A. Fish, and J. Sydow. 2007. Interorganizational Networks at the Network Level: A Review of the Empirical Literature on Whole Networks. *Journal of Management* 33(3):479–516. ▶ https://www.researchgate.net/publication/254801923_Interorganizational_Networks_at_the_Network_Level_A_Review_of_the_Empirical_Literature_on_Whole_Networks.

[PH14]   Porter Michael, and James E. Heppelmann. 2014. How Smart, Connected Products Are Transforming Competition. *Harvard Business Review*.

[Por85]   Porter, E. 1985. *Competitive Advantage: Creating and Sustaining Superior Performance*. New York: Simon and Schuster.

[PT17]   Paustian, M., and L. Theuvsen. 2017. Adoption of precision agriculture technologies by German crop farmers. *Precision Agriculture* 18:701–716.

[Sch12]   Schumpeter, J. 1912. *Theorie der wirtschaftlichen Entwicklung. Nachdruck*. Berlin: Duncker & Humblot. ISBN-13 : 978–3428117468.

[SEW+10]   Strecker Otto, Anselm Elles, Hans-Dieter Weschke, and Christoph Kliebisch. 2010. *Marketing für Lebensmittel und Agrarprodukte*, 4th ed. DLG.

[Str20]   Strategyzer 2020. *Business Model Canvas*. ▶ https://www.strategyzer.com/canvas/business-model-canvas. Accessed 20 Nov 2020.

[SSM16]   Sydow, J., E. Schüßler, and G. Müller-Seitz. 2016. *Managing Inter-organizational Relations – Debates and Cases*. London: Palgrave / Macmillan Publishers, Red Globe Press.

[TA17]   Thul, M.J., and A. Altherr. 2017. Digitale Transformation der Landwirtschaft. In *Management der digitalen Transformation: Interdisziplinäre theoretische Perspektiven und praktische Ansätze*. Eds. V. Lingnau, G. Müller-Seitz, and S. Roth, 223–234. Munich: Vahlen. ISBN-13 : 978-3800655403.

[TB18]   Tidd, J., and J. Bessant. 2018. *Managing Innovation: Integrating Technological, Market and Organizational Change*, 6th ed. Hoboken: Wiley. ISBN-13 : 978-1118360637.

[Tee10]   Teece, D. J. 2010. Business models, business strategy and innovation. *Long Range Planning* 43(2–3):172–194. ▶ https://www.sciencedirect.com/science/article/abs/pii/S002463010900051X.

[Thi19]   Thielen, T. 2019. *Netzwerkbasierte Geschäftsmodellinnovationen: Untersuchung des Innovationsprozesses von Geschäftsmodellen auf Whole Network Level vor dem Hintergrund von Industrie 4.0*. Düren: Shaker, ISBN-13 : 978–3844068719.

[TL18]   Täuscher, K., and S. Laudien, 2018. Understanding platform business models: A mixed methods study of marketplaces. *European Management Journal* 36(3):319–329. ▶ https://www.researchgate.net/publication/316667830_Understanding_Platform_Business_Models_A_Mixed_Methods_Study_of_Digital_Marketplaces.

[Tid18]   Tidd, J., and J. Bessant. 2018. *Managing Innovation: Integrating Technological, Market and Organizational Change.*, 6th ed. Hoboken: Wiley. ISBN-13 : 978-1118360637.

[Xar20]   Xarvio. 2020. *Digital farming*. Electronically published under the URL: ▶ https://www.xarvio.com/gb/en.html. Accessd 26 March 2020.

[Yar18]   Yara. 2018. *Yara übernimmt trecker.com*. ▶ https://www.yara.de/news-veranstaltungen/news/yara-uebernimmt-trecker.com/. Accessed 3 Feb 2021.

[ZA07]   Zott, C., and Amit, R. 2007. Business model design and the performance of entrepreneurial firms. *Organization Science* 18(2):181–199. ▶ https://www.researchgate.net/publication/228956182_Business_Model_Design_and_the_Performance_of_Entrepreneurial_Firms.

[ZA08]   Zott, C., and R. Amit. 2008. The fit between product market strategy and business model: Implications for firm performance. *Strategic Management Journal* 29(1):1–26. ▶ https://onlinelibrary.wiley.com/doi/10.1002/smj.642.

[ZA10]   Zott, C., and R. Amit. 2010. Business model design: An activity system perspective. *Long Range Planning* 43(2–3):216–226. ▶ https://faculty.wharton.upenn.edu/wp-content/uploads/2012/05/businessModelDesign_Amitzott_LRP2010.pdf.

Technology Perspective

Thomas Herlitzius, Patrick Noack, Jan Späth, Roland Barth, Sjaak Wolfert, Ansgar Bernardi, Ralph Traphöner, Daniel Martini, Martin Kunisch, Matthias Trapp, Roland Kubiak, Djamal Guerniche, Daniel Eberz-Eder, Julius Weimper and Katrin Jakob

Contents

3.1 Efficient Systems Engineering for Automation and Autonomous Machines – 111
3.1.1 Introduction – 111
3.1.2 Characteristics of Agricultural Machinery Development – 112
3.1.3 Mechanization and Automation as Drivers of Productivity on the Farm – 115
3.1.4 Paradigm Shift from "Bigger, Faster, Wider" to Sustainability, Robotics, and Autonomy – 116
3.1.5 Challenges of Autonomous Systems – 118
3.1.6 Summary and Outlook – 120

3.2 **Precision Farming – 120**
3.2.1 Introduction and History – 121
3.2.2 Technology – 122
3.2.3 Applications – 125

3.3 **Safe Object Detection – 127**
3.3.1 Introduction – 127
3.3.2 V-Model – 129
3.3.3 Challenges in Safe Surround Sensing – 129
3.3.4 Solutions for Safe Object Detection – 132
3.3.5 Conclusion and Outlook – 137

3.4 **Interoperability and Ecosystems – 137**
3.4.1 Introduction – 138

© The Author(s), under exclusive license to Springer-Verlag GmbH, DE, part of Springer Nature 2022
J. Dörr and M. Nachtmann (eds.), *Handbook Digital Farming*,
https://doi.org/10.1007/978-3-662-64378-5_3

3.4.2	A Reference Architecture for Integrated Open Platforms and Key Components for Interoperability – 139	
3.4.3	A Lean, Multi-Actor Approach for Ecosystem Development – 142	
3.4.4	Conclusions and Future Development – 144	
3.5	**Artificial Intelligence – 145**	
3.5.1	AI in the Agriculture Context – 145	
3.5.2	Capturing the Environment – 146	
3.5.3	Data Exchange and Shared Understanding – 146	
3.5.4	Interpretation, Analysis, and Decision Support – 147	
3.5.5	Getting Smarter: Machine Learning – 148	
3.5.6	Artificially Intelligent Robots – 151	
3.5.7	Economics of AI – 152	
3.5.8	Outlook: Individualized Optimization – 153	
3.6	**Agricultural Data and Terminologies – 153**	
3.6.1	The Data Landscape in the Agricultural Sector – 154	
3.6.2	A Global Data Space—Achievable or Wishful Thinking? – 157	
3.6.3	Controlled Vocabularies, Thesauri, and Ontologies – 159	
3.6.4	Conclusion and Outlook – 162	
3.7	**The Role of Geo-Based Data and Farm-Specific Integration: Usage of a Resilient Infrastructure in Rhineland-Palatinate, Germany – 162**	
3.7.1	Introduction – 162	
3.7.2	Overview – 163	
3.7.3	Site-Specific Resilient and Climatic Smart Farming – 164	
3.7.4	The GBI as an Example for Infrastructural Resilience – 166	
3.8	**Technology Outlook – 170**	
3.8.1	Introduction – 170	
3.8.2	Key Enablers for Digital Agriculture – 170	
3.8.3	Autonomous Machinery for Tomorrow's Agriculture – 175	
3.8.4	Digital Twin: The Farmer as a Factory Manager – 178	
3.8.5	Democratization of Agriculture Through Digitalization – 179	
	References – 180	

Technology Perspective

3.1 Efficient Systems Engineering for Automation and Autonomous Machines

Thomas Herlitzius

Abstract

The history of AgMachinery technology shows a shift toward methodological and analytical approaches. Nevertheless, research into the manifold material laws is still inadequate. The goal of development was and is the continuous increase of productivity. What is new is the motivation for more sustainable production methods. In the next twenty years, machine concepts will change in the direction of highly automated, flexible equipment systems that can be used collaboratively and semi-autonomously, and thus, the direct correlation between productivity and the size of a machine will become less important.

3.1.1 Introduction

In contrast to industrial production processes, crop production takes place under outdoor conditions and on stationary crop sites. The individual process steps are separated from each other both, spatially (different fields) and temporally (growth cycle), and are subject to changing conditions of soil, terrain, and weather. In general, the treatment and processing of natural substances are characterized by a multitude of parameters, which cannot be defined and measured exactly and which exhibit great variance. As with any production process, the aim in agricultural production is also to achieve end-to-end plannability of effort and results in order to produce high yields with the desired and constant quality. The focus on a few specially bred crops and machine-oriented cultivation processes is, to a certain extent, continuously changing, which is characterized by constant evolution, diversity, and a fluid balance of fauna and flora. To the extent that crop production is intensified by artificial fertilizers, chemical crop protection, and more powerful, but also larger and heavier machines, the intervention into existing ecosystems and the resulting changes become more pronounced. On the positive side, there is unprecedented productivity in agricultural production at low cost and with high quality. At the same time, however, we also see the undesirable side effects, such as nitrification of groundwater, decline in biodiversity, and harmful compaction in the soil, to name just a few examples. Besides breeding and fertilization / crop protection, the mechanization and automation of work processes is the main driver of the productivity achieved. The economically justified demand of agriculture was and is the continuous increase of productivity in order to control and, if possible, further reduce process costs. The development departments of the AgMachinery manufacturers have so far successfully responded to this by further increasing the size of the machines, which can operate on a site-specific basis due to automation and achieve a reduction in effort and operating resources by increased efficiency. The quite reasonable and timely demand for reducing the undesirable side effects of agricultural production at the same time as increasing productivity in order to move from a predominantly economically driven agriculture to a better balance in the sustainability triangle is, although latently present, not yet a primary requirement for technology development today. This is also unlikely to change as long as additional costs for more sustainable production methods and plant-friendly technology cannot be passed on as quality characteristics in the value chain. Consumers, however, have been fed a romanticized and diffuse ideal image of agriculture by the food industry and the retail trade, and innovations are difficult to integrate into this image. It is therefore (for all participants in the new value creation networks) a matter of showing that careful treatment of nature and animals is also possible on the basis of technological progress [VHS06].

3.1.2 Characteristics of Agricultural Machinery Development

The main focus of AgMachinery development is on the functional theories of the working elements for material processing, where the work result per time and area as well as the energy demand are the most important optimization criteria. Despite extensive work in recent decades, research into the manifold material laws is still inadequate. Simplified models supported by measured process parameters are still characteristic in this field. ◘ Figure 3.1 illustrates the change in AgMachinery development to methodological–analytical approaches, which have always been used in traditional engineering sciences.

Today, empirical theories with different foundations allow estimations, which, however, still have to be validated by experimental analysis. For AgMachinery processes, a universal analysis does not appear to be appropriate in the near future, neither theoretically nor in terms of measurement technology [KSS11].

In the last 30 years, powerful virtual development tools have been added to the engineers' toolbox. Although they can narrow the gap to a complete description of machine functions, they are still far from closing it. In particular, moving critical development issues forward improves development efficiency and reduces iterations and the number of test machines while lowering development costs and / or increasing the quality of the result at the same time (see ◘ Fig. 3.2). Virtual product development is used for component design, simulation of lifetime and functionality, as well as for statistical test planning. Well-known tools that are based on the three-dimensional parametric CAD modeling world are the finite element methods (FEMs), multi-body simulation (MKS), or flow and particle simulation (CFD / DEM), to name just a few.

However, there are also clear limits to simulation tools for function development, as they do not adequately represent the diversity of natural substances and the calculation parameters can rarely be derived directly from material parameters (if these

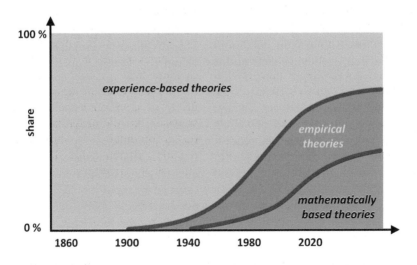

◘ **Fig. 3.1** Development trend for the theoretical foundation of AgMachinery according to [KSS11]

Technology Perspective

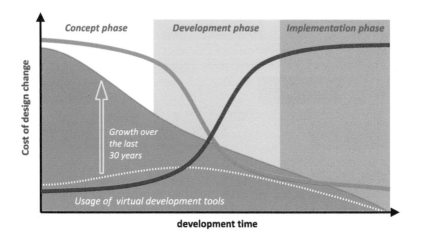

Fig. 3.2 Increasing use of virtual development tools

are even known) and therefore require additional transfer functions to be created. The simulation world, which is already idealized in itself, is additionally simplified and based on assumptions in order to optimize computing time or to be able to perform computations at all. This results in a multitude of error sources that can be controlled the better the more development experience can be built up in a development team. Based on such experience, it is possible to optimize and control the problem-specific balance between simulation, laboratory experiments, and field experiments.

Today, digitalization technologies are opening up completely new perspectives for agriculture to find acceptable solutions for the seemingly unavoidable contradiction between profitability and sustainability, while at the same time realizing an increase in value creation. Digitalization technologies are an outstanding example of "technology push," where a market is sought for disruptive innovations based on visions and trends. Technology push means developing a technology out of an opportunity, whereby an inherent risk exists that the market requirements will be missed or misjudged. In contrast, "technology pull" means relying on the clearly expressed needs of known customer segments and markets. This is low risk, but the disadvantage is that it produces mostly incremental innovations that do little to create differentiation and uniqueness in the market. Again, it is important that there is a balance between pre-development and product development, so that trends are not missed, but a disciplined development process can be followed that delivers innovative as well as reliable and cost-effective products (see ◘ Fig. 3.3).

One characteristic of current product innovations is that new features increasingly deliver cross-product benefits, such as bidirectional data exchange with Farm Management Systems (FMIS) or the coordination of neighboring machines. Following the long phase of component and machine automation, we have now entered the era of process automation, where the individual machine becomes the actuator of an FMIS. From this point of view, "machine intelligence" is limited to machine-internal process optimization, while the specifications for site-specific processing come from the FMISs, which realize a new level of automation above the machine level with the corresponding added value. According to this definition, it is not correct to speak of

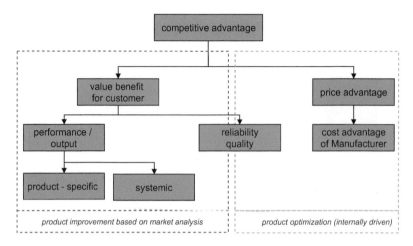

Fig. 3.3 Product development strategies

"smart tractors", as these only support the attached implements and their embedded machine intelligence. From the perspective of machine development for tractors, implements, and self-driving machines, there are therefore new requirements that need to interact with the level of process and farm automation in the best possible way.

The wide variety of influencing factors that will shape technology development in the next 20 years is depicted in ◘ Fig. 3.4. The results of an unpublished expert survey by AVL in 2018 show that there is little doubt that digitalization technologies and electrification will play a major role. The development of drones and robotics,

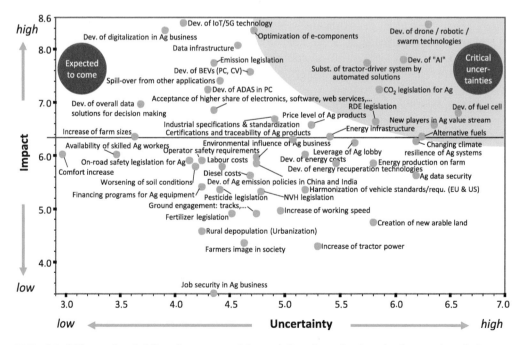

Fig. 3.4 Effects and probability of occurrence of factors influencing technology development in agriculture

the replacement of operators by autonomization, and the use of AI technologies are also seen as highly effective for the successful transformation of agriculture, but are fraught with great uncertainties, as implementation timeframes, user acceptance, and economic viability are hard to predict and must be demonstrated starting in market niches (see ▶ Sect. 1.5). The resource-rich, globally active, large AgMachinery manufacturers have a problem with this approach, as risks and markets are very well understood and niches do not provide acceptable business cases, which can prevent such disruptive innovations for a long time in strongly process-oriented organizations. In the survey, the traditional development topics such as chassis that protect the ground or increased working speed and engine power were shown to have little effect. This is in line with the finding that the limitation in size and weight in the known machine concepts progressively increases the effort required to realize further productivity gains in relation to user benefits.

3.1.3 Mechanization and Automation as Drivers of Productivity on the Farm

The steadily increasing efficiency of today's machines and implement systems in agricultural machines has contributed significantly to modern crop production reaching high yield levels. This has enabled industrialized countries to achieve security of supply with high-quality food at low cost. However, the risks and side effects of high-performance technology and cultivation methods are becoming visible and measurable today (e.g., soil compaction, partial overfertilization, declining biodiversity). Against this background, the requirements for new machine systems or even concept changes are extremely high, the constraints to be considered are complex,

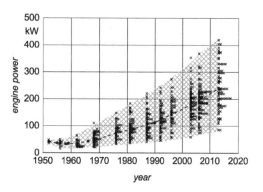

Fig. 3.5 Development trend of the installed engine power of combine harvesters, exemplified by the German market [Böt13]

and isolated machine developments are only useful for niches. The field of AgMachinery development is thus becoming even more complex, and in addition to finding solutions for the technical and economic requirements, it must also assume responsibility for the interaction between agriculture and the ecosystem with the objective of achieving closed cycles.

Automation helps to ensure that the technically installed process capacity can actually be utilized, but it has not been able to stop or reverse the trend toward ever greater growth in size. The continuous increase in productivity with a simultaneous widening of the gap between the lower and upper performance classes is characteristic. The observed trend in growth over time is illustrated in ◘ Fig. 3.5 using the example of the combine harvester.

In the upper segment, growth is five times stronger at 6.3 kW/a than in the lower segment at 1.4 kW/a. The widening gap between the lower and the upper power segments (1990: 200 kW vs. 2020: 350 kW) is an indicator that the high productivity of large machines cannot be implemented economically in smaller farm structures and that there is and will continue to be demand in the lower power segment. An adaptation of farm structures and field sizes to higher machine performance oriented toward high utilization of the technology has so far taken place only to a limited extent, and its acceptance in society is declining today. In addition to the increasing development effort required to comply with the dimensional and axle load limits specified by road traffic regulations, manufacturers are making further efforts to counteract the diversity of parts caused by the wide spread of power classes by using platform strategies. However, with the increasing spread of the model range and the growing number of variants, this is only successful to a limited extent. For the reasons mentioned above, there is a general expectation among experts that machine concepts will change over the next twenty years in the direction of machine systems with electrified drive systems that are highly automated and can be used flexibly and semi-autonomously [Böt15, KBS+13, GMU+17, BKA13, GB07, USW+17, HF17]. Productivity will then no longer be determined by machine size alone, but will become configurable for the specific use case through the number of units used when the power of the individual systems is reduced. The paradigm of continuous growth in the productivity of a unit (machine), which has been valid since the beginning of mechanization in agriculture, could be replaced by modular machine systems with smaller units and a high level of automation, which would make productivity scalable again and better adaptable to the process chains. This is another reason why the development focus of AgMachinery manufacturers has been on process automation since the 1990s. Despite the availability of vehicle platforms with sophisticated automation features as a prerequisite for process automation and autonomous driving, there is still a great need for development in the acquisition of information on soil and plant properties. The current state of development in ambient and process sensor technology is still inadequate to enable complete process automation. Digitalization technologies will penetrate all areas of agricultural production, which will enable a further increase in the level of automation.

3.1.4 Paradigm Shift from "Bigger, Faster, Wider" to Sustainability, Robotics, and Autonomy

Since around 2015, at Agritechnica in November, it has been clearly evident that the comprehensive digitalization of operational processes with a variety of products is starting to arrive in practice. Since then, several studies of autonomous machines have been presented, but all of them do not yet appear ready for production and are currently aimed at niche applications in order to get the technology established.

The productivity of AgMachinery is directly linked to the size of the process area of a particular function. In the past, productivity increases were mainly achieved by increasing the process area represented by working width, machine-internal process channel width, or working speed. When it comes to tillage, crop protection, fertilization, and harvesting, the corresponding storage volumes have also increased. As machines at the upper performance band increasingly reach the reasonable or legal upper limits on weight and dimensions, further required productivity increases must be achieved without a further increase in weight and dimensions. Two seemingly mutually exclusive ways to meet and reduce axle loads and dimensions become apparent as follows:

— A further increase in functional power density through intensive functional development and the use of fiber composites to reduce the specific weight. High development costs and additional material costs for fiber composites lead to a progressively deteriorating cost–benefit ratio, with the costs increasing more

Technology Perspective

than the benefits, which makes machine amortization even more difficult.
- Well-known machine concepts are shifting toward highly automated, flexibly operating machine systems with hybrid energy sources and distributed drive architectures [Wit19, BSW+05]. Multi-machine operation or autonomous systems mean that labor costs do not increase and machines can be smaller again. Thus, the number of platforms and the variety of parts can be reduced, which would be a basic prerequisite for disruptive ideas currently under discussion, such as Spot Farming or Patch Cropping with site-specific and extended crop rotations [WUH+17].

Whereas in the first case, functional development has to be intensified and new material knowledge has to be implemented—from calculation via validation and manufacturing to recycling—the focus in the second case is on process automation, platform and module strategies, and the best possible (HMI) support for the new operator role. On the one hand, the time has come to make a decision because both paths require considerable development resources and involve risks and therefore require lead time. On the other hand, there are efforts to further push existing machine concepts to the limit of what is feasible in order to postpone the time for a decision so that the trends become more visible and can be assessed better. The basic development strategy question of defining productivity in a machine or in a machine system is additionally influenced by the possibilities and trends of digitalization technologies and the demand for best possible support of process and farm automation (Smart Farming). Under this aspect, the development of Ag-Machinery must expand its focus from the individual product to the production system of a farm, which will cause major problems especially for large manufacturers with their machine-focused, thoroughly optimized, and highly meshed product development processes. If we consider the ability of manufacturers to successfully establish Smart Farming technologies on the market and superimpose their customers' understanding of the applications on their own digitalization competence, the following diagram emerges (see ◘ Fig. 3.6).

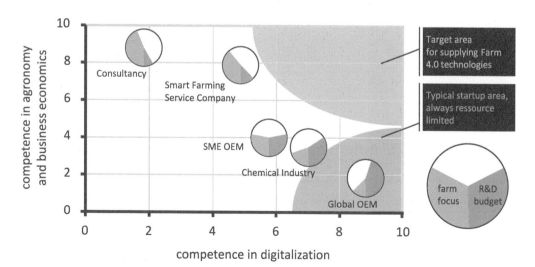

◘ **Fig. 3.6** Strengths and weaknesses of technology providers and service providers in a Smart Farming application world

The resource-rich market leaders lack the application background to cooperate with FMIS. The approach of integrating FMIS functionality on the machine or around the machine suffers from the inherent problem that the levels of automation become mixed and that data flows and decision chains become non-transparent and inhomogeneous. The smaller companies, on the other hand, lack the foresight and the vision of digital networking, even though they are much closer to the users and also benefit much more directly from successful customers, due to the respective players' focus on users with their products.

Today, there are only a few areas where money can be made directly with digital tools. Most manufacturers are (still) in the investment phase and cross-subsidize. Here, added value for the respective providers is more likely to be generated through "customer loyalty" or "customer transparency". This can be seen quite clearly in the Innovation and Digitalization segment of BayWa, where Digital Farming and e-business activities are bundled: With revenues in the amount of €5.6 million in the first half of 2020, EBIT of €-6.8 million was generated [Bay21].

No one is yet in the "sweet spot" of digitalization competence and user understanding, and the question for everyone is whether and how to get there and which internal and external obstacles must be overcome to get there, or which new companies and business models will emerge here.

3.1.5 Challenges of Autonomous Systems

The use of autonomous systems makes sense specifically when automation simplifies work in agriculture and increases the sustainability of agricultural production. Besides numerous advantages for autonomous machines such as workload relief or more sustainable management through digital recording and evaluation of nutrient cycles, there are also disadvantages that must be understood, such as high investment costs, failure probabilities due to changing environmental factors, legal uncertainties, or farm structures not designed for autonomous AgMachinery. When considering autonomous agricultural systems, the evaluation of the benefits must be differentiated according to application areas, and labor and social aspects must be taken into account. In principle, autonomy is never about replacing the labor force, but about the changed role away from routine activities and toward management and problem-solving competence.

In agriculture, the use of autonomous agricultural systems is currently the subject of a controversial debate. In the next few years, automated and autonomous AgMachinery is expected to become much more widespread in practice. Therefore, it is necessary to analyze the manifold open questions regarding benefit, reliability, safety, responsibility, and legal security (see ▶ Sect. 1.7). To increase the usability of the technology, systemic solutions as well as practice-oriented system integration are important. Completely autonomous systems are not feasible in the open field in the short to medium term and might not even be desirable. In contrast to road traffic, which defines six levels on the way to autonomy, AgMachinery technology must use a derived approach of its own because a distinction must be made between different states of driving and process functions (see ◘ Tables 3.1 and 3.2) and because the transitions between the individual levels will be gradual.

The discussion about autonomous machines has become so central in recent years because, on the one hand, the necessary technologies are gradually becoming available and, on the other hand, because, as described above, the potential is seen to meet the current challenges regarding AgMachinery far better than is possible with

Technology Perspective

Table 3.1 Six levels of autonomous driving based on the SAE Standard J3016

Monitored operation	1 Driver only	
	2 Assisted	Machine navigates and provides diverse degrees of processing component automation
	3 Partial Automation	
Non-monitored operation	4 Conditional Automation	Machine operates autonomously, operator on standby
	5 High Automation	Operator not required for certain periods
	6 Full Automation	

existing machine concepts. The technology drivers are, in particular,
- sensor and actuator technology,
- image processing and AI,
- wireless communication technologies,
- electrified drives,

while the application-relevant drivers are
- climate change and changing crop cultivation systems including crop rotations,
- process cost management and profitability,
- trust in and acceptance of autonomous technology by farmers, and
- policy framework and regulations, including constraints regarding fertilizer application and crop protection.

◘ Figure 3.7 shows an overview of how the trend toward autonomous machines will continue over time and for specific markets according to the various categories in ◘ Table 3.2.

While fully human-controlled systems will remain similar to the way they are today, significant changes are to be expected for the other three classes. Assisted systems

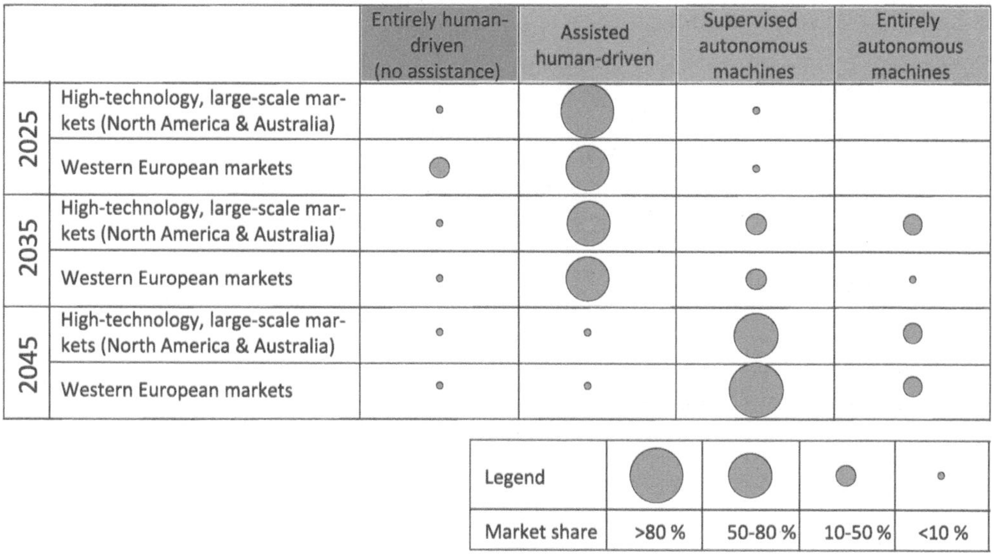

		Entirely human-driven (no assistance)	Assisted human-driven	Supervised autonomous machines	Entirely autonomous machines
2025	High-technology, large-scale markets (North America & Australia)	○	●	○	
	Western European markets	○	●	○	
2035	High-technology, large-scale markets (North America & Australia)	○	●	○	○
	Western European markets	○	●	○	○
2045	High-technology, large-scale markets (North America & Australia)	○	○	●	○
	Western European markets	○	○	●	○
Legend		●	●	○	○
Market share		>80 %	50-80 %	10-50 %	<10 %

◘ **Fig. 3.7** Expert survey on development trends in machine autonomy [DFH+19]

Table 3.2 Revised levels of autonomy for AgMachinery in the ISO 18497-1 standard

Levels of autonomy	Manual Non-Automated	Partially automated	Semi-autonomous	Autonomous
Functions	Non-Automated			
		Automated		
Modes	Manual mode			
			Autonomous mode	

will be able to support more complex actions and will be equipped with more cameras and more safety functions. The class of supervised autonomous machines will continue to evolve, with decreases in the requirements on the presence and qualifications of the supervisor over time and an increase of field- and application-specific autonomous actions, all the way to more universal actions. The class of fully autonomous machines is evolving from pure demo systems via systems that act autonomously on the field based on defined plans once they are there, to systems that receive a rough plan from other systems and then act fully autonomously.

3.1.6 Summary and Outlook

Supporting sustainable production systems in agriculture is not a primary requirement for technology development. This will not change much as long as additional costs for more sustainable production methods and plant-friendly technology cannot be passed on as added value in the value chain. In the last five years, it has become clearly evident that the comprehensive digitalization of operational processes with a variety of products is starting to arrive in practice. Since then, several studies of autonomous machines have been presented, but all of them do not yet appear ready for series production and currently aim at niche applications in order to get the technology established.

Under this aspect, the development of AgMachinery must expand its focus from the individual product to the production system of a farm, which will cause major problems especially for large manufacturers with their machine-focused, thoroughly optimized, and highly networked product development processes. Completely autonomous systems are not feasible in the open field in the short to medium term and might not even be desirable.

In general, it can be concluded that there will be a clear trend toward autonomous agricultural systems in the long term, but the speed with which farmers will adopt autonomous systems will differ significantly between the different markets. The adoption of autonomous technology will be spearheaded by North America and Western Europe.

3.2 Precision Farming

Patrick Noack

Abstract

The following section gives an introduction to the history as well as the state of art of Precision Farming and ends with an outlook on future developments. It also provides an insight into the core technologies behind the concept of Precision Farming.

3.2.1 Introduction and History

Precision farming is a term that has been coined the 1990s initially focusing on plant production. In the following years, it has been extended to cover additional fields. This section has a strong focus on plant production and will not cover the applications of precision farming in other areas like e.g., livestock farming. The focus of Precision Farming lies on optimizing agricultural production based on data and information in opposition or in addition to management based on knowledge and tradition. The optimization may target economical, ecological, or social aspects and is therefore closely related to sustainability.

During the last years, different terms have been coined to describe information-based farm management. Precision agriculture was the first term already introduced in the 1990s. Later on, information or data-based agriculture has been labeled as Smart Farming, Digital Farming or Farming 4.0. None of the above-mentioned terms have been defined clearly enough to justify a distinction. From a practical point of view, all concepts supply digital toolsets which help to support the production of food, feed, raw materials, and biofuels by providing decision support or the automation of processes.

One of the initial drivers for precision farming was yield mapping. In the late 1980s and early 1990s, scientists and the agricultural industry started integrating mass flow sensors in combine harvesters along with satellite navigation technology (GNSS). The yield data collected in the field were processed in geographic information systems (GIS), resulting in yield maps revealing the local differences in yield on a subfield level. The concept of variable rate application (VRA) evolved out of the maps supplying information on local yield variation. VRA implies that fields are subdivided into management zones, which are treated according to local deficiencies. The term is closely related to site-specific management. Additionally, the emergence of GNSS technology provided the ability to determine the location of tractors and implements and to control the application rate at a given position in real time.

Initially, the concept of site-specific management and variable rate application struggled with both technical and agronomical hurdles. On the technical side, the development of electronics in agriculture was still in an early phase, resulting in premature systems and limitations with respect to computing power. At the same time, the lack of standards for connectors and data protocols led to proprietary and monolithic solutions not sharing common components like displays or sensors (GNSS). The agronomical challenges were founded on unclear strategies with respect to managing low and high yielding zones. Some agronomists suggested to reduce the management intensity in areas with low yield, arguing that less input is needed to maintain plant growth. Others suggested to increase the intensity with the aim of increasing yield and making use of the yield potential. The underlying issue is that plant growth is driven by many different factors like the content of different nutrients in the soil, soil pH, water availability, soil compaction, soil temperature, and other meteorological measures and that even under same conditions different crops and breeds will react differently to the environmental factors listed above.

In early 1990s and early 2000s, site-specific management has only been integrated into the farm management by early adopters. Apart from the fact that electronic components were improved and the standardization of the communication between controllers and terminals on tractors and implements [ISO17] went forward helping technology for site-specific management to become more reliable and versatile, new applications based on GNSS technology were introduced into the market. Starting off with so-called parallel swathing systems that indicated the optimal driving direction for minimizing gaps and overlaps while working the field by LEDs, steering technology entered the market as of the beginning of the new decade in 2000. The first automatic steering systems were already operating based on RTK-GNSS and featured 2,5 cm absolute accuracy when steering a tractor automatically along predefined tracks. Despite the high price, the systems were quickly accepted by farm managers and machine operators as their benefits (reduce overlaps and gaps) were independent of crop, breed, nutrient content, soil moisture, and meteorological conditions.

Automatic steering systems were initially also expensive because the retrofit integration into existing tractor hydraulics was complicated and thus expensive. As a result, so-called steering assistance systems have been introduced. They apply electric motors to interface with the tractor steering wheel or the steering column reducing the time for installation dramatically. Today, most modern tractors are equipped with interfaces for automatic steering systems ex-factory which—along with other factors—has been resulting in a substantial price reduction. An additional application based on GNSS technology is section control (SC). With increasing boom widths, the manual actuation of single sections at the headland of fields has been introducing an increasing inaccuracy with respect to longitudinal overlapping and gaps when applying agrochemicals. Section control systems monitor and actuate the activity of single boom sections (or even nozzles). The systems monitor the area which has already been covered by the application (coverage map). Sections travelling over an area that has already been covered are automatically switched off and back on when re-entering untreated areas. Again, section control was successfully introduced and well accepted by farmers because their operation is independent of agronomical factors like plants, soil, and weather. The system has also been adopted to control fertilizer spreaders, seeders, and even hoeing machines.

Meanwhile, both, automatic steering systems and section control systems with partly standardized interfaces (ISO 11783) are state of the art and are being adopted even on medium-sized and small farms. This has also removed most of the economic entry barriers for site-specific management as most of the components needed to implement variable rate technology like GNSS receivers and standardized terminals are already present on tractors and have been paid for. In addition, the availability of spatial information on plant growth and plant nutrient content has become available at a much lower or even no cost due to the availability of high-quality satellite data (Sentinel Mission) and unmanned aerial vehicles (UAV) with multispectral cameras.

3.2.2 Technology

Precision farming technology is a set of tools, most of which have not been specifically designed for the application in agriculture.

3.2.2.1 GNSS and Other Positioning Devices

When it comes to plant production, GNSS is the most important tool as it provides accurate information on absolute position, time, speed, and heading [Zog09]. However, GNSS was initially developed for military purposes in the late 1970s and early 1980s by the Americans and the Russians in parallel. GNSS receivers were the midwife for precision farming when enabling yield mapping to reveal the extent of variability within fields. They also play a key role for variable rate application, section control, and automatic steering systems. In addition, GNSS sensors are applied for field boundary mapping, soil sampling, field scouting, the documentation of tasks, and the autonomous operation of UAV. It becomes clear that the application of GNSS is fundamental to the implementation of precision farming in crop production.

Apart from GNSS, relative positioning devices such as lasers, ultrasonic sensors, and tactile sensors are being used to control the path of travel of machinery or the height of implements (e.g., sprayer booms).

3.2.2.2 GIS and FMIS

The second tool are geographic information systems (GIS) which are generally designed to display, create, edit, transform, and merge spatial information in both vector format and raster format. They may be installed on (mobile) PCs/laptops, servers or mobile devices such as tablets or mobile phones. If installed on mobile devices, they are applied for mapping and location-based data collection. A company called ESRI (Environmental Systems Research Institute [E20]), founded in 1969, was and is one of the main drivers for the development and deployment of GIS. As the name states, GIS were initially developed to understand and study the spatial nature of environment in general.

The ability to process raster data in GIS is the basis for analyzing and processing satellite imagery and aerial images acquired with UAV and airplanes. Other incoming data are yield maps, soil maps, sensor data, and task data collected during field operations. In opposition to imagery, most of this data is handled as vector data which consists of geometries (points, lines, polygons) and information related to the geometries mostly in the form of databases. GIS are used to create and export data, e.g., in the form of application maps for variable rate application. In this case, the database contains desired target rates for given locations, which are to be processed when travelling in the field for controlling the application rate of implements.

GI technology is closely related to Farm Management Information Systems (FMIS). This term describes software that is designed to store and analyze data on farms, fields, machinery, tasks, inputs, and work force. It may be described as an Enterprise Resource Planning software (ERP) for agricultural operations. These software products need spatial awareness and thus GI functionality if the spatial variability is to be taken in account for optimizing the farm, machine operations, the use of resources, or a dedicated production system. FMIS have traditionally been operated on local computers and laptops. Today, the trend goes toward server- and cloud-based solutions, which can be accessed from different devices and even machines and implements with an online connection to a communication network.

3.2.2.3 Wired and Wireless Communication Systems

The concept of precision farming implies that devices such as controllers, sensors, actuators like hydraulic valves, and electric motors need to communicate based on wireless or wired communication channels. At the same time, controllers have to access data sources for reading or writing data and information.

The communication between devices in a tractor, in an implement or in a self-propelled machine like a harvester, may be implemented on a proprietary basis. However, if machinery from different manufacturers exchange data through a communication network, standardization is a key for seamless data exchange. Based on a German initiative [D97], the ISO 11783 [ISO17] standard has been developed to define a communication protocol, size and layout of connectors, functionalities (section control, variable rate, task documentation), data formats (ISO XML) and universal terminals (UT) which provide a brand-independent user interface to implements and the above-mentioned functionalities. The standardization is key for controlling and optimizing processes during farming operations and enables multiple use of components and functionalities independent of brand and model of tractor and implement. The development and continuous improvement and extension of the ISO 11783 have helped to make the application of precision farming successful and to raise the acceptance of farm managers and machine operators.

Data exchange with Farm Management Information Systems (FMIS) has traditionally been performed with data storage devices such as CF-, SD cards, or USB sticks. The devices were used both, for importing data like application maps for variable rate application as well as exporting data like task data log files documenting field operations for later use in the FMIS.

Today, the trend goes toward the wireless transmission of data, mainly based on mobile communication networks. On the one hand, this accelerates the data exchange, especially when machinery is being operating in the field over a longer period without returning back to the farm premises. This is especially of great relevance when it comes to transmitting data that is time critical like nitrogen application maps or the real-time position of vehicles in harvest chains. On the other hand, issues related to defective storage devices and data conversion can be avoided when transferring data directly from a tractor or implement to a server or cloud service.

3.2.2.4 Sensor Technology

Apart from GNSS and sensors applied in standard mechatronic applications (angle encoders, inductive sensors), image and spectral sensors play the most important role in agriculture. Images from ordinary RGB cameras are applied for detecting defective grains in combine harvesters and for identifying weeds and bugs in the field. Attached to UAVs, they have also proven to be helpful when it comes to evaluating crop cover or plant density. Stereo cameras are being applied for steering and control applications (e.g., detection of swath height during forage harvest).

The application of spectrometers or imaging sensors that measure the reflection, transmission, or absorbance of light in the range beyond red (near infrared) is widely applied for determining the chemical composition of materials. Multispectral sensors with sensitivity to near-infrared wavelengths have been applied for remote sensing to acquire satellite images as of the late 1970s. Despite the low resolution, satellite images have helped to understand the variation of plant growth in both, natural ecosystems, and agricultural crop production systems. Remote sensing has evolved with a growing number of satellite systems, decreasing costs, and increasing spatial and spectral resolution. Today, remote sensing is a key technology in precision farming. It has been supplemented by the rapid progress in the realm of UAV, which can carry multi- or even hyperspectral cameras, resulting in images with a much higher resolution than satellite images while being mainly independent of cloud cover.

Mounted on tractors, NIR sensors are being applied for deriving the nitrogen content of crops based on the reflection of sunlight or artificial light sources. These systems are known as nitrogen sensors. They help to adapt the application rate of fertilizer spreaders or sprayers online according to the nutrient status of the plants. NIR sensors are also increasingly used for monitoring the dry matter content and other chemical properties of feed, energy plants and small grains during harvest in combines and more commonly in self-propelled forage harvesters. Some of these systems may even be used in the cowshed for analyzing the quality of grass and corn silage for optimizing the feed ration. During the last years, NIR has also entered the analysis of soil properties. Attached to implements or stand-alone systems, NIR spectrometers continuously monitor the content of organic matter and other chemical properties of the soil either for adapting seed rate or seed depth online or for mapping and later consideration and the compilation of application maps.

The understanding of soil moisture distribution and the ability of the soil to hold water becomes more and more important when facing the challenges of climate change. Mobile sensor devices measuring the geo-electric or the geo-magnetic conductivity of the soil help to visualize the spatial distribution of the water-holding capacity. The resulting data and maps can be used offline to adapt seeding density and to control irrigation systems (variable rate seeding and variable rate irrigation). In opposition to the above-mentioned mobile sensors delivering relative spatial distribution of soil water content, stationary sensors measuring the soil electric conductivity or applying more sophisticated methods like frequency domain reflectometry (FDR) or time domain reflectometry (TDR) provide absolute soil moisture measures based on calibrations for different soil types. These systems propagate electrical or magnetic field in the soil and determine the soil moisture based on the frequency shift caused by the soil water or the time for an echoed signal to return to its origin.

3.2.3 Applications

3.2.3.1 Steering and Autonomous Vehicles

The most prominent application of precision farming in crop production is automatic steering systems. They help minimize overlaps and gaps during all field operations and result in even distribution of fertilizer, AgChemicals, and seed. It is also reported that operating speed is generally higher and the time for turning can be decreased when the driver is being supported by steering systems. The systems help to reduce driver fatigue, extend the working hours into times of poor vision (fog, dust, night), and release a substantial amount of attention of the driver for monitoring implements and system parameters of the operation process resulting in higher operation quality. Besides that, steering systems are a door opener for new approaches to farm fields like controlled traffic farming, strip tillage, intercropping, and cross-hoeing. Automatic steering systems have been paving the way toward autonomous agricultural vehicles. The technical feasibility of autonomous tractors and robots has been proven in numerous research projects (see ► Sect. 3.1). The dissemination is very slow due to legal constraints and liability issues (see ► Sect. 1.7).

3.2.3.2 Documentation and Mapping

The documentation of tasks performed in the field is the basis for later analysis in FMIS and optimization. The ISO 11783 standard provides a basis for logging all relevant parameters during field operations, e.g., position, speed, fuel consumption, and application rate. Additional data can be collected with dedicated sensors or sample mapping missions aiming at logging data which is relevant for plant growth or reflects plant growth development (yield, nutrient content, electric conductivity, soil organic matter, plant cover). The data pool may be enriched by crop scouting data collected in the field. The data pool or data lake provides the basis for later analysis of costs, quantities, and qualities of yield as well as environmental impact of farming: it provides the basis for Big Data Analytics and the application of artificial intelligence to agricultural data.

3.2.3.3 Implement Control

Precision farming provides various methods for optimizing the operation of tools on implements, in self-propelled machinery, or even for having implements control actuators on tractors (tractor implement management, TIM).

Section control is a very straightforward approach originally designed to control sections or even single rows or nozzles on sprayers, seeders, and fertilizer spreaders for avoiding longitudinal overlaps and gaps. This mechanism helps to reduce the consumption of resources, saves costs, and helps to distribute material (seed, fertilizer, agrochemical) evenly in the field. However, the potential of section control has not yet been fully exploited. Section control systems may be applied when it comes to switching sprayers or fertilizer spreaders automatically off when they approach protected landscape elements like hedges, rivers, and lakes. Additionally, a current trend is to apply section control when it comes to leaving single tracks or parts of the field free for ecological purposes (nesting birds, bee pasture).

Variable rate technology is designed to control the application rate of implements according to application maps or sensor values when travelling over the field.

This approach is key for reducing costs and unintended leaching of nutrients and AgChemicals satisfying both, the economic welfare of farmers and the requirement to reduce negative impact on the environment. A wide variety of data is available as input for the creation of application maps, such as soil maps, soil sensor maps, UAV images, satellite images, and meteorological models. Numerous research projects have proven that local models are able to produce viable application maps for local conditions [ZSJ+10]. Artificial intelligence is already starting to help creating regional or even global models that process application maps with less effort and in a reproducible manner.

Another promising approach is the automation of tractors based on implement controllers. This concept has been labelled as tractor implement management (TIM). The background is that sensors on the implement can collect data which help to improve the efficiency or the quality of the current operation. However, in traditional tractor-implement configurations, the implement controller lacks the ability to manipulate the actors residing in the tractor, e.g., tractor speed, three-point-hitch position or hydraulic valve actuation. TIM provides a CAN-based protocol based on the ISO 11783 standard that enables the implement controller to send commands to the tractor to adapt the above-mentioned parameters. TIM has proven to increase the efficiency of farming operations by supporting the operator in various demo projects.

3.2.3.4 Outlook

One of the main challenges in plant production is reducing the application of AgChemicals. Due to the high pressure from ecologists, politics, and the consumer side, it is very likely that precision farming solutions related to the reduction of AgChemicals or alternative methods for detecting and erasing weeds and bugs will be available in the near future.

Artificial intelligence and Big Data science may be tools that help to support the development. They will also help to extract valid information, decision support or data for the desirable features for further reducing the risks caused by decreasing work force in agricultural production. This fields needs regulations and a proper concept for liability (see ▶ Sects. 1.6 and 1.7).

Last but not least, man–machine interfaces need to be improved in order to maintain the link between the unquestioned value of human knowledge, intuition, and tradition on the one hand and the potential of data mining and automation on the other hand.

3.3 Safe Object Detection

Jan Späth and Roland Barth

Abstract

At the Agritechnica 2019, one could see a notable variety of autonomous machines. Yet, seeing the number of autonomous machines in the field, it becomes obvious that there is still a way to go until a fully autonomous machine is operating at the field. However, what we do see, are more and more functions which enable, for instance, tractors to work almost autonomously. RTK-based steering is only one example that points out the path in which the AgMachinery sector is leading. In order to continue the path of more and more automation, safe surround sensing is, among other challenges (e.g., legal aspects, social aspects, see ▶ Chapter 1), the key challenge. In this section, it shall be outlined what are the technological challenges related to safe surround sensing and how they can be overcome.

3.3.1 Introduction

There is a plethora of reasons why farmers need to be more efficient in the upcoming years. As described in ▶ Sect. 1.2, from a European perspective, we do see more and more upcoming regulations governing the use of fertilizer and plant protection products, an increasing price pressure for land and labor, and a continuing decrease of arable land while at the same time the world population is increasing. As described in previous chapters of this book, Digital Farming and with this the automatization of agricultural processes can be one tool to support farmers in this difficult environment.

Seeing the numerous concept studies of different tractor manufacturers at the Agritechnica 2019, it becomes obvious that manufactures are collecting experience with developing (semi-)autonomous machines. This is nothing new. In fact, in 1962, Cornelius Siegling had already developed an autonomous plowing tractor [FRK+15]. But the fact that these are all prototypes underlines also that there is still a way to go until the autonomous tractor will be reality in the field.

However, even though, there is not yet one autonomous tractor working fully autonomous in the field, does not mean that the functions which come with such a tractor are not yet implemented on the machines. Typical features like plant detection, automatic steering, and row detection have already been in place for years and show that even though there is no autonomous tractor yet, one can observe a great variety of autonomous functions being implemented at AgMachinery nowadays.

To understand the gap between an autonomous working tractor and autonomous functions, a closer look at the Society of Automotive Engineers (SAE) Standard J3016 for automated driving is helpful [SAE20]. The standard differentiates between 5 different levels of autonomous driving (see ◘ Fig. 3.8). On the least automated level (SAE Level 1), the driver has assisting features like blind spot warning, but must steer, brake, accelerate, and always supervise possible features. At the highest automated level (SAE Level 5), the driver might sit in the driver seat, but the driver is not required to steer, break, or accelerate, not even to interfere. It is assumed that the system will perform its task completely autonomously at all times, on all terrains and under all different circumstances, independent of unplanned events.

Following this classification, we witness a time of being just between SAE level 3 with functions like GPS-guided steering and SAE level 4 with, for example, systems where one tractor is leading others (e.g., Feldschwarm from John Deere [Fel20] or Guideconnect from Fendt [Agc20]). Yet, to bridge this gap between SAE level 3 and SAE level 4, a safe, reliable surround

	SAE LEVEL 0	SAE LEVEL 1	SAE LEVEL 2	SAE LEVEL 3	SAE LEVEL 4	SAE LEVEL 5
What does the human in the driver's seat have to do?	You are driving whenever these driver support features are engaged - even if your feet are off the pedals and you are not steering			You are not driving when these automated driving features are engaged - even if you are seated in "the driver's seat"		
	You must constantly supervise these support features; you must steer, brake or accelerate as needed to maintain safety			When the feature requests, you must drive	These automated driving features will not require you to take over driving	
	These are driver support features			These are automated driving features		
What do these features do?	These features are limited to providing warnings and momentary assistance	These features provide steering OR brake/ acceleration support to the driver	These features provide steering AND brake/ acceleration support to the driver	These features can drive the vehicle under limited conditions and will not operate unless all required conditions are met		This feature can drive the vehicle under all conditions
Example Features	automatic emergency braking blind spot warning lane departure warning	lane centering OR adaptive cruise control	lane centering AND adaptive cruise control at the same time	traffic jam chauffeur	local driverless taxi pedals/steering wheel may or may not be installed	same as level 4, but feature can drive everywhere in all conditions

Fig. 3.8 SAE levels of driving automation; source: own creation based on Shuttleworth [Shu20]. Blue: driver activity, grey: automated driving activity

sensing in a rough environment is certainly one crucial enabler. Seeing the status quo, this is an extremely difficult task, especially if the whole system shall be commercially feasible (what we should assume). So far, no machine manufacturer has been able to take this huge step. Figure 3.9 underlines the challenges: look closely, do you recognize the man standing next to the machine?

The aim of this section is to contribute to the debate on safe surround sensing for AgMachinery by outlining the challenges related to safe surround sensing, by providing possible solutions to the challenges, and by giving a short outlook on how the industry might develop. To structure the different challenges (and later the different solutions) in a comprehensive way, we will use a simplified V-Model. This model encompasses

Fig. 3.9 Object detection in the field

all important steps of the development of a (software) system and is thus a good guidance to ensure the different challenges are considered at different points in time. The following part of this section gives will give a short outline of this model.

3.3.2 V-Model

The V-Model is designed to guide development processes in an encompassing way by differentiating between the different development steps [FKS+09]. In this section, we simplify the V-Model and differentiate between four different steps: system understanding, system architecture, implementation and testing/validation (see ◘ Fig. 3.10 below). For further details regarding the V-Model and how you can use it for project management see [FKS+09].

In the first phase, the so-called system analysis phase, it is essential to understand the system to specify the final system as detailed as possible. Having a user perspective and being able to answer questions such as for what, how, when, why, and under what conditions the future user will work with the system is essential in this phase. A sophisticated specification of the system can only be developed once these questions are answered completely and hence a deep system understanding is reached. For this purpose, it makes sense to break the system down into different functional units / subsystems (components), which are defined very precisely in the specification.

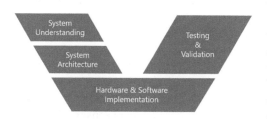

◘ Fig. 3.10 Simplified V-Model Source: ITK Engineering

Based on the precise specifications, the system architecture is derived. The central questions during this task are what software and hardware are needed to meet the criteria specified before. Moreover, the different interfaces between the hardware and the software components need to be specified. At the end of this phase, the whole system should be ready to be implemented.

The implementation is the next step. Once the different components have been implemented, the testing begins, starting with the single hardware/software modules. The next step is the integration testing. In this context, the single components are integrated and tested to check for instance problems regarding the interaction between the different components. In the last part of the testing, the whole new system is validated.

There are obviously strong interactions between the different development steps not only in the order in which they are described here, but also between steps on the left and right side. For instance, those who write down the specifications need to consider that these specifications must be testable in a coherent way.

3.3.3 Challenges in Safe Surround Sensing

Each of the steps described above is associated with great challenges when it comes to the development of a safe object detection system for the off-road terrain. In the following part of this section, these challenges will be described.

In line with the V-Model, we assume that the desired product would be a safe object detection system which can be integrated into different types of machines. Then the definition of safe in the context of the variety of use cases of different machines would be the first challenge. How can this be further specified? As noted by [DFH+19], there is a plethora of different regulations out there, but not a coherent definition/guideline being applicable in this context. This makes it extremely difficult to develop a safe system.

Another quite significant challenge is to specify all the different (environmental) conditions surrounding the system. Things like different weather conditions, different terrains, different growth phases of plants, different disturbing objects (e.g., animals, bushes, pedestrians) make the specification task a quite challenging one. This variability of the surrounding environment, leading to an immense number of different cases, is certainly a significant challenge, which interferes with the development of a safe object detection at different stages of the V-Model.

Assuming an adequate specification for the topics above has been achieved, when it comes to the system architecture the next challenge is already ahead. Especially by focusing on the agricultural context it is quite hard to find the components which would be necessarily needed to build up a safe system. As discussed above, we are facing an environment with a high variability. Based on this, we do have demanding requirements for the system as it must work under (very) rough environmental conditions, such as dust, frost, vibration and in a landscape with great varieties (different soil types, different crops, different slopes, different disturbing factors).

Facing the technical dimension of the problem, it is worth investigating the different technologies that can be used for safe object detection: stereo camera, LiDAR, and Radar. A stereo system consists of two cameras, arranged next to each other at a fixed distance. Due to the slightly different perspectives, depth information can be calculated. For this, both images are horizontally aligned, and corresponding image points are searched. The depth information is obtained by the difference in distance of the corresponding image points [KVR+09].

LiDAR is based on the emission and reception of light. Laser beams are emitted instead of diffuse light. The distance is calculated based on the elapsed time between emitting and receiving the beam. Two types can be distinguished: (1) mechanical LiDAR uses rotating parts for beam steering and (2) solid-state LiDAR, which uses techniques without mechanical elements for beam steering (e.g., an optical phased array) [PYC+17]. The radar sensor emits radio waves and receives the reflected signals. Similar to LiDAR, there are two different techniques for measuring distance: (1) pulse radar measures the time of flight of a radio wave pulse and (2) the frequency-modulated continuous wave (FMCW) radar measures distance by analyzing the phase shifts. Frequency modulation is used to cope with ambiguities. Relative motion of a target causes a Doppler shift in the frequency of the transmitted radar waves, which can be used to measure the relative target speed [CDC+17].

Based on the different principles how these technologies function, there are differences in the usage (advantages) of the different sensor types. We included ◘ Table 3.3, to underline that there are many points to be considered. However, discussing each of them would not be feasible in this context. Yet, the clear take away is that it is a complex topic and that most likely there will be more than one sensor-type needed to establish a safe object detection.

Once we accept that it is most likely that we need more than one sensor type to design a safe object detection, we must accept that we will have different (data) outputs from different sources. In order to fuse these different data outputs into one concrete recommendation, e.g., "no object near the machine", a sensor fusion has to happen, which we here define in line with [Elm02] as "the combination of sensory data or data derived from sensory data such that the resulting information is in some

Technology Perspective

■ Table 3.3 Assessment of different sensor types; source: [Eym19]

Specification		Camera	Radar	LiDAR
Distance	Range	Good	Very good	Very good
	Resolution	Good	Very good	Good
Angle	Range	Very good	Good	Very good
	Resolution	Very good	Moderate	Good
Classification	Velocity resolution	Moderate	Very good	Good
	Object categorization	Very good	Moderate	Good
Environment	Night	Moderate	Very good	Very Good
	Rain/Clouded	Moderate	Very good	Good

sense better than would be possible when these sources were used individually".

However, when a sensor fusion is required, the control device processing this information must have considerable processing power. This requirement, combined with the rough environmental conditions, leads to a challenge, that is more a general problem of AgMachinery manufactures, namely the relatively small piece numbers which are needed by the manufacturers in comparison with automotive quantities, resulting in the problem that it is difficult to get such sophisticated components (like sensors or control devices) for an economically feasible price.

Focusing on the sensor fusion, another challenge is coming from the software perspective. As described by [FHSR+20], there are three challenges when it comes to sensor fusion:
- "What to fuse: What sensing modalities should be fused, and how to represent and process them in an appropriate way?
- How to fuse: what fusion operations should be utilized?
- When to fuse: at which stage of feature representation in a neural network?".

Another extremely critical point related to the missing experience related to autonomous driving is the lack of training data which can be used to train the whole system (in case one wants to work with Artificial Intelligence, which is more or less state of the art). [FHSR+20] conclude that there are considerable limitations regarding the data diversity (e.g., too few data sets) and regarding the data quality (e.g., spatial and temporal misalignments of different sensors).

Besides these technological problems, there are economic problems linked to them. It is very likely that the safety problem discussed already in the specification section will continue and might even evolve further. Especially as the technology turns out to be more and more complex (and costly), the trade-off between finding an economically feasible, functioning system and assuring safety is getting more and more difficult.

Regarding the implementation of the system, we do have to consider that we are working with the rhythm of nature and do have a very seasonal influenced development cycle. Hence, engineers sometimes have only a very short (not easily movable) timeframe to implement a system out-

Table 3.4 Challenges during the different phases of the development process; source: own creation

Development process	Challenge
System specification	How to operationalize safety
System specification	How to operationalize all the different environmental conditions
System architecture	Economical feasible component selection
System architecture	Sensor fusion
System architecture	Trainings Data availability and quality
Implementation	Seasonality
Testing	Availability of testing data
Testing	Seasonality

side at the field. Of course, one could argue that due to the climatic differences between countries there might be another window of opportunity to implement a system in another country slightly later—this is certainly valid. However, it must be noted that this is related to quite some logistic challenges (e.g., moving big equipment over considerable distances), which means higher development costs.

When it comes to the testing phase, we do see a comeback of several challenges indicated already in the development phase. The first one to be mentioned is the lack of data, which can be used to test against the development of the system. As already discussed above, there is hardly any training data in sufficient quality and quantity available. Of course, when we want to test different components we can use the same dataset, but it is also a necessity to test the system with different data to see whether the components/system is behaving in the way it is designed to.

Another (already discussed) challenge is the seasonality, resulting also in challenges regarding the system testing. Again, the possible timeslots for working outside in the field are relatively short, making testing a considerable challenge.

Summarizing this section, there is a considerable long list of different challenges (see Table 3.4). However, when going through the reasons behind these challenges, we do have quite some similarities which can be grouped: high complexity, high (development) costs, seasonality, and the unclear legal situation. These similarities give us hope for the next section, in which we will describe solutions to overcome these problems.

3.3.4 Solutions for Safe Object Detection

After having described the different challenges, it is now the right moment to focus on possible solutions. The question mark regarding operationalizing safety is a big task. Nevertheless, as shown by [JWS19], there are ways to deal with the complexity. The authors applied the IEC TS 62998-1:2019 [IEC19] for safety-related sensors designated for outdoor use to design an architecture for a reliable object detection for autonomous mobile machines. In this context, they established a process similar to the process of the V-Model discussed above (see Fig. 3.10) [JWS19]. There are two points to be highlighted: the first is the iterative design of the process, phasing in the lessons learned from the prototyping back into the system architecture. By doing this,

Table 3.5 Challenges and their possible solutions during the different phases of the development process; source: own creation

Development phase	Challenge	Solution
Specification	How to operationalize safety	Applying related norms Make safety an inherent part of the development process Cooperation with the professional (insurance) association
Specification	How to operationalize all the different environmental conditions	Applying worst cases
System architecture	Economical feasible component selection	Usage of Simulation
System Architecture	Sensor fusion	Usage of Simulation
System Architecture	Trainings Data availability and quality	Usage of AI
Implementation	Seasonality	Usage of Simulation
Testing	Availability of testing data	Usage of AI
Testing	Seasonality	Usage of Simulation

one can be assured that the fast feedback will save time and hence money.

The other notable point here is that they implemented safety measures at every step, making the safety assessment an integral part of the whole development process (see Fig. 3.11), combining functions and norms. Addressing complex topics such as safety from the beginning onwards throughout the whole development phases is a wise decision as it helps to minimize the risks of discovering unintended safety problems during the development or, even worse, once the product is released. A good example of such a safety measure would be the integration of a runtime verification scheme. This is a safety measure, which is constantly "checking for the correctness, i.e., the correct implementation of a given specification, of the system" [FSA17]. Once this safety guard detects an unspecified behavior of the system, processes start to push the system into a safe state.

Besides the IEC TS 62998-1:2019 norm, another valuable source of important information can be found in the DIN EN ISO 18497 [DIN19] on AgMachinery and tractors—Safety of highly automated agricultural machines—Principles for design. Especially Sect. 5 is an interesting source of information as it specifies the requirements for a test obstacle. In Fig. 3.12 (Dimensions in millimeters), you will find more specified information on the dimensions. Moreover, according to the DIN EN ISO 18497 the test obstacle shall be filled with water to represent the composition of the human body, material must be plastic, e.g., polyethylene with a matt surface and the color must be olive green also with a matt surface.

Another valuable way to do deal with the problem of operationalizing safety is to work in close cooperation with a professional (insurance) association. In Germany, for instance, there is the Institute for Occupational Safety and Health of the German Social Accident Insurance (IFA) which can assist and approve developments.

When focusing on the multitude of different environmental conditions, a possible solution is to focus on the worst-case conditions [JWS19]. To do this in a comprehensive way, objects or situations are grouped

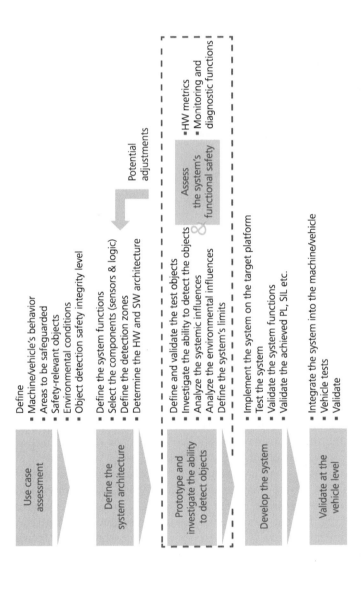

Fig. 3.11 Example of a safe object detection process; source: ITK Engineering

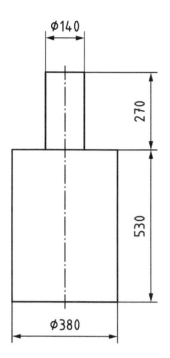

Fig. 3.12 Dimensions in millimeters of test object; source: [DIN19]

By transferring the reality into the virtual world, we can make (quick) progress at different phases of the development. In the specification phase, it is possible to define the system in a digital environment and simulate the outcomes. The results of this simulation can then be used to improve/validate the specification. Also, the problems related to testing can be mitigated, as a test algorithm in an early stage with synthetic data can be generated, reducing time and effort. This idea is in line with what has been discussed above, regarding the iteration between prototyping and the actual system.

Especially when it comes to building up a complex system, selecting (and comparing) different sensor setups at different positions with different fusion technologies, a sophisticated digital twin can save time and financial resources as it allows a faster feedback. For instance, as shown in Fig. 3.13, Raytracing technology is used to simulate LiDAR, radar, and ultrasound dispersal in such a way that the emitted signal is abstracted via geometrical beams and their dispersal in the 3D environment is calculated to the point where their reflections bounce back to the sensor [JWS19].

The idea of transferring the real world into the virtual world is typically associated with the use of artificial intelligence (AI). There is a wide debate on the definition of AI. [KBK+09] illustrate this with the example of four different definitions which can be found in an English dictionary. We do not want to be too involved in this debate but understand AI in this special context as an instrument developed to solve a problem by being able to learn. For systems that require visual perception systems, AI algorithms and specifically deep neural networks are the state of the art. These systems learn from data and can hence achieve tasks for which an explicit specification is unfeasible (open context).

based on their similar characteristics. Then, out of this group the most difficult object/situation needs to be identified. Next to that, the sensors are also tested under the least favorite conditions (e.g., LiDAR sensors must detect diffuse reflective objects with just 1.8% reflectance). If the system works under these difficult conditions, it can be assumed that it works also in less extreme scenarios.

Regarding the challenges of high development costs, high complexity, and the seasonality, one possible solution is the usage of digital twins. Seeing that there is no universal definition of digital twins, we do understand a digital twin in line with [GS12] as "an integrated multiphysics, multiscale, probabilistic simulation of an as-built vehicle or system that uses the best available physical models, sensor updates, fleet history, etc., to mirror the life of its corresponding [...] twin".

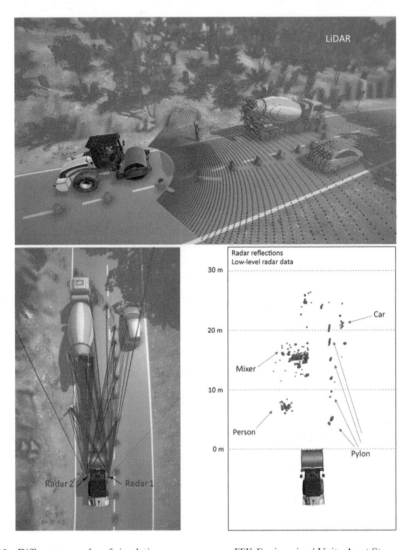

Fig. 3.13 Different examples of simulating sensors; source: ITK Engineering | Unity Asset Store

Yet, as mentioned above, an AI algorithm for learning from data requires a great amount of (different) training data. To increase this amount of training data, data augmentation can be used, which we describe in line with [CZM+19] as an "effective technique to increase both the amount and diversity of data by randomly "augmenting" it". An example of such a data augmentation would be style transfer, which can be defined as technique in which an "image of a certain domain A (..) is translated into the style of a different domain B" [JHS+18]. Such a style transfer can be used, for instance, to transform an image created on a sunny day to an image under rainy conditions, having more and more variation of data. However, assuring safety for these complex AI systems is an unsolved problem, as there is no standard or regulation for using AI yet.

3.3.5 Conclusion and Outlook

In the previous sections, we have discussed the different challenges and possible solutions regarding safe object detection along the V-Model. Thereby we have identified several challenges which need to be addressed. Fortunately, many of the challenges identified in this section have similar root causes, among others considerable complexity, high development costs, and the difficulties regarding seasonality. We have shown that there needs to be a mixture of different methods and techniques to address these challenges.

We have highlighted that safety thinking and connected to that safety measures should be integrated as cornerstone of the development. To do this, it makes sense to consider IEC TS 62998-1:2019 and DIN EN ISO 18497, to work closely with the employers' liability insurance association, to think (and test) worst-case scenarios, to use the digital twin, and to use AI, for example, to capture more data. These are demanding methods and technologies. Using them to solve the challenges demands interdisciplinary teams. Of course, managing such interdisciplinary teams/projects is another considerable challenge that needs to be considered.

How will this field further evolve? As indicated above, the number of interdisciplinary teams will certainly increase. But besides that? As discussed above, the hardware requirements are very demanding whereas at the same time the number of units needed is relatively small (in comparison to the automotive sector), leading toward high hardware costs. It would be no surprise to us if more and more software will be used in the future to replace (new) hardware where possible. However, we see the automotive industry is taking action and adjusting its organizational set-up in line with the growing importance of software (e.g., Car.Software.org by Volkswagen) and anticipate similar developments for the agricultural sector as well. This in turn means that the software and hardware, which is currently under development, require a solid foundation, so that new features can be added in the future.

Regarding the future of safe object detection, it is difficult to predict how it will further evolve. Partly due to legal issues and the manufacturers' economic calculations, it might still take a while until we encounter a tractor operating autonomously in the field with a safe environment detection. To state it a bit catchier: based on what we have discussed above regarding simulation, we will first see a digital automation of development processes, then we will see an automation at the field level.

3.4 Interoperability and Ecosystems

Sjaak Wolfert

Abstract

The digital transformation of agriculture is taking place in a System of Systems (SoS) context in which farm management is supported by a plethora of hardware devices and software systems from various vendors that seamlessly have to co-operate with each other. This co-operation between systems is often referred to as interoperability and can be defined at different levels: from hardware, data to applications and business processes. Poor interoperability hampers adoption and value creation in supply chains. To improve interoperability, it is important to develop and foster Integrated Open Platforms based on a reference architecture with Minimal Interoperability Mechanisms, reusable components, semantic interoperability and service- and data monetization. Technical development should go hand-in-hand with ecosystem development, embedded in a lean multi-actor use case approach in which structural organizations (e.g., standardization bodies, farmer's organizations) play a crucial role in promoting and sustaining interoperability. In the future, Digital Innovation Hubs at a regional and local level can enhance interoperability.

3.4.1 Introduction

The digital transformation of agriculture is an ongoing development in which farmers and related farm workers are challenged by a plethora of digital systems in all kinds of forms ranging from small separate sensors for temperature, humidity, etc. to complex GPS-based mapping and application systems that are often integrated in existing equipment such as tractors or sprayers (see ▶ Sect. 3.2). Although the number of such devices and systems is overwhelming, it is generally acknowledged that the adoption rate is still lagging behind its true potential (see ▶ Sect. 1.5). One of the main barriers—often perceived as complaints by farmers—is that various devices and systems are poorly co-operating with each other. Porter and Heppelmann [PH14] nicely describe this development by the **Systems-of-Systems (SoS)** concept. This means that nowadays in virtually all industries and sectors digital systems can be considered as part of a larger system and have to be compatible with other systems and seamlessly co-operate with each other as it if was one system. This has several implications for suppliers of subsystems and to be competitive, interoperability has become a key asset.

In ICT terms, seamless co-operation between different systems is referred to as **interoperability**. Interoperability can be defined at four different, interdependent integration levels [Gia04, WVV+10]. First, hardware must be enabled to connect with each other either physically (e.g., by a plug) or remotely (e.g., by Bluetooth). Second, data sharing should be possible by common data definitions and standards, also (semantic interoperability). Third, software applications which are often put on top of digital devices and systems should be aligned so that different applications can work together as if it was one aligned system (see SoS). Finally, it is important to realize that all these devices and systems should facilitate complete business processes at the farm, so it helps if these processes are defined in a standardized manner. It is this final process level where the Systems-of-Systems approach becomes important. In a farm management context, this implies that a farmer is doing his daily business integrating various systems ranging from hard technical systems on a machine to market information systems, weather information, human resources, customer relations, etc. For example, when spraying a crop, market information is required to know whether certain chemicals can be applied or not and how. Weather conditions must be checked and sometimes have to be recorded for the sake of documentation and certification procedures. At larger farms, this task has to be appointed to a farm worker or an external contractor. The chemical has to be bought from a supplier. In the past, various independent systems were used to support all these processes, but to increase the adoption rate and acceptance by farmers and from a competitive point of view, it is important that these systems are co-operating in a Systems-of-Systems setting in which interoperability at all levels is improved.

The example above illustrates how different systems, often from different, multinational vendors are involved in the farm management process. Ideally, all these vendors and systems should be tuned to each other at all levels of interoperability. This is actually happening in various ways, amongst others in common standardization organizations (e.g., AEF [AEF21], AgGateway [AgG21], GS1 [GS121]) and other alliances (e.g., AIOTI [AIO21]). However, this involves complex, technical negotiation processes in which competitive advantages sometimes work counterproductive. Hence, there is a common plead for development and fostering of vendor-independent **Integrated Open Platforms** [KWS+16].

The objective of this section is to describe how such IOPs are currently taking shape and which interoperability mechanisms and standards are playing a role. The success of these IOPs relies on the critical mass of systems and their end-users that will adopt them requiring a development of **ecosystems** of developers, users and all kinds of other related actors. First, a reference architecture for IOPs in digital agriculture will be described including current key components for interoperability. Then an approach for related ecosystem development will be proposed that is currently being deployed in ongoing projects. Finally, some recommendations for future development will be provided.

3.4.2 A Reference Architecture for Integrated Open Platforms and Key Components for Interoperability

As described in the previous section, digital systems for farm management support involve many different processes, subsystems, applications, data, and devices. In the end, every task requires a specific piece of software, specific devices, data, etc., and even every different farm context requires tailored software. However, many underlying layers of a final piece of software are not specific at all, but generic for all kind of different purposes and sectors. For example, a sensor measuring the temperature in a field is basically the same as for measuring the temperature in an office or factory. Or a geographical coordinate in a field is not different from a coordinate in a city. Hence, it would be very inefficient to develop a specific standard for all these different purposes for agriculture as it will be the same for many other sectors. This was the reason for the European Commission in 2010 to start the Future Internet Public–Private Partnership (FI-PPP) in which various sectors were challenged to build applications on a common software architecture [FIP21]. Agriculture was also part of this programme, especially represented by the SmartAgriFood and FIspace project [VWB+16, WSG14, KGS+14]. Currently, the IoF2020 project, which focusses particularly on the Internet of Things (IoT) in farming and food, is building on the results of the previous projects [VWB+17]. To promote interoperability between various systems in agriculture, using generic components that are also used in other sectors, IoF2020 has developed a reference architecture for Integrated Open Platforms as depicted in ◘ Fig. 3.14.

The reference architecture leverages the IoT-A Architectural Reference Model [BBB+13], basically reflecting the four interoperability integration layers as described in the introduction. The main layers from bottom to top are:

- **Physical device layer:** Different IoT devices and AgMachinery deployed in the field, that are capable of sensing their environment and generating data of interest for digital farming applications.
- **Connectivity layer:** Enables two-way transmission of the data produced by devices between these layers.
- **IoT service layer:** Exposes the raw data generated from devices to upper layers in the architecture through different application-level transport protocols based on different paradigms (publish/subscribe, request/response, etc.). In addition, it offers interfaces that allow to communicate with devices for management or actuation purposes.
- **Mediation layer:** Transforms raw data coming from devices or other external services, into curated, harmonized, and

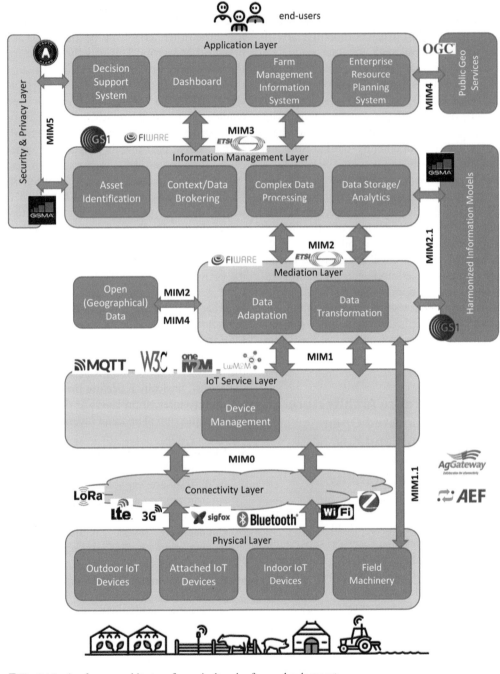

Fig. 3.14 A reference architecture for agricultural software development

possibly aggregated data that can be exposed to data processing algorithms or analytics. This layer is also capable of sending actuation commands to the IoT Service Layer.
- **Information management layer:** The main component of this layer is usually a data hub (which could be a context broker) which enables the publication, consumption, and subscription of all the information relevant to a digital farming solution. The information present at this layer, which can be current or historical, may have been aggregated from different sources, not only IoT. In addition, this layer may offer complex event processing, storage, or analytics services, which can generate insights, prescriptions, or predictions.
- **Application layer:** Contains different digital smart farming applications that could be used by stakeholders, particularly farming professionals. They include, but are not limited to, systems related to decision support (DSS), farm management (FMIS), dashboards or enterprise resource planning (ERP systems).

A cross-cutting layer on **Security and Privacy** is defined, aimed at guaranteeing secure access to information and devices, while respecting the privacy of farmers and exploitations. In addition, other external entities play a relevant role, namely:
- **Open Data Providers:** Represented, for instance, by databases offering open data in the agricultural domain (pests, disease, weather historical data, etc.) or services publishing certain contextual data such as weather forecasts, weather alerts or weather observations. Satellite data/image publication platforms or geo-services, which provide geospatial data, are also under this scope.
- **Harmonized Information Models:** Define the structure and representation of the information to be managed.
- **Public Geo Services:** Offering public geospatial data related to agricultural assets (for instance fields), frequently coming from geo-information systems owned by public authorities.

Minimum Interoperability Mechanisms (MIMs) are representing a minimal, common ground that shall help developers as well as end-users to learn about the current state of the art when aiming at the digitization of agri-food processes as well as facilitating openness and access to data and services. It is expected that if developers comply with these minimum requirements, it will increase the openness of solutions, foster competition, and encourage solution providers to focus on their unique selling propositions, increasing quality, decreasing time to market and finally also aiming at a better cost–benefit ratio for both solution providers as well as end-users. The MIMs identified in this reference architecture and depicted in ◘ Fig. 3.14 are:
- **MIM0:** Connectivity enabler for IoT Devices and AgMachinery. Multiple communications technologies can be considered as its basis, including traditional wireless short range (Wi-Fi, Bluetooth, IEEE 802.15.4, etc.), Machine to Machine (M2M) powered by global telco networks (3G/4G/5G), or long range IoT networks specifically designed for IoT (LPWA).
- **MIM1:** Enabling the exposition of the data and services offered by IoT Devices through well-known programmatic interfaces. MQTT, OMA Lightweight M2M, oneM2M and W3C Web of Things are the main technology enablers available in the industry today.
- **MIM1.1:** Enabling bi-directional transmission of data between AgMachinery and the upper layers that deal with information management. ISOBUS, ADAPT, EFDI and ETSI NGSI-LD are key enabling and emerging technologies to realize this MIM.

- **MIM2:** Enables (i) the transformation, aggregation, harmonization and publication, as context information, of harmonized data coming from IoT Devices, AgMachinery or other sources of information (open data portals, web services providing contextual data, etc.) and (ii) exposes a unified way to send commands and to mediate with IoT Devices or AgMachinery, regardless the interface exposed by the IoT Service Layer or the physical machinery. FI-WARE, NGSI-LD and GS1 are key enabling technologies here.
- **MIM2.1:** Main enablers of this MIM are Harmonized Information Models that allow to publish digital farming information following the same meta-model, data representation formats and conventions (units of measurement, etc.). This is key when it comes to portability of solutions at the data layer. FI-WARE Data Models (a superset of the GSMA Harmonized Data Models) and GS1 are specifications of common information models.
- **MIM3:** Provide access to all the data of interest to digital farming applications, including, but not limited to, real time data, historical data or analytics results. In addition, it allows the subscription to data changes and to publish new data coming from the application layer. FIWARE, ETSI NGSI-LD and GS1 are under this scope.
- **MIM4:** Enables the Application and Mediation Layers to consume public Geo-Services offering open geospatial data, enriching the digital farming applications with geospatial data and off-the-shelf visualizations. OGC WFS and WMS play an important role here.
- **MIM5:** A cross-cutting interoperability point that facilitates the secure interchange of information between the different layers and actors. The GSMA IoT Security Guidelines and the traditional security technology stacks (TLS, DTLS, PKI, etc.) or protocols (particularly OAuth2) are under this scope.

A more detailed description of this reference architecture and MIMs can be found in several deliverables of the IoF2020 project providing a detailed step-by-step approach how it can be implemented [VST+19, CIV+18]. More technical descriptions can be found via the IoF2020 website [IOF21] and GitHub, including reference examples of implementations in various agricultural sectors [Git21].

This reference architecture and several implementations of its layers and components should be considered as a vast basis for the development of digital systems and software in agriculture that will improve interoperability between systems at all relevant levels. However, this requires that the architecture is really supported and used by developers of these systems and that it is constantly updated to cater for new innovations and developments. This is not just a technical issue but involves organizational challenges including several stakeholder groups. The next section will describe how this challenge can be approached.

3.4.3 A Lean, Multi-Actor Approach for Ecosystem Development

Based on experiences in various projects a lean, multi-actor approach was developed that is characterized by use cases representing a mix of agricultural sectors but also specific processes (e.g., fertilizing, crop protection) [VWB+17, VST+19]. Each use case involves all relevant actors from the whole value chain, including system developers and end-users. While every use case is developing its own specific digital solution, it is searched for synergies between the different use cases to promote reuse of standards, components, etc., that come together in the reference architecture. This approach is visualized in ◘ Fig. 3.15.

Each use case solution has an architecture that is an instance of the reference architecture as described in the previous section. It involves several IoT components that are systematically brought together in the IoT catalogue [IoT21]. This catalogue serves as a basis for component reuse by other use cases. Implementations of the use case IoT systems are carried out within a lab environment so that configurations and instances of systems and components become available as references and reuse by others. Finally, the IoT systems are deployed resulting in services and data becoming available for reuse by other use cases for other purposes. For example, if one use cases has developed a service to collect detailed weather information at a specific site, another use case that needs the same data could reuse this service and doesn't have to invent this wheel again. At the same time, this will improve interoperability between systems and in practice it sometimes also means that end-users such as farmers do not have to pay twice for the same underlying services.

Although ◘ Fig. 3.15 and its description suggests a linear approach, in reality a much more iterative approach is followed, taken from the current trend of lean start-up methodology based on minimum viable products [Rie11]. In that approach, it is key that intermediate results are tested and discussed by all involved actors. Therefore, it is called the lean multi-actor approach. The bottom of ◘ Fig. 3.15 indicates the involvement of various other projects and organizations. As already described in the previous section, the reference architecture is also not re-invented from scratch, but builds on existing ones. At the same time, projects are temporary constructions by definition so it is important that results are sustained by more structural (global) organizations such as AgGateway, GS1, and FIWARE. They can also strongly promote reuse of components toward a larger user community.

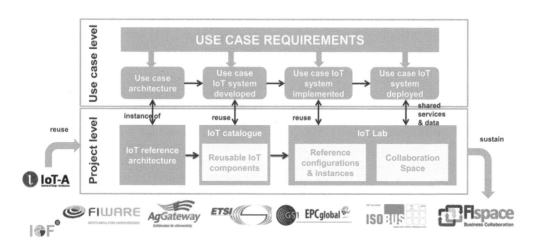

◘ **Fig. 3.15** The lean-multi-actor approach based on use cases ensuring reuse and interoperability at project level

3.4.4 Conclusions and Future Development

The digital transformation of agriculture is taking place in a System-of-Systems context in which adoption and value creation depends on the seamless interoperability between various digital solutions. To facilitate this, it is important to develop and foster integrated open platforms based on a reference architecture with minimal interoperability mechanisms, reusable components, semantic interoperability and service- and data monetization. Ecosystem development should go hand-in-hand with technical development, embedded in a lean multi-actor use-case approach in which structural organizations (e.g., standardization bodies, farmer's organizations) play a crucial role in promoting and sustaining interoperability.

The projects behind this approach have developed a vast ecosystem of various actors ranging from developers, intermediates and end-users such as agronomists and farmers. Some of them have a global scope, especially the mentioned standardization organizations and associations and thus potentially a large impact. Still this ecosystem that is developed by large, centrally led projects are only a fraction of the potential ecosystem.

At the technical side, new European projects are leveraging the interoperability architecture and infrastructure. The ATLAS project is building an open, distributed, and extensible data Interoperability Network, based on a micro-service architecture, which will offer a high level of scalability from a single farm to a global community [ATL21]. The DEMETER project is focusing on interoperability as the main digital enabler, extending the coverage of interoperability across data, services, platforms, M2M communication, and online intelligence but also human knowledge, and the implementation of interoperability by connecting farmers, advisors, and providers of ICT solutions and machinery [DEM21]. With regard to data, several other initiatives can be mentioned. Most of them have a multiple domain objective with specific focus or working groups on agriculture. The International Data Space Association [IDS21] aims to guarantee data sovereignty by an open, vendor-independent architecture for a peer-to-peer network, which provides usage control of data from all domains. In Germany, the Fraunhofer institute tries to develop a specific Agricultural Data Space [ADS21], to make available all the data that is necessary for making decisions in agriculture. In the GAIA-X project [GAI21], representatives from politics, business and science from France and Germany, together with other European partners, create a proposal for the next generation of a data infrastructure for Europe: a secure, federated system that meets the highest standards of digital sovereignty while promoting innovation.

At the innovation ecosystem side, the concept of Digital Innovation Hubs (DIHs) is currently rising in various domains and sectors, similar to the earlier Future Internet programme by the European Commission [RS18]. A DIH refers to a local or regional ecosystem through which any business can get access to the latest knowledge, expertise and technology to test and experiment with digital technology relevant to its products, processes, or business models. A DIH also provides the connections with investors, facilitates access to financing and helps to connect users and suppliers of digital solutions across the value chain. The recently started EU-funded project SmartAgriHubs [SAH21] tries to set up and foster a European-wide network of DIHs for agriculture, to enhance the Digital Transformation for Sustainable Farming and Food Production [WMB19]. Concurrently, a vast network of competence centers (CCs) is being developed that are connected to one or more DIH(s), which provides an opportunity to promote the reference architecture and standards for interoperability for Digital Farming. It is expected that through this ecosystem development representing a large part of the targeted user community—especially at a local, regional level—interoperability between various digital systems will be promoted and improved.

■■ Acknowledgements

The contents of this section are largely based on the work that is done in work package 3 of the project 'Internet of Food and Farm 2020' (IoF2020) supported by the European Union's Horizon 2020 research and innovation programme under grant agreement no. 731884.

3.5 Artificial Intelligence

Ansgar Bernardi and Ralph Traphöner

Abstract

As a computer science discipline, AI works on systems which interact with their environment, process complex information, draw non-trivial decisions from data, and pursue useful goals with a certain degree of autonomy. AI technology contributes to Digital Farming by offering intelligent assistance in a complex environment, comprising environmental observation, flexible data sharing and cooperation, decision support, machine learning solutions, and intelligent robotics application. Ongoing developments promise situation-aware and individualized solutions, high degrees of autonomy, and data-driven insights and optimizations.

3.5.1 AI in the Agriculture Context

The wide application field "agriculture" poses numerous challenges for AI solutions: plant production in open fields or indoors and animal farming on pasture or in barns ask for AI support, ranging from location-specific and individual diagnostics and activity control to farm-wide or cross-regional monitoring, strategic planning and recommendations. Plant production in open fields covers wide areas and heavy machinery for, e.g., hay, grain crops, potatoes, or beets. Wine and fruit production pose special and individualized requirements. Vegetables show quick turna-

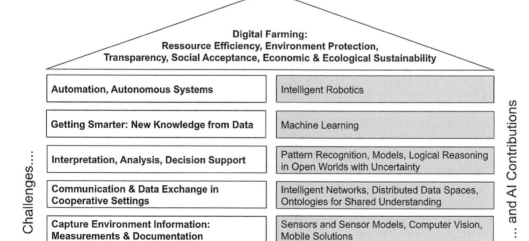

◘ Fig. 3.16 AI contributions toward Digital Farming

round cycles and the transgression toward in-house/greenhouse production modes; short-rotation plantations mark the transition to forestry. In animal farming, on the other hand, the need for careful considerations of the individual is obvious for both ethical and economic reason.

Furthermore, agriculture is characterized by strong cooperative settings in farm operation (with providers of seeding material and agro-chemistry, consulting services, and all kinds of contracted services) and the subsequent product processing which results in both, food and bio raw materials. Supporting relevant integration, communication, logistics, and economical processes request an additional AI support.

In summary, AI in agriculture pursues intelligent assistance in a complex environment comparable to modern industrial production settings, but with the additional challenge resulting from dealing with living creatures and the full dynamics of nature.

3.5.2 Capturing the Environment

Any IT support in agriculture is faced with the need to obtain valid input data about the tasks and objects under consideration. Capturing information about the environment via technical sensors is often the first step toward AI support in agriculture.

Building the sensors themselves is not the task of AI—physics and engineering know-how realize the technical gadgets to capture numerous signals from the agricultural world and transform them into IT-processable electric signals which serve as input to AI systems. Widely used sensors in agriculture nowadays offer data about soil and weather, provide positional information, and gather optical images even beyond the visible light spectrum—see ▶ Sects. 3.2 and 3.3 for more details.

AI comes into play as soon as we start to interpret the signals received from sensors, build the respective models of the environment, and start to trigger actions based on our insights.

Relevant AI technology starts with the semantic modeling of individual sensors: Formalisms like SensorML [OGC21] enable detailed descriptions of type, quality and characteristics of a sensor which lays the basis for reliable interpretation by subsequent IT systems. **Pattern Recognition** and **Computer Vision** are widely applied AI technologies to process sensor data. Sensors can be used to directly control precision farming operations or enable multiple analysis tasks up to mobile "labs" which measure chemical compositions of crops or manure based on near-infrared (NIR) signals.

Besides the operation-oriented applications based on sensor modeling and pattern recognition from sensor data, automated documentation of the measurements taken is an important agricultural application in itself which eases documentation tasks, helps to fulfill governmental regulations, and contributes to much-required transparency in many ways.

3.5.3 Data Exchange and Shared Understanding

Data acquired by various sensors and systems in the field, the barn or elsewhere during agricultural work processes will be valuable only if humans and/or technical systems will use such data to get insights and to act in some meaningful and useful way. As modern agriculture involves the cooperation of many participants, the first challenge after data acquisition is the communication, exchange, and mutual understanding among relevant, cooperating partners in agricultural work settings. The traditional approach to reach such shared understanding is standardization; relevant examples are ISOBUS, AGROVOC, or INSPIRE.

Beyond standards (which are expensive to establish), the rich AI tradition of knowledge representation and reasoning facilitates automatic solutions where logic-based formalisms open up a new way to handle disperse data formats and vocabularies: Once participants in a data exchange are able to formally define their respective concepts and terms, computer algorithms can automatically decide whether some data represents an instance of a particular concept and can calculate whether different expressions have identical meaning. Thus, the mapping between data from different sources can be realized automatically.

The necessary tools are well-known and already widely adopted in the worldwide web as we know it, see in particular the Resource Description Framework RDF [SR14], RDF Schema language RDFS [BG14] for object definitions, Web Ontology Language OWL [HKP+12] with higher logical expressive power, or the Simple Knowledge Organization System SKOS [IS09] which is used for AGROVOC. By using such **Semantic Technologies**, current activities combine the various approaches into widely usable meta models which shall effectively support the automatic interoperation and allow to provide many relevant resources (like e.g., lists of approved pesticides or fertilizers available). The most prominent activity toward this end is the development of the standardized Agricultural Data Space [ADS19], see also ▶ Sect. 3.4.

3.5.4 Interpretation, Analysis, and Decision Support

Utilizing AI in farming aims at increased yields, better use of resources, e.g., water, optimal nutrition, minimal application of plant protecting agents and many more. Working toward these objectives requires numerous decisions to be taken by the farmer every day. Here the role of AI is to augment the human decision making and to automate routine tasks.

If, for example, the agricultural crop land is scanned regularly by a drone, then the collected image data must be processed to identify areas with increased weed or plants with diseases. This cannot be done manually by the farmer. We instead use AI to classify every square meter of land for weed and plant diseases. Based on those findings, appropriate treatments will be proposed and evaluated with respect to sustainability. Such proposals may consider further data such as soil composition or weather forecasts.

In the above example, the AI system tackles three sub-tasks. It first interprets data to detect situations that require action and then analyzes those situations to determine its nature. Finally, the latter plus related additional information are the basis for evaluation to propose appropriate action and to predict its possible result. Hence, decision support is divided into identifying the need for a decision, classifying it to propose a decision and then evaluate to predict the outcomes.

AI offers five general approaches to implement these steps, i.e., connectionism, symbolism, Bayesian, analogy, and evolutionism [Dom15]. Although the biggest distinction between these approaches is the underlying learning paradigm (see Sect. 3.5.5: Getting Smarter: Machine Learning), we briefly put them into the context of the decision support process.

Connectionism mimics the function of natural nervous systems and brains. These so-called artificial neural networks (ANNs) are well suited for the identification and classification steps in the decision support process, especially in situations where sensory perception abilities are required. The xarvio Scouting App [Xar20] is an example application that provides image analysis to identify weeds, classify insects, recognize diseases and more.

Symbolism, i.e., symbolic AI, together with the connectionist approach, is one of the oldest AI technologies dating back to the 1950s [NS56]. It applies the principles of **logic** where facts and rules express explicit knowledge. Deduction derives new facts from known facts by applying rules and induction derives new rules from facts. Formerly, rule-based expert systems were the most widely known representatives of this kind. Nowadays, they are semantic technologies, and their knowledge representation is known as **Knowledge Graphs**. The application of symbolic AI in the decision support process is for classification and evaluation. In case of structured or semi-structured data, e.g., database records, so-called decision trees serve the classification purpose well. Evaluation is often carried out best with symbolic AI when explicit knowledge must be applied.

Bayesian systems apply conditional probabilities to determine decisions, e.g., if there is a set of symptoms then such a system calculates the probabilities of the presence of diseases. Therefore, estimates of the probability of a disease under the condition that a specific symptom is present are required. There are also machine learning approaches available to acquire those estimates. In the decision support process Bayesian methods are well suited for classification and evaluation tasks for structured and semi-structured data. As such they compete with symbolic AI. The latter is preferred if explicit knowledge is available while the former succeeds if probability estimation is easy because of experience or statistical data. A typical application of Bayesian systems in agronomy is the prediction of crop yield [GAP16].

Analogical reasoning uses similarity measures to identify past situations that are closest to a current situation and then adapts and applies the knowledge from the past to the present one. The best-known analogical technologies are **Case-Based Reasoning** (CBR) and **Recommender Engines**. Recommendation is a standard component of e-commerce sites such as Amazon or Netflix. Here past buying habits are analyzed to suggest new products for purchase. CBR has many use cases, e.g., diagnosis and recipe management in industry, and is applicable for the classification and evaluation step in decision support. Analogy is also the simplest machine learning approach and is best suited whenever data is rare or sparse.

Finally, **evolutionism** is the application of genetic algorithms. **Genetic algorithms** generate models and problem solutions with random variations. Of those only the fittest survive and form the next generation for iterating the algorithm. This approach mimics the natural evolution and its success strongly depends on the right choice of the fitness function. Its practical use is still limited, though active research is still ongoing.

All those approaches have strengths and weaknesses, and none is universally applicable. ANN performs compression and generalization of input data. Therefore, they are computationally efficient classifiers but may deliver strange results when presented with exceptional input data. CBR systems on the other hand do not compress data at all and can handle exceptions well, but are computationally more expensive. Hence, it is often advantageous to combine different approaches to tackle real-world problems, e.g., routine cases are best dealt with ANN, while CBR is strong for exceptional cases. However, the choice also depends on the actual availability of data and domain knowledge.

3.5.5 Getting Smarter: Machine Learning

The programming of decision systems which explicitly represent human domain knowledge and scientific insight is hard and cumbersome manual work. Machine learning, as an important subfield of artificial intelligence, promises a remedy: While traditional programming concentrates on telling the computer in detail what to do (and thus enabling it to perform complex and repetitive tasks automatically and without failure), a machine learning system is targeted toward facilitating the computer to change (i.e., improve) its own behavior in reaction to observed input data. So, instead of being programmed, a machine learning system shall learn how to act by some training process, i.e., by processing a set of training data. Ultimately, a machine learning system will thus generate an appropriate model out of the training data.

Consequently, machine learning systems are applied in situations where the details and laws of the reality under observation are not (yet) known or cannot be expressed in some traditional program, but where a sufficiently large data set represents this reality.

A typical example for such problems is the recognition of some object of interest in pictures provided by some camera (see also ▶ Sect. 3.3). Describing in detail what shall constitute, e.g., a particular plant and then building a program which identifies these plant characteristics within the pixel array delivered by a camera is simply not feasible in practice. However, providing a large amount of example pictures of the plant of interest in various surroundings as well as counterexamples is much easier.

Machine learning has developed a rich variety of technical approaches for enabling the computer to derive a generalized behavior from observed data. In recent years, the so-called **artificial neural networks** (ANN) have reached a new level of performance which proved to outperform many of the more symbolic, more traditional approaches tried previously.

The idea of ANN is quite old (e.g., the original papers by Rosenblatt [Ros58] date back to the 1950s) and inspired by our knowledge about human brain functionality: A brain neuron combines input signals and distributes an output signal to other neurons once a certain threshold has been reached. ANN mimics this functionality by using weights (to represent and modify the significance of input data) and some activation function (which calculates the output of the artificial neuron based on the weighted input data). Many artificial neurons and their interconnection (the network architecture) facilitate the effective handling of complex recognition tasks. The training process adjusts the internal weights such that the network creates the intended correct output for the respective input data. Building and training complex ANN architectures with many internal levels realize so-called deep learning.

The power of current ANN approaches and the high public interest in their capabilities (in public opinion, ANN and deep learning are often perceived as the one and only machine learning approach) is result of progress in three key areas:

- New and powerful ANN architectures, developed and improved in world-wide open-source exchange and discussion.
- Availability of large amounts of training data, often due to collaborative efforts on the Internet and the availability of cheap sensors.
- Huge computational power of GPU programming as the predominant means for ANN calculations.

Ultimately, machine learning tasks can be described as solving classification problems: classify whether some data point is correct or incorrect with respect to the question (e.g., does the object in the picture show the plant to be recognized—or not?), whether some data point is within or without the expected behavior (e.g., does a machine signal show normal behavior or is it an indicator of imminent breakdown?), or whether some potential outcomes contribute to an intended optimization (e.g., learn how to predict the best next step in a strategy game). Given the training data, the machine learning system approximates a function which realizes the intended classification.

Successful application of machine learning approaches heavily depends on the availability of suitable data. The training method—or type of machine learning—determines the actual data requirements.

The so-called **supervised learning** relies on an available set of training data which contains pairs of input data and corresponding results; it is clearly marked ("labeled data") whether the result is correct (positive example) or incorrect (negative example) for each data pair. During training, the input–output data pairs and their assessments are fed into the ANN; the internal parameters of the ANN are adapted accordingly. It is crucial to avoid that the system adapts too closely to the training data: An ANN that exactly replicates the training data set but cannot handle slightly different input data is of no use when confronted with new input from reality—this pitfall is known as "over-fitting".

Many applications for pattern-recognition type tasks can be solved using supervised learning—the impressive progress in image analysis and object recognition often are based on such training with relatively huge sets of labeled training data.

A word of caution is in place here: As the ANN creates a function based on the given training data, the quality of the result (i.e., the correctness of the learned model) depends entirely on the quality of the training data. Machine learning is data-driven and thus governed by the laws of traditional statistics. Hence, it is prone to the risk of blindness for rare events and subject to hidden bias in the training data. Correctly labeled, representative and bias-free training data in sufficient quantity can be very difficult to obtain! (see Sect. 3.5.7: Economics of AI).

The so-called **Unsupervised Learning** approaches improve this to some extent. Here, labeled training data are not needed from the start. Instead, classification algorithms, clustering methods and similar mathematical tools try to find regularities in the incoming data and identify peculiar deviations. Such approaches are well-suited to supervise continuous data streams and time series of data and find indications of abnormal behavior, e.g., for warning of impending failures in machines as a tool in predictive maintenance scenarios. Optimization of work procedures and machine parameters is another interesting application area. Finding the right algorithms and configurations for obtaining useful results in the application scenario is the core challenge of such approaches and requires significant domain knowledge.

So-called **reinforcement learning** is an interesting approach to explore unknown areas: Instead of consuming large training data sets, a learning system is configured in a way which facilitates the generation of new data configurations (e.g., by systematic variation of possible input values and calculating outputs). If it is possible to assess whether the generated output becomes "better" with respect to the intended goal, the system can learn an optimal performance by itself. The spectacular success in complex strategy games ("alphaGO" [Dee21]) quite recently was an example of successful reinforcement learning.

Applications of machine learning systems in agriculture are manifold, and numerous research and development activities investigate new opportunities. Relevant examples comprise, among others:

− The identification of plants and pests in the field. Smartphone apps [Isi21] connected to a machine learning-based service already achieve good recognition results and market penetration.
− Monitoring and assessment of animal behavior in stable and field, using various sensors like cameras and collars with acceleration sensors. The machine learning

system identifies patterns which indicate health, stress, heat or general well-being of individual animals [KM13].
- Evaluation of signals in mobile lab technology. Near-infrared sensors, combined with powerful signal analysis, offer interesting opportunities to measure composition of biological fertilizers (manure) or harvest details.
- Evaluation of camera imagery from drones, satellites and alike. Together with multi-spectrum imagery, the potential benefits seem to cover any imaginable ground assessment and monitoring task.
- Dynamic observation and prediction of time-series data, including optimization of machine operation, complex scheduling, or even market-related economic predictions.
- Autonomous behavior of vehicles, drones and alike.

Ultimately, machine learning approaches promise to uncover complex pattern in multidimensional data spaces, thus helping us to detect new knowledge like yet unknown dependencies and cause–effect relations in the interaction between environmental factors and plant growth. While such approaches will create only hypotheses, strictly speaking, it is nevertheless expected that such automatic data mining approaches will help to cope with current and impending challenges posed by climate change and the need for environmental protection while still feeding a growing world population—if we succeed to mobilize the cross-enterprise, cross-situation and cross-country data transfer needed for such endeavors.

Furthermore, learning new insights from complex data poses additional challenges to the explainability and transparency of machine learning systems. An insight hidden in some mysterious black box and visible only as unintelligible weight figures from some ANN structure is not helpful to be used to gain new knowledge in the agricultural domain—we ultimately aim for better understanding of the agricultural system using terms which are suitable for humans.

3.5.6 Artificially Intelligent Robots

Robotics and AI are separate fields, though sometimes mixed up. Robots are physical machines that interact with the physical world via sensors and actors.[1] These machines are programmable and autonomous. Industrial robots, e.g., perform manufacturing tasks in a deterministic and repetitive manner within a controlled environment. Most of these robots are not "intelligent", i.e., they cannot adopt to a changing environment or dynamically plan the execution of a task.

AI augments robots with perception and planning capabilities. Instead of just sensing the presence of an object and moving it, the robot recognizes the object and executes tasks based on the object's type. Such artificially intelligent robots adapt to changing environments. This is a prerequisite for many tasks in agriculture since they do not take place in a controlled environment such as a factory. Examples of artificially intelligent robot applications (beyond autonomous driving) are:
- Fruit harvest is a manual task due to the fruit's susceptibility to pressure and skin injuries. Robots equipped with computer vision systems and pressure-sensitive manipulators locate fruits, evaluate its ripeness, pick ripe fruits and sort with respect to quality classes.
- Grain harvesting is an example of a task which requires collaboration of differ-

---

[1] Hence, robotic process automation is not about robots, but just software.

ent systems. A harvester stores the collected grain temporarily, i.e., it must be unloaded when its storage is full without interrupting the harvesting process. Tractor-drawn trailers collect the grain from the harvester by navigating in parallel at the same speed to the harvester. This task can be executed autonomously by communication and coordination between the harvester and the tractor.
- Plant protection as described in a previous section based on individual images can be automated by autonomous robots. The robot captures plant images with a computer vision system, classifies these pictures with respect to diseases and applies the appropriate amount of chemical treatment.
- Weed removal is a labor-intensive task and has been mostly replaced by the application of herbicides. Robotics can reverse this trend. So-called weed-picking robots navigate agricultural areas autonomously. A computer vision system enables them to distinguish weed from cultural plants and to remove the weed with specialized actuators without harming other plants.
- Autonomous robot platforms can be utilized to collect field data. These data logging robots facilitate the measurement of plant individual soil nutrition, humidity, etc., or the systematic visual inspection of plants. Hence, data collection from the field can be executed almost without human intervention.

There are robots in use for livestock farming as well as in arable farming. However, milk robots or feeding systems are not artificially intelligent robots. They are traditional industrial automation systems.

3.5.7 Economics of AI

The economics of AI in agriculture does not differ fundamentally from other successful AI application areas. Any business that relies on decisions under uncertainty will benefit from good predictions. AI automates and augments decisions with predictions, which increases productivity [BRS19]. This increase manifests in higher yields, less consumables, and less labor. AI requires investments as well, i.e., tangible assets such as robotic machinery and IT infrastructure, and intangible assets. The latter are data and knowledge. AI will be economically successful when the productivity gain outweighs the investments.

The cost of data and knowledge is often the unknown in this equation. The acquisition of data may require a long period of time, e.g., many growth periods, or a large variety of crop samples from a whole region. Its use may require many experiments and learning by doing. To mitigate the risk of this unknown, it is necessary to apply the micro-perspective of the single agricultural business and the meso-perspective of the agricultural sector as a whole. An individual farmer will not be able to collect thousands of images of weed and crop diseases, but if all farmers share this effort, then large data sets can be collected fast. This approach is called crowd sourcing and e.g., part of the business model of xarvio [Xar21].

AI has the potential to enable new business models that would not have been possible without this technology. The semantic interoperability of data in conjunction with knowledge graphs, e.g., facilitates the transparency of supply chains. Any bottle of milk could be traced back to the cow that produced its content to verify the production and health state. However, to become a viable business model, this requires that consumers or the food industry is paying for this information and farmers will obtain better prices for their products. The same is true for services that exploit open data such as weather or geo information. If their utilization leads to actual cost reductions, they will be adopted and deliver a return on investment.

As any technology, AI will be successful if it satisfies an actual need, no matter whether this is a significant financial gain or an increase in work–life balance. A fully autonomous seeding machine will not be a game changer for a farmer who spends two working days a year seeding. However, the milking robot is a true success story because it frees up time of the farmer and offers quiet Sunday mornings.

3.5.8 Outlook: Individualized Optimization

AI application in agriculture will lead to new solutions where automated systems exhibit new degrees of autonomy and dynamic adaptation to situation-specific individual circumstances. On the long run, this will allow to realize fine-grained optimizations and individualized operations up to the level of single plants and individual animals—while satisfying many detailed and individualized customer requests as well. Hence, modes of work such as mechanical field work, which have been abandoned because of the cost of manual labor, become economically viable again, due to the availability of autonomous robots. When swarms of small robots can assess and treat plants individually, environmental-friendly mechanical weed removal (instead of heavy chemistry) becomes attractive again. When reliable data become easily accessible and are shared for mutual benefit, AI algorithms will detect new insights and realize fine-grained optimization opportunities.

Actually, there are already initiatives to systematically acquire and utilize such data, e. g. Google's Mineral project [Xco21]. Increasingly interlinked networks of systems will help to combine highly individual, situation-specific measures with global maximum efficiency and improved traceability and transparency, leading to sustainable ecological and economic benefits and social acceptance.

3.6 Agricultural Data and Terminologies

Daniel Martini and Martin Kunisch

Abstract

Agricultural production today requires decisions to be made under increasingly demanding constraints, taking into account a wide variety of influencing parameters and the impact on profitability, environmental protection and animal welfare, food safety, sustainability, and more. It has become an information-intensive business, and achieving a systemic view requires the integration of data from multiple sources. Existing datasets include data from public sources such as data on crop varieties or crop protection products, satellite images, or spatial vector data, as well as data from private sources and data obtained on the farm, for example, from AgMachinery or sensor data. Currently, data are delivered in a number of different formats, which leads to difficulties with regard to interoperability. One of the key questions is whether format and interface specification mechanisms, as implemented in most standardization initiatives, are capable of overcoming these barriers. Semantic technologies provide mechanisms for the self-description of data using globally unique identifiers and a generic data modeling mechanism that can be applied to virtually any dataset. This includes the use of ontologies and defined terminologies. This section introduces examples of agricultural data with content summaries and formats, presents technology requirements for integration, and uses examples of existing resources and technologies to show how approaches for achieving this integration might look like.

3.6.1 The Data Landscape in the Agricultural Sector

3.6.1.1 Content Scope, Data Categories, and Data Types

Today, we live in a world of growing amounts of data in almost every field. Another important characteristic of the agricultural sector is that it represents an intersection of different disciplines. Current problems in agriculture—for example in the area of efficient use of nutrients and energy [KPV+19] or risk management with regard to climate variability and change [HTA+15]—require a systemic approach: To successfully leverage data, e.g., in simulation models or computational models, data from different sources must be retrieved and integrated. How efficient such cross-source retrieval and integration can be accomplished depends on how data is represented at the level of services. A targeted approach involves obtaining an overview of the data landscape, of the requirements for agriculture from which needs for data evaluation arise as as well as of the resulting challenges first. After that, suitable data representation mechanisms and methods for specific problem areas can be derived from this.

The task of agriculture is the purposeful production of food, raw materials, and energy using natural resources such as soil, plant material, and livestock. Accordingly, agricultural sciences have links to biology, but also to earth sciences. Production is influenced by weather and climate, so meteorology plays a role as well. Ever since the advent of mechanization and automation in agriculture, topics from engineering sciences such as mechanical engineering and electrical engineering are also touched upon. Findings in chemistry have enabled a better understanding of plant and animal nutrition as well as chemical crop protection and treatment of animals with pharmaceuticals. In addition, production is the source of income for farms and a part of the value creation for the economy, so economic aspects are important. The goal of sustainability also requires the overarching consideration of all subsectors in the sense of fulfillment of the United Nations Sustainable Development Goals [UN15] (see also ▶ Sect. 1.3).

From this breadth of content and the aforementioned disciplines, relevant data categories for agriculture are derived:

— Agriculture-oriented data on resources used: plant species and varieties with growth and quality characteristics, animal breeds and their performance and breeding values, AgMachinery and its technical data, fertilizers and their nutrient contents, crop protection products and pharmaceuticals with their compositions, active ingredients, and areas of application, etc.
— Data on the influencing environment: weather and climate data, but also data on hydrology, geology, and in the landscape context with structures worthy of protection, such as hedgerows and landscape or water protection areas, etc.
— Data on the market environment and statistical data: produced quantities and harvest statistics, price time series of agricultural products, current supply situation of the population such as self-sufficiency rates, etc.

These data can be fed into other software systems, for example for statistical evaluation or processing using artificial intelligence algorithms. Decision makers in agriculture such as farm managers, but also policymakers as well as representatives of upstream and downstream sectors act as consumers of the evaluations and raw data and make decisions on this basis. However, agriculture is increasingly producing data itself as well. In addition to the established recording processes of accounting and documentation, machines in the field log data during work execution, such as application rates and consumption, and milking systems record milk quantities and quality parameters for individual animals. In line with the heterogeneity of the disciplines as well as the myriad of entities associated with the data categories and the increasing role of farmers also as data producers, there is a certain range of data types to be considered: For example, spatial categorization often plays a role, so geospatial data such as polygonal features, points, etc. and their respective annotations with technical data are highly significant. The time dimension is often also relevant, so time series are common for market data and agricultural statistics, but also for on-farm data such as milk yield curves. For earth observation data, which has entered the scene in recent years, both the spatial and the temporal dimensions are relevant. In addition, there also exist simple, structured data, which may be stored in relational databases—e.g., farm input Master Data. However, information for farmers is not only available as structured data. Agriculture as a discipline also has strong qualitative–descriptive features. A considerable proportion of specialist information is therefore available in unstructured text corpora as reports, scientific articles, or legal texts. In addition, there is the experiential knowledge of the farmers, which is currently rarely recorded formally.

3.6.1.2 Data Formats and Standards

In accordance with the use cases considered by the agricultural sector and the disciplines involved, a heterogeneous landscape of data formats and standards has emerged, of which we only pick out a few characteristic examples here. In particular, the standard ISO11783 (ISOBUS) originates from the agricultural sector, but so do initiatives such as ISOagriNet. In addition, formats and standards from the geodata sector are also relevant for agricultural software, such as the service and format specifications of the Open Geospatial Consortium (OGC), which have been incorporated into the ISO191xx series of standards and are an essential foundation of the Infrastructure for Spatial Information in the European Community (INSPIRE). In the following, we will briefly discuss these examples.

The ISO11783 standard—"Tractors and machinery for agriculture and forestry—Serial control and communications data network" [ISO19] currently has 14 parts. Its core application areas are data connections between tractors and agricultural and forestry implements as well as the exchange of work orders and recorded data with management systems in agriculture and forestry. Part 10 specifies a data format based on the eXtensible Markup Language (XML), a standard originally designed for marking up and structuring documents such as manuals in the IT sector [BPS+08]. The central element is the so-called Task, which describes a work process in the field. Linked to this element is data about implements, workers/drivers, fields on which the cultivation takes place, etc. The ISO11783 standard also defines its own encoding for geodata, which can be used to create application maps and tracks for automatic steering systems. The format is designed such that the work order can be transferred to a machine and records can be transferred back to the FMIS. During this process, sensor data such as PTO speeds, fuel consumptions, applied rates, harvested quantities, etc., are recorded together with the associated time and location stamps. The standard also defines a compact binary coding for this log data.

Based on the ISOBUS for arable and permanent crop farming, the ISOagriNet initiative attempted to develop a similar system for livestock farming, i.e., essentially for the exchange of data and the networking of equipment for animal husbandry (e.g., climate computers, automatic feeders, milking systems, etc.). This resulted in the Agricultural Data Interchange Syntax (ADIS) and the Agricultural Data Element Dictionary (ADED). The system is therefore also known as ADIS/ADED. This work has found its way into standardization via the ISO standards 11787 [ISO95], 11788 [ISO00] (both now withdrawn) and 17532 [ISO07]. ADIS/ADED includes its own format specification, which is based on messages in a text format derived from database records. Based on this, possibilities were later also created for exchanging specialist content from livestock farming via web services and web interfaces common in this environment, such as XML or JSON [Bra17]. The standard has not gained widespread use across the full range of its originally envisioned use cases, but it is being used in specific areas. For example, it has practical significance for milk performance testing and breeding value estimation. Furthermore, the protocol forms the basis for the interfaces of the HI-Tier Platform.

Besides these specifications developed in the agricultural sector, established geodata standards play an important role for the particularly relevant category of geospatial data. The development of specifications not only for geodata formats, but also for operations on geodata in databases, catalog and map services is being driven by the Open Geospatial Consortium (OGC). The OGC has also contributed a number of their standards to ISO (ISO 191xx series of standards). However, specifications are also available freely as Open Standards [Ogc21]. A number of development teams have developed compliant software implementations available as open source. The Geography Markup Language (GML) [Por12] has been developed as a format for the exchange of geospatial data. It is based on the so-called feature property model, which at its core transforms the concept known from printed maps of geo-object types or "layers" (settlements, water bodies, streets, natural landmarks, etc.) marked with specific colors or symbols into an object-oriented data model. Geometries (points, polylines, polygons) to draw on a map are properties of these objects. This kind of modeling provides a representation well-suited for spatial analysis, but requires the object types with their properties to be determined to the greatest possible extent before setting up a system ("a priori"). The Working Group of the Surveying Authorities of the States of the Federal Republic of Germany (AdV) has therefore presented comprehensive object type catalogs and schemas, for example [Adv21].

In addition to the aforementioned ISO standards and OGC specifications, a number of other interface standards and data formats are in use, which will not be discussed in detail here. Moreover, several information services exist that are relevant to agriculture, where data is embedded into an application or is merely offered for download in simple ad-hoc or proprietary data formats.

3.6.1.3 Status Quo of Data Interoperability

From the perspective of a geospatial data standardization stakeholder, the standards developed in this context offer a high degree of interoperability. Nevertheless, agricultural stakeholders often complain about poor data interoperability. How can we explain this apparent contradiction?

Standards developer communities have so far mostly considered the problem of interoperability only within their scope and their use cases. It is possible to provide specific services that are useful for farmers based only on data sources designed in compliance with the standards of a single developer community. However, given complex interrelationships and influence paths, there is potential in merging different internal and external datasets. These also touch on various disciplines mentioned above and are therefore in many cases coded and retrievable according to incompatible specifications.

It must be questioned whether the obvious solution approach—that the degree of standardization is not sufficient and that the efforts of the individual initiatives and the communication must be intensified—is actually promising. The amount of data collected and the accessibility of different data sources is currently increasing significantly over time. This further increases the heterogeneity of the agricultural data landscape. A solution based on agreements between developer communities appears rather unrealistic due to the rather limited "data rates" in the exchange of information between people [COD+19], especially since, as bilateral agreements are to be made on a case-by-case basis, the number of the necessary coordination processes is approximately the square of the number of stakeholders. Bundling of coordination processes can be helpful here, but again leads to other problems—e.g., with regard to ensuring the ability of committees to work when the number of participants is too large.

3.6.2 A Global Data Space—Achievable or Wishful Thinking?

The outlined problem of the lack of scalability of common standardization processes leads to the question of whether there are other options of creating a global, interoperable data space. The decisive factor here is whether it is possible to find approaches that keep the coordination effort as low as possible—for example, by merely agreeing on common basic principles and replacing some of the aspects commonly described in textual specifications with machine-readable descriptions and then leaving their interpretation to computers.

3.6.2.1 The FAIR Principles and the Role of Linked Data

The FAIR principles were designed in the context of a framework for European research data management, but they also address in a generalized way precisely those difficulties that arise from some of the characteristics of the current data landscape described above: Given growing data volumes, heterogeneous sources and application requirements, and distributed providers, how can data be provided in a **F**indable, **A**ccessible, **I**nteroperable, and **R**eusable way [WDA+16]. They do not limit themselves to formulating these principles, but also provide guidance on implementation as well as examples of technologies and formats that meet the requirements (see [Gof21]). It can be shown that the FAIR principles can be understood as necessary conditions for the achievement of the goals underlying these four keywords. This means, for example, that in order to achieve interoperability, care must be taken to ensure that formal languages are used for representing knowledge (Principle I1), vocabularies are used that themselves also fulfill the FAIR criteria (Principle I2), and links are present in the data and metadata to other (meta)datasets (Principle I3).

In the following sections, only two aspects will be examined in more detail: the use of globally unique, persistent identifiers and considerations regarding suitable data models.

Existing technology stacks can be evaluated with regard to the fulfillment of the FAIR criteria. With the W3C Semantic Web recommendations such as the Resource Description Framework (RDF)

[SR14], the Web Ontology Language (OWL) [HKP+12], and the Linked Data Platform (LDP) [SAM15], technologies exist in which almost all FAIR principles are either already explicitly anchored in the specifications or can be implemented with reasonable effort.

Linked Data is often associated with Open Data or even equated with it. However, the specifications do not make any statement about data usage rights and how to handle them. In his 5-Star Model for Open Data [Ber10, 5st21], which also builds on Linked Data, Tim Berners-Lee does propose the use of open licenses, but there is no technical specification or restriction on this at all. The FAIR principles take up data usage rights aspects in Principles R1.1 and A1.2: R1.1 requires that clear, machine-readable license terms be assigned to the data—machine readability is particularly important for search engines and registries, so that users can immediately recognize the extent to which the data provided can be used in their context. Principle A1.2 clarifies that "accessibility" does not necessarily mean "free" or "open", but rather that protocols for data retrieval should also support authentication and authorization, if required. Accordingly, the frequently expressed wish of farmers and private-sector companies that third-party access should only take place in a controlled manner can also be taken account in this environment.

3.6.2.2 The Need for Globally Unique Identifiers

To enable communication that is free of misunderstandings, unique identifiers must be defined for specific elements, things, or concepts about which data is to be exchanged in order to ensure that sender and receiver refer to the same concept when interpreting the data. The focus may be on the coding of classifications and attributes (crops, types of objects, parameters recorded, etc.) or of individuals (animals, fields, farms). In agriculture, the codes for the first group include, for example, various systems of crop codes created over the years from the perspectives of different use cases or regional coverages, such as the crop codes of the European Plant Protection Organisation (EPPO), crop codes of the variety approval authorities, and crop numbers of the IACS (Integrated Administration and Control System of the EU) procedure. The variety of different codes for quite similar, partly even identical concepts illustrates one of the core problems in this field: As soon as data is to be transmitted from one application domain to another, or is to be merged across domains, such code systems must be translated into each other with great effort. For example, one application goal could be to merge and evaluate information on varieties such as resistances with data from crop protection advice. Either the user has to do a translation manually when inputting the data, or developers have to provide transfer mechanisms. An additional challenge in this context is that systems usually only overlap in part or combine concepts at different levels (e.g., crop group codes in IACS).

The use of identification and code systems which fulfill the requirements for global uniqueness and persistence in accordance with FAIR Principle F1, is therefore one of the most important basic requirements for data interoperability. In addition, A1 demands that it should also be possible to retrieve data via the assigned identifier using open standard protocols. This can only be realized with reasonable effort by relying on consolidated, widely used standards with appropriate support by software tools, such as Uniform Resource Identifiers (URIs) [BFM05]. It should not only be possible to identify data objects or individuals; core concepts of data representations such as entities, object classes, or data fields should also be assigned unique identifiers. These are the prerequisite for reusing individual elements from data standards and recombining them into new data formats.

3.6.2.3 A Suitable Data Model

Existing systems and standards are based on different paradigms and design patterns. Common data management systems often build on the relational model. Web services often use hierarchically organized document formats such as XML or JSON to serialize data. Depending on the underlying use case, data is organized into formats or tables in different ways. Relations can, for example, be modeled in different directions: For example, the relation between the driver and the machine can be represented as: "Driver X drives Machine Y" or as "Machine Y isDrivenBy Driver X". In strictly hierarchical data structures, these different views lead to incompatible representations. In order to achieve interoperability, data must be structured according to an abstract data model where different use-case-driven views no longer play a role, at least in the "external view", i.e., on the level of an interface. In view of the increasingly fluid data landscape, a suitable model must therefore be extensible in addition to being generally valid. Ideally, data from new sources can be added to existing datasets without time-consuming programming of transformations.

If a data model has this degree of generality and extensibility, it can be used not only to represent the actual domain-oriented data, but also to describe these domain-oriented representations themselves—the separation between schema and data that is otherwise common in databases or with JSON and XML can be dispensed with. Descriptive metadata can thereby be provided and retrieved in the same way as the data itself. The a priori knowledge needed to use data sources is minimized and both developers and software agents can first of all exploratively retrieve important descriptive information such as existing object classes, properties, data types, and relations existing among the data in order to then use the same mechanisms—technically possibly even via the same API—to request the data itself. This means that not only data is coded, but also parts of the knowledge required to interpret the data. Semantic technologies are the basis of such a knowledge representation, and the explanations to FAIR Principle I1 name some language specifications that meet the necessary requirements on generality, extensibility, and capability for self-description, e.g., RDF or OWL.

3.6.3 Controlled Vocabularies, Thesauri, and Ontologies

For practical use in data services, vocabularies and ontologies can be developed with the help of the above-mentioned methods and technologies. Vocabularies are oriented to the necessary minimal formal language scope for data exchange, i.e., they provide simple classes and attributes. For semantic descriptions of technical terms that occur both in structured datasets—e.g., in text fields—and in more unstructured text corpora, electronic thesauri and knowledge organization systems are available. These also include relations between concepts such as subject relatedness or synonymy.

Ontologies can be seen as a further semantic expansion stage. Gruber defines ontologies in the information-technical sense as follows: "An ontology is a description (like a formal specification of a program) of the concepts and relationships that can formally exist for an agent or a community of agents." […] "Ontologies are often equated with taxonomic hierarchies of classes, class definitions, and the subsumption relation, but ontologies need not be limited to these forms. Ontologies are also not limited to conservative definitions—that is, definitions in the traditional logic sense that only introduce terminology and do not add any knowledge about the world. To specify a conceptualization, one needs to state axioms that do constrain the possible interpretations for the defined terms." [Gru93].

Ontologies include aspects of vocabularies and thesauri such as classes, terms, relations, and properties, but can also go beyond this to include statements about relationships and rules for deriving new knowledge. Classification criteria can also be described: They specify which conditions must be fulfilled for an object to belong to a particular class. So-called inference engines or reasoners can then be used to decide whether the criteria are met, based on appropriately provided data about the objects in question. In the agricultural sector, this function could be used, for example, to describe whether farming practices on a certain piece of land meet requirements of organic farming. Due to the complexity of the underlying regulations, however, the complexity of the descriptions in an ontology would also be high.

3.6.3.1 AGROVOC

Currently, the most comprehensive thesaurus for the agricultural sector is AGROVOC, which is provided and maintained by the United Nations FAO [CSM+13]. It was first published as a book in three languages in the early 1980s. Nowadays it is delivered electronically as Linked Data in RDF format. It uses the so-called Simple Knowledge Organization System (SKOS), an RDF vocabulary whose central class is a concept that can be equipped with preferred and alternative text labels and organized into super-/sub-concept hierarchies [IS09]. ◻ Figure 3.17 uses the concept "Lettuce" to illustrate these references. Currently, AGROVOC contains about 40,000 concepts with about 800,000 labels in up to 40 languages. It is managed and further developed by the FAO's Agricultural Information Management Standards Group with the support of a globally distributed team of editors who jointly input new concepts and translations via the VocBench platform.

AGROVOC is currently used to support various use cases. Since labels in different languages are available for concepts, AGROVOC can be used, for example, to implement multilingual search engines. In

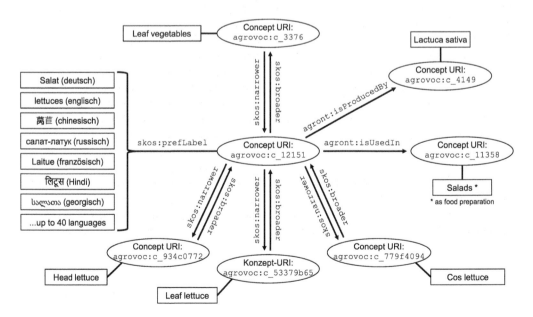

◻ **Fig. 3.17** Graphical representation of the concept "Lettuce" with a selection of its text labels, its embedding in the terminological hierarchy, and relations to other concepts

addition, AGROVOC can be used to automatically tag texts or web pages, e.g., to describe the topic or keywords frequently occurring in a text. This information can then be used for navigation, for example via topic trees. Various organizations also use the concepts of AGROVOC to tag datasets with metadata describing, for example, which parameters are included in a dataset, where the data was collected, or which organisms or environmental conditions play a role in a dataset.

In addition to various access paths for research and processing by humans [Fao21a], AGROVOC is also provided as Linked Open Data in a machine-readable form. AGROVOC concept URLs are retrievable in the sense of RESTful web services; moreover, a public access point is available for queries using the standardized query language SPARQL [W13]. These APIs enable software developers and the computer systems developed by them to specifically query or use the data contained in AGROVOC. Integration of the complete AGROVOC into one's own systems is also possible by downloading the dataset. AGROVOC concepts and labels are tagged with URLs as unique identifiers and can therefore also be reused in RDF vocabularies and ontologies, for example to uniquely define terms in data exchange systems or annotate classes and data fields with multilingual text descriptions.

3.6.3.2 Ontology Developments in the Agricultural Sector

Fully developed ontologies have been used only to a limited extent in the agricultural domain. For AGROVOC, relations between concepts that are typical for an ontology were added some time ago as AgrOntology [Fao21b]. An example are "causative relations", which describe cause–effect relations: produces/isProducedBy, hasDisease/isDiseaseOf etc. On this basis, agricultural knowledge is coded in a computer-readable way and can then be used by expert systems (see also ▶ Sect. 3.5.4), for example. The AgrOntology relations are currently undergoing a comprehensive revision aimed at simplification and more consistent use.

In the meantime, ontologies have found their way into the experimental systems of large or international agricultural research institutions. As part of the Platform for Big Data in Agriculture, CGIAR has designed the Agronomy Ontology (AgrO) [ABL17], which is used to automatically generate digital field books that subsequently allow data from different testing facilities to be merged. A mapping to the concepts of AGROVOC exists for AgrO; it is therefore an example of the interplay between thesauri and ontologies.

What such an interplay of thesauri, ontologies, and datasets can look like in concrete terms is currently also being demonstrated in the EU project DEMETER. Deliverable 2.1 [Rou20] of the project describes a series of data models, outlines ontology-driven interoperability mechanisms, and describes some ontologies developed in that context. Within the FOODIE Cloud (foodie-cloud.org), a data service is provided that builds on the technologies and methods described above and links and integrates a series of open data sources, for example on land use and transport, in an interoperable way with the help of ontologies and thesauri [Foo21].

In related fields such as bioinformatics, ontologies are already in wider use. The Open Biological and Biomedical Ontology (OBO) Foundry [Obo21] provides infrastructure components and tools for managing ontologies as well as ontologies themselves on a range of biological topics. Prominent big data services such as UniProt [Uni21] also use ontologies and semantic technologies to describe and deliver their data as well as query mechanisms such as the above-mentioned SPARQL.

3.6.4 Conclusion and Outlook

Due to its heterogeneity and its interfaces with other disciplines, the agricultural sector faces considerable challenges when it comes to data integration and the creation of overarching data spaces via interoperable services. Standardization mechanisms based on effort-intensive coordination processes that have been customary up to now are no longer effective in the current environment. Instead, basic abstract principles and targeted methods must now be used to ensure findability, accessibility, interoperability, and reusability (FAIR) in the provision of data and standards. This includes a more comprehensive formal description of data—meaning that knowledge on how to interpret data needs to be encoded. Semantic technologies, vocabularies, and ontologies are tools that, on the one hand, make use of these methods and, on the other hand, can serve as building blocks of a future provision of data that is better usable even without comprehensive a priori coordination and specifications.

3.7 The Role of Geo-Based Data and Farm-Specific Integration: Usage of a Resilient Infrastructure in Rhineland-Palatinate, Germany

Matthias Trapp, Roland Kubiak, Djamal Guerniche, Daniel Eberz-Eder and Julius Weimper

Abstract

Historically, authorities collected multiple agricultural data types, e.g., protection zones, field boundaries. This can be relevant for a more farm and field-specific application of digital farming measures. This section describes an example how farmers can access this data, developed by the public agricultural advisory service in Rhineland-Palatinate. Specific use cases are outlined and discussed in this section.

3.7.1 Introduction

The German Federal Ministry for Food and Agriculture (BMEL) pointed out in 2020 on the website, that "climate change mitigation, adaptation to climate change, food security and the production of renewable resources, including wood, are closely inter-related. While agriculture and forestry are affected by climate change, they are, at the same time, an important part of the solution. Arable farming produces the great majority of basic foodstuffs and animal feed and therefore represents the foundation for our daily food. The past decades have seen a significant increase in output. This guarantees a secure, high-quality supply of food. However, the high level of productivity in arable farming also entails challenges in terms of environmental protection, nature conservation, economic factors and social acceptance. Furthermore, existing conflicts of interests need to be resolved" [BME20] (see also ▶ Sect. 1.4).

Additionally, the Thünen Institute on behalf of the BMEL formulated 2020 that "Agriculture shall produce high-grade food and renewable resources in a reliable manner. In future, agricultural production will face increasing pressure to release fewer greenhouse gases and air pollutants. International agreements on climate change mitigation and air pollution control oblige Germany to reduce emissions" [Thü20].

Therefore, to provide continued food security, agriculture must use the limited arable land more efficiently. However, to prevent shifting problems into the future, the required transformation is only feasible if sustainability aspects are considered. Moreover, with a strong social and political awareness of global change, sustainability but also animal welfare is increasingly influencing the discourse [UB15]. A wide range of new technologies and concepts are currently developed to support farmers to manage their land not only more sustainably but also more efficiently. The following section will provide a brief overview of the developments to date. Moreover, exemplary, related solutions in the form of a resilient infrastructural and site-specific smart farming concept are outlined.

In the position paper of the digital association BITKOM on arable farming strategy 2035, the following overarching statements are formulated [Bit20a]:
- Digitization is a central element of the future of (smart) farming and should be part of all recommendations for action
- digital technologies are already available—they must now be brought more strongly into the area (a wider implementation practice)
- sustainable agriculture is only possible with a digital administration
- there has to be a holistic view of sustainability

In its article "Digitalpolitik Landwirtschaft" (Digital Policy for Agriculture), the BMEL describes "Precision Farming" and "Smart Farming" as technical developments that have been tested and used in agricultural engineering practice for more than two decades [BME20]. According to a survey by DLG - Agrifuture Insights [Bit20b], the acceptance of farmers using new technologies and digital information is already very high in Germany [DLG18].

Therefore, with smart farming integrating the possibilities of the "Internet of things" (IoT), there are abundant opportunities. This means using technologies not only to gain higher production yields with less input, but also using them to optimize administration processes, monitoring, reporting, even at the landscape level to integrate a more holistic approach.

RLP AgroScience has worked in the framework of digital landscape analysis and precision agriculture on different orders of (spatial) scales for over 20 years. This section will provide an overview of the last 25 years of research performed by RLP AgroScience in the context of smart farming. The current efforts will be explained in more detail, especially the establishment of a mixed private–public solution as a step toward to a resilient smart farming concept, the so-called Geoboxinfrastructure (GBI). Consequently, the main focus of this section is the description of existing communication infrastructure and spatial resources from authorities, and integrating these with geo-information to create a resilient infrastructure to support farmers in their daily work. For us, a view of the entire diversity of individual agricultural holdings is important. This is why we want to create a platform that allows every farm to retrieve relevant public information for their existing farmed areas. This platform provides an introduction to the diversity of precision and smart farming, and raises awareness of the benefits of business-relevant geo-based data. The previously mentioned GBI is presented as an example of such an infrastructural resilient smart farming concept in Germany's state Rhineland-Palatinate.

3.7.2 Overview

Since the beginning of the 1990s, GPS (global positioning system) has been available for non-military use in Europe. The first applications in agriculture began to use the positioning data, with the first GPS-enabled harvester being able to measure yield and position. For the last 25 years, precision farming was mainly triggered by site-specific approaches, focusing primarily on optimizing fertilizer and yield (fertilization on withdrawal) and guiding the lane assistance of tractors [RNT97, SLS+93].

In 1998 and 1999, a first on-farm research trial was performed by varying the amount of fertilizer within one parcel on discrete georeferenced strips of approximately 1 hectare in size. Each strip was analyzed with respect to soil information and the different slopes generated from terrain models. This trial, supported by a GPS-enabled John Deere tractor and harvester, led to a better understanding of the correlation between soil quality, topography, and fertilizer quantity on the final yield. The results were published in a PhD Thesis by Trapp [Tra03].

The combined German research projects "Preagro I" and "Preagro II" (2000 to 2008, respectively) were one of the largest research projects in the context of precision agriculture in Germany. In the first phase, important scientific and technical fundamentals were developed on over sixteen farms. In the second funding phase, the project focused primarily on the information management on these farms. Additionally, the possibilities of integrating the entire value chain up to the consumer level were investigated. One of the main results of the entire Preagro project was to show what is technically feasible, taking small-scale differences into account in management and developing rules to include these site differences in the overall design. This allows the farmers to optimally adapt their cultivation to their respective sub-areas within their fields, basing everything on plant cultivation and ecological criteria [Pre20].

Subsequently, the development of a geo-data-infrastructure (GDI) based on the INSPIRE-process (see ▶ Sect. 3.6) and the processing of specialized geodatabases was also improved by authorities and research institutes. In 1997, the implementation of the Land Parcel Information System (LPIS) in the framework of the Integrated Administrative Control System (IACS) by the EU was a crucial milestone for the agricultural sector. One of the first application was the "Olive-GIS" project in Greece (2003), where all olive trees were counted based on digital orthoimages to aide in funding decisions [GGA+07].

Overall there are two approaches which drove the establishment of these technologies in the EU framework of agriculture. Firstly, the EU-wide gathering and management of geo-data and information for administration and controlling, and secondly, the optimization of farming practices on a single plot of land.

3.7.3 Site-Specific Resilient and Climatic Smart Farming

In the research project "KlimLand" (Impact of climate change on agriculture and viticulture in Rhineland-Palatinate 2013), the RLP AgroScience developed a geo-database for all agricultural parcels in Rhineland Palatinate (RLP, a federal state of Germany), with a huge amount of additional information [TTK13] (◘ Fig. 3.18).

Additionally, indicators were calculated based on the site-specific geo-data such as the climatic resilient index, which combines the results of climatic parameters with the topographical and soil information of each plot (◘ Fig. 3.19).

Farmers and consultants were therefore able to make decisions on site-specific information not only for one plot, but for whole landscapes and regions. Unfortunately, due to missing technological capabilities at the time this information was gathered, it could not be used adequately by farmers and the authorities. Additionally, such static geo-information needs to be updated regularly to remain accurate. The information was mostly used for research projects, and not integrated into the operative process chains in the administrations.

Nowadays, technological development enables just-in-time information management. However, due to climate change and increasingly varying local weather conditions, there is a need to be able to react earlier to unex-

Technology Perspective

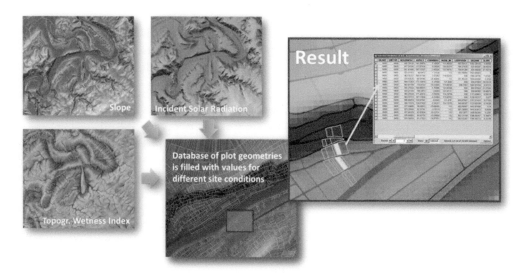

Fig. 3.18 Transfer of values of continuous raster information into the database of discrete objects [TTK13]

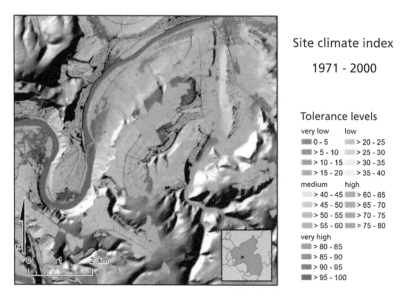

Fig. 3.19 Site climate index for the period 1971–2000 on agriculturally used sites in the Middle Moselle [TTK13]

pected weather events [STK+18]. Microclimate data recorded specifically for each plot by local sensors can be used to extend weather forecasting systems, allowing measures to be implemented with greater temporal and spatial precision. Algorithms based on artificial intelligence (AI) use and process all information from near and remote sensing sources to create zone-specific management plans. RTK-GPS enables the precise control of autonomous machines and robots down to the centimeter.

However, without a communication and geo-data infrastructure, the meaningful use of the data and model outputs cannot be used by farmers or advisors in their operations.

Therefore, the GBI will be presented as an existing and working example on the level of an EU member state (Germany) and on the scale of a federal state (Rhineland Palatinate). The purpose of this section is not the technical description, but possible applications which require such an infrastructure.

3.7.4 The GBI as an Example for Infrastructural Resilience

The motivation behind designing the GBI was triggered by the question "How can the resilience of a digital agriculture be strengthened through usable and interactive systems?"

The overall objective of the GBI is the development of a digital and resilient infrastructure. It includes, "in addition to map- and table-based presentation of farm data, geo-forms for placing orders and encrypted communication with other users. By combining the GBI with a site passport, in which all relevant public geo-data are bundled and processed, new approaches and findings can be developed and implemented in the future. This is in addition to the pre-screening of requirements regarding fertilization, plant protection, biodiversity, etc., also at the landscape level" [RSE+18]. One of the key points in the project is the focus on a resilient and decentral data management. Therefore, it is very important to reduce the strong dependency of operational data from central servers. In order to reduce the risks from different blackout scenarios, we focus on a hybrid digital architecture where the most important operation data are stored decentralized on an edge device while getting online information like dynamic weather data. This concept of a Resilient Edge Computing (REC) is based on the principal of offline first. Through the implementation of the Open Horizon Framework, it is possible to manage thousands of decentral edge devices which serve as a decentral data storage [RKS+20] (Fig. 3.20).

The GBI includes three components: the GeoBox-Viewer (GBV) as a visualization tool, the messenger for communication, and the site passport (field atlas) as a local storage system on the farm. The geo-data services of the state and the authorities that are relevant to the farms are made available to them free of charge; this includes both on-farm storage and distribution, as well as combined online and offline use on mobile devices. With this approach, relevant basic data for smart farming can be stored in a standardized way at the farm level. This also ensures data sovereignty and thus the value creation remains on the farms.

Spatial management requirements are becoming increasingly complex. Therefore, the need for valid official geographic and technical information is increasing. For example, in order to promote non-chemical crop protection methods, additional rule-based knowledge must be prepared in information portals via modern cognitive services so that it can be accessed by farmers in a location- and situation-specific manner. In the case of sovereign tasks, it is a challenge for the federal states to be able to reach all farms in the various regions equally with digital and geo-data-based information services.

An example of such geo-data-based information services will be delivered by the ongoing research project SOFI (Smart Soil Information for Farmers, BMEL). Sensor data-based map services for soil-conserving cultivation and environmentally compatible fertilization are generated and published in the GBV as a mapping service.

Technology Perspective

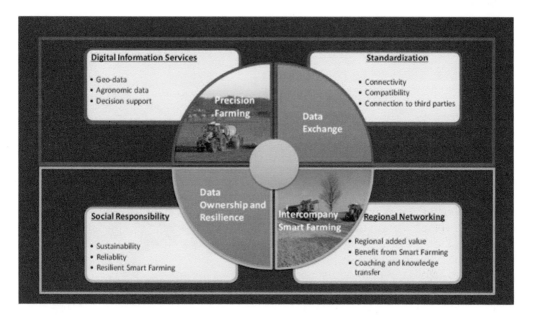

Fig. 3.20 Overview of the Geobox infrastructure as part of the Digital Agrarportal (DAP) of Rhineland-Palatinate [RSE+18, DLR21c]

Using weather data from agrometeorological stations and soil information about field capacity and layer thickness, the potential soil moisture will be modeled daily for all agricultural sites and polygons of the available soil data. The modeling results are classified into different classes, from very dry to very moist, which can be translated into classes describing the trafficability of soils to avoid soil damaging by compaction [BLE21]. To derive susceptibility of soils to compaction and conversely derive trafficability, several approaches have been published (see, e.g., overviews compiled by [Kuh19]). In the SOFI-Project, some of the published approaches were implemented as exemplary model combinations and checked for their applicability to existing datasets. So far, an approach based on derived trafficability according to simulated soil moisture seems promising. It allows for fast computation to estimate soil texture and soil moisture-dependent trafficability and can be implemented efficiently based on existing datasets. The advantage of this approach in the current situation is the comparable small set of necessary input data and high computational efficiency for whole landscape units in combination with the soil moisture model outlined above. This is especially a practical benefit as it allows comparable fast "delivery" of information to possible users [Kli20].

Two examples for the calculation of potential soil moisture in Rhineland-Palatinate are shown in the following figures, one with respect to the drought in 2019, and the other showing the year 2020, where there was more precipitation (Fig. 3.21):

As can be seen clearly in Fig. 3.21, on the left side the model results show that nearly all agricultural parcels in Rhineland Palatinate calculated on March 03, 2019, were classified as dry. This means the water capacity in the upper soil layer was not sufficient to support plant growth. Fig. 3.21 on the right side, the calculation made for March 03, 2020 shows a completely different picture. Due to a higher precipitation, the water capacity in the upper soil layers was sufficient to support plant growth.

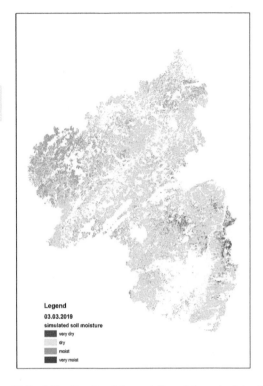 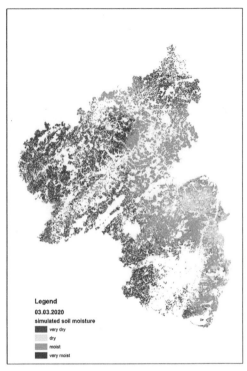

Fig. 3.21 Results of the modeling of the potential soil moisture for one day in 2019 (left) and 2020 (right)

The temporal resolution of the modelled geo-information is currently a single day. The plan is to model each entire federal state at night, and provide farmers and advisors the datasets they can use to plan their daily tasks in the morning. The spatial resolution is shown in the following figure for selected parcels located on the agricultural lead farm "Neumühle", which is part of the research project "Experimentierfelder Südwest" [DLR21a] and whose data will become part of an open data farm (◘ Fig. 3.22):

The combination of this location-based data enables, for the first time, the derivation of spatiotemporally high-resolution real-time maps as a service for farmers, contractors, and machinery rings. The result is a set of flexibly usable geo-data sets that can be integrated both into expert and consulting tools in web-based media (apps), and into existing, partly vehicle-based, agricultural software solutions in order to optimize control processes there. Through the close connection to existing public portals, such as the Digital Agricultural Portal in RLP (Rhineland-Palatinate), the results can be made available quickly and efficiently to all participating farmers and inter-farm organizations.

Verification and validation of these modeled results are necessary for consequent refinement and improvements. Currently, most validation is discontinuous, either spatially (classical measurement stations) or temporally (farm-scale field campaigns). Gathering "ground truth" data is, in both cases, costly. Therefore, low cost and easy-to-use sensors which measure soil moisture, temperature, and humidity will be installed in field to verify the modeled results. To automate the sensoring as far as possible and with respect to resilience, LoRaWan-enabled sensors will be used. The

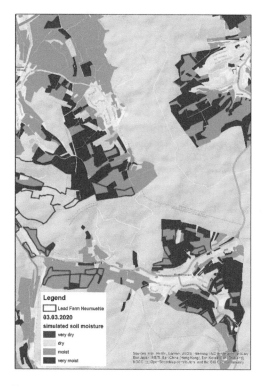

Fig. 3.22 Detailed map section of modeled soil moisture at the Lead Farm Neumühle to show the spatial resolution of the current simulation approach

sensor network will not be placed haphazardly, but rather with respect to terrain analysis and the agrometeorological stations to optimize the network and the selected sites. The storage in the open-source platform "thingsnetwork" ensures free access to the sensor data [Lora20, The21].

The GBI and especially the communication component "Messenger" with geo-formulations enable farmers and advisors to send back information directly and integrate them immediately into the visualization component GBV.

Consequently, the next step is establishing a landscape data space as a hybrid system; integrating cloud computing methods with a local and decentralized data storage. Enabling the farmers to get sovereignty on their own data, combined with the possibilities to share the data on a secure platform leads to a cooperative concept, adding value to different fields of application and scales.

Additional applications to be investigated include pesticide management and risk mitigation on a landscape level through applications like PAM3D or research results like the RiskMin-approach [TDS+20, Zep20]. In agriculture, various plant protection products are subject to slope restrictions as part of the approval process. The aim is to avoid the discharge of pesticides into adjacent water bodies caused by surface runoff and soil erosion. To date, no instruments exist to reliably and objectively determine the slope. The planning and implementation of pesticide applications is the responsibility of the farmer, according to good professional practices. The goal of the PAM3D project was to build a web service that enables the farmer to fulfill slope requirements in an automated and traceable way [Zep20]. Biodiversity is the basic resource maintaining and supporting ecosystem services and functions. It is present at different levels in the complexity of a given landscape. Therefore, when dealing with biodiversity related aspects, it is important to consider the whole landscape instead of only analyzing single fields or elements. The European Food Safety Authority (EFSA) published different opinions and scientific papers in the context of recovery and recolonization of populations. In their opinion, a clear relationship between landscape and risk is formulated. Therefore, in the research project RiskMin, the effects of different kinds of risk mitigation strategies were simulated on a landscape level [TDS+20].

All these examples show how an infrastructure like the GBI can be used to improve impacts at the landscape level and how to manage the challenges of farming today. This is especially important for small-scale landscapes and farms. They need user-friendly information and free easy to use services that work on different devices, including mobile devices. The main challenge of such an approach is to scale it to the country, European, and global level. In order to solve these challenges, the focus of the GeoBox project is not to create a new standard for data, but use existing global standards like FAIR principles, Linked-Data, etc. At present, a major challenge is how to define the "hand-over" between public and commercial offers.

3.8 Technology Outlook

Katrin Jakob

Abstract

Continued technical development of sensors, AI algorithms, computing power, batteries, robotics and in field communication will ultimately culminate in the creation of digital twins, a complete digitalization of the physical agricultural production chain from demand to production and sales, removing inefficiencies, increasing traceability, thereby improving production economics and sustainability. Technologies developed by start-ups are on the forefront of this 4th Agricultural Revolution. At the same time, access to digital technologies democratizes agricultural production in developing countries by allowing access to resources not available before.

3.8.1 Introduction

Digitalization creates the opportunity to catapult agricultural production into its 4th revolution, allowing for informed and unbiased decisions, removing unknowns, increasing agility, and thus creating more effective Production Systems while reducing its environmental footprint. However, adoption of digitalization in agriculture has been slow due to many inefficiencies. Incompatibility of different data platforms and file formats (see ▶ Sects. 3.4 and 3.6), no apparent ROI for the farmer (see ▶ Sect. 1.5), high costs, connectivity issues, and real-time availability in the field, scalability, and insufficient AI are some of the factors that have hampered broad implementation of digital technologies in agriculture.

Digital technologies, challenges, and how to overcome these have been discussed in the previous sections. Start-up companies are critical in developing new technologies and bringing them to market. At the end of 2020, there have been more than 1,600 start-ups in the field of digital agriculture [Hal20]. Despite the initial shortcomings, digitalization will transform agriculture in the next decades. Insufficient manual labor resources, compliance and regulatory policies and environmental impact of agricultural production will speed up development and adoption of digital technologies. Changing customer behavior is also affecting digitalization. Consumers today are becoming more aware of food production practices and are asking for transparency in farm production and supply chains and traceability of sustainable practices. Digitalization has opened a new toolbox for farmers like no other before.

3.8.2 Key Enablers for Digital Agriculture

3.8.2.1 Sensors: The Eyes, Ears, and Noses of Tomorrow's Farmer

Sensing and monitoring devices on the ground, on machinery, and in the air have taken a leap with the advent of IoT, AI, and wireless data transmission capabilities. For digital agriculture, sufficient information of crop, soil and machinery status is needed in real time. Soil and plant parameter mapping, yield monitoring, variable rate seeding, fertilizing, and spraying, weed mapping, topography and boundaries and

Technology Perspective

machine guidance systems are some of the applications these data are applied to. Technical advances in optical, chemical, electrical, mechanical, organic and ultrasonic sensing are opening new ways how plant, soil, machinery and weather parameters will be monitored in the future.

Nanomaterials and nanosensors can play a role for in situ sensing, allowing for high spatial and temporal resolution. Nanotubes materials have been tested in plants and on plant leaf surfaces to measure stress-related and other plant molecules [GLK+15, WGK+17], acoustic sensors measure pests based on the sound of their chewing and movement [MSP15], (Agrint Sensing Solutions [Agr20a]) and electrical noses sense plant and disease specific volatiles given off by the plant when attacked by diseases or insects [CIA+19, SWC17]. Measuring signaling molecules directly in the plant allows for earlier detection of stress hence enabling a timelier response to counteract stress, preventing yield loss. As an example, different fluorescent proteins can be attached to nutrient, drought, or pathogen stress signaling molecules and the signal of plants in the field can be captured with a hyperspectral camera (InnerPlant, [Inn21]). Printed electronics have become a means to produce biodegradable, chip-less sensors measuring nitrogen and moisture content. Widespread dispersal of these sensors in the soil would allow for unique data collection [ADM+2020].

Increasing water scarcity and regulation push innovation in soil moisture sensors. Electrochemical, mechanical, dielectric soil moisture, and airflow sensors are used to measure soil parameters and added wireless data transmission capabilities allow for remote data collection. To offset soil disturbance when installing soil sensors, CropX has developed a geometric shape for its sensor to prevent soil disturbance providing more accurate readings [Cro21]. Uniquely, Groguru has developed soil sensors that are installed below tilling depth and can stay in the field for 5 years [Gro20]. Teralytic has taken soil measurement to the next level by adding nitrogen, phosphorus and potassium measurement capabilities to its wireless soil probe [Ter20]. Arable has developed an acoustic sensor to measure rainfall with the highest accuracy [Usp18].

Developments in gamma radiation, microwave, and optics sensing and analytics allow for remote sensing from machinery, satellites, or drones. SoilOptix's topsoil mapping system, mounted on machinery, is based on gamma radiation measuring nutrient properties and soil texture characteristics plus plant available water, and bulk density [Red20]. Vandersat uses microwaves to predict soil moisture from satellite data at a 100 m $\times$ 100 m resolution at 10, 20 and 40 cm soil depth [Van20].

A leap in optical sensor, camera, and machine/deep learning algorithms development has allowed for monitoring, assessing, and calculating critical plant and soil parameters based on RGB, hyperspectral and multispectral imagery in combination with thermal and other sensors. The addition of LiDAR and synthetic aperture radar technology has increased image quality and precision. With these improved images, datasets and deep learning algorithms, predictive data are becoming more robust. Slantrange's sensors are based on six spectral bands with True Color RGB and feature an integrated dual-antenna RTK GPS, LIDAR rangefinder, and extended Kalman filter navigation solution for precise individual plant size and shape definition [Sla2020]. Persistence data mining employs NIRS hyperspectral sensors mounted on a tractor to identify organic matter, soil type, moisture, and nutrients and micronutrients [Per20]. Phenospex's plant eye combines 3D laser vision with spectral image to calculate plant parameters [Per20]. Veri's Technologies provides "on the go" soil mapping with sensors mounted on a sled, including an optical sensor beneath the crop residue mapping organic matter [Ver20]. Croptimal deploys a spectroscopy solution based on energy-dispersive X-ray fluorescence (ED-XRF) technology for soil and plant analysis reducing analysis time from days to about 50 min [Cro20a].

Due to continued chip technology and battery development, sensors are becoming smaller, more versatile, and last longer. Particularly, improved optical sensing, computer vision and AI-based modelling and analytics will allow for remote data collection and precise prediction of soil, field and crop conditions, from nutrient status to water content without direct measurements. Data transmission via LoRaWan from sensors to base stations to the cloud has been evolved as the most common data transmission for smart field sensor technologies and will likely be a viable option for some time as connectivity in the field remains problematic. Edge devices for real time-in field communication and AI added analytics, satellite capabilities and improved battery life will continue to boost remote sensing.

3.8.2.2 Artificial Intelligence

Technology follows roughly an exponential growth in data gathering volume, analytical capabilities to transform that ocean of data into models and determine model-based actions. Through digitalization, immense datasets will be produced, and model generation, calibration, and actionable conclusions will bess mined by artificially intelligent systems (see also ▶ Sect. 3.5). At the same time, the cost of access to the systems will go down exponentially. This will allow smaller businesses to utilize such systems, especially as business models for access become more per-use based rather than equipment purchase based.

Agricultural production is an uncontained, complex system of the soil and crop environment and weather, exposed to constantly changing conditions, and hence remains difficult to manage. In addition, demand, supply, and market prices change constantly. Artificial intelligence will be key to understanding the system's complexity and to converting data into precise action at the right place at the right time, and along the entire Ag production chain from farming and farm management to distribution and sales. The application of AI will have unprecedented impact on how crops will be produced, stored, distributed, and sold. Inefficiencies and uncertainty along the production chain can be removed and replaced with accuracy and resource efficiency not seen before. AI in the agriculture market is projected to grow from about USD 1.0 billion in 2020 to USD 4.0 billion by 2026, at a CAGR of 25.5% between 2020 and 2026 [Mar20a].

AI and deep learning capacity depends, among other factors, on the availability of computation power and training data. Computer chip and code development will continue and improve processing power, reduce power consumption, and improve computational power, keeping following Moore's law. Developing domain-specific architectures, which are tailored to specific tasks, is one way to improve computing power. GPUs have been developed for graphics processing in computer gaming and have been proven to be well suited for machine learning applications. Tensor processing unit chips have been built by Google specifically for deep neural network inference tasks [WWB+19] and have started helping to decipher biology's complexity [Cal20]. Crop production happens in specific ecosystem including soil, crops, microbiomes, weather and more, and entails several biological systems for which present computing capabilities might not be sufficient. The advent of quantum computing can provide further opportunity to decode these complex systems to improve crop production efficiency (PsiQuantum, [Psi20]). Enhancing prediction of weather, and pest and disease pressure will be crucial to mitigate crop losses from biotic and abiotic stress and save inputs and resources.

OneSoil and Blue River both are using convolutional neural networks on Nvidia

GPUs for their robotics technology. One limitation of this technology is the need for large data sets to train these networks. OneSoil manually marked out boundaries on nearly 400,000 fields for training data. Now its algorithms can automatically create boundaries from satellite data [Mar19a]. Blue River used over a million images for network training [Mar19b] to enable their "See&Spray" technology for weeding. One Shot Learning using a Siamese Network has come into focus asserting to minimize the number of training images. The process involves a classification of images based on probability of the presence of an image creating classes, and each class only requires one training sample [KZS15].

Optimized computer vision in combination with new sensors and advanced AI algorithms will allow for applications, which we might not even imagine yet, like hyper-precision application of crop inputs, a spot on a leaf to destroy a fungus or aphid, or harvest a fruit that has the perfect ripeness, even if a leaf or branch is in the way of the harvesting device. Vision Robotics enables robots to autonomously create three-dimensional models of vines which are used for correct pruning, and in lettuce for correct thinning [Vis20]. Through advanced image processing and AI, Taranis achieves 8 cm resolution in the field based on areal images allowing spotting an insect on a leaf [Sfs20].

Earlier AI analytics platform solutions to predict plant and soil parameters were siloed, difficult to use, provided no actionable insights and only offered single applications. The platform of the future needs to bring these silos together and provide a comprehensive set of solutions from planting recommendations and stand measurements to irrigation, fertilizer, disease and harvest management and more. Each specific area will continue to develop and plug and play platforms both for hardware and software, which can seamlessly integrate farmers knowledge, the most advanced sensor, new data sets, crop models and AI algorithms will be required.

Microsoft's Azure FarmBeats provides a platform for customers to build their AI and machine learning models leveraging agriculture datasets across different data providers from sensor data to drone and satellite data [Mic20]. Stream.ML provides a market platform for ML [Sml19]. DroneDeploy's aerial maps can be downloaded into different apps at John Deere's, Agremo, Climate's FieldView and others [Dro20a]. Arable's Mark can add third-party soil sensors, while its disease modeling data can be integrated into BASF's xarvio platform for fungicide recommendations [Bur19]. Arable's sensor data are being integrated into Netafim's automated precision irrigation and fertigation cloud-based platform as well as into their crop model algorithms [Ara20]. Regrow's FluroSense platform offers ordering of high-resolution satellite data that can be integrated into their decision platform. In addition to integration of sensor data, crop models and other data, this platform establishes two-way integrations with major agronomy-focused farm management systems (Proagrica, Agrian, Agworld) [Reg21]. Farmwave's CORE analysis details data from sensors, machinery, and software and is designed for compatibility with agriculture SaaS platforms [ABB+19]. Draganfly's Quantix-Mapper provides non-proprietary image formats from its drone images in open file format that can be integrated in mapping and modelling tools [Dra20].

Importantly, NEVONEX is closing the missing link between digital platforms from different vendors and backend integration into farmers' machinery. The company offers a retrofit kit for machines to integrate digital services chosen by the farmer which can then easily be accessed on any device [Nev21]. A consolidation of IoT Ag companies is also expected to successfully compete in the market and bundle services and products delivered to the farmer.

3.8.2.3 Real-Time Communication in the Field: 5G, Edge Computing, Satellites

High-volume, real-time data transmission in the field without latency is needed in precision agriculture for real-time alerts and execution via actuators on irrigation, machines, and robotics as well as for real-time video and mapping analysis from drones and planes. Connectivity in rural areas remains an obstacle. Approaches to solving this problem are next-generation cellular networks, edge computing, and universal provision of broadband internet via satellites.

Roll out of 5G has begun worldwide in 2018 but remains slow. Building coalitions with telecom and edge device technology companies to bring 5G to rural areas might be an accelerator for faster access. The US Rural Cloud Initiative aims to cover 3.8 Mill km^2 with an edge cloud network based on private LTE networks and a 5G Network using open radio access network technologies and solutions [Tri20]. Similarly, China Telecom worked with partners to enable 5G and smart farming in the rural Zhejiang Province [Hua20]. Facebook, Google, Intel and Qualcomm and international VCs have invested over 20bn USD to support building out online service and extend into rural areas in India [Sin20]. International telecom companies are testing 5G in Latin America [Lop20]. In Africa, first 5G tests started in 2018 and a commercial 5G is available in South Africa [Aja19]. While Europe had planned to be fully 5G ready in 2020, it is lagging as only 11 member states have 5G roadmaps [Sol20].

Interestingly, there will be shortcomings of 5G for IoT and AI applications because it has not been developed with these applications in mind and research into 6G has begun to offset these. One of the main aspects of 6G will be AI involved communication as well as Space-Air-Ground-Sea integrated communication [ZYX19]. This technology will be more suitable for powerful IoT and M2M applications.

It is expected, that even with expansion of 5G, its data transmission capabilities will not be sufficient for the ever-increasing data volume. Hence, AI will be moving to edge devices, complementing server and cloud AI. Data is collected, stored, refined, and send back out in near real time reducing latency from milliseconds to microseconds. Such setup allows sensors and machinery connecting to the farm network without depending on a remote server or cloud enabling real-time autopiloting of machinery and other automated farm operations. As an example, FreeWave's wireless M2M communication system enables any number of smart agriculture applications in the field. It captures data from sensors and robots and a base station receiver can send real-time data corrections to machinery in the field for navigation and other tasks [Fre18].

Decreasing satellite and launch costs and focus on nanosatellites cruising in low orbit (600–800 km) for low latency are promising developments to economically provide broadband access via satellites. Sensoterra's soil sensors are planning to use Hiberband's Low Power Global Area Network of nanosatellites at 600 km above Earth to regulate irrigation in crops [Sen20]. SpaceX's Starlink mission aims at providing worldwide broadband access with the goal to fly as many as 42,000 satellites [Eth20]. As of January 2022, over 2,042 Starlink satellites have been employed and people can sign up globally to receive internet service. Even more exciting for autonomous and AI-based vehicles is the outlook that Starlink wants to connect to moving vehicles, albeit large vehicles first [Bro21a], and the satellite dish can go with the customer to any place [Bro21b]. Oneweb plans to install about 48,000 low orbit satellites and Amazon's project Kuiper received approval to install 3,236 low orbit internet satellites [One20, Ama20]. The race for broadband internet availability via low-orbit satellites will be advantageous for the farming industry and would help to faster adapt digital technologies, particularly in areas where cellular access is not feasible for a foreseeable future.

3.8.3 Autonomous Machinery for Tomorrow's Agriculture

3.8.3.1 Satellites for Sensing and Mapping

Satellites offer unmatched consistent imagery over drones and planes for wide areas but have lacked hyper precision resolution, were too expensive to be regularly updated with the latest sensor technology and costly to send them into orbit. Building a satellite can easily cost a few 100 million USD depending on the technology and launch into orbit used to cost from 50–400 million USD and more [Glo20]. However, significant changes are coming to the satellite industry. AI now enables better analysis of historic and present satellite data and satellites can be miniaturized, are becoming less heavy and can be launched into orbit on a regular basis. The satellite race has already started, and competing companies offer satellite data and analytics software that can observe Earth 24/7 or provide services for satellite data analytics. Capella Space plans to offer hourly coverage of every point on Earth, rendered in sub-meter resolution and provide imagery on demand within 30 min [Cap20] and BlackSky Global promises to revisit most major cities up to 70 times a day. Applications for agriculture will follow. SpaceX has launched a rideshare tool for Falcon 9 launches starting at $1 million for payloads ranging up to 200 kg. Latitudo40 and Ursa Space provide service for satellite data analytics that combines artificial intelligence and machine learning to analyze date from networks of satellites [Lat20, Urs20].

CubeSat's 10-cm cube with less than 1.3 kg weight is used by Planet to significantly enhance their satellites. The latest launch of 44 new satellites in January 2022 on SpaceX's Falcon9 brings the overall number of Planet satellites in orbit to 127, and continuous expansion will provide world class, high-resolution images. Satellite Vu intends to launch an infrared satellite sensor that can rate the energy performance of every building on the planet, and this technology might also be applicable to agriculture [Kol20]. ChipSats of the size of a stamp powered by solar energy have also been deployed, forming a swarm of cracker-sized nanosatellites at an altitude of 300 km and the prospect is, that people build their own ChipSat tailored to their interests [Bat19]. In summary, satellite technology has become accessible through which vast amount of satellite data will be available at affordable prices.

3.8.3.2 Unmanned Aerial Vehicles and Drones

The arrival of drones pushed IoT in agriculture to new heights but short battery life, flight regulations, connectivity issues, inadequate drones, high costs, insufficient sensors and analytics, and no immediate ROI for the farmer hindered quick adoption. Many features have been improved and flight regulations for AG drones in the US have been lifted which has rekindled interest from VCs for further investments in the space (DroneDeploy, Aerobotics, AgEagle, Taranis). Drones today carry self-calibrating multispectral sensors, are autopiloting, have longer flight times, and became more affordable. The market potential for agricultural drones in the next 5 years is forecasted to have an exceptional growth of 35.9% CAGR from 2020 to 2025, raising from 1.2 billion USD to 5.7

billion [Mar20b]. Taranis has managed a new level of efficiency and resolution, capturing 0.3 mm/pixel resolution images of fields from planes and drones at a speed of ~40 ha (100 ac) in 6 min [Tar20]. A fully autonomous drone fleet is available from American Robotics. Fixed wing drones are promising higher efficiency compared to quadcopters [Ili19]. Drone-based datasets might need to be sharable for future plug & play analytics and management platforms to compete in the marketplace, as seen with Draganfly's and DroneDeploy's data sets.

Drones for area monitoring have advantages over satellites. They can more easily be updated and equipped with the latest sensors, adjusted faster than satellites, provide for better resolution, and are cheaper. However, advances in satellite data capture and analysis and falling prices for launches are making satellite data competitive with drone data which will likely lead to an increase in satellite data use. Application of drones in the future might focus less on area monitoring tasks but on specific plant monitoring, like fruit counting (Aerobotics, [Aer20]) and use of drones for specific work tasks. Spraying and fertilizing by planes is not suited for hyper-precision but drones will be able to deliver hyper-precision for seeding, spraying, pollination, harvest and other tasks. Today, variable rate application via drones is possible for fertilizer, pesticides, and herbicides (DJI Agras MG-1, [Dji20]). Rantizo provides exchangeable seeding and spraying tanks [Ran20]. XAG's drones are flying in swarms for spraying [Gan19] and Dropcopter drones pollinate orchards [Dro20b]. Tevel Aerobotics follows fruit development in apple orchards over the season and had planned to steering the drone to the apple at harvest based on data acquired during growth [Tev20]). Drones will replace working tasks of heavy machinery that are more effectively executed than by robots on the ground. They can also access areas that might otherwise not be accessible by ground machinery.

3.8.3.3 Autonomous Field Machinery and Robots

Sensor and IoT technology advances, and particularly progress in computer vision, AI, real-time or near-real-time data analytics via edge devices and the cloud have become enablers of autonomy and robots in the field. This development has also been spurred by significant shortages of manual labor. As we get closer to being able to map every plant and leaf and provide plant and leaf specific treatments, machinery is needed that can deliver with the same precision as we map to the plant and leaf within centimeter or even millimeters. Visibility throughout the production chain, enabled by blockchain, and renewed interest in different crop production systems like intercropping, and optimized applications are other factors that have stimulated robotic development in agriculture.

Hurdles in developing and adopting robots in the field have been manifold. Robustness, battery life, precision, and speed as well as the economics are some of the factors that have made the adoption of robots in the field slow. However, this is changing rapidly. Like the drone market, the robotics market is expected to grow significantly, from 4.6 billion USD in 2020 to 20.3 billion USD in 2025, at a CAGR of 34.5% [Mar20c].

The traditional symbol of agriculture, the tractor, is transforming into a robot that comes in many forms. Two companies have been leading the development of a fully autonomous, electric, smart tractor, Monarch Tractors and Ztractor. Machines of both companies are equipped with sensors, cameras and AI technology to sense and analyze on the go which might replace direct sensors in the field in the future. Monarch's tractor provides 40 HP [Mon20], which is at the lower end of traditional tractor's power [Joh20] but an impressive start into a new era of tractor development. Other companies have been developing robots that are replacing the tractor in the traditional sense and are specialized for tasks and tool platforms.

There are 4 trends evolving in robots for Ag: (1) highly specialized autonomous robots that are focused on one specific task, (2) a robotics platform, where one autonomous base station can carry exchangeable tools for different tasks, (3) autonomous robots that can be managed in swarms or fleets and can either be specialized or flexible with different tool mounting options, (4) a combination of base stations serving robots and drones.

(1) The first robotic development approaches during the last decade have been specialized for weeding, spraying, trimming, and harvesting, and mostly in high value crops and orchards where manual labor has been in high demand. Farmwise has been one of the first companies solving the weeding problem in vegetables to reduce manual labor, based on computer vision combined with mechanical weed removal. Carbon Robotics thrusted itself into this market breaking autonomous weeding records by weeding 15–20 acres per day. Also based on Nvidia's chip system, and AI vision models, weeds are eliminated by high energy lasers [Car21]. Manual strawberry harvesting is being replaced by robots developed by companies like Agrobot, Advanced Farm Technologies, Harvest Croo Robotics, and Dogtooth technology (greenhouse). Harvest Croo Robotics claims to replace up to 30 workers with one robot [Paq19]. FFRobotics is developing autonomous robots for apple harvesting. Field tests have shown difficulties ranging from rugged terrain, low speed to correct object recognition. Fieldwork Robotics is in early trials to harvest raspberries, which should allow for harvest of other fruits once the technology has been developed for such a delicate fruit [Wil20]. A completely autonomous spraying robot for orchards has been developed by GUSS which can work in a fleet and up to 8 robots handled by one person from a PC or tablet [Gus20]. Robotic phenotyping in the field is being approached by Earthsense/Terrasentia and Acuris monitoring corn, soybean and other crops and kiwi, respectively [She20, You18]. Similarly, Google is working on a ground bot that can count pods and seeds in soybeans [Gra20].

(2) To make robots more flexible and less expensive, robotic platforms are developed which enable different tools to be mounted for different agricultural tasks from seeding, weeding, pruning and spraying to harvesting different crops. Semesis is in early development of a platform executing sowing, weeding, spraying and fertilizing [Sem20]. Thorvald's autonomous modular platform from Saga Robotics works for fruits and vegetables and performs UV-treatment, picking fruits and vegetables, phenotyping, in-field transportation, cutting grass for forage, spraying and data collection/crop prediction [Sag20]. A kiwi harvester has also been developed by Robotics Plus [Rob20]. Etarob's autonomous platform in vegetables not only enables adding tools for field tasks like mechanical weeding, spraying and others but can also be combined with a logistics and packaging station, expanding its use to after harvest technologies [Ail20].

(3) Swarm Farm and Rabbit Tractors provide robots that can be tailored to different tasks and work in swarms [Swa20, Mar20e], similar to Fendt's Xaver model project [Fen20]. Xaver's most recent development shows a specialized sowing robot [Val20]. As robots are usually smaller, swarming is an effective way to cover large areas with high precision. One advantage of smaller robots is their flexibility to adjust to different farming practices which might evolve in the future, like intercropping or agroforestry.

(4) A combination of an autonomous robotic base station/edge device with drones for apple harvest has been employed by Tevel Robotics [Tev20]. Taking it a step further, Yanmar's SMASH system entails

a mobile base with a robotic arm, a drone and a ground station to execute tasks from crop monitoring to weeding, taking soil samples and more [Cla20].

Agricultural robotics is just at the beginning. Robotic arms, imaging technology, AI, and analytical platforms will continue to evolve and eventually become more standardized and at the same time more flexible. As an example, AI will not only handle to recognize weeds but there will be modules trained for many other Ag applications from recognizing a disease on a leaf to vegetables and fruits for harvest. These modules are interchangeable and can be combined with an appropriate standardized robotics part for pruning, harvesting and other tasks. This will bring efficiencies in robotic equipment production, which will bring down costs and accelerate adoption.

Specialty crop robots are leading this revolution as the economics for robots are most favorable in this sector due to the high labor cost which can be up to 70% of the overall production cost. The Global Harvest Automation Initiative by the California organization Western Growers aims to automate 50% of harvest across the fresh produce industry within 10 years. Such an initiative brings together farmers, Ag organizations and technology companies which can significantly streamline development and adoption preventing technology developers from inventing the wheel twice and use robotics and field expertise already available. Row crop production, already highly automated, will still benefit from further digitalization and autonomy from production to sales. Here, application of seed, fertilizer and other inputs will be applied highly localized at the right time, with the right amount, saving up to 80% or more on cost for certain inputs, i.e., pesticides.

Breeding and biotechnology will also play a role in the transformation to autonomous agriculture. Crops and trees might need to be adjusted to field robots. Plant or tree density, row width and crop architecture from branching to fruit location affect how a robot can access a field and a crop. Despite progress in breeding and genetic engineering through CRISPR, developing new crop varieties takes several years and it would be advantageous to know how the robots of the future look like to be able to start breeding today. Robotic standardization can help to make this process faster.

Robot technology can be less heavy to reduce soil compaction, more precise to reduce water, chemical and fertilizer applications, execute new tasks, i.e., pollinating, replace manual labor, and can be deployed 24/7. Robotics, Farming, or Weeding -as-a-Service (RaaS, FaaS, WaaS) are evolving as new business models and will make access to new robot technologies easier for the farmer. Another positive effect of this technology development, including AI and other digital technologies, is, that younger generations have taken an interest in agriculture helping to bring digital technologies into the field and usher agriculture into its 4th revolution.

3.8.4 Digital Twin: The Farmer as a Factory Manager

Advances in digitalization will culminate in the creation of digital twins for the agricultural production, a digital replica of the physical production chain including all steps from customer demand to production to supply chain to end customer sales, minimizing any inefficiency in the entire production and supply chain. A digital twin enables optimizations of all processes through simulation of multifarious scenarios which allows its users to make informed decision and adapt quickly to changes along the production chain [DDK+20]. The combination of a digital twin with blockchain technology can ensure secure accessibility and authentication of users, traceability, and data immutability in the system [HSJ+20].

Research projects to create a digital twin of tomato or for the N Cycle in a farming operation are underway [Kni20] while Agronomeye has made its goal to completely digitize one's farming operation [Agr20b]. Taiwan has started a research project on digital twins in agriculture and aquaculture hoping it can increase production efficiency of most of its small farm holders by 30% [Tec19].

Digital twins already exist in industry sectors like manufacturing but a digital twin in agriculture is an ambitious goal. Greenhouse operations will become fully digitized faster than farm operations, given that they operate in a controlled environment. While building a digital twin, production, distribution, and sales will become more and more independent from direct human interaction. The farmer of tomorrow will be able to oversee the farming operation on a control platform and will be working with digital agronomists to capture the highest value from his farming operation while optimizing inputs in a sustainable, environmentally friendly, and economic production system.

3.8.5 Democratization of Agriculture Through Digitalization

Digital technologies are catapulting agricultural production into a new era of accessibility, openness, and traceability. While the food production system has been inaccessible and centralized in the hands of a few large companies, increasing freedom of information enables in principle wider access to resources and technology and increased efficiency of agricultural production in countries which didn't have direct access to technology before.

Not surprisingly, developed countries are leading the adoption of digital technologies, followed by Brazil [Mar20d]. Digitalization of the entire production process and supply chain will eventually lead to full transparency of agricultural production which in turn will define what the consumer will buy and how policies in agriculture will be shaped.

Digitalization will positively impact food production in developing countries where millions of small farm holders, the main agricultural producers, lack knowledge and access to financial and other agriculture related services for crop production. Mobile technology offers these farmers access to information, water, seed, fertilizer, markets and more including access to money/credit bringing new empowerment to small farm holders and rural communities. Knowledge-sharing platforms like WeFarm, PEAT/Plantix, or CropDiagnosis [Wef20, Pla20, Cro20b] are easily accessible via mobile and are used by over 500,000 small-scale farmers around the world, including Africa and India. Bloom is an integrated technology for Asia and Latin America providing small-holder farmers via mobile phone platform with an 'all in' platform from seed recommendation to sales of their products [Blo20]. Access to these platforms will allow for more sustainable production practices, prevent food loss, and increase income. Once the farmers are becoming more productive and grow their operations, they will be able to increasingly incorporate additional digital technologies like renting a tractor (Hello Tractor, [Hel20]) or drones (AcquahMeyer Dronetech, [Amd20]). Rent a service business models might enable farmers to use technology they cannot afford otherwise. This can be a leap in improvement of a small farmers life, develop rural communities, improve agricultural practices, and make them sustainable. In addition, farmers can work together to negotiate better prices for service or materials, and consequently, farm consolidations are expected to grow to achieve economy of scale [Pra20]. There remain obstacles. Internet access in rural areas, affordability of data, literacy and digital literacy require continued improvements to drive adoption of digital technologies. Local governments and NGOs together with the private sector need to step up to further improve training and access to digital resources. Impact focused investment will play an increasing role in developing countries. The venture capital firm Omnivore provides a bold vision for India, which could serve as a model for other developing countries [Pra20].

References

[5st21] 5-star Data. ▶ https://5stardata.info/de/. Accessed 26 March 2021.

[ABL17] Aubert, Céline, Pier Luigi Buttigieg, Marie-Angelique Laporte, Medha Devare, und Elizabeth Arnaud. 2017. *CGIAR Agronomy Ontology*. ▶ http://purl.obolibrary.org/obo/agro.owl. Accessed 5 March 2021.

[ABB+19] Adams, Chuck, Charles Bassham, Frank Benoit, Brian Bull, Chris Chan, Craig Ganssle, Bryan DiMatteo, Tom Hyatt, Steve Hyland, Noriaki Kono, Krishna Padmanabha, and Nick Palczynski. 2019. *Farmwave White Paper*. ▶ https://www.farmwave.io/whitepaper Accessed 10 Aug 2020.

[ADM+2020] Atreya, Madhur, Karan Dikshit, Gabrielle Marinick, Jenna Nielson, Carson Bruns, and Gregory L. Whiting. 2020. (Poly(lactic acid)-based ink for biodegradable printed electronics with conductivity enhanced through solvent aging. *ACS Appl. Mater. Interfaces 2020* 12(20):23494–23501.

[Adv21] *Arbeitsgemeinschaft der Vermessungsverwaltungen der Länder der Bundesrepublik Deutschland (AdV)*. ▶ http://www.adv-online.de/GeoInfoDok/broker.jsp?uMen=d3b70780-c5f2-bc61-f27f-31c403b36c4c. Accessed 4 March 2021.

[ADS19] Agricultural Data Space Whitepaper. *Eine Veröffentlichung des Fraunhofer-Leitprojekts "Cognitive Agriculture"*. Fraunhofer IESE, November 2019. ▶ https://www.iese.fraunhofer.de/content/dam/iese/de/dokumente/innovationsthemen/COGNAC_Whitepaper_ADS2019.pdf Accessed 11 March 2021

[ADS21] *The Agricultural Data Space is to make available all the data that is necessary for making decisions in agriculture*. ▶ www.iese.fraunhofer.de/en/innovation_trends/SmartFarming/cognitive-agriculture/agricultural-dataspace.html. Accessed 12 Feb 2021.

[AEF21] *The Agricultural Industry Electronics Foundation*. ▶ www.aef-online.org. Accessed 12 Feb 2021.

[Aer20] Company Aerobotics. ▶ www.aerobotics.com. Accessed 13 March 2022.

[Agc20] AGCO. 2020. *Fendt Guide Connect*. ▶ https://www.fendt.tv/home/fendt-guideconnect_1408.aspx. Accessed 16 Dec 2020.

[AgG21] *AgGateway: a global, non-profit organization with the mission to develop the resources and relationships that drive digital connectivity in global agriculture and related industries*. ▶ www.aggateway.org. Accessed 12 Feb 2021.

[Agr20a] Company Agrint. ▶ www.agrint.net. Accessed 13 March 2022.

[Agr20b] Company Agronomeye. ▶ www.agronomeye.com.au. Accessed 13 March 2022.

[Ail20] Company Ai.Land. ▶ www.a-i.land/en. Accessed 13 March 2022.

[AIO21] *The Alliance for Internet of Things Innovation with the mission to contribute to a dynamic European IoT ecosystem*. ▶ http://www.aioti.org/. Accessed 12 Dec 2021.

[Aja19] Ajao, Oluniyi D. 2019. *The State of 5G in Africa in 2020*. ▶ https://tech.africa/5g-africa/. Accessed 13 March 2022.

[Ama20] Amazon. ▶ https://blog.aboutamazon.com/company-news/amazon-receives-fcc-approval-for-project-kuiper-satellite-constellation. Accessed 10 Aug 2020.

[Amd20] Company AcquahMeyer. ▶ www.amdronetech.com. Accessed 13 March 2022.

[Ara20] ▶ https://www.arable.com/2020/01/31/netafim-arable-how-high-quality-data-enables-precision-irrigation-easily-affordably-and-globally/. Accessed 13 March 2022.

[ATL21] *ATLAS project that will build an open, distributed and extensible data Interoperability Network, based on a microservice architecture which will offer a high level of scalability from a single farm to a global community*. ▶ www.atlas-h2020.eu. Accessed 12 Feb 2021.

[Bat19] Bate, Tom. 2019. *Inexpensive chip-size satellites orbit Earth*. ▶ https://news.stanford.edu/2019/06/03/chip-size-satellites-orbit-earth/ Accessed 13 March 2022.

[Bay21] Bay, Wa. *Halbjahresbericht-2020*. ▶ https://www.baywa.com. Accessed 18 March 2021.

[BBB+13] Bauer, M., M. Boussard, N. Bui, F. Carrez, C. Jardak, J. De Loof, K. Magerkurth, S. Meissner, A. Nettsträter, and A. Olivereau. 2013. *Deliverable D1. 5—Final architectural reference model for the IoT v3. 0. Internet of things architecture (IOT-A)*.

[Ber10] Berners-Lee, Tim. 2010. *"Is your Linked Open Data 5 Star?"*, World Wide Web Consortium (W3C). retrieved on 26 Mar 2021 from: ▶ https://www.w3.org/DesignIssues/LinkedData.html#fivestar. Accessed 28 March 2021.

[BFM05] Berners-Lee, Tim, Roy Fielding, and Larry Masinter. 2005. *"Uniform Resource Identifier (URI): Generic Syntax"*, Internet Engineering Task Force. ▶ https://tools.ietf.org/html/rfc3986. Accessed. 5 March 2021.

[BG14] Dan Brickley, and R. V. Guha. (Eds). „RDF Schema 1.1.", W3C Recommendation 25 February 2014.▶ https://www.w3.org/TR/rdf-schema/. Accessed 13 March 2022.

[Bit20a] ▶ https://www.bitkom.org/Bitkom/Publikationen/Bitkom-Stellungnahme-zur-Ackerbaustrategie-2035. Accessed 13 March 2022.

[Bit20b] ▶ https://www.bitkom.org/Presse/Presseinformation/Schon-8-von-10-Landwirten-setzen-auf-digitale-Technologien.

[BKA13] W. Bangert, A. Kielhorn, R. F. A. Albert, P. Biber, and S. Grzonka. 2013. *"Field-Robot-Based Agriculture"*, in Land.Technik, AgEng. Hannover.

[Ble21] Federal Office for Agriculture and Food (BLE). ▶ https://www.ble.de/DE/Projektfoerderung/Foerderungen-Auftraege/Digitalisierung/Machbarkeitsstudie/Machbarkeitsstudie.html. Accessed 13 March 2022.

[Blo20] ▶ https://www.linkedin.com/company/blooom-democratizing-the-future-of-farming/about/. Accessed 13 March 2022.

[BME20] The Federal Ministry of Food and Agriculture (BMEL). ▶ https://www.bmel.de/EN/topics/farming/climate-stewardship/climate-stewardship_node.html. Accessed 13 March 2022.

[Böt13] Böttinger, S. 2013. *Stand und Tendenzen der Mähdrusch-Entwicklung*, in VDI-MEG Kolloquium Mähdrescher 12./13. September 2013, Stuttgart/Hohenheim.

[Böt15] Böttinger, S.. 2015. Mähdrescher. *Jahrbuch Agrartechnik* 27:158 – 170.

[BPS+08] Bray, Tim, Jean Paoli, C. M. Sperberg-McQueen, Eve Maler, and Francois Yergeau. 2008. *"Extensible Markup Language (XML) 1.0 (Fifth Edition)"*, World Wide Web Consortium (W3C).▶ http://www.w3.org/TR/xml/. Accessed 5 March 2021.

[Bra17] Bray, Tim. 2017. RFC8259: The JavaScript Object Notation (JSON) Data Interchange Format. *Internet Engineering Task Force (IETF)*. ▶ https://tools.ietf.org/html/rfc8259. Accessed 5 March 2021.

[Bro21a] SpaceX plans Starlink broadband for trucks, ships, and planes [Updated]. ▶ https://arstechnica.com/information-technology/2021/03/spacex-plans-starlink-broadband-for-cars-boats-and-planes/. Accessed 7 May 2021.

[Bro21b] Dishy McFlatface to become "fully mobile", allowing Starlink use away from home. ▶ https://arstechnica.com/information-technology/2021/04/dishy-mcflatface-to-become-fully-mobile-allowing-starlink-use-away-from-home/. Accessed 7 May 2021.

[BRS19] Brynjolfsson, Erik, Daniel Rock, and Chad Syverson. 2019. Artificial Intelligence and the Modern Productivity Paradox: A Clash of Expectations and Statistics. In: *The Economics of Artificial Intelligence: An Agenda, Agrawal, Gans, and Goldfarb*.

[BSW+05] Blackmore, S., B. Stout, M. Wang, and B. Runov. 2005. *Robotic agriculture – the future of agricultural mechanization?*. In Fifth European Conference on Precision Agriculture.

[Bur19] Burwood-Taylor, Louisa. *BASF's Xarvio partners with Arable to improve fungicide recs in lead up to outcome-based pricing*. ▶ https://agfundernews.com/basfs-xarvio-partners-with-arable-to-improve-fungicide-recs-in-lead-up-to-outcome-based-pricing.html. Accessed 13 March 2022.

[Cal20] Callaway, Ewen. 2020. "It will change everything': DeepMind's AI makes gigantic leap in solving protein structures. *Nature* 588(10):203

[Cap20] *Company Capella Space*. ▶ www.capellaspace.com. Accessed 13 March 2022.

[Car21] *Company Carbon Robotics*. ▶ www.carbonrobotics.com. Accessed 4 May 2021.

[CBR+20] Castrignano, A., G. Buttafuoco, A. M. Raj Khosla, D. Moshou, and O. Naud. 2020. Agricultural Internet of Things and Decision Support for Precision Smart Farming. *Academic Press*. ▶ https://doi.org/10.1016/B978-0-12-818373-1.12001-3.

[CDC+17] Cooper, K. B., S. L. Durden, C. J. Cochrane, R. R. Monje, R. J. Dengler, and C. Baldi. 2017. Using FMCW doppler radar to detect targets up to the maximum unambiguous range. *IEEE Geoscience and Remote Sensing Letters* 14(3):339–343.

[CIA+19] Cui, Shaoqing, Elvia Adriana Alfaro Inocente, Nuris Acosta, Harold M. Keener, Heping Zhu, and Peter P. Ling. 2019. *Development of fast E-nose System for early-stage diagnosis of aphid-stressed tomato plants. Sensors (Basel)* 19(16):3480. ▶ https://doi.org/10.3390/s19163480.

[CIV+18] Cantera, J. M., J. S. Issa, P. van der Vlugt, S. Klaeser, T. Bartram, A. Kassahun, I. Neira, and T. Milin. 2018. D3.3 Opportunities and barriers in the present regulatory situation for system development, in: IoF2020 (Ed.), *IoF2020 project deliverables*.

[Cla20] Claver, Hugo. 2020. *Yanmar develops modular robotic platform for agriculture*. ▶ https://www.futurefarming.com/Machinery/Articles/2020/4/Yanmar-develops-modular-robotic-platform-for-agriculture-572599E/. Accessed 13 March 2022.

[COD+19] Coupé, Yoon Mi Oh, Dan Dediu, and Francois Pellegrino. 2019. Different languages, similar encoding efficiency: Comparable information rates across the human communicative niche. *Science Advances* 5(9). ▶ https://doi.org/10.1126/sciadv.aaw2594. Accessed 5 March 2021.

[Cro20a] ▶ https://www.croptimal.com/technology. Accessed 13 March 2022.

[Cro20b] *Company CropDiagnostics.* ▶ www.cropdiagnosis.com. Accessed 13 March 2022.

[Cro21] ▶ https://cropx.com/tutorial/preferential-flow/. Accessed 13 July 2021.

[CSM+13] Caracciolo, Caterina, Armando Stellato, Ahsan Morshed, Gudrun Johannsen, Sachit Rajbhandari, Yves Jaques, Johannes Keizer. 2013. The AGROVOC Linked Dataset. *Semantic Web* 4(3):341–348. ▶ https://doi.org/10.3233/SW-130106. Accessed 5 March 2021.

[CZM+19] Cubuk, E. D., B. Zoph, D. Mane, V. Vasudevan, and Q. V. Le. 2019. Autoaugment: Learning augmentation strategies from data. In *Proceedings of the IEEE conference on computer vision and pattern recognition* (pp. 113–123).

[D97] DIN 9684-1:1997-02. 1997. *Landmaschinen und Traktoren - Schnittstellen zur Signalübertragung - Part 1: Punkt-zu-Punkt-Verbindung.* Beuth Verlag GmbH.

[DDK+20] Denis, Nicolas, Valerio Dilda, Rami Kalouche, and Ruben Sabah. 2020. *Agriculture supply-chain optimization and value creation.* ▶ https://www.mckinsey.com/industries/agriculture/our-insights/agriculture-supply-chain-optimization-and-value-creation. Accessed 13 March 2022.

[Dee21] *Deepmind's AlphaGo.* ▶ https://deepmind.com/research/case-studies/alphago-the-story-so-far. Accessed 11 March 21.

[DEM21] *The DEMETER project is a large-scale deployment of farmer-driven, interoperable smart farming-IoT (Internet of Things) based platforms, delivered through a series of 20 pilots across 18 countries (15 EU countries).* ▶ www.h2020-demeter.eu. Accessed 12 Feb 2021.

[DFH+19] Dörr, J., B. Fairclough, J. Henningsen, J. Jahić, S. Kersting, P. Mennig, ..., and F. Scholten-Buschhoff. (2019). *Scouting the Autonomous Agricultural Machinery Market*, IESE Report No. 041.19/E, Kaiserslautern, Germany.

[DIN19] DIN EN ISO 18497:2019-08 Agricultural machinery and tractors – Safety of highly automated agricultural machines – *Principles for design (ISO 18497:2018).* Beuth

[Dji20] ▶ https://www.dji.com/mg-1. Accessed 11 July 2020.

[DLG18] DLG e. V. 2018. *"Digitale Landwirtschaft – Chancen. Risiken. Akzeptanz. Ein Positionspapier der DLG"*, DLG e. V. (ed.).

[DLR21a] *Dienstleistungszentrum Ländlicher Raum Rheinhessen-Nahe-Hunsrück (DLR).* ▶ http://ef-sw.de/. Accessed 13 March 2022.

[DLR21b] *Dienstleistungszentrum Ländlicher Raum Rheinland-Pfalz (DLR).* ▶ https://www.dlr.rlp.de/Digitales-AgrarPortal/DAP. Accessed 13 March 2022.

[DLR21c] *Dienstleistungszentrum Ländlicher Raum Rheinland-Pfalz (DLR).* ▶ https://www.dlr.rlp.de/Digitales-AgrarPortal/GeoBox-/Das-Projekt. Accessed 13 March 2022.

[Dom15] Pedro Domingos: *The Master Algorithm.* Basic Books, 201.

[Dra20] ▶ https://draganfly.com/products/quantix-mapper/. Accessed 13 March 2022.

[Dro20a] *Company DroneDeploy.* ▶ www.dronedeploy.com. Accessed 13 March 2022.

[Dro20b] *Company Dropcopter.* ▶ www.dropcopter.com. Accessed 11 July 2020.

[E20] ESRI. (2020). *About Esri.* Retrieved from ▶ https://www.esri.com/en-us/about/about-esri/overview.

[Elm02] Elmenreich, W. 2002. An introduction to sensor fusion. *Vienna University of Technology, Austria* 502:1–28.

[Eth20] Etherington, Darrell. 2020. *SpaceX launches 58 more Starlink satellites and 3 planet Skysats for first rideshare launch.* ▶ https://techcrunch.com/2020/06/13/spacex-launches-58-more-starlink-satellites-and-3-planet-skysats-for-first-rideshare-launch/. Accessed 13 March 2022.

[Eym19] Eymann, G. (2019). *Automatisiertes Fahren: Sensortechniken im Check.* ▶ https://www.vdi.de/news/detail/automatisiertes-fahren-sensortechniken-im-check. Accessed 13 March 2022.

[Fao21a] *Food and Agriculture Organization of the United Nations.* ▶ http://www.fao.org/agrovoc/access. Accessed 5 March 2021.

[Fao21b] *Food and Agriculture Organization of the United Nations.* ▶ http://www.fao.org/agrovoc/agrontology. Accessed 5 March 2021.

[Fel20] Feldschwarm 2020. *John Deere.* ▶ http://www.feldschwarm.de/index.php/partner/john-deere. Accessed 13 March 2022.

[Fen20] ▶ https://msc.fendt.com/modules/create_pdf/pdf2/7398_web_en_2020-07-31_12-08-48.pdf.

[FHSR+20] Feng, D., C. Haase-Schütz, L. Rosenbaum, H. Hertlein, C. Glaeser, F. Timm, ..., and K. Dietmayer. (2020). *Deep multi-modal object detection and semantic segmentation for autonomous driving: Datasets, methods, and chal-*

lenges. *IEEE Transactions on Intelligent Transportation Systems*.

[FIP21] *The Future Internet Public-Private Partnership (FI-PPP) is a European programme for Internet innovation*. ▶ https://www.fi-ppp.eu/. Accessed 12 Feb 2021.

[FKS+09] Friedrich, J., M. Kuhrmann, M. Sihling, and U. Hammerschall. 2009. *Das v-modell xt*, 1–32. Berlin: Springer.

[Foo21] *FOODIE Metaphactory*. ▶ https://metaphactory.foodie-cloud.org/resource/:Start. Accessed 26 March 2021.

[Fre18] Freewave. 2018. *Florida Orchard Deploys Autonomous Tractor for Precision Agriculture Research*. ▶ https://www.freewave.com/wp-content/uploads/2018/12/case-study-florida-orchard-autonomous-tractors.pdf. Accessed 13 March 2022.

[FRK+15] Fehrmann, J., A. Ruckelshausen, R. Keicher, and K. Weidig. 2015. Autonomer Plantagen-Pflegeroboter für den Obst-und Weinbau. *ATZoffhighway* 8(3):32–43.

[FSA17] Feth, P., D. Schneider, and R. Adler. (2017, September). *A conceptual safety supervisor definition and evaluation framework for autonomous systems*. In International Conference on Computer Safety, Reliability, and Security (pp. 135–148). Springer, Cham.

[GAI21] *GAIA-X, a project initiated by Europe for Europe with the aim to develop common requirements for a European data infrastructure*. ▶ www.data-infrastructure.eu/GAIAX. Accessed 12 Feb 2021.

[Gan19] Gan, Joe. 2019. *XAG get its drones upgraded with tie-ups with Bayer and Huawei*. ▶ https://agfundernews.com/xag-taps-on-bayer-and-huawei-to-help-its-drones-take-off.html. Accessed 10 Aug 2020.

[GAP16] Gandhi, Niketa, Leisa J. Armstrong, and Owaiz Petkar. 2016. Predicting Rice Crop Yield using Bayesian Networks. In: *2016 International Conference on Advances in Computing, Communications and Informatics (ICACCI)*, Jaipur, pp. 795–799

[GB07] Griepentrog, H. W., and B. S. Blackmore. 2007. *Autonomous Crop Establishment and Control System*, in Agricultural Engineering, Hannover.

[GGA+07] González-Jiménez, Javier, Cipriano Galindo, Vicente Arevalo, and Gregorio Ambrosio. 2007. Applying Image Analysis and Probabilistic Techniques for Counting Olive Trees in High-Resolution Satellite Images. In *Advanced Concepts for Intelligent Vision Systems, Lecture Notes in Computer Science*, eds. J. Blanc-Talon, W. Philips, D. Popescu, P. Scheunders, Vol 4678.

Springer, Berlin. ▶ https://doi.org/10.1007/978-3-540-74607-2_84.

[Gia04] Giachetti, R. E. 2004. A Framework to Review the Information Integration of the Enterprise. *International Journal of Production Research* 42(6):1147–1166.

[Git21] *A number of datamodels for the agri-food sector*. ▶ https://github.com/smart-data-models/dataModel.Agrifood. Accessed 12 Feb 2021.

[GLK+15] Giraldo, Juan P., Markita P Landry, Seon-Yeong Kwak, Rishabh M Jain, Min Hao Wong, Nicole M Iverson1, Micha Ben-Naim, and Michael S Strano. 2015. A ratiometric sensor using single chirality near-infrared fluorescent carbon nanotubes: Application to in vivo monitoring. *Small* 11(32):3973–3984.

[Glo20] ▶ https://globalcomsatphone.com/costs/. Accessed 13 March 2022.

[GMU+17] Gaus, C., T. Minßen, L. Urso, T. de Witte and J. Wegener. 2017. *Mit autonomen Landmaschinen zu neuen Pflanzenbausystemen. Abschlussbericht FKZ 2814NA012*. Braunschweig: BMEL (BÖLN).

[Gof21] *GO FAIR Initiative*. ▶ https://www.go-fair.org/fair-principles/. Accessed 4 March 2021.

[Gra20] Grant, Elliott. 2020. *Embracing the complexity of nature*. ▶ https://blog.x.company/embracing-the-complexity-of-nature-45afc5bf5573. Accessed 17 July 2020.

[Gro20] *Company Groguru*. ▶ www.groguru.com/products/. Accessed 13 March 2022.

[Gru93] Gruber, Thomas R. 1993. A translation approach to portable ontology specifications. *Knowledge Acquisition* 5(2):199–220, ISSN 1042-8143. ▶ https://doi.org/10.1006/knac.1993.1008. Accessed 5 March 2021.

[GS12] Glaessgen, E. H., and D. S. Stargel. 2012. The Digital Twin Paradigm for Future NASA and U.S. Air Force Vehicles. In 53rd Struct. Dyn. Mater. Conf. Special Session: Digital Twin, Honolulu, HI, US.

[GS121] *A not-for-profit organisation that develops and maintains global standards for business communication*. ▶ www.gs1.org. Accessed 12 Feb 2021.

[Gus20] *Company GUSS*. ▶ www.gussag.com. Accessed 17 July 2020.

[Hal20] Hall, Christine. 2020. *Agtech Sector blooms as more dollars and startups rush in*. ▶ https://news.crunchbase.com/news/agtech-sector-blooms-as-more-dollars-and-startups-rush-in/. Accessed 13 March 2022.

[Hee13] Heege, H.J. 2013. *Precision in Crop Farming*. Netherlands: Springer. ▶ https://doi.org/10.1007/978-94-007-6760-7.

[Hel20] Company hello tractor. ▶ www.hellotractor.com. Accessed 13 March 2022.

[HF17] Herlitzius, T., and J. Fehrmann. 2017. *Gutachten Stand und Tendenzen der Roboteranwendungen im Bereich der Pflanzen- und Tierproduktion*. Wissenschaftlicher Dienst des Deutschen Bundestages, Dresden

[HKP+12] Hitzler, Pascal, Markus Krötzsch, Bijan Parsia, Peter F. Patel-Schneider, and Sebastian Rudolph (eds.): *OWL 2 Web Ontology Language Primer (Second Edition)*, W3C Recommendation 11 December 2012. ▶ https://www.w3.org/TR/owl2-primer/. Accessed 11 March 2021

[HSJ+20] Hasan, Haya R, Khaled Salah, Raja Jayaraman, Mohammed Omar, Ibrar Yaqoob, Sasa Pesic, Todd Taylor, and Dragan Boscovic. *A Blockchain-based Approach for the Creation of Digital Twins*. ▶ https://doi.org/10.1109/ACCESS.2020.2974810, IEEE Access.

[HTA+15] Haigh, Tonya, Eugene Takle, Jeffrey Andresen, Melissa Widhalm, J. Stuart Carlton, and Jim Angel. 2015. *Mapping the decision points and climate information use of agricultural producers across the U.S. Corn Belt. Climate Risk Management* 7:20–30, ISSN 2212–0963. ▶ https://doi.org/10.1016/j.crm.2015.01.004. Accessed 5 March 2021.

[Hua20] Huawei. 2020. *Using 5G to revolutionize farming*. ▶ https://www.cio.com/article/3564550/using-5g-to-revolutionize-farming.html. Accessed 13 March 2022.

[IDS21] *The International Data Spaces Association (IDSA) is on a mission to create the future of the global, digital economy with International Data Spaces (IDS), a secure, sovereign system of data exchange in which all participants can realize the full value of their data*. ▶ www.internationaldataspaces.org. Accessed 12 Feb 2021.

[IEC19] IEC TS 62998-1:2019 *Safety of machinery – Safety-related sensors used for the protection of persons*. VDE.

[Ili19] Iliaifar, Amir. 2019. *Fixed wing drones vs quadcopters: A project comparison*. ▶ https://www.sensefly.com/blog/fixed-wing-drones-vs-quadcopters/. Accessed 13 March 2022.

[Inn21] Company InnerPlant. ▶ www.innerplant.com. Accessed 4 May 2021.

[IOF21] *IoF2020 has developed and tested a series of IoT components and solutions*. ▶ https://www.iof2020.eu/results/technology-resources/scientific-community. Accessed 12 Feb 2021.

[IoT21] *The IoT Catalogue : the one-stop-source for Internet of Things (IoT) knowledge, innovations and technologies, aiming to help IoT stakeholders (developers, integrators, advisors, end-users, etc.) to take the most advantage of the Internet of Things for the benefit of society, businesses and individuals*. ▶ https://www.iot-catalogue.com/. Accessed 12 Feb 2021.

[IS09] Isaac, Antoine, and Ed Summers (eds.). 2009. SKOS Simple Knowledge Organization System Primer. *W3C Working Group Note* 18 August. ▶ https://www.w3.org/TR/2009/NOTE-skos-primer-20090818/. Accessed 11 March 21.

[Isi21] *isip Rübenblatt Scan*. ▶ https://www.isip.de/isip/servlet/isip-de/apps. Accessed 11 March 21.

[ISO95] International Organization for Standardization. 1995. *ISO-11787: achinery for agriculture and forestry—Data interchange between management computer and process computers—Data interchange syntax*.

[ISO00] International Organization for Standardization. 2000. *ISO-11788: Electronic data interchange between information systems in agriculture—Agricultural data element dictionary*.

[ISO07] International Organization for Standardization. 2007. *ISO-17532: Stationary equipment for agriculture—Data communications network for livestock farming*.

[ISO17] ISO/TC 23/SC 19. 2017. ISO 11783. *Tractors and machinery for agriculture and forestry—Serial control and communications data network*. Retrieved from ▶ https://www.iso.org/standard/57556.html.

[ISO19] International Organization for Standardization. 2019. *ISO-11783: Tractors and machinery for agriculture and forestry — Serial control and communications data network*.

[JHS+18] Junginger, A., M. Hanselmann, T. Strauss, S. Boblest, J. Buchner, and H. Ulmer. 2018. *Unpaired high-resolution and scalable style transfer using generative adversarial networks*. arXiv preprint ▶ arXiv:1810.05724.

[Joh20] Company John Deere. ▶ https://www.deere.com/en/tractors/compact-tractors. Accessed 9 Dec 2020.

[JWS19] Jakobs, S., A. Weber, and D. Stapp. 2019. Zuverlässige Objekterkennung für autonome mobile Arbeitsmaschinen. *ATZheavy duty* 12(2):46–51.

[KBK+09] Kok, J. N., E. J. Boers, W. A. Kosters, P. Van der Putten, and M. Poel. 2009. Artificial intelligence: Definition, trends, techniques, and cases. *Artificial intelligence* 1:1–20.

[KBS+13] Karner, J., M. Baldinger, P. Schober, B. Reichl, and H. Prankl. 2013. Hybridsysteme für die Landtechnik. *Agricultural Engineering* 68(1):22–25.

[KGS+14] Kaloxylos, A., A. Groumas, V. Sarris, L. Katsikas, P. Magdalinos, E. Antoniou, Z. Politopoulou, S. Wolfert, C. Brewster, R. Eigenmann, and C. Maestre Terol. 2014. A cloud-based Farm Management System: Architecture and implementation. *Computers and Electronics in Agriculture* 100:168–179.

[Kli20] ▶ https://innovationstage-digital.de/uploads/tx_bleinhaltselemente/Innovationstage_2020_-_KlimAgrar__Nachlese_.pdf.

[KM13] Krieter, Joachim, and Bettina Miekley. 2013. Perspektiven sensorgestützter Expertensysteme in der Tierhaltung. In: *Steuerungselemente für eine nachhaltige Land- und Ernährungswirtschaft - Stand und Perspektiven*. KTBL-Tage. KTBL-Schrift Nr. 500, S. 100–107. Kuratorium für Technik und Bauwesen in der Landwirtschaft e.V., Darmstadt, April 2013.

[Kni20] Jan Knibbe, Willem. 2020. *WUR is working on Digital Twins for tomatoes, food and farming*. ▶ https://www.wur.nl/en/newsarticle/WUR-is-working-on-Digital-Twins-for-tomatoes-food-and-farming.htm. Accessed 13 March 2022.

[Kol20] Kolemann, Lutz. 202. *Satellite Vu: Next Gen Infrared Constellation to help resolve Global Warming*. ▶ https://www.linkedin.com/pulse/satellite-vu-next-gen-infrared-constellation-help-resolve-lutz/. Accessed 13 March 2022.

[KPV+19] Koppelmäki, Kari, Tuure Parviainen, Elina Virkkunen, Erika Windquist, Rogier P. O. Schulte, and Juha Helenius. 2019. *Ecological intensification by integrating biogas production into nutrient cycling: Modeling the case of Agroecological Symbiosis*. *Agricultural Systems* 170:39–48. ▶ https://doi.org/10.1016/j.agsy.2018.12.007. Accessed 26 March 2021.

[KSS11] K. Krombholz, Stockach, and R. Soucek. 2011. Geschichte der Landtechnik. In *Jahrbuch Agrartechnik*, ed. R. M. H.-H. Harms, 23. Münster: DLG Verlag.

[Kuh19] Kuhwald, Michael. 2019. *Detection and modelling of soil compaction of arable soils: From field survey to regional risk assessment*. Dissertation for the award of the doctorate of the Faculty of Mathematics and Natural Sciences of the Christian-Albrechts-Universität zu Kiel. Kiel.

[KWS+16] Kruize, J., J. Wolfert, H. Scholten, A. Kassahun, and A. Beulens. 2016. A reference architecture for farm software ecosystems. *Computers and Electronics in Agriculture* 125:12–28.

[KVR+09] Klappstein, J., T. Vaudrey, C. Rabe, A. Wedel, and R. Klette. 2009, January. Moving object segmentation using optical flow and depth information. In *Pacific-Rim Symposium on Image and Video Technology* (pp. 611–623). Berlin: Springer.

[KZS15] Koch, Gregory, Richard Zemel, and Ruslan Salakhutdinov. 2015. *Siamese Neural Networks for One-shot Image Recognition*. Proceedings of the 32 nd International Conference on Machine Learning, Lille, France, JMLR: W&CP volume 37

[Lat20] *Company Latitudo40*. ▶ www.latitudo40.com. Accessed 13 March 2022.

[LM20] Lorenz, F., and K. Münchhoff. 2015. *Teilflächen bewirtschaften: Schritt für Schritt*. Frankfurt: DLG-Verlag.

[Lop20] Lopez, Marianna. 2020. *The state of 5G rollout in Latin America*. ▶ https://www.contxto.com/en/mexico/5g-latin-america. Accessed 13 March 2022.

[Lora20] *LoRa Alliance*. ▶ www.lora-alliance.org. Accessed.

[Mar19a] Martin, Scott. *Dig In: Startup OneSoil tills satellite data to harvest farm AI*. ▶ https://blogs.nvidia.com/blog/2019/04/15/startup-onesoil-tills-satellite-data-to-harvest-farm-ai-gpu/. Accessed 13 March 2022.

[Mar19b] Martin, Scott. 2019. *Goodwill Farming: Startup harvests AI to reduce herbicides*. ▶ https://blogs.nvidia.com/blog/2019/05/02/blue-river-john-deere-reduce-herbicide/. Accessed 13 March 2022

[Mar20a] Markets & Markets. April 2020a. *Artificial Intelligence in Agriculture Market by Technology (Machine Learning, Computer Vision, and Predictive Analytics), Offering (Software, Hardware, AI-as-a-Service, and Services), Application, and Geography – Global Forecast to 2026*. ▶ https://www.marketsandmarkets.com/Market-Reports/ai-in-agriculture-market-159957009.html. Accessed 13 March 2022.

[Mar20b] Markets & Markets. 2020b. *Agricultural Drones Market*. ▶ https://www.marketsandmarkets.com/Market-Reports/agriculture-drones-market-23709764.html?gclid=Cj0KCQjwsuP5BRCoARIsAPtX_wFe5nZscAQV0fsvshxkWS0blO0eGceIsIY-eBrp72Y72W2iLNzSLtQaAmwEEALw_wcB. Accessed 13 March 2022.

[Mar20c] Markets & Markets. 2020c. *Agricultural robot market*. ▶ https://www.marketsandmarkets.com/Market-Reports/agricultural-robot-market-173601759.html. Accessed 13 March 2022.

[Mar20d] Marktes & Markets. 2020d. *Smart agriculture market*. ▶ https://www.marketsandmarkets.com/Market-Reports/smart-agriculture-market-239736790.html. Accessed 13 March 2022.

[Mar20e] Martin, Scott. 2020. *Field Day: AI Startup Cultivates Robo Tractors for 'Swarm Farming' Disruption.*. ▶ https://blogs.nvidia.com/blog/2020/04/23/rabbit-tractors-swarm-farming/. Accessed 13 March 2022.

[Mic20] ▶ https://azure.microsoft.com/de-de/blog/democratizing-agriculture-intelligence-introducing-azure-farmbeats/. Accessed 13 March 2022.

[Mon20] *Company Monarch*. ▶ www.monarchtractor.com. Accessed 13 March 2022.

[MSP15] Martin, Betty, S. Maflin Shaby, and M.S. Godwin Premi. 2015. Studies on acoustic activity of red palm weevil the deadly pest on coconut crops. *Procedia Mater. Sci* 10:455–466.

[Nev21] Company NEVONEX. ▶ www.nevonex.com. Accessed 4 May 2021.

[Noa18] Noack, P. 2018. *Precision Farming – Smart Farming – Digital Farming: Grundlagen und Anwendungsfelder*. Wichmann.

[NS56] Newell, Allen, and Herbert A. Simon. 1956. The Logic Theory Machine: A Complex Information Processing System. *The RAND Corporation, Report P-868*. ▶ http://shelf1.library.cmu.edu/IMLS/MindModels/logictheorymachine.pdf. Accessed 11 March 2021.

[Obo21] *The Open Biological and Biomedical Ontology (OBO) Foundry*. ▶ http://obofoundry.org. Accessed 5 March 2021.

[OGC21] *Sensor Model Language*. ▶ https://www.ogc.org/standards/sensorml. Accessed 11 March 2021.

[Ogc21] *Open Geospatial Consortium*. ▶ https://www.ogc.org/docs/is. Accessed 4 March 2021.

[One20] ▶ http://www.spaceref.com/news/viewpr.html?pid=55744. Accessed 13 March 2022.

[Paq19] Daniel, Paquette. 2019. *Farmworker vs. robot*. ▶ https://www.washingtonpost.com/news/national/wp/2019/02/17/feature/inside-the-race-to-replace-farmworkers-with-robots/. Accessed 13 March 2022.

[Per20] Company Persistence Data Mining. ▶ www.persistencedatamining.com/news/3. Accessed 13 March 2022.

[PH14] Porter, M.E., and J.E. Heppelmann. November 2014. How Smart, Connected Products are transforming competition. *Harvard Business Review* 2014: 65–88.

[Phe20] Company Phenospex. ▶ www.phenospex.com. Accessed 13 March 2022.

[PL17] Pedersen, S. M., and K. M. Lind. 2017. *Precision Agriculture: Technology and Economic Perspectives*. Springer International Publishing.

[Pla20] Company Plantix. ▶ www.plantix.net/en/. Accessed 13 March 2022.

[Por12] Portele, Clemens. 2012. *"OGC® Geography Markup Language (GML) — Extended schemas and encoding rules", Open Geospatial Consortium*. ▶ http://www.opengis.net/spec/GML/3.3. Accessed 5 March 2021.

[Pra20] Prabhakar, Umang. *Omnivore Vision 2030: report* ▶ https://www.omnivore.vc/wp-content/uploads/2020/09/Vision-2030-report-08092020.pdf.

[Pre20] ▶ http://preagro.auf.uni-rostock.de/preagro_mops/. Pre Agro Verbundprojekt. Accessed 13 July 2021.

[Psi20] ▶ www.psiquantum.com/. *PsiQuantum*. Accessed 12 Oct 2020.

[PYC+17] Poulton, C. V., A. Yaacobi, D. B. Cole, M. J. Byrd, M. Raval, D. Vermeulen, and M. R. Watts. 2017. Coherent solid-state LIDAR with silicon photonic optical phased arrays. *Optics letters* 42(20):4091–4094.

[Ran20] ▶ https://rantizo.com/products/. Accessed 11 July 2020.

[Red20] Redmund, Stephen. 2020. *Gamma radiation mapping: A system for topsoil mapping and variable rate nutrient application*. ▶ www.croplife.com/precision/gamma-radiation-mapping-a-system-for-topsoil-mapping-and-variable-rate-nutrient-application. Accessed 13 March 2022.

[Reg21] Company Regrow. ▶ www.regrow.ag. Accessed 4 Mai 2021.

[Rie11] Ries, E. 2011. *The lean startup: How today's entrepreneurs use continuous innovation to create radically successful businesses*. Currency.

[RKS+20] Kuntke, Franz, Christian Reuter, Wolfgang Schneider, Daniel Eberz, and Ansgar Bernardi. 2020. Die GeoBox-Vision: Resiliente Interaktion und Kooperation in der Landwirtschaft durch dezentrale Systeme. In *Mensch und Computer 2020 - Workshopband*. Bonn: Gesellschaft für Informatik e. V., eds. C. Hansen, A. Nürnberger, B. Preim. ▶ https://doi.org/10.18420/muc2020-ws117-407.

[RNT97] Resch, Hans Norbert, Thomas Nette, and Matthias Trapp. 1997. GIS zur Unterstützung des „Precision Farming" – Kostenoptimierung und Trinkwasserschutz-. *Geo-InformationsSysteme (GIS)* 3:10–13, [Rob20]. Company RoboticsPlus. ▶ www.roboticsplus.co.nz. Accessed 13 March 2022.

[Ros58] Rosenblatt, F. 1958. The perceptron: A probabilistic model for information storage and organization in the brain. *Psychological review* 65 (6): 386.

[Rou20] Roussaki, Ioanna. (ed.). D2.1 Common Data Models and Semantic Interoperability Mechanisms – Release 1. *DEMETER Project Consortium*. ▶ https://h2020-demeter.eu/wp-content/uploads/2020/10/DEMETER_D21_final.pdf. Accessed 26 March 2021.

[RS18] Rissola, G., and J. Sörvik. 2018. *Digital Innovation Hubs in Smart Specialisation Strategies*. Luxembourg: Publications Office of the European Union.

[RSE+18] Reuter, Christian, Wolfgang Schneider, Daniel Eberz, Markus Bayer, Daniel Hartung, and Cemal Kaygusuz. 2018. Resiliente Digitalisierung der kritischen Infrastruktur Landwirtschaft – mobil, dezentral, ausfallsicher. In *Mensch und Computer, Workshopband, Gesellschaft für Infor-*

matik e. V.,, ed. R. Dachselt, G. Weber, 623–632. Dresden, Germany. ▶ https://dl.gi.de/bitstream/handle/20.500.12116/16930/Beitrag_330_final__a.pdf.

[SAE20] Society of Automotive Engineers. 2020. *Levels of driving automation.* ▶ https://www.sae.org/news/2019/01/sae-updates-j3016-automated-driving-graphic. Accessed 3 Dec 2020.

[Sag20] Company Saga Robotics. ▶ www.sagarobotics.com. Accessed 13 March 2022.

[SAH21] *SmartAgriHubs is a European-funded project that aims to realise the digitalisation of European agriculture by fostering an agricultural innovation ecosystem dedicated to excellence, sustainability and success.* ▶ www.smartagrihubs.eu. Accessed 12 Feb 2021.

[SAM15] Speicher, Steve, John Arwe, and Ashok Malhotra. 2015. Linked Data Platform 1.0. *World Wide Web Consortium (W3C).* ▶ https://www.w3.org/TR/ldp/. Accessed 5 March 2021.

[SCK18] Shannon, D.K., D.E. Clay, and N.R. Kitchen. 2018. Precision Agriculture Basics. *American Society of Agronomy Crop Science Society of America Soil Science Society of America.* ▶ https://doi.org/10.2134/precisionagbasics.

[Sem20] ▶ https://www.semesis.ch/en/products. Accessed 13 March 2022.

[Sen20] ▶ https://www.sensoterra.com/en/product/connectivity/Alternative-Communication-Options/. Accessed 13 March 2022.

[Sfs20] Successful Farming Staff. 2020 *Artificial Intelligence spurs rel-time scouting say terranis Officials.* ▶ https://taranis.ag/2020/07/22/artificial-intelligence-spurs-real-time-scouting-say-taranis-officials/. Accessed 13 March 2022.

[She20] Sheikh, Knvul. 2020. *A growing presence on the farm: robots.* ▶ https://www.nytimes.com/2020/02/13/science/farm-agriculture-robots.html. Accessed 13 March 2022.

[Shu20] Shuttleworth, J. (2019). *SAE Standards News: J3016 automated-driving graphic update.* ▶ https://www.sae.org/news/2019/01/sae-updates-j3016-automated-driving-graphic. Accessed 13 March 2022.

[Sin20] Singh, Manish. 2020 *Google invests 4.5 billion in India's reliance Jio platforms..* ▶ https://techcrunch.com/2020/07/15/google-invests-4-5-billion-in-indias-reliance-jio-platforms. Accessed 13 March 2022.

[Sla2020] ▶ https://slantrange.com/production-agriculture/. Accessed 13 March 2022.

[SLS+93] Schmidt, Frank, Ruth Lütticken, Thilo Steckel, and Hans Norbert Resch. 1993. In *Anwenderorientierte Weiterentwicklung von Informations- und Kommunikationsstrukturen in der Landwirtschaft*, eds. E. Schulze, Bs Petersen, H. Geidel, 197–200. Referate der 14. GIL.

[Sml19] ▶ https://www.prweb.com/releases/stream_technologies_launches_stream_ml_deep_learning_platform/prweb16674265.htm. Accessed 13 March 2022.

[Sol20] Solomon, Gabriel. 2020. *The state of European connectivity. How ready are we for 5G?.* ▶ https://www.ericsson.com/en/blog/2020/7/the-state-of-european-connectivity-how-ready-are-we-for-5g. Accessed 13 March 2022.

[SR14] Schreiber, Guus, and Yves Raimond (eds.): "RDF 1.1 Primer", W3C Working Group Note 24 June 2014. ▶ http://www.w3.org/TR/rdf11-primer/. Accessed 23 Nov 2020.

[STK+18] Samaniego, Luis, Stephan Thober, and Rohini Kumar et al. 2018. Anthropogenic warming exacerbates European soil moisture droughts. *Nature Clim Change* 8:421–426, ▶ https://doi.org/10.1038/s41558-018-0138-5.

[Swa20] ▶ www.swarmfarm.com, *company Swarm-Farm.* Accessed 13 March 2022.

[SWC17] Sun, Yubing, Jun Wang, and Shaoming Cheng. 2017 Dec. Discrimination among tea plants either with different invasive severities or different invasive times using MOS electronic nose combined with a new feature extraction method. *Computers and Electronics in agriculture.* 143:293–301. ▶ https://doi.org/10.1016/j.compag.2017.11.007.

[TDS+20] Trapp, Matthias, Marc Deubert, Lucas Streib, Björn Scholz-Starke, Martina Roß-Nickoll, and Andreas Toschki. 2020. Simulating the Effects of Agrochemicals and Other Risk-Bearing Management Measures on the Terrestrial Agrobiodiversity: The RISKMIN Approach. In *Landscape Modelling and Decision Support, Innovations in Landscape Research, Springer*, eds. W. Mirschel, V. Terleev, K. O. Wenkel. Cham. ▶ https://doi.org/10.1007/978-3-030-37421-1_23.

[Tec19] TechNews. ▶ https://technews.tw/2019/10/28/iii-digital-twin-solutions-for-smart-farming/.

[Ter20] ▶ https://order.teralytic.com/products/soil-probeAccessed 13 March 2022

[Tev20] *Company Tevel Aerobotics Technologies.* ▶ www.tevel-tech.com. Accessed 13 March 2022.

[The21] The Things Industries. ▶ https://www.thethingsnetwork.org. Accessed 13 March 2022.

[Thü20] ▶ https://www.thuenen.de/en/ak/, Thünen-Institut. Accessed 2 Nov 2021.

[Tra03] Trapp, Matthias. 2003. *Geodatenmanagement zur standortangepassten Ressourcenoptimierung in der Landwirtschaft*, Trierer Bodenkundliche Schriften, Band 6, ISBN: 3-9807099-5-7.

[Tri20] Trilogy Team. 2020. *Rural cloud initiative spearheading the digital transformation of rural America*. ▶ https://trilogynet.com/news/rural-cloud-initiative-spearheading-the-digital-transformation-of-rural-america-2/. Accessed 13 March 2022.

[TTK13] Trapp, Matthias, Gregor Tintrup gen. Suntrup, and Christian Kotremba. 2013. Auswirkungen des Klimawandels auf die Landwirtschaft und den Weinbau in Rheinland-Pfalz. In *Schlussberichte des Landesprojekts Klima- und Landschaftswandel in Rheinland-Pfalz (KlimLandRP)*, ed. RHEINLANDPFALZ, KOMPETENZZENTRUM FÜR KLIMAWANDELFOLGEN, 170. Teil 3, Modul Landwirtschaft

[UB15] adelphi / PRC / EURAC. 2015. *Vulnerabilität Deutschlands gegenüber dem Klimawandel*, Umweltbundesamt, Climate Change 24/2015, Dessau-RoßlauBMEL.

[UN15] United Nations. 2015. *Transforming our world: the 2030 Agenda for Sustainable Development. Resolution of the General Assembly dated 25 September 2015, A/RES/70/1*.

[Uni21] *UniProt Consortium*. ▶ https://www.uniprot.org/. Accessed 5 March 2021.

[Urs20] ▶ www.ursaspace.com. Accessed 17 July 2020.

[Usp18] US patent US 9,841,533 B2.

[USW+17] Underwood, J., A. Wendel, B. Schofield, and L. McMurray. 2017. Efficient in-field plant phenomics for row-crops with an autonomous ground vehicle. *Field Robotics* 34(6):1061–1083.

[WUH+17] Wegener, J. K., L.-M. Urso, D. v. Hörsten, T.-F. Minßen, and C.-C. Gaus. 2017. Neue Pflanzenbausysteme entwickeln – welche innovativen Techniken werden benötigt? *Landtechnik* 72(2).

[Val20] Vale, Steven. 2020. *New look for Xaver Fendt robot*. ▶ https://www.profi.co.uk/news/new-look-fendt-xaver-field-robot. Accessed 13 March 2022.

[Van20] Vandersat Specification sheet. 2020. ▶ https://vandersat.com/data/soil-moisture/. Accessed 13 March 2022.

[Ver20] ▶ https://veristech.com. Accessed 13 March 2022.

[VHS06] Vierboom, C., I. Härlen, and J. Simons. 2006. Akzeptanz organisatorischer und technologischer Innovationen in der Landwirtschaft bei Verbrauchern und Landwirten. *Schriftenreihe Organisatorische und technologische Innovationen in der Landwirtschaft*, 21.

[Vis20] company Vision Robotics. ▶ www.visionrobotics.com. Accessed 5 July 2020.

[VST+19] Verdouw, C., H. Sundmaeker, B. Tekinerdogan, D. Conzon, and T. Montanaro. 2019. *Architecture framework of IoT-based food and farm systems: A multiple case study. Computers and Electronics in Agriculture* 165:104939.

[VWB+16] Verdouw, C., J. Wolfert, A. Beulens, and A. Rialland et al. 2016. Virtualization of food supply chains with the internet of things. *Journal of Food Engineering* 176:128–136.

[VWB+17] Verdouw, C.N., S. Wolfert, G. Beers, H. Sundmaeker, and G. Chatzikostas. 2017. *IOF2020: Fostering business and software ecosystems for large-scale uptake of IoT in food and farming, in The International Tri-Conference for Precision Agriculture in 2017*, ed. W. Nelson, 7. Hamilton.

[W13] The W3C SPARQL Working Group. 2013. SPARQL 1.1 Overview. *World Wide Web Consortium (W3C)*. ▶ https://www.w3.org/TR/sparql11-overview/. Accessed 5 March 2021.

[Wef20] Company WeFarm. ▶ www.wefarm.co. Accessed 15 Aug 2020.

[WDA+16] Wilkinson, Marc D., Michel Dumontier, Ijsbrand Jan Aalbersberg et al. 2016. The FAIR Guiding Principles for scientific data management and stewardship. *Sci Data* 3:160018. ▶ https://doi.org/10.1038/sdata.2016.18. Accessed 5 March 2021

[WGK+17] Min Hao Wong, Juan P Giraldo, Seon-Yeong Kwak, Volodymyr B Koman, Rosalie Sinclair, Tedrick Thomas Salim Lew, Gili Bisker, Pingwei Liu, and Michael S Strano. 2017. Nitroaromatic detection and infrared communication from wild-type plants using plant nanobionics. *Nat Mater* 16(2):264–272.

[Wil20] Willimas, Alan. *Fieldwork Robotics completes initial field trials of raspberry harvesting robot system*. ▶ https://phys.org/news/2019-05-fieldwork-robotics-field-trials-raspberry.html. Accessed 17 July 2020.

[Wit19] de Witte, T. 2019. Wirtschaftliche Perspektiven autonomer Kleinmaschinen im Ackerbau. *Journal für Kulturpflanzen*, 71(4):95–100.

[WMB19] Wolfert, S., L. Mira da Silva, G. Beers, D. Pompeu Pais, J. Anda Agarte, M. Lora Lozano, and N. Molina Sanz. 2019. SmartAgriHubs. Connecting the dots to foster the digital transformation of the European agri-food sector – highlighting the Portuguese innovation ecosystem. *Cultivar* 16:45–53.

[WSG14] Wolfert, J., C.G. Sørensen, and D. Goense. 2014. A future internet collaboration platform for safe and healthy food from farm to fork. In *Global Conference (SRII), 2014 Annual SRII*, 266–273. San Jose, CA, USA: IEEE.

[WVV+10] Wolfert, J., C.N. Verdouw, C.M. Verloop, and A.J.M. Beulens. 2010. Organizing information integration in agri-food – a method based on a service-oriented architecture and living lab approach. *Computers and Electronics in Agriculture* 70(2):389–405.

[WWB+19] Yu (Emma) Wang, Gu-Yeon Wei, and David Brooks, Benchmarking TPU, GPU, and CPU Platforms for Deep Learning, 2019. ▶ arXiv:1907.10701v4 [cs.LG].

[Xar20] *The Assistant in Your Pocket.* ▶ https://www.xarvio.com/gb/en/products/scouting.html. Accessed 11 March 2021.

[Xar21] *Xarvio Scouting App.* ▶ https://seedworld.com/facts-xarvio-scouting-app/. Accessed 11 March 2021.

[Xco21] *Mineral.* ▶ https://x.company/projects/mineral/. Accessed 11 March 2021.

[Zha15] Zhang, Q. 2015. *Precision Agriculture Technology for Crop Farming.* CRC Press.

[You18] Young, Ashton. 2020. *Kiwi agritech startup flourishes with homegrown legal marketplace.* ▶ https://bizedge.co.nz/story/kiwi-agritech-startup-flourishes-homegrown-legal-marketplace. Accessed 17 July 2020.

[Zep20] ZEPP, Informationssystem Integrierte Pflanzenproduktion e. V. (ISIP). 2020. Julius Kühn-Institut, John Deere European Technology Innovation Center (ETIC), Kuratorium für Technik und Bauwesen in der Landwirtschaft e. V. (KTBL). In *Hangneigungsauflagen sicher einhalten", Rheinische Bauernzeitung*, eds. S Estel, K Albrecht, C Federle, B Golla, B Kleinhenz, D Martini, A Aurelia Maria Moanţă, Zvonimir Perić, Tanja Riedel, and Manfred Röhrig, 24–25.

[ZSJ+10] Zhang, X., L. Shi, and X. Jia et al. 2010. Zone mapping application for precision-farming: A decision support tool for variable rate application. *Precision Agriculture* 11:103–114. ▶ https://doi.org/10.1007/s11119-009-9130-4.

[Zog09] Zogg, J.-M. 2009. Essentials of Satellite Navigation. ▶ https://www.u-blox.com/sites/default/files/products/documents/GPS-Compendium_Book_%28GPS-X-02007%29.pdf. Accessed 13 March 2022.

[ZYX19] Zhao, Yajun, Guanghui Yu, and Hanqing Xu. 2019. *6G Mobile Communication Network: Vision, Challenges and Key Technologies (in Chinese)*. Sci Sin Inform, ISSN 1674-7267, Pre-published, 10.1360/N112019-00033. ▶ http://engine.scichina.com/doi/10.1360/N112019-00033.

Agronomy Perspective

Jörg Migende, Johannes Sonnen, Sebastian Schauff, Julian Schill, Alexa Mayer-Bosse, Theo Leeb, Josef Stangl, Volker Stöcklin, Stefan Kiefer, Gottfried Pessl, Sebastian Blank, Ignatz Wendling, Sebastian Terlunen, Heike Zeller, Martin Herchenbach, Fabio Ziemßen and Wolf C. Goertz

Contents

4.1 The Development of Agricultural Distributors into Solution Providers: Who is Helping Farms to Successfully Apply Smart Farming? – 195
4.1.1 Status of the Use of Digital Technologies – 195
4.1.2 Reasons for the Limited Use of Smart Farming Technologies by Farmers – 195
4.1.3 Common Features of these Obstacles – 196
4.1.4 The Term "Solution" – 196
4.1.5 Overcoming the Obstacles – 197
4.1.6 Who Can Help to Overcome the Obstacles? – 198
4.1.7 From Product Seller to Solution Provider – 199
4.1.8 Summary and Outlook – 200

4.2 Cross-Manufacturer Data Exchange Interoperability as a Basis for Efficient Data Management in Agriculture – 200
4.2.1 Introduction – 200
4.2.2 Technology Development—History, Important Actors and Current Projects – 202
4.2.3 Market Development—Current Status – 207
4.2.4 Conclusion and Outlook – 209

4.3 E-Commerce and Logistics – 210
4.3.1 Types of Digital Distribution Channels – 210
4.3.2 Differentiation Strategy of ag.supply – 212

© The Author(s), under exclusive license to Springer-Verlag GmbH, DE, part of Springer Nature 2022
J. Dörr and M. Nachtmann (eds.), *Handbook Digital Farming*,
https://doi.org/10.1007/978-3-662-64378-5_4

4.3.3	Requirements for the Online Trade of Agricultural Input Goods – 212	
4.3.4	Effects of Digital Distribution Channels on Agricultural Contribution Margin Accounting – 213	
4.3.5	Sustainability Effects of Digital Distribution Channels – 214	
4.3.6	Outlook – 214	
4.4	**The Digital Eco-System of Sustainable Farming: Agricultural Insurance as a Glue – 214**	
4.4.1	Introduction – 214	
4.4.2	When Agricultural Insurance is Parameterized, Their Digital Loss Assessments Result in Immediate Pay-Outs – 215	
4.4.3	Typical Insurance Covers Along the Agricultural Supply Chain – 216	
4.4.4	Insurance as a Glue in a Digital Ecosystem of Sustainable Farming – 217	
4.4.5	Digitally Enabled Risk Analysis and Mitigation—Farming Risk Profiles – 218	
4.4.6	Enabling More Sustainable Production: Risk Profiles as Ally – 218	
4.4.7	Conclusion – 219	
4.5	**Soil and Seed Management – 220**	
4.5.1	Introduction – 220	
4.5.2	Methods of Soil Management – 220	
4.5.3	Increasing the Performance of Soil – 221	
4.5.4	The Role of Crop Rotation – 222	
4.5.5	The Impact of Climate Change – 222	
4.5.6	Seed Management – 223	
4.5.7	Outlook – 225	
4.6	**Nutrient Supply: From Whole Fields to Individual Plants – 226**	
4.6.1	Introduction to Metering and Spreading – 226	
4.6.2	Determination of Machine Settings – 227	
4.6.3	GPS-Based Automation Systems – 228	

4.6.4	Field-Zone-Specific Nutrient Identification and Supply – 229	
4.6.5	Pneumatic Fertilizer Spreaders – 230	
4.6.6	Organic Fertilization – 231	
4.6.7	Conclusion and Outlook – 232	

4.7   Crop Protection: Diverse Solutions Ensure Maximum Efficiency – 233
4.7.1   More Than 50 Years of Modern Plant Protection Technology – 233
4.7.2   How Did the Plant Protection Technology Commonly Used Today Come About? – 233
4.7.3   Which Progress can be Expected in the Next 5 Years? – 235
4.7.4   What are the Long-Term Prospects 2025–2030? – 239
4.7.5   Conclusion: Future Crop Protection will be More Specific and Diverse – 240

4.8   Weather and Irrigation – 241
4.8.1   Introduction – 241
4.8.2   Current Status and Tool Kits Availability for Mitigation of Risks for Farmers – 242
4.8.3   Key Components and Technologies Presently Available – 243
4.8.4   Future Development – 245

4.9   Harvest Sensing and Sensor Data Management – 246
4.9.1   Introduction – 246
4.9.2   Vehicle-Based Sensing Systems – 247
4.9.3   Remote Sensing Support – 250
4.9.4   Data Quality Management (Post Correction) – 250
4.9.5   Automation – 251
4.9.6   Outlook: Ubiquitous Sensing and the Autonomy Challenge – 252

4.10   Direct Agricultural Marketing and the Importance of Software: It's not Possible Without Digitalization! Which Software Solutions Help to Digitalize the Administrative Work Steps of Direct Agricultural Marketing? – 253
4.10.1   Direct Agricultural Marketing—Blessing or Curse? – 253
4.10.2   The Evolution of Direct Agricultural Marketing – 254

4.10.3	Current Software Solutions for the Digitalization of Administrative Work Steps in Direct Agricultural Marketing – 256
4.10.4	Future Developments of Direct Agricultural Marketing – 257

4.11	**Challenges and Success Factors on the Way to Digital Agricultural Direct Marketing: "We Do Not Need a Homepage" – 258**
4.11.1	Introduction – 259
4.11.2	Status Quo of Direct Marketing – 259
4.11.3	Challenges of Digital Direct Marketing – 260
4.11.4	Success Factors of Digital Direct Marketing – 260
4.11.5	COVID-19 and the Digitalization – 262
4.11.6	Conclusion – 263

4.12	**Digitalization in the Food Industry – 263**
4.12.1	How Digitalization is Changing Food Sales – 263
4.12.2	More Transparency: Close to the Customer and Yet Far Away – 264
4.12.3	"Direct to Consumer": Not New, but Different – 264
4.12.4	E-Food Startups: New Models, Processes, and Infrastructure – 265
4.12.5	New Distribution Channels: "Tiny Stores, Dark Stores, Ghost Stores" – 265
4.12.6	Agile Approach: Learning from the Start-ups and Joining in – 266
4.12.7	The Digital Path to "B2B2C" – 266

4.13	**Artificial Intelligence and Sustainable Crop Planning: Better Planning and Less Waste Through Digital Optimization – 267**
4.13.1	Artificial Intelligence—Mystery or Helpful Tool – 267
4.13.2	Why Don't You just Go Sustainable? – 268
4.13.3	Thinking Vegetable Production and AI Together – 268
4.13.4	Conclusion – 271

References – 271

Agronomy Perspective

4.1 The Development of Agricultural Distributors into Solution Providers: Who is Helping Farms to Successfully Apply Smart Farming?

Jörg Migende

Abstract

Farms realize the potential of digital technologies to increase their production efficiency and operate in a more environmentally friendly way. The use of such technologies in farming practice, however, is still low. This is due to the high complexity of their implementation and use. In addition, the users are insecure regarding the issue of economic feasibility and the use of their data. Those distributors and service providers that offer both agricultural supplies and agricultural machinery are particularly well placed to help users overcome these obstacles. They combine many of the required competencies under one roof. However, their organizational and operational structure needs to evolve from that of a pure product seller to a solution provider.

4.1.1 Status of the Use of Digital Technologies

More complex digital solutions that offer decision support and thereby optimize agricultural processes are only used by very few farms to date. In a recent study by the industry association Bitkom in Germany, only 9% of the respondents indicated that they are using data-driven systems. At the same time, however, 81% of the respondents believed that digitalization is an opportunity to increase their production efficiency. A share of 79% of the respondents stated that digitalization leads to more environmentally friendly production methods [Bit20].

These more complex digital solutions will be referred to as Smart Farming technologies in the following. Here the question arises which difficulties farms see in the introduction of Smart Farming technologies and how these can be overcome.

The explanations presented below are based on the long-term experience of the author and his teams at BayWa AG and FarmFacts GmbH. Special thanks go to Dr. Josef Bosch and Dr. Wolfgang Angermair, as well as to VISTA GmbH.

4.1.2 Reasons for the Limited Use of Smart Farming Technologies by Farmers

Surveys of German farms on the biggest obstacles to the use of Smart Farming technologies on farms have produced consistent results across surveys [Bit20, GSE18, BHR16] (see also ▶ Sect. 1.5).

The cited surveys point first to the lack of compatibility of the individual system elements and the high application complexity. These difficulties lead to errors in use and are frustrating due to the large amount of time needed for installation and operation. Investments already made in machines, sensors, as well as hardware and software turn out to be incompatible. In practice, these are then not suitable for ensuring seamless data transmission to enable process control.

The high level of investment required and the uncertainty in terms of economic feasibility are also among the most common challenges for farm managers.

There is also consensus on concerns regarding insufficient data protection, lack of data security, and unclear sovereignty over the generated data. The area of data sovereignty also includes the fear of more possibilities for government control [Bit20].

4.1.3 Common Features of these Obstacles

If positive influences of Smart Farming technologies with regard to sustainability, economic feasibility, and quality of agricultural products are proven in research, the desired effects will only arise if many farm managers use these technologies on their farms. This is why the above-mentioned obstacles in agricultural practice must be overcome. They are based, among other things, on the complexity of Smart Farming technologies and a lot of insecurity among the users.

4.1.3.1 Complexity

A farm is the sum of individual management and production processes. The farm itself is in turn embedded into other processes that connect it with the upstream and downstream sectors of agribusiness and society. For example, it is part of the supply chain for food and energy. Positive and negative influences of agricultural production on natural resources affect ecosystems.

A process always consists of individual elements that are mutually dependent and influence each other. Smart Farming technologies are a useful tool when they improve one or more processes. For example, field-zone-specific fertilization helps to optimize nutrient management in terms of fertilization effort, plant yield, and avoidance of negative environmental effects.

It is the job of the farm manager to organize, control, and permanently optimize these processes. This is a complex task, in which Smart Farming technologies can provide support. However, Smart Farming technologies also consist of several elements. One example is field-zone-specific seeding, whose effect results, among other things, from the optimal combination of variety and crop rotation, the findings from soil analysis, suitable machine technology and IT, as well as smooth data transmission. The use of Smart Farming technologies is therefore also a complex process.

4.1.3.2 Insecurity

Smart Farming technologies can trigger insecurity for the user, especially if they include digital elements. Several reasons can be identified for this:

Some farms cite a lack of IT knowledge [BHR16]. In addition, there are the doubts regarding data protection, data security, and data sovereignty already mentioned in ▶ Sect. 4.1.2.

The aforementioned doubts regarding economic viability also foster insecurity. Particularly with regard to economic viability, it should be noted that farming takes place predominantly outdoors. The benefits of Smart Farming technologies in crop cultivation do not occur uniformly in every year and can only be assessed over the course of several years. This is also true for the amortization of investments. Finally, the complexity of controlling and analyzing the processes on a farm, as described above, should not be forgotten.

The complexity of the decision and the application, and the insecurity in the use of Smart Farming technologies must be reduced for the farm manager in order to promote the use of these technologies.

4.1.4 The Term "Solution"

Smart Farming is therefore to be understood as process control. Processes are defined and implemented to achieve one or more goals. Fertilization, for example, serves to promote plant growth and maintain soil fertility. At the same time, however, the fertilization process must minimize undesirable nutrient discharge and meet the legislative requirements. Good process control therefore solves production problems.

Hence, it is too short-sighted to speak of a Smart Farming technology as a "product". In the following, the term "solution" will be used. Sale and support of Smart Farming technologies are therefore not "product sales" but "solution sales". Using this approach, the product seller becomes a "solution provider". The task of the solution provider is to eliminate the insecurity of the farm manager and manage the complexity on his behalf.

For the farm manager, the use of complex Smart Farming technologies must become easier. Reliability of use must be guaranteed at all times, since agricultural production processes in crop cultivation often take place in narrow time windows (temperature, precipitation, wind, length of day, logistics capabilities, supply contracts).

4.1.5 Overcoming the Obstacles

In this section, we present ways to overcome the three major obstacles: lack of compatibility, insecurity regarding the cost/benefit ratio, and low confidence in data protection and data security.

4.1.5.1 Compatibility

Compatibility should be possible not only between the machines of one manufacturer, but also between different manufacturers. Beyond the machinery, smooth data transmission into and out of the Farm Management Information System is also needed (see ▶ Sect. 3.4). Compatibility with sensors and partner systems in upstream and downstream agribusiness and public authorities as well as with external data sources make an integrated system landscape complete. Collaborations across agmachinery manufacturers, such as the agrirouter (see ▶ Sect. 4.2) and the NEXT Machine Management of FarmFacts GmbH in collaboration with the Agriculture Application Group (AAG), point in the right direction in this regard.

In practice, however, it is evident that even if compatibility is largely ensured on individual farms, system failures do happen in day-to-day work. Here, specialists must be available on site or remotely to enable the user to continue working. These tasks are increasingly being taken over by AgMachinery workshops, which is also reflected in a significant change in the requirements for the job profile of "agricultural and construction mechatronics technician". In combination with software support geared towards the requirements of agricultural production processes, rapid assistance can be guaranteed in the event of technical problems.

In troubleshooting activities, crop consulting also plays an important role. Together with the user, such a consultant analyzes and corrects deviations from the goals when using Smart Farming technology. In the author's personal experience, an organizational connection and direct communication between the crop consulting on the one hand and data model and software development on the other hand has proven effective here. This means, for example, that plant growth models are constantly improved through feedback from the field.

However, compatibility goes far beyond machine systems. A Smart Farming solution must also fit the natural conditions, the existing production systems, the expectations of the farm manager, and the legislative requirements. In addition, the local wireless data transmission options must also be considered.

Thus, guaranteeing compatibility requires a holistic support and consulting approach. In the next section, we will discuss who can fulfill this as a partner of the farm.

4.1.5.2 Insecurity Regarding the Cost/Benefit Ratio

It is crucial that this ratio is not determined exclusively on the basis of general studies, but in the context of the concrete individual farm. The following questions (see ▶ Sect. 2.4) have to be answered there:

- Does the solution have effects on plant development, working time, and sustainability, and how large are these?
- How does the solution affect the costs and the yield of the respective process?
- Is the solution technically feasible in agricultural practice so that the farmer can continue to use it stably and as autonomously as possible in the future?
- Does the solution guarantee fulfillment of third-party legal or qualitative requirements on the farm?

Answering these questions can be challenging, as it is unrealistic to conduct a scientifically sound study in the investing business prior to every investment decision. An experienced external solution provider should provide decision-making support, take responsibility for the implementation, and ensure application. In this regard, it is not enough for the solution provider to have only technical competencies. Rather, a high level of crop-related knowledge is also necessary.

The exchange of experience with professional colleagues or independent test reports on Smart Farming technologies could also help farm managers make decisions. Vocational training and continuing education provide the framework for farm managers to become informed decision makers regarding the use of state-of-the-art Smart Farming technologies.

To make decisions, farm managers can use agile project management methods together with their solution providers. In concrete terms, this means gradually moving from less complex pilot projects to larger process changes. However, in order to avoid aimless "trial and error", it is important to develop a target picture for the use of Smart Farming technologies beforehand. Realistic and honest expectation management on the part of the solution provider helps to avoid frustration on the part of the farm manager.

4.1.5.3 Data Protection and Data Security

The topic of data security, data protection, and data sovereignty has already been discussed in ▶ Sect. 1.7.

It is not just about clear legal regulations that can help to reduce the insecurity of the user. Transparency on the part of the solution provider in the collection and usage of the data is another important element. However, it also takes informed users who know their rights and understand digital business models.

A personal relationship of trust based on a long-standing business relationship between the user and the solution provider helps to overcome insecurities and establish more Smart Farming technologies on a farm.

If the user wants a high degree of data sovereignty, the solution provider's ability to pursue data-driven services for free of charge is limited. Examples in this regard include personalized advertisement and cross-selling. This, in turn, has an impact on the pricing of the respective Smart Farming technology.

4.1.6 Who Can Help to Overcome the Obstacles?

Let us illustrate the complexity of a Smart Farming solution with the example of "field-zone-specific inorganic fertilization". In the author's experience, more than ten technological components and services are required just for the initial installation and calibration of field-zone-specific inorganic

fertilization on a farm. This requires at least the expertise and active participation of the following specialists:
- Soil sample service
- Crop cultivation consulting, including assistance in complying with the Fertilization Ordinance
- AgMachinery sales, including steering systems and possibly sensors
- AgMachinery service, including data transmission
- Fertilizer distributor for the provision of customized fertilizers and fertilizer mixtures
- Software provider(s)
- Data provider(s), for example for application maps

If organic fertilization is included, the complexity increases even more. But the multitude of specialists mentioned above does not make it easier for the user. A solution provider should therefore combine all competencies in their organization. It is then their task to coordinate the different special functions in such a way that the farm manager does not need to do this. This coordinating function also helps in the event of disruptions during practical use.

For this reason, manufacturers of AgMachinery and supplies such as fertilizers, seeds, and pesticides are only suitable for this role to a limited extent. They are usually focused on one product area, which covers only one part of the agricultural production process. The approach of cooperating as a digital platform provider with different companies and thereby covering all process steps is also only practicable to a limited extent, as the farmer values human contact when it comes to advice and troubleshooting. The complexity in the interaction among the individual process components still cannot be controlled only remotely.

Consequently, companies that have both, expertise in AgMachinery and in crop cultivation, have an advantage in this respect. It is crucial for the solution providers to have a dense network of mechatronics technicians for AgMachinery as well as crop cultivation consultants in rural areas, so that they can also do business in operative terms. This is a major difference to purely digital Smart Farming platforms.

Businesses from the cooperative farm supplies and AgMachinery sectors, for example, are ideally placed for this. If such businesses have already built up their own Smart Farming expertise, they are predestined to be solution providers. Examples are the German BayWa AG with its subsidiaries FarmFacts GmbH (Farm Management Information Systems and data connectivity solutions) and VISTA GmbH (remote sensing data and data modeling), as well as the French INVIVO Group with its subsidiary SMAG.

Employees' long-standing relationships with customers can also strengthen trust in data security and data sovereignty. This protects the user, as data misuse would jeopardize the long-standing customer relationship.

Other approaches for businesses to act as solution providers for farmers are collaborations among businesses, such as that of Raiffeisen Warenzentrale Köln with FarmFacts GmbH. Collaborations can also be a way for businesses that do not wish to make extensive investments in their own Smart Farming expertise. Things are also made easier for a farm if the solution provider cooperates with contractors or machinery rings.

4.1.7 From Product Seller to Solution Provider

Agricultural distributors are still largely product-oriented and organized into clearly demarcated departments. Departmental thinking must be overcome and coordinating functions must be created. Professional change management facilitates the necessary change in attitude among employees.

In order to now become a solution provider, the willingness to cooperate among the various specialists must be supported by management, suitable communication channels must be implemented, for example via CRM systems, and new coordinating positions have to be created within the company. These "Key Account Managers Smart Farming" also serve as contact persons for the customer at the same time.

Joint pilot projects between more product-oriented and more solution-oriented employees, in particular, help to test new ways of working together and implement them positively in the sales and service organization. This is also a good method to build collaborations between independent businesses. Methods such as Design Thinking or business model canvas can be of assistance in the development of such pilot projects (see ▶ Sect. 2.5).

4.1.8 Summary and Outlook

In this section, we have demonstrated that solution providers can play a crucial role in the sustainable implementation of Smart Farming in agricultural practice. Investments in new employee competencies, in new positions such as "Key Account Manager Smart Farming", but also in technological and organizational innovations are already being made by agribusiness companies.

However, Smart Farming technologies must also lead to sustainably positive revenue models for solution providers. Otherwise, solution providers will be unable to fulfill their important function of helping farms use these technologies successfully.

4.2 Cross-Manufacturer Data Exchange Interoperability as a Basis for Efficient Data Management in Agriculture

Johannes Sonnen

Abstract

This section focuses on data exchange and interoperability from a farm operation perspective, incl. the description of selected commercial data exchange solutions, key associations and research activities in the field. Starting with the description of the most important developments, this section ends with a drafted big picture for data exchange and services from a farmer perspective.

4.2.1 Introduction

Access to data and its practice-oriented interpretation is an innovation driver for more sustainable agriculture. The increasing complexity of agricultural practices as well as the growing amount of data require new data management concepts. These include data exchange platforms, databases, data management systems, and user applications (e.g., Farm Management Information Systems, FMIS).

A key problem to date has been the poor or non-existent compatibility and interoperability of the software and hardware offered for agriculture, as many manufacturers offer closed systems (see ▶ Sect. 3.4). Continuous optimization of agricultural production processes (e.g., automatic documentation, more targeted and thus reduced application of seeds, fertilizers, and pesticides) along the entire value chain (see ◘ Fig. 4.1) requires new features such as cross-manufacturer data evaluations, process documentation, sustainability certificates, and decision support.

Agronomy Perspective

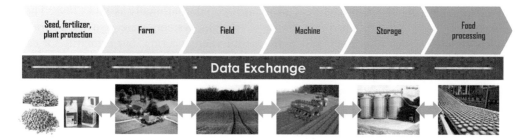

Fig. 4.1 Agricultural value chain—from seed to food

In order to be able to offer these features demanded on the market, cross-manufacturer data exchange takes on a central role, as machinery and agricultural software products from different manufacturers are used in agricultural production chains. Standardized data exchange formats (e.g., as described in ISO 11783/10) have long been considered the appropriate means for cross-manufacturer data exchange. For example, many machines can export and import the ISOXML data format via USB stick (or similar storage media). However, in recent years, different folder structures and standards not being fully adhered to have led to a high level of frustration among users as the desired cross-manufacturer data exchange has not been possible consistently. In addition, agricultural software companies, in particular, have not embraced these standardized formats, but have rather also relied on closed solutions in combination with consumer hardware. This makes them independent of the proprietary web interfaces developed by AgMachinery manufacturers for Internet of Things (IoT) applications. For providers of agricultural software, connecting to and operating many different web interfaces, in particular, has often proven to be too time-consuming and thus too cost-intensive a solution for interacting with machines (see **Fig. 4.2**). Therefore, many providers of agricultural software have developed closed, self-controllable systems and are offering them on the market. These systems are more or less dependent on the discipline of the operators, as they are an integral part of data collection and processing.

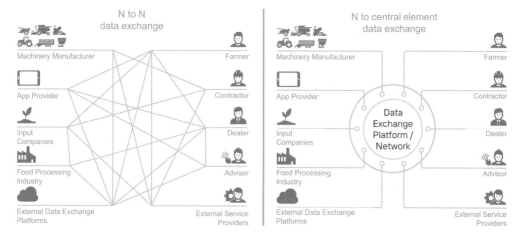

Fig. 4.2 N to N versus N to central data exchange element

Large AgMachinery manufacturers also offer their own self-contained solutions consisting of existing machines and agricultural software functions. Until now, valuable agronomic and machine data could therefore not be used across manufacturers by end users, farmers, and contractors.

To enable data exchange between the individual actors in an agricultural production chain, the need for a centralized, web-based platform or standardized network for farmers and contractors to exchange data has greatly increased.

Such a central element (see ◘ Fig. 4.2) or network for cross-manufacturer data exchange should fulfill all currently known requirements along the value chain. A universal data exchange platform for farmers and contractors, through which AgMachinery and agricultural software from different manufacturers can be interconnected, requires a high degree of flexibility to enable the integration of new elements. In addition to flexibility, such a system must also fulfill all data security and data sovereignty requirements. The provider must be able to guarantee that only the users have access to their data and can define data streams and their usage themselves, and can change these at any time.

4.2.2 Technology Development—History, Important Actors and Current Projects

In the context of digitalization, cross-manufacturer connectivity of AgMachinery and agricultural software applications is becoming increasingly important. One prerequisite for these developments, which are now referred to as "Agriculture 4.0", was a uniform data bus system. This had been developed by Bosch in the form of the CAN bus in the 1980s already, together with the chip manufacturer Intel. CAN stands for Controller Area Network and was later standardized internationally as ISO 11898. These CAN bus systems were also used in agricultural engineering, but primarily for controlling the machines and/or for communication between the tractor and the implements. At that time, the focus was not yet on data management.

Today, the expectations not only of farmers and contractors are strongly influenced by their experience with applications in the consumer sector and by knowledge about available technologies. This results in a list of requirements of the above-mentioned users for an agricultural data management system:

– Open and cross-manufacturer data exchange between machines and software products without conversion problems to optimize and increase the efficiency of agricultural production processes with a central data exchange platform
– Free choice of machines and agricultural software applications and the possibility of individual arrangement and configuration of a farm's own data management ecosystem
– High attractiveness and quality of the data exchange platform to increase acceptance
– Possibility to use third-party services (apps) and the resulting added values
– Automatic collection, interpretation, and evaluation of agronomic process data
– Availability of information at any place and any time on today's standard device market
– Use of regionally available software solutions that meet state-specific requirements (agronomic, legal, economic), e.g., with regard to documentation obligations
– Time-limited access control by the data producer (farmer/contractor) to their own data with exclusive data sovereignty for the end user

Agronomy Perspective

– Self-determination for storage of usage data, and transparent overview of data flows

This has resulted in some concrete solutions already being used in agricultural practice, which will be described below.

4.2.2.1 Commercial Solutions
The agrirouter Platform

In early 2014, six AgMachinery manufacturers decided to collaborate in the area of data management and to convince other AgMachinery manufacturers to participate in the newly founded initiative. The result of this development, which took the form of a consortium, is the cross-manufacturer and web-based data exchange platform agrirouter, which has been available for free use since March 2019. agrirouter transports data (similar to the way the postal service transports letters and packages), but does not store data. The data transport is carried out in secure form and, just as with the postal service, the data packages are not opened to analyze their content.

With the increasing need to interconnect different participants in agricultural production processes, the complexity grows. The technology approach described here enables the user to easily set up their own customized data network. In this way, farmers and contractors both in conventional farming and in organic farming thereby always keep control over their data in their network in a central place.

Machines can be connected directly or indirectly via existing, proprietary systems (e.g., telemetry systems). Technically, direct connection is made via a communication unit located on the machines. Existing market machines can be retrofitted and connected at any time using retrofittable, certified communication units. Indirect connection of machines is also possible and takes place via the manufacturer-specific platforms (these often have the prefix "my" in their name, e.g., "my-grimme"). Software applications, often Farm Management Information Systems (FMIS, see ◘ Fig. 4.3), or third-party apps (e.g., producers of Application Maps), as well as software appli-

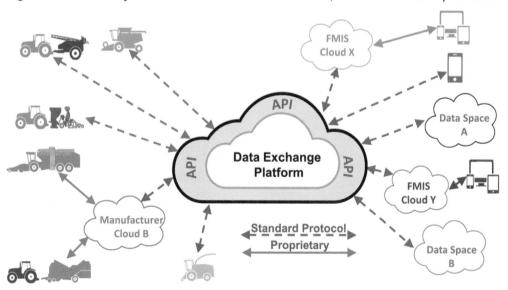

◘ Fig. 4.3 Design of a data exchange ecosystem, exemplified by agrirouter

cations of AgMachinery manufacturers can be connected by the farmer/contractor according to their needs.

With this web application, which can be configured individually by each user and can be adapted time and again, data from AgMachinery, which is currently collected by a multitude of sensors, or data from Agricultural Software products can be transported to other connected software applications via the data exchange platform in accordance with the routes (rules) set by the end user. The data from all machines used in a production process thus forms the basis for the documentation, the desired increase in efficiency and quality, as well as cost minimization in the future design of the farmer's production processes.

A data exchange platform enables the cross-manufacturer use of valuable agronomic and machine data that is demanded by the market. The farmer can use not only the data from their own machines, but also the data from the machines of their contractor, provided that consent for the release of such data has been given. Software applications selected by the end user can also exchange data with each other; e.g., application maps can be created for the application of fertilizer or pesticides. Such a map would then be transferred via the platform to the machine applying the product.

The end user can set up their own on-farm ecosystem (see ◘ Fig. 4.3), consisting of different apps and machines that exchange data with each other according to the routes (rules) the farmer has set individually. With agrirouter, the desired interoperability has taken a big step forward.

No comparable concept or product is currently available on the market. Globally, developments in this area are also moving in the direction of cross-manufacturer and open systems. In August 2020, John Deere announced that further interfaces have been integrated into its Operations Center management platform and that with the "MyTransfer" app [pro20], the possibility has been created to transfer and use data from the machine displays of the most common manufacturers. Like agrirouter, John Deere thereby now also enables direct connection of machines (terminals) in addition to indirect connection via manufacturer-specific clouds.

The DataConnect Solution
DataConnect is an initiative launched in 2019 by the companies John Deere, Claas, and 365Farmnet. In this cooperation, data exchange between company-internal telemetry platforms is to be realized. The scope of exchangeable data currently comprises five agronomic values. Other manufacturers may join DataConnect, but they must be operating their own manufacturer platform (see manufacturer cloud in ◘ Fig. 4.3). With DataConnect, farmers and contractors can view and process their machines in the platform of their choice and do not need to switch between systems.

Some AgMachinery manufacturers offer external partners the possibility to connect to their own telemetry platform via an interface (API). The APIs, and thus the scope of the data, vary from manufacturer to manufacturer. Market acceptance among agricultural software providers is rather low.

The NEVONEX Platform
Under the name NEVONEX, Robert Bosch GmbH [Bos20] is offering a platform for digital services since the Agritechnica 2019 (see ◘ Fig. 4.3, right part). With the help of this application environment, digital services can be loaded from a feature store (similar to an app store) onto hardware installed on AgMachinery. This can be a piece of additional retrofitted hardware or a factory-installed piece of hardware. Using these hardware components, farmers can extend the functionality of their machines with digital services (assistance systems). For example, a digital service could ensure compliance with wa-

ter protection zones when applying pesticides. The hardware used is connected to the NEVONEX Feature Store via a cellular connection to enable "downloading" of digital services. After installation, respective activation on the hardware, the digital services can use the cellular connection for example to obtain external data (in our example, the local maps for the water protection zones). In 2022, a connection between the data exchange platform agrirouter and the services platform NEVONEX will be made possible in this way with the service "agrirouter connector". This will enable customers to realize their data exchange as well as the operation of digital services using a single piece of hardware. In the example above, an application map can be used as the basis for the service ensuring compliance with water protection zones. The pesticide sprayer then automatically complies with the water protection zones and also varies its application quantity (e.g., liquid fertilizer).

The JoinData/DjustConnect Network
In the livestock farming sector, an initiative called "JoinData" exists in the Netherlands [Joi20], which has established a data exchange network mainly in dairy production. This static network can be configured and used by the end customer.

A similar approach to optimizing dairy production is being pursued by the initiative "DjustConnect" of the research institution ILVO [ILV20] for the Belgian region of Flanders. ILVO stands for multi-disciplinary, independent research and specialized services in all areas related to agriculture, fishery, and food in Flanders. ILVO, an internationally recognized scientific institute, is part of the regional government of Flanders.

4.2.2.2 Associations
Agricultural Industry Electronics Foundation (AEF)

Seven international manufacturers of agricultural equipment and two associations founded AEF on 28 October 2008 [AEF18]. AEF is an independent international organization and provides resources and know-how to foster the use of electronics and electrical equipment in agriculture. Currently, eight manufacturers and three associations are engaged as premium members in AEF, along with another 200 members, to:
- Improve cross-manufacturer compatibility of electronic and electrical components in agricultural equipment
- Ensure transparency in compatibility issues

A central point of the joint work is the introduction of international electronic standards, but not the development of products. Furthermore, AEF promotes the development and introduction of new technologies. In addition to the ISOBUS topic, EFDI (Extended Farm Management Information System Data Interface), electric drives, camera systems, high-speed ISOBUS, and wireless field communication have been added as new topics at AEF in recent years. As a representative of the Ag-Machinery industry, AEF is a partner in the EU research project ATLAS [EC20a].

AgGateway
AgGateway is an initiative founded in North America in 2005, which has gradually expanded to Latin America and Europe. AgGateway Europe, as a subsidiary of AgGateway Global, is working to optimize the interoperability and traceability of field operations as well to simplify the use of soil and weather data [AEF18]. In addition to standardizing data formats in crop production, the Ag-Gateway Europe initiative is also active in standardizing data interoperability in the area of livestock production and processing. The standards developed by AgGateway can also be included in the future as additional data ex-

change formats in data exchange platforms such as agrirouter or cloud solutions such as DataConnect. Moreover, all solutions must be able to add market-relevant data exchange formats along the agricultural value chain at any time in the future.

4.2.2.3 Research Activities

EU Research Project ATLAS
Within the framework of the EU research project ATLAS [ATL20], different companies in the AEF are working on implementing data exchange possibilities via a standardized network. This would allow future data exchange between platforms such as agrirouter and DataConnect. According to the author's assessment, two data management worlds could emerge in the coming years, which can be compared with the consumer world. On the one hand, there will continue to be the closed John Deere world around the "Operations Center", which, like Apple's iOS, allows selected third-party providers dedicated access to data. On the other hand, there will be the open "agrirouter" world, which, similar to Google's Android operating system, will give all providers along the agricultural value chain open access to data exchange via the central agrirouter data exchange platform (see ◘ Fig. 4.3). The data exchange network conceptually developed in the EU research project ATLAS is intended to serve as a bridge between the two worlds in the future in order to enable the end customers, farmers, and contractors to engage in a complete exchange of data across all manufacturers.

Research Projects in the GAIA-X Initiative
With GAIA-X, representatives from government, industry, and research from France and Germany, together with other European partners, are jointly developing a proposal for the design of a next-generation data infrastructure in Europe. The goal is a secure and interconnected data infrastructure that meets the highest standards of digital sovereignty and promotes innovation. An open and transparent digital ecosystem shall allow data and services to be made available, merged, and shared in a trustworthy manner [BMW21a]. In the future, GAIA-X will give users access to a broad, relevant, and specialized portfolio of products and services from cloud providers and thus enable them to use customized solutions. In this context, GAIA-X offers full transparency through self-description and certified data protection as well as regulatory criteria of the products and services offered [BMW21a].

There are several focal areas within the GAIA-X project. One domain focuses on the area of agriculture, within which the requirements for the shared data infrastructure are to be analyzed in the context of use cases. In the use case project Agri-Gaia, for example, an AI ecosystem for the agricultural and food industry, which is dominated by small and medium-sized enterprises, is to be researched and developed on the basis of GAIA-X. For this purpose, an innovative B2B platform will be realized that provides industry-specific adapted AI components as easy-to-use modules and brings together users and developers of AI algorithms. Agri-Gaia closes the circle of sensor data acquisition in the field, training of algorithms on appropriate servers, and continuous updating/optimization of the algorithms. Appropriate interfaces and standards will be developed to create a cross-manufacturer infrastructure for the exchange of data and algorithms [BMW21b]. During the project, cross-domain requirements will be evaluated regularly based on the results of the individual use cases and aggregated in a superordinate layer. In the future, the user will thus only need one central access, for example, to login to the authorized GAIA-X area based on their rights. Domain-specific data access solutions thus become unnecessary.

4.2.3 Market Development—Current Status

A key finding from various research projects and from the market requirements is that the existing cross-manufacturer cooperation between AgMachinery companies in the areas of mechanics, hydraulics, and electronics needs to be expanded to include the area of data management. Although the self-evident compatibility between tractor and implements from various manufacturers enables the exchange of data between these two machines that is necessary for control (ISOBUS, e.g., the CCI—Competence Center ISOBUS e. V., see [CCI20]), it does not allow for data exchange of this combination of machines with the outside world.

Until now, large agricultural technology corporations have often offered proprietary software solutions (i.e., solutions geared only to their own products), which allow only the connection of manufacturer-specific machines. In the meantime, more and more manufacturers have recognized that customers demand cross-manufacturer solutions, because on farms worldwide, farmers and contractors practically always use machines from different manufacturers, so-called mixed fleets.

The joint development of a cross-manufacturer data exchange platform, respectively a data exchange network, (see ◘ Fig. 4.3) has therefore been and continues to be a natural step from a market and above all from a user perspective. However, especially Central European markets will only accept such possibilities for exchange if they are developed by a community, so that they can be considered manufacturer-neutral and do not originate from a single manufacturer.

To date, farmers and contractors have lacked a solution approach that enables individual networking of the products (machines and software) used in their respective production processes and individual control of the desired data exchange.

As there is no central access via the existing manufacturer-specific solutions, many individual interfaces from the part of the agricultural software and hardware providers to the manufacturers must be set up and maintained with great effort. Valuable know-how of these providers is thus not usable as far as the farmer is concerned. A comparison with popular smartphone app platforms suggests itself here: They have long since managed to offer all software providers the possibility of getting their services to customers. Providers of agricultural software applications or telemetry Retrofit Hardware confirm that the chosen central, neutral, and open approach is the preferred one. From the perspective of agricultural software and hardware providers along the value chain (see ◘ Fig. 4.1), the market potential of a cross-manufacturer, open data exchange platform is considered to be very high.

In addition to the established manufacturers of agricultural software, other companies—some of them start-ups—have offered and continue to offer new software products, mostly for very specific applications. Many of these software tools require cross-manufacturer access to agronomic as well as machine data in order to support farmers in their efforts to optimize their production processes.

A data exchange platform, respectively network, designed according to the above criteria enables the step-by-step use of data available today and in the future along the entire agricultural value chain.

Based on the big picture (see ◘ Fig. 4.4), the integration of the individual elements shows the short-term and future possibilities of data exchange. For example, farmers and contractors will not only be able to exchange data between software products and AgMachinery of different manufacturers, but will also be able to

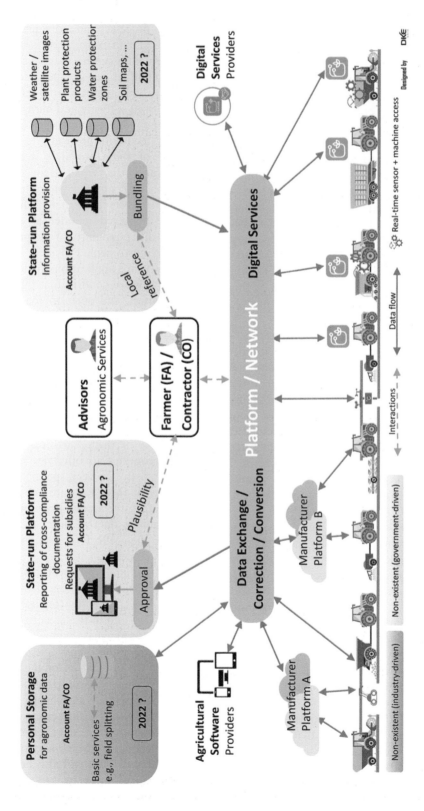

Fig. 4.4 Big picture of agricultural data and services management 2022

store this data centrally and personally in the future. In addition, data from the public sector can be included in farmers' decision-making in the future, and communication with government institutions can also take place digitally in the future. This integration will allow workloads to be reduced and the quality of the exchanged data to be improved significantly. All these elements help the farmer to gradually optimize and possibly reorient their production processes.

Regarding the approach of bundled and region-specific provision of publicly available data (see ◘ Fig. 4.4 top right), it should be noted that this is already being implemented or available in some German federal states. In Germany, the final report of a "Feasibility study on state-run, digital data platforms for agriculture" commissioned by the German Federal Ministry of Food and Agriculture (BMEL) was published at the end of 2020 [BLE21]. In parallel, under the term "Common European Agricultural Data Space" [EC20b], the EU member states are currently working on evaluating objectives and work packages derived from these and on facilitating the exchange of agricultural data along the entire value chain in both industry and research across states. The project is still in its early stages and is driven mainly by the European Commission directorate DG CNECT in cooperation with DG AGRI.

4.2.4 Conclusion and Outlook

The market potential of open data exchange platforms, respectively future networks, to increase interoperability is very high because the concept takes into account all market participants along the value chain and enables open access to all stakeholders. In addition, all participants can offer their end customers, the farmers and contractors, the possibility to exchange data with all other participants using a single interface. The end users are offered a large portfolio for the individual exchange of data for their agricultural value chain. This increases the likelihood that the concept of an open data exchange platform will lead to high market penetration and possibly market leadership.

From the author's point of view, a variant similar to the above-mentioned parallel existence of two relevant smartphone operating systems and applications programmed for these is also a possibility. A serious difference between the current situation in the consumer sector and AgMachinery is the ATLAS project described above, which intends to build bridges between the operating system worlds without questioning their existence. As in road traffic, one would agree on a common data traffic regulation, while the two system providers would apply their individual approval regulations.

From the perspective of agricultural software or hardware providers, this results in the decisive advantage—just as in the consumer world—of not having to serve an unknown number of interfaces, but preferably only one or a few. If, for example, an agricultural software provider wants to offer a mobile app solution to the market in the future, they only need to integrate the agrirouter and proprietary solutions like the John Deere Operations Center interface into their solution.

The comparison with the app stores of alternative operating systems, which we have used several times in this section, is also important because the expectations of all stakeholders involved in agricultural production processes are shaped by the experiences they have made there in terms of compatibility, ease of use, and data protection. This is particularly true for the generation of the so-called digital natives.

If the market participants along the value chain manage to build bridges across operating systems and proprietary boundaries, not only farmers and contractors as well as the providers and manufacturers involved, but also consumers will benefit

4.3 E-Commerce and Logistics

Sebastian Schauff and Julian Schill

Abstract

In the past, the market for agricultural input goods was characterized by local, stationary traders. However, this market is undergoing change with the ongoing digitalization. Farmers are no longer dependent on local traders, they can now buy almost any product online. On the one hand, existing retail chains offer their products in their own online store, and on the other hand new start-ups have emerged which have entered this market. Against the background of the growing online market for agricultural input goods, this section provides an overview of various digital distribution channels established in the market. In addition, the requirements for the trade of agricultural input goods, effects on the contribution margin accounting of farmers and sustainability effects are shown. The section ends with an outlook.

4.3.1 Types of Digital Distribution Channels

Looking at the digital distribution channels established on the market, three different types can be derived.

Type 1: Digital Marketplace
A digital marketplace is characterized by the fact that a large number of different retailers offer products on one platform. Customers can use this digital marketplace to find out about the products on offer. In the event of a purchase decision, the retailer and customer can complete the transaction directly. Digital marketplaces offer traditional retailers the opportunity to make their products available with little effort. On the customer side, such marketplaces offer the advantage of being able to obtain information about a product and its price from different retailers. This transparency leads to a price competition between the retailers due to the comparability of prices. Another advantage of the digital marketplace is that it corresponds to the model of digital consumption. The disadvantage of this type of marketplace is that the platform operators usually do not provide any guarantees for the processing of the transaction and the shipment. Furthermore, platform operators give no guarantee for the quality and authenticity of the traded products. This creates the risk that counterfeit products are sold and the customer does not notice this.

Type 2: Reverse Digital Marketplace
The reverse digital marketplace corresponds in many respects to the digital marketplace. The main difference, however, is that customers publish a demand and retailers make offers for it. This means that a customer publishes a demand on the platform and various retailers can respond to it. The advantage of this type of marketplace is that customers may receive offers below the market price. For retailers, this principle represents a high effort. On the one hand they have to search for suitable demands, on the other hand they have to make an offer for every single demand. Furthermore, there is no guarantee that this offer will be accepted and that a transaction will take place. There is also the risk of an imbalance between offers made and successful transactions. As a result, this marketplace loses its attractiveness for retailers and fewer offers are submitted. This is due to the fact that many customers use this marketplace to compare the prices of local traders. Another disadvantage of the reverse digital marketplace is that it is generally only attractive for large lot sizes. For small batches or individual parts, the sales process represents too much effort.

Type 3: E-Commerce

E-commerce differs from the other two types in that a retailer offers products for sale on his own platform. Customers can use this site to find out about the products and conclude a transaction directly. The transaction is also processed via this platform. This type of distribution has the advantage that the customer can purchase all products from one retailer. This retailer is responsible for the quality and authenticity of the products offered. The retailer guarantees the customer that the transaction will be processed correctly and that the complete logistics process of his order will be handled. Another advantage of e-commerce is the high degree of standardization of the ordering, payment and logistics processes. This results in a more predictable shopping experience for the customer at an online retailer, in which the customer can build trust, similar to a local retailer. A disadvantage of e-commerce platforms is that the customer cannot compare prices on the different sites. In order to compare prices, the customer has to make an additional effort and find out about comparative prices on other platforms. On an e-commerce platform, there is therefore no price competition, rather the prices are set by the provider.

At this point it should be noted that there are not only these three types of digital distribution channels (◘ Fig. 4.5). There are also hybrid forms, which, for example, pursue a business model based on e-commerce but also offer a digital marketplace. Furthermore, there are cooperative approaches that cover all areas of purchasing for a closed circle of members. However, for this to work, the farmer must initially purchase membership as an entry barrier or use the provider's agricultural management software.

After the different characteristics of digital distribution channels have been shown, a short overview of the most prominent representatives of the respective types on the German agricultural market will now be given. It is worth mentioning that especially companies in the field of e-commerce have their roots in the classic catalog business or are spin-offs of established companies. In contrast, current marketplace pro-

Digital distribution channels		
Type 1: Digital marketplace	**Type 2: Reverse digital marketplace**	**Type 3: E-commerce**
• Many dealers offer products • Many customers request products • Dealer ≠ Platform provider	• Many customers post demands • Many dealers submit offers • Dealer ≠ Platform provider	• One dealer offers products • Many customers demand products • Dealer = platform provider
Advantages: + Traditional retailers can offer products online without much effort + Several suppliers for one product + Transaction can be carried out directly + Price war + Corresponds to the model of private consumption	**Advantages:** + Chance of offers below market prices	**Advantages:** + Direct settlement of transactions + Guaranteed fulfillment of the order + Highly standardized processes + Corresponds to model of private consumption
Disadvantages: − Usually no guarantee for payment processing and shipping	**Disadvantages:** − High effort to draft the demand − High effort for submitting offers − No direct transactions − Used for comparison with local market prices	**Disadvantages:**

◘ Fig. 4.5 Three types of digital distribution channels

Table 4.1 Prominent representatives of digital distribution channels		
Type 1: Digital marketplace	**Type 2: Reverse digital marketplace**	**Type 3: E-commerce**
– Cropspot – House of Crops	– Agrando – Agrimand	– ag.supply – BayWa Shop – Schlepperteile – Agridirekt

viders have entered the market as start-ups (Table 4.1).

On an international level, the Farmers Business Network (FBN) in the USA is worth mentioning. Started as a company that wanted to provide farmers with data to help them optimize their purchasing and product selection, it now operates its own e-commerce platform, FBN Direct. Another example is Agrofy, which operates on the South American market. A prominent representative on the European e-commerce market is the French company Agriconomie. They can be seen as an established start-up, which is active on the French and Belgian market. A further international view shows that there are efforts in many countries to establish digital distribution channels.

4.3.2 Differentiation Strategy of ag.supply

As e-commerce is an international trend and the author is a co-founder of ag.supply, the following deep dive focusses his experiences. The company ag.supply was founded in 2018 and is a start-up that operates an e-commerce platform for agricultural input goods. The company, founded by Eric Schüßler and Sebastian Schauff, offers approximately 3 million products online in the areas of spare parts for AgMachinery, seeds, crop protection, and fertilizers. The products are currently available in Germany and the Netherlands.

In a growing e-commerce market ag.supply wants to differentiate itself from other competitors by the product range offered and its quality as well as by qualified consulting.

The product portfolio includes all types of agricultural input goods. For example, other competitors offer a range of 30,000 to 50,000 spare parts. This is sufficient, in order to make many repairs with wearing parts possible. The company ag.supply wants to cover the product need of all repairs and offer original parts as well as high-quality reproduction parts.

Beside a broad product portfolio, the qualified customer support is of great importance. Depending on their request, customers are referred to an agricultural engineer or an AgMachinery mechanic for advice. This is intended to ensure a high quality of advice and to provide advice that is independent of manufacturers.

For ag.supply it turned out that the factors supply ability, quality, consultation, and logistics are the most important factors, which affect the purchase decision of the farmers. The factor price is to be regarded thereby only as hygiene factor.

4.3.3 Requirements for the Online Trade of Agricultural Input Goods

The digital distribution of agricultural input goods presupposes that these goods are tradable online on the one hand and that they are also logistically capable on the other.

The ability to trade products online depends heavily on whether the customer is willing to buy them online. In the classical sense, the farmer obtained all products from local traders. Since the emergence of the first digital distribution channels, however, there has been a trend away from local trade towards digital marketplaces or e-commerce. However, this trend is not equally strong for all product categories. Today, for example, there is greater online demand for spare parts for AgMachinery than for crop protection or seeds.

Logistics capability incl. all efforts involved in shipping certain products. This effort is relatively low, for example, for the shipping of spare parts. Spare parts can be shipped by courier, express and parcel services without additional effort. Pesticides pose a greater challenge for shipping, as they are usually hazardous goods. The shipping is however possible with a small additional expenditure. However, the dispatch of fertilizers, which is characterized by large lot sizes, is problematic. Here the ordered quantity of fertilizer has to be divided into the logistic sizes TL (Truckload) and LTL (Less than Truckload). TL means that the load corresponds to the capacity of the vehicle, usually a truck. In contrast, the load in the LTL range is smaller than the available capacity. In both cases the same transport costs must be considered. Therefore, orders of fertilizers, which correspond to the capacity of the vehicle, are realizable. However, orders that only use a part of the available capacity are difficult to realize. The reason for this is that the transport costs per transported unit of fertilizer are significantly higher than the transport costs in the TL area.

In summary, it can be said at this point that products are logistically viable if they can either be shipped using the standardized courier, express and parcel market or the shipment is large enough to fill a truck. The latter is difficult to implement due to the structure of German agriculture with a large number of small farms or part-time farmers.

4.3.4 Effects of Digital Distribution Channels on Agricultural Contribution Margin Accounting

In the contribution margin calculation, the farmer usually only considers the product costs. Marginal savings are possible through online procurement. However, it does not consider the time savings that a farmer experiences through online trading (similar to other Digital Farming related cost centers, see ▶ Sect. 2.4). In the classical sense, an order is processed as follows: The farmer calls his local dealer and inquiries about the availability of a product. If the dealer does not have this product in stock, it must be ordered. One or two days later the farmer has to drive to the dealer, pick up the product and drive back to his farm. The additional purchase costs are about 30€ to 40€ per purchase. The time required is also not to be neglected. However, if he orders the product digitally, the product is delivered directly to his farm without any further effort. If one looks at the orders of ag.supply, then 40% of the purchases are transacted by App. That means that the farmer can take care of the purchase with his cell phone for example during the field work with GPS driving control. Thus, substantial savings in procurement and working time are created. Depending on the product group, savings of up to 15% can be generated. Another advantage is that customers are no longer bound to opening hours. Customers can therefore place orders after hours or on weekends. This is a particular advantage for part-time farmers.

4.3.5 Sustainability Effects of Digital Distribution Channels

In reference to ag.supply experiences, sustainability topics are addressed on the side of the farmers only rarely. Customers are focused on products and their quality. Another problem is that many farmers regard express delivery as standard and demand it. Compared to standard shipping, this is much more harmful to the climate. However, by deciding to purchase products online, farmers indirectly enable sustainability measures that ag.supply is pursuing. For ag.supply the greatest possible potential for sustainability lies in the saving of logistics stages. The classic trade structure is as follows: A product is produced by a manufacturer, shipped to a wholesaler who in turn ships it to a local retailer from whom the farmer purchases the product. When products are shipped from one logistics stage to the next, they are repackaged and transferred. By skipping individual logistics stages these processes could be saved. The company pursues the goal of sending products directly from the manufacturer or wholesaler to the customer and thus skipping subsequent logistics stages. To make this possible, the networking of the various partners is of utmost importance.

4.3.6 Outlook

The importance of digital distribution channels for agricultural input goods will continue to grow in the coming years. From a current perspective, there is an infrastructural oversupply in the form of multi-level supply chains and excess storage capacities for seeds, crop protection and fertilizers. The question must be answered as to how processes and the associated infrastructure can be optimized and made more efficient for the benefit of farmers. It is already becoming apparent today that significant efficiency gains can be achieved by skipping logistics stages.

In addition to efficiency, the focus is increasingly shifting to the farmer as a customer. If the customer can choose from a variety of digital providers, the choice will be the most customer-friendly one. Providers will therefore have to make a great effort to be attractive to customers.

Strong market growth will increase the number of start-ups trying to penetrate this market. The environment is an important factor for the establishment of start-ups. Currently, the supply of venture capital for start-ups in this area in Germany is considered weak. In order to enable the foundation of new start-ups, a rethinking of venture capital providers is required at this point. After all, in order to meet the farmers' customer requirements, agile start-ups are needed that can operate free of established business processes.

4.4 The Digital Eco-System of Sustainable Farming: Agricultural Insurance as a Glue

Alexa Mayer-Bosse

Abstract

This section describes agricultural insurance solutions incl. different indices, parameters and risks and why Digital Farming solutions benefit from combination with insurance solutions and vice versa. The agronomic focus is on climate-based production risks targeting e.g., yield, soil, and weather such as precipitation or temperature.

4.4.1 Introduction

Farming is all about quality inputs, sophisticated AgMachinery, market opportunities, appropriate financing, and sound agronomical expertise to benefit best from natural resources. In recent years, digitalization has been added to the game. On the one hand, digitalization makes agricultural production more efficient, while on the other hand it equips the enterprise for the future and unlocks new potential. On top of using high-tech machinery and managing complex data, and data platforms, farmers must find suitable financial instruments—from supplier credits to bank loans with appropriate collaterals, from leasing agreements to all types of insurances. Even with all this effort one constant in agriculture remains: risk.

Farmers have been experts in managing risks, ever since crops have been grown or animals raised to market for profit. Risks in crop farming are all about time, operation, price, and weather. Livestock risks are all about health, mortality, and price. New risks have been recognized by the introduction of innovative technologies in business decisions such as autonomous in-field driving [SB+21], sensor technology, and artificial intelligence [BT21] in business decisions. In the future Digital Farming creates new risks in terms of cyber or data security.

There are numerous solutions for insuring farm risks. Insurance solutions for losses that might occur while using Digital Farming systems, i.e., autonomous driving (see ▶ Sect. 3.3), services [MR20] in the agricultural sector have to be analyzed case by case, depending on the amount of data produced. Crop yield risks are however unique and one of the most complex ones to quantify. Crop losses are increasing and occur more regularly, all over the globe. They increase when, for example, conventional practices lose their effectiveness, whether as a result of resistance to certain AgChemicals or as a result of restrictions imposed by the EU fertilizer regulation. In particular due to climate change, losses in agricultural production are on the rise. As adverse weather conditions persist locally for longer, farms subsequently experience too much or too little rain with devastating losses [Fa+18]. Of particular interest for farmers is how losses in crop production can be better assessed, mitigated and evaluated through digitalization, especially with respect to climate risks.

In this section, digital crop insurance is discussed from various perspectives. Insuring risk and evaluating it is looked at independently. Firstly, crop insurance is a valuable tool to mitigate the financial impact of climate risks. Digital technologies allow these risks to be parameterized in insurance products. Due to its digital nature of these modern insurance products, all partners in the agricultural supply chain can assist in financing production risks. It is also stated that crop risks are complex and expensive. Lastly, the calculation of risk profiles gives a monetary benchmark on how climate risk-resilient a farmer can produce.

4.4.2 When Agricultural Insurance is Parameterized, Their Digital Loss Assessments Result in Immediate Pay-Outs

Traditionally, agricultural insurance policies cover crop losses. Why would insurance then have a new role in the age of Digital Farming? Hail insurance, for example, is an important support for farmers and will remain that. Yet it is precisely the digital treasure of data available that makes new forms of insurance possible. Agricultural insurance is called Parametric Index Insurance if it can use suitable proxies for agricultural production and its risks. Losses are usually quantified as a shortfall in crop yield. Alternative concepts also capture defaults in weather metrics, biomass production margin, or soil moisture. Basically, these proxy factors and indices estimate agricultural risks without measuring the actual damage on the field. For this, index insurance does not require any technical risk inspection and no individual claims settlement. Both, underwriting and loss assessment, are exclusively data driven. For example, climate risks such as drought and heat waves [Ta21] can be determined by a weather index based on temperature to assess the pollination failure caused by a heatwave. If during the pollination phase of about three weeks the daily temperature in a particular area exceeds, e.g., 33 °C for a predefined number of days, the insurance product would pay out. It is fair to mention that index insurance policies are criticized for their basis risk. In this heat index example, it could happen that pollination is hindered even though the weather station has not exceeded the agreed daily temperatures. In this situation the basis risk would result in no pay-out because the threshold value was not met. But here too, there is an advantage of digitalization: the more risk relevant data is digitally recorded, the more the basis risk can be steadily reduced.

For the development of such a parametric cover, historical and especially georeferenced risk related data are key. At the core of all risk analyses are the digitally available Georeferenced Field Boundaries. With georeferencing, all satellite imagery and localized weather information can be assigned to regions relating to the field being studied. Crop rotations, historical yields and farming practices are digitally tracked and recorded. The calculation of the insurance premium and the pay-out must also follow the same parametric logic. Immediately after the reading of the recorded temperature, the compensation is determined. If the temperature exceeds the chosen 33 °C, payments are initiated directly.

Particularly in the case of drought insurance, a policy that requires a physical loss assessment that has to be conducted after sudden events such as hail or storms is not practical or cost-effective. Drought is a long-lasting and widespread weather phenomenon that would require a large number of loss adjusters in a short period of time just before harvest. In order to avoid these bottlenecks, Index Insurance is the right choice, which also saves costs in the loss assessment.

4.4.3 Typical Insurance Covers Along the Agricultural Supply Chain

De-risking the supply chain from the production risk is based on geo-referenced index insurance, where index types and risks are matched (◘ Table 4.2).

There already exist many bilateral insurance agreements in the agricultural supply chain. Farmers around the world insure themselves against crop shortfall, for example based on an area yield index. Agribusinesses cover their own crop production related risks. Seed suppliers select a precipitation index for germination protection, which they stick on a seed bag. Crop protection companies offer either a satellite-based vegetation index or a remotely sensed soil moisture index to their clients. Traders offer producers yield index insurance based on yield statistics. Machinery manufacturers provide drought index insurance. Food and beverage processors insure their supplies against the processing risk due to weather-related reduction in crop quality. These types of index covers can also be structured in a cost-efficient manner, when an agribusiness company purchases one policy for many enrolled farmers.

Agronomy Perspective

Table 4.2 Examples of parametric index insurance types and climate risks

Index	Parameter	Climate risk	Production risk
Area yield index	Yield statistics	Various perils	Crop shortfall
Vegetation index	Biomass	Drought	Crop shortfall
Weather index	Temperature	Heat stress	Pollination
Weather index	Precipitation	Drought	Germination
Weather index	Precipitation	Excess rainfall	Process quality
Soil moisture index	Soil moisture	Drought	Germination
Soil moisture index	Soil moisture	Excess moisture	Harvest time

4.4.4 Insurance as a Glue in a Digital Ecosystem of Sustainable Farming

Digital agricultural platforms provide analyses of in-field technical devices and sensor data of machinery in order to support the farmer with agronomical decisions. With this digital intelligence, the path to new production methods with risk-reducing characteristics and risk monitoring is short. It is interesting to see how digitally enabled index insurance can bring even more benefits to the ecosystem.

Insurance serves as a glue when bilateral agreements merge into a multilateral system and all members benefit from it. While de-risking the supply chain remains the primary objective of insurance, additional benefits are generated for further services and transactions. For example, a growing market of accurately collected data serve as compelling value propositions for various stakeholders to engage within such a digital ecosystem. Agribusinesses—whether seed or crop protection providers—are expanding their offering to farmers by taking out index insurance with the farmer being the sole beneficiary of a pay-out. Traders offer producers pre-financing with insurance embedded in an off-taking agreement. Machinery manufacturers see a competitive advantage in granting an insured lease instalment forgiveness in the event of drought. And food producers enforce a certain quality of agricultural products by insuring their quality risk on behalf of the farmers (◘ Fig. 4.6).

◘ **Fig. 4.6** Climate risk insurance entry points along the agricultural production cycle

A look at the various value propositions reveal that supply chain actors embedding insurance can unlock further business potential. Since the producer is compensated for all covered operating costs in periods of drought, the discount for a seed bag in the form of insurance encourages the farmer to buy it. In turn, the farm provides the digital field data via the platform to the players. In this way, sales increase and co-financed production risk (by various agribusiness companies), ultimately puts the farmer in a creditworthy position with his supplier and bank.

When applying for production credit to purchase inputs, the farmer's credit application is backed by an index insurance. Using digital intelligence, a bank could consider additional lending to increase or even replace asset-backed loans by assessing the production potential and providing farms specific index insurance. Each year, insurance cover can only be renewed if the data is up-to-date and reflects information such as crop rotation and newly planted fields, if it is digitally accessible. In turn, the insurance provides additional quality assurance for the data. It becomes clear how important accurate data is for all parties involved.

The basis for all this are digital risk profiles modelled per region, per field, and per crop. It is feasible on behalf of the farmers as digital crop insurance can be sliced and diced accordingly.

4.4.5 Digitally Enabled Risk Analysis and Mitigation—Farming Risk Profiles

Digital technologies [Mi21] are a blessing and radical gamechanger for the sector. They also play an important role in risk assessment and mitigation, both in terms of agronomic and financial risks. Farming practices must respond to emerging risks and digital tools assists in doing so—whether in avoiding risks by applying an optimal amount of fertilizer according to the Growth Stages or in reducing losses through the precise application of crop protection.

Above all, digital systems have the potential to boost management practices as smart risk profiles are a major positive spinoff of digitalization. Next to crop rotations, historical yields and farming practices, soil data can be assessed and updated, be it nutrients, texture or organic matter content. With all this information, a complex risk profile is constructed, whereby past weather events, probability modelling and predictive analytics give shape to specific loss scenarios. As the probability, that adverse events will increase damages, risks can be expressed as expected monetary (negative) value [Usm20]. An example might be a 25% crop shortfall event. By calculating the insurance premium, this risk profile premium gives an annual monetary value that should be subtracted from the farmer's contribution margin.

4.4.6 Enabling More Sustainable Production: Risk Profiles as Ally

Extreme droughts are worrying. Local heavy rainfall is also becoming more frequent as a result of climate change. A dreaded scenario is that the high amount of rainfall causes the soil in the fields to erode within a short time. Besides the loss of fertile soil, public infrastructure is damaged resulting in the rebuilding of roads, and private households have to replace damaged cellars and foundations. According to an EU study [Pa+18], up to 0.43% of harvest productivity is lost annually due to severe erosion, with a total damage of around 1.25 billion Euros. A number of agronomic and financial measures can be taken to manage this risk:

1. increasing the water infiltration and storage capacity of the soil is the agronomical approach;
2. purchasing insurance for the full loss exposure would be the financial measure.

Digital Farming technologies are an enabler in addressing these measures. As soon as farm managers rethink their practices to embrace soil smart farming, they may need to invest in other AgMachinery and adapt their inputs accordingly. With digitally recorded data, farms are then well equipped to monitor the impact of changed farming practices on soil quality, also known as regenerative agriculture.

Following this thought, insurance serves several functions. On the one hand risk transfer (insurance), on the other hand the economic perspective of regenerative agriculture [SL20]. The issue here is what savings in risk costs could be achieved through increased water infiltration capacity in the soil. By comparing insurance premiums of different farming practices, the practices more vulnerable or more resilient and therefore more sustainable e.g., after of a severe weather event, can be identified.

A weather index-based risk profile helps to assess the risk of soil erosion by estimating the likelihood of a heavy rainfall event in a certain region. According to the German Umweltbundesamt [UBA20], if a rainfall of more than 20 l/m² in a period of two days can already cause soil erosion. The risk profile of a rainfall index can be valued at about 150 €/ha per year, calculated for a producer's digitally georeferenced fields and measured at this local weather station. By significantly increasing the water storage capacity, a farmer might only want to insure the remaining risk of soil erosion. If he or she were to choose a threshold of 60 l/m² rainfall—as this scenario is less likely—the insurance premium would only be around 30 €/ha. It is obvious that the calculated cost savings of 120 €/ha per year can be invested in equipment to improve soil quality.

This perspective of soil erosion and its risk profile is interesting. Isn't there a public interest in co-financing the insurance premiums to incentivize the farmer converting to a nature positive farming, both to start and to continue? Such a funding would be linked to conditions imposed on farms aimed at reducing the more frequent erosion risks through appropriate soil management. Verification from the farmers' point of view can be provided by digital agricultural technologies. In terms of a drought coverage: Isn't there a public interest in co-financing insurance premiums for catastrophic drought events, and at the same time creating incentives for farmers to invest in increasing water storage capacity in order to be more resilient in drier years?

4.4.7 Conclusion

As climate risks in agriculture become more common, farmers are increasingly challenged to stand their ground in order to sustain. This opens a lot of opportunities of combining Digital Farming systems with risk profiles, and eventually with insurance. If de-risking of the supply chain is co-financed by the various stakeholders, farmers are supported in their business for the sake of all. Addressing production risks is complex but at the same time has a powerful impact.

Other emerging risks in Digital Farming are also on the insurance industry's agenda in the future, be it risks of autonomous driving or services in artificial intelligence. The risk of breaching environmental regulations or traceability guarantees is also being looked at.

One of the unnoticed levers of insurance is the skill to evaluate risk profiles. The parametric insurance approach gives a monetary value to resilience towards climate risks. The often-discussed internalization of external climate costs can be operationalized in agricultural production thanks to the digital treasure trove of data. Climate risks farmers are exposed to get a real cost tag which has to be considered.

Digital Farming creates a platform to combine agricultural risk mitigation (Parametric Insurance) with climate change adaptation (soil resilience), and in due course with climate change mitigation (soil carbon sequestration).

4.5 Soil and Seed Management

Theo Leeb and Josef Stangl

Abstract
This section provides an overview of the historical development, influencing factors, as well as current optimization approaches, and future developments of agricultural soil and seed management.

4.5.1 Introduction

Soil and seed management are subject to great changes. 50 years ago, farmers knew their site in great detail and on a very small scale. Regarding soil management in field zones, they were able to make decisions based on many years of experience. For example, on a clay hilltop, an additional pass with a Power Harrow proved positive for good emergence of the seeds. In other places, the plow had to be placed flatter to avoid bringing dead soil to the surface. In the area of seed management, too, a lot of know-how was available on farms regarding seed rate or placement depth.

This experience was passed on from generation to generation and continuously optimized. When farms started to grow by acreage, this growth was mainly realized through leased land. The site-specific characteristics of these newly cultivated areas could hardly be transmitted and passed on, meaning that this information was largely lost. Increasingly powerful technology compensated for this loss of information. In practice, different soil management systems were established.

4.5.2 Methods of Soil Management

Currently, we distinguish three different methods in soil management:
- No-till, where there is no or almost no mechanical intervention in the soil
- Conservation tillage, where Stubble Cultivator technology is mostly used and where the degree of intensity can vary greatly
- Inversion tillage with the plough or similar turning implements

With no-till, the tillage tasks are taking over by bioturbation, root penetration, and self-loosening of the soil. This is suited particularly well to sites with a functioning and well-established Macrofauna. However, a consistent water supply throughout the year is required. Thus, this system is rarely, if ever, found on sandy sites or dry sites with black soil. In addition, soils that tend towards natural compaction are less suitable for no-till. Soils with an angular skeleton, such as sandy soils, tend towards natural compaction. Sites with high water retention capacity in combination with extreme winter waterlogging also tend to have compaction zones and natural compaction. If, on the other hand, the soil consists of more than 20% clay, it loosens itself by shrinking and swelling, which is positive for root penetration in the soil. This process is becoming increasingly important as due to the advancing climate change, the frost effect on the soil structure becomes less reliable for loosening the soil in winter.

In Central Europe, no-till only plays a very minor role. In the majority of cases, the soil is tilled. Mechanical tillage is divided into conservation and inversion methods. In conservation soil management, cultivators with tine technology or other cutting implements are used. In this system, the organic matter, such as straw or intermediate crops, is mixed into the tillage hori-

◻ **Fig. 4.7** HORSCH Tiger 4AS with SteelFlex Packer working in an intermediate crop

zon. The mixing of the organic matter and the tillage by the tines, which breaks the soil mostly at predetermined breaking points, leaves a load-bearing soil that quickly stabilizes again. Depending on the tillage intensity, some of the organic matter remains on the surface, which significantly reduces the risk of wind and water erosion. For these reasons, conservation tillage has become widespread in practice in recent decades (◻ Fig. 4.7).

The plough stands for inversion tillage. In this method, the soil is inverted by turning it sideways in furrow slices. In this process, breaking the furrow slices or even pouring the soil stream at the plow blade is desirable to leave the loosest possible soil. Inverting the topsoil of the field safely spills seeds, plants growing at the surface, and organic matter. What remains is weed-free, brown soil—"a clean table".

4.5.3 Increasing the Performance of Soil

In addition to this high-performance soil management, increasingly optimized fertilization in combination with crop protection has improved yields at many sites. This was one of the success factors that enabled farmers to increase crop yields from the 1980s onward, although animal husbandry was abandoned on many crop farms and the importance of organic fertilization on large areas declined.

Around the turn of the millennium, however, the first farms realized that the performance of the soils would not be sustainable in the long term without organic fertilization. Soil biology and soil physics became more of a focus. In recent years, the pressure to increase fertilizer efficiency has intensified the need to look more closely at soil biology and soil physics and to place them at the center of fertile soils.

These changes will greatly affect soil management. The classical doctrine of soil management—starting shallow and then working the soil progressively deeper in the second and subsequent phases—is being expanded. The cultivation of intermediate crops is becoming an important measure in soil management strategies, despite challenges like establishing the intermediate crop and creating optimal conditions for the succeeding crop. The intermediate crop residues (roots and plants) must be made digestible for soil biology through tilling. The emergence of volunteer plants in the succeeding crop—in the future also without the use of total herbicides—must be avoided, without compromising seed bed quality or water drainage in the soil.

In the course of soil optimization, the following approaches are used in practice, partly supported by digital tools:

− In shallow stubble cultivation, the savings or optimization potential seems to be very low due to the low control depth.
− Selecting the working depth depending on the amount of straw (e.g., derived from the combine's yield recording) can be an approach to optimize straw incorporation.
− Variable choice of the working depth during basic tillage to carve out compaction zones in order to increase the root space is a promising approach on deep soils. However, there is currently

still a lack of data to create an application map. Depth control as a function of Tractive Power requirements is also being tested. Watch out is that deeper control requires more power, thus stronger machines, which can lead to increased soil compaction
- Depth control by means of sensors/cameras with a view of the topsoil in order to produce a homogeneous working pattern.

4.5.4 The Role of Crop Rotation

In the future, however, soil management will have to change much more than just varying the depth by field zone. It will be geared to a much greater extent to objectives determined largely by the main crop and by crop rotation. Unfortunately, as crop rotation only exists in theory on many farms due to delayed harvest date of the previous crop, the weather, economic or political changes, long-term planning is very difficult. Nevertheless, crop rotation elements should be built up under new aspects. The intermediate crop preceding corn or sugar beet is already firmly established, but can be further optimized according to the specific site. Corn requires nitrogen much later than sugar beets. For this reason, an intermediate crop with a bigger C:N ratio with slower N release and less intensive crushing of the intermediate crop by tillage is conceivable for corn. In addition, corn does not require as fine a seedbed as sugar beets. However, an intermediate crop that is well integrated into the crop rotation can do more than store and release nutrients. Deep-rooted species can loosen the soil and stabilize its structure. Through deep roots, organic matter and thus carbon reaches the subsoil, where humus can be built up in the long term, one of the key measures in Carbon Sequestration. The roots of the intermediate crop are more important for soil fertility than the aboveground plants. Undersown Crops are an alternative option as well.

The future challenge is to develop field and field zone adopted crop rotation elements that can be combined independent and flexible to maintain manageability on farm level.

September to September could serve as a crop rotation period. Example ideas for crop rotation elements:
- Canola with an Undersowing of clover, first tillage with knife roller and the possibility of a fall crop, e.g., cereal after cultivator sweep
- Cereals with stubble fall, deep tillage followed by canola seeding—open soil with sun/UV radiation once in the crop rotation
- Change from winterization to summarization—plenty of time for intermediate crops with a wide range of objectives: a) optimization of nitrogen release for the succeeding crop; b) optimization of humus build-up/soil fertility; and c) optimization of water balance

There are certainly other approaches and combinations as well, derived from e.g., site requirements, farming practices, commercially viable crop spectrum.

4.5.5 The Impact of Climate Change

In view of climate change, not only before mentioned Carbon Sequestration is gaining importance, but from a short-term yield protection, the item "optimization of water balance" will be one of the most important goals in proper soil management. Most studies on climate change assume almost constant annual precipitation in many regions, but the distribution of precipitation over the year is changing. In crop farming, we increasingly have to plan for droughts lasting for months at a time. The exciting question is therefore: With which soil management system can we save water from rainy months and keep it available for plants during dry periods? This goal could possibly be reached with deep loosening, an intermediate crop with low water consumption, and a mulch layer that is tolerable for seeding.

In order to reconcile deep loosening, integration of intermediate crops, and storage of water with safe emergence of the crop, soil management and seeding technology must be coordinated. If soil management, including compaction, does not match the seeding system, safe and uniform emergence is hard to achieve under difficult conditions. However, uniform emergence is the foundation for good yields. Uniform stands are stronger in suppressing weed in both, organic and integrated crop cultivation, and all other subsequent production steps are easier to perform. The key to uniform stands is seed management.

4.5.6 Seed Management

The crucial factors in seed management are:
- **Seed rate:** Depending on sowing date, variety, and location, it significantly influences stand density and consequently the required weed suppression.
- **Colter pressure:** This has an influence on root development. If it is too high, it will be harder for the roots to penetrate the side walls of the seed furrow that are compressed by the sowing discs. If it is too low, the placement depth cannot be kept uniform by the seeding unit.
- **Placement depth:** It needs to be deep enough that the seed gets enough water to germinate, but also only deep enough that nothing stands in the way of optimal crop development. For optimum placement depth, good coordination between tillage, including compaction, and the setting of the seeding technique is imperative.

These three factors depend very much on soil differences. Since the site properties differ on a small scale on the sections of a field, it is necessary to either find a good compromise in the settings or to control the adaptation via an application map. The basis for the generation of a map is basic information such as soil type maps or analyses of satellite maps, on the basis of which the yield potential can be derived on a small scale. In cereal cultivation, seed rates are currently still mostly assigned to the respective zones by the farm manager. They often revise the zoning manually afterwards based on their expert knowledge.

The variation in seed rate is influenced by the following factors:
- Condition of the seedbed:

– Low emergence expected	=> increase in seed rate
– Well-prepared seedbed	=> decrease in seed rate

- Yield potential of field zones:

– Weak areas	=> decrease in seed rate
– Zones with good yield	=> increase in seed rate
– Places with weak Tillering	=> increase in seed rate
– Places with good Tillering	=> decrease in seed rate

In addition, as already mentioned, the sowing date and the site-specific production technique are taken into account when deciding on the quantity of seeded grains/m^2. In particular, expected plant losses caused by insect or slug damage should be mentioned here. But conscious adjustment of the seed rate due to calculated losses as a result of mechanical or chemical plant protection influence the seed rate. Some farms already use variable rate seeding maps—and have made good experiences with these.

However, currently only few farms are using them and therefore we have to ask why field-zone-specific seed rate control for cereal crops is not used more intensively in practice, although the technology has been available on many farms for more than ten years. Another way to influence stand density during seeding is to regulate colter pressure and the pressure on the press roller. Currently, ways are being sought to vary colter pressure, compaction, or placement depth using fixed sensors, independent of chart-based systems. For example, there exist approaches for using online sensor technology to regulate seeding depth and embedment via a wide range of parameters.

Currently, the following parameters are under discussion:
— Using the color of the soil (humus content) to regulate seed rate and sowing depth
— Optimizing seeding depth via soil moisture sensors
— Other parameters such as pH, clay content, water retention capacity, etc. are being examined and control functions are being sought (◻ Fig. 4.8).

The situation is different for single-seeded crops. For corn, there has recently been support from various growers' houses or even ready-made chart recommendations for selecting the correct seed rate for a specific site, taking into account relevant basic data. To maintain the optimum target spacing of plants in the row, seeders with large working widths control the metering devices differently across the working width when cornering, thus ensuring the same seed rate everywhere. Optimization of seed embedment is achieved by means of variable colter pressure regulation. This takes place depending on the penetration resistance of the seeding unit to the soil. The automatic control prevents too much or too little colter pressure—both have a strong negative influence on development in the juvenile stage of the corn. Too much colter pressure compacts the seed furrow walls, so in the juvenile stage, the roots grow along the seed furrow and do not develop optimally. Too little colter pressure results in the grain lying in "loose" soil and delayed emergence under dry conditions. Due to the easy handling and the clearly recogniz-

◻ **Fig. 4.8** HORSCH Pronto 6 DC in grain sowing with variable seed rate

able benefits, this technology has been in high demand for new corn planters in recent years.

4.5.7 Outlook

Improvements in precision seeding will continue to be made in the coming years. The reasons for this are the limited water supply and the resulting variety adaptation towards Dent Varieties, which are less forgiving of seeding mistakes. In addition, changes in fertilizer regulation demand more care in the use of nutrients for our crops. The fertilizer efficiency of nitrogen and phosphorus is coming under increasing scrutiny. Both contact fertilization in the seed furrow and underground fertilization next to the seed can provide solutions for increasing fertilizer efficiency. Especially under cool conditions, fertilizer near the seed will force the juvenile development of the crop. This type of targeted fertilization acts as a development accelerator not only in corn, but also in other crops. Rapid juvenile development allows our crops to compete more effectively against weeds. In fertilization, the need to integrate seed technology and soil management is becoming increasingly apparent. Soil cultivation implements offer the possibility to place the fertilizer even deeper as a deposit or mix it more broadly into the soil. Attracted by the nutrients (ammonium and phosphate), the plant roots grow towards these deposits and not only utilize the deposited nutrients very effectively in this area, but also root faster into deeper soil and thus exhibit improved drought tolerance. Scientific trials in recent years have shown the potential of contact and underground fertilization in seeding or deep fertilization in soil management (◘ Fig. 4.9).

In summary, soil management will have to perform additional tasks, such as saving water from rainy months for dry periods—which are becoming more frequent due to climate change, dealing with intermediate crops in terms of establishing them and working them in, and avoiding Volunteer Plants in the main crop when the use of total herbicides might no longer be allowed. Soil management, the choice of intermediate crops, and the seeding of the crop must be consistently coordinated. However, if

◘ Fig. 4.9 HORSCH Focus 6 TD in rapeseed sowing + deep fertilization

crop rotation is not considered as well, this alone will not be sufficient to significantly improve soil fertility. In the past, the benefits from coordinated crop rotation on positive resultant yields were often subordinated to short-term economic decisions. These goals can only be achieved if the knowledge about soil biology, soil chemistry, and soil physics is reweighted and newly interlinked. The synergy effects from the individual disciplines must be strengthened and converted into yield. This will certainly not make production easier, but we sense strong interest on the part of farmers in these topics and the opportunities they offer.

There are clear benefits of bringing soil and seed management measures back to the resolution it had 50 years ago. To translate this level of accuracy to individual fields and field zones while keeping the scale, will require more data sources with higher resolution as well as additional decision support and variable application features, one of the key challenges and opportunities for Digital Farming.

4.6 Nutrient Supply: From Whole Fields to Individual Plants

Volker Stöcklin

Abstract

Digitalization has already led to many sensor and control systems being used today for the precise and needs-based supply of nutrients to plants. In mineral fertilization with Twin Disc Spreaders, in particular, sensors for measuring the mass flow have become established. GPS-based control systems with section control and field-zone-specific application have also become established following the introduction of the ISOBUS. Small-scale identification of nutrient needs is enabled by new soil sensors, nitrogen sensors, and high-resolution satellite maps. The increased use of pneumatic spreaders with spreading systems that can switch off individual manifolds and spread different quantities will make it possible in the future to respond to the needs of individual plants even in very small field zones.

4.6.1 Introduction to Metering and Spreading

In order to protect the environment and optimally supply the plants (see ▶ Sect. 1.4), fertilization must be carried out precisely and needs-based. In recent years, digitalization in mineral fertilization has made great progress. In the most widely used twin disc spreaders, sensor systems are available for the two important machine functions—metering and distribution—with which the entire work process can be controlled and automated.

To control the metering, i.e., to determine the amount of fertilizer per time, weighing technology with Shear Force Sensors is used on the one hand. These shear force sensors are used to measure the weight loss in the hopper, which is strongly affected by dynamic influences and must be filtered with large time constants. On the other hand, EMC systems with torque sensors for mass flow control are also very widely used. EMC systems use the effect of the Coriolis force. By measuring the torque at the drive of the spreading disc, the mass flow can be calculated directly via a characteristic linear curve for each type of spreading disc if the speed is known (see ◘ Fig. 4.10). The advantages of EMC over weighing technology lie in the possibility to determine the mass flow separately for each side left/right very dynamically and with high precision, independent of vibration or slope influences. Torque sensors, e.g., based on the magnetostrictive measuring principle, are used for this purpose.

Control systems for optimizing the spread distribution, such as AMAZONE ARGUS TWIN (see ◘ Fig. 4.11) or RAUCH AXMAT DUO, are based

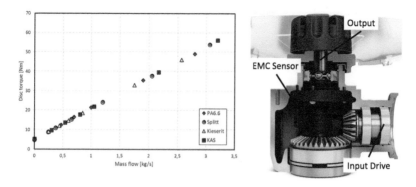

☐ **Fig. 4.10** EMC mass flow curve of different spreading materials at constant speed [Sto11], throwing disc drive with magnetostrictive EMC torque sensor [Rau20]

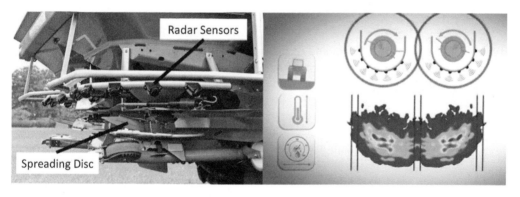

☐ **Fig. 4.11** Spread pattern monitoring with AMAZONE ARGUS TWIN [Ama20a]

on radar sensors that detect the circumferential distribution of the fertilizer directly at the spreading disc and can control it to a specific setpoint value of the flight direction for each type of fertilizer, type of spreading disc, and working width. In this way, the spread pattern can be kept constant and uniform in the event of changes in fertilizer properties due to air humidity, temperature, and particle size distribution.

4.6.2 Determination of Machine Settings

In order to ensure optimal settings for the fertilizer spread, the machine manufacturers empirically determine setting data for each type of fertilizer and each working width and make them available to the operators. In the past, this setting data was published in a book. Today, smartphone apps are used to select the type of fertilizer and the setting data, and allow wireless transmission of the setting parameters to the machine. To check fertilizer distribution on the field, collecting vessels are placed perpendicular to the driving direction, the fertilizer is collected, and the quantities in the vessels are compared. Alternatively, collection mats are laid out and the fertilizer on the mat is photographed using a smartphone app (see ☐ Fig. 4.12) and the optimized machine settings are calculated automatically.

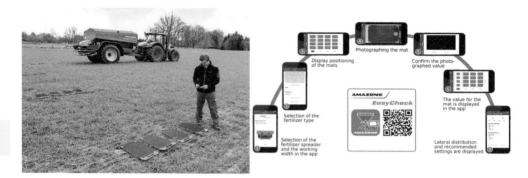

◻ **Fig. 4.12** Checking lateral distribution with AMAZONE Easy Check [Ama20b]

4.6.3 GPS-Based Automation Systems

Typically, fields are not uniform and rectangular in shape, but have various special areas. In the case of non-parallel tramlines, uneven tramline spacing, or right-angled and sloping headlands, the spread patterns must be adjusted in order to achieve the most uniform fertilizer distribution possible.

For this purpose, GPS-based section control systems such as SULKY Econov or RAUCH VariSpread are used to control lateral distribution. To change the spread pattern, the spreading disc speed, the drop point, and the application rate can be adjusted.

In the machine control unit, the correlations of the previously mentioned setting parameters are assigned to the part-width sections and can be called up at the corresponding position in the field and automatically set by the machine.

Due to the large spatial extension of the spread pattern, e.g., 50×30 m with a working width of 28 m, areas with over- or undersupply can arise in the headland. According to [Thu11], these areas can be minimized by automatically calculating the switch-on or switch-off distances in the headland, depending on the flight characteristics of the fertilizer, the working width, the driving speed, and the application rate (see ◻ Fig. 4.13). These calculation algorithms are implemented and available as commercial products, e.g., in AMAZONE SwitchPoint or RAUCH OptiPoint.

As fertilizer savings of >10% can be achieved with GPS-based systems for fertilizer spreaders, these systems have now become widespread. Typically, the machines are equipped with the necessary actuator technology when they are shipped from the manufacturer, so that only a software license for the operator terminal needs to be purchased and activated to use them.

Another control system for minimizing external environmental influences on the spread pattern is the AMAZONE WindControl system. Here, an anemometer mounted on the fertilizer spreader records the speed and direction of the wind and automatically balances the spread pattern by adjusting the spreading disc speed and the drop point. Especially for difficult wind locations, this system can significantly extend the time window for using the disc spreader.

In the case of the RAUCH HillControl system, an inclination and yaw rate sensor records the position and the change in position of the spreading discs. Based on this, calculation algorithms are used to determine optimized machine settings for drop point, application rate, and disc speed for each spreading disc separately, depending on the flight characteristics of the fertilizer, the working width, and the slope inclination. This calculation is carried out fully

Agronomy Perspective

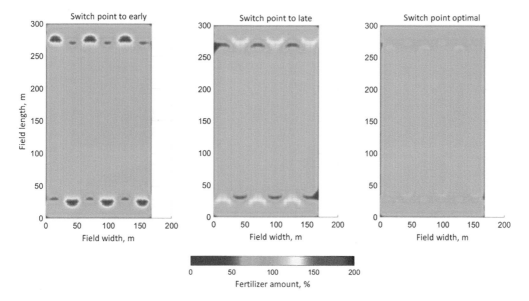

Fig. 4.13 Comparison of field distribution without/with optimized switch intervals [Thu11]

automated while driving. This ensures consistently good lateral fertilizer distribution on slopes and in highly variable field topologies, such as crests and dips.

4.6.4 Field-Zone-Specific Nutrient Identification and Supply

With the introduction of the ISOBUS and the Task Controller functionality, the use of application maps and the documentation of the nutrients applied have become very easy. In the past, data was transferred from Farm Management Information Systems (FMIS) to the ISOBUS system on the tractor or implement via USB stick. In the meantime, GSM communication modules have also become established and the data can be transferred to various end points using platforms such as agrirouter or Data-Connect (see ▶ Sect. 4.2). The farmer can store and manage their data in an FMIS. Here, all historical data about the respective field and the processes carried out, respectively planned, is also available. In addition, live telemetry data from the work processes, i.e., the states and process data of tractors with implements, can be displayed. The result is a comprehensive view of all information required for crop farming in the form of a digital field.

This digital field in turn forms the basis for determining the nutrient needs of the plants, taking into account the requirements of the respective crop and complying with the thresholds of the fertilizer regulation. Soil samples must also be included for this purpose.

A study on soil and plant variability [Sto11] with regard to the optimal size of the sections for precision farming applications showed variability in the soil of less than 20 m and even less in the plant population in analyses performed on Danish farms. This requirement on the resolution was not met in the past with standard soil analysis, where representative samples were taken every 3 to 5 ha. With newly developed soil sensors, e.g., the smart spade of STENON or the colter-mounted soil sensors from SOILREADER, high-resolution online measurements are possible at any point in the field within a few seconds.

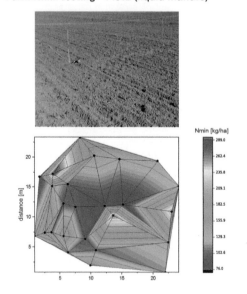
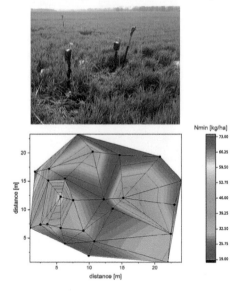

Fig. 4.14 Spatial distribution of Nmin in the soil with organic/mineral fertilization [Ste19]

This provides a very detailed picture of the nutrient composition in the soil, which can vary greatly within a few meters (see Fig. 4.14).

Nitrogen sensors that measure the requirements online have already been available for several years. Besides the sensors from YARA, Greenseeker from Farmfacts, OptRX from AgLeader, or CropSpec from Topcon, which only measure the current conditions of the plants, ISARIA from Fritzmeier has established itself with a new approach to the use of yield potential maps. Due to high acquisition costs, complex operation, and benefits that are not always measurable directly, these sensors have not yet been able to establish themselves on a broad scale.

An alternative is to analyze plant populations with satellite data, e.g., on the basis of the NDVI indicator. The Sentinel 2 satellites provide optical multispectral data for this from 440 to 2200 nm. In addition, radar spectral data from the Sentinel 1 satellites is now available, which makes it possible to also determine moisture and nutrients in the soil. Although the raw data is provided free of charge by ESA, formats processed for and usable by end users are only available for a fee from various service providers on the internet. While resolutions of 20×20 m were common in the past, more recent satellite and remote sensing systems have made it possible to get data at a much higher resolution of less than 1 m. [Dar19] was able to demonstrate that higher resolution of the data leads to different characteristics in the field.

4.6.5 Pneumatic Fertilizer Spreaders

The next big step in improving nutrient distribution will be the renaissance of pneumatic fertilizer spreaders (see Fig. 4.15). They make it possible to spread fertilizer over much smaller areas. Many of the control systems mentioned above for twin disc spreaders are not required for pneumatic spreaders, as the narrow spacing of the outlet manifolds of approx. 1 m makes the dis-

Fig. 4.15 Example of a pneumatic spreader

tribution almost insensitive and much more robust to external influences such as wind, slope inclination, or even poor, varying fertilizer quality.

In boundary spreading, too, pneumatic spreaders with their small overlaps offer a much more sharply defined spread pattern than disc spreaders. By depositing the fertilizer almost on a line, part-width section control can also be realized much more accurately. New developments, such as the RAUCH MultiRate metering system, enable pneumatic spreaders to follow target maps with high precision, with a resolution of 1 m (see **Fig. 4.16**), and thus to cover the crop-specific nutrient requirements almost down to the individual plant.

4.6.6 Organic Fertilization

In contrast, there is still some potential for further technological development in organic fertilization. Although it is true that weighing systems are nowadays used to measure the mass flow of solid manure spreaders and to control metering via the feed rate of the scraper floor, more ad-

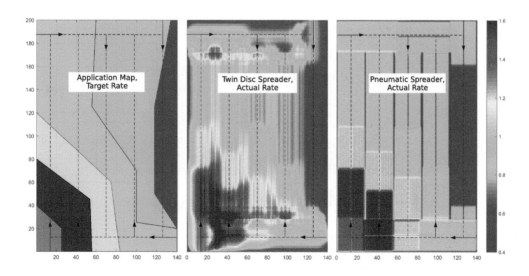

Fig. 4.16 Comparison of target/actual map for twin-disc and pneumatic spreaders [Sto19]

vanced control systems for automating the work process and the distribution are only available to a limited extent.

In the application of liquid manure, volumetric flow meters have been joined by Near-Infrared Spectroscopy (NIRS) and Nuclear Magnetic Resonance (NMR) sensor systems when it comes to determining the nutrient content of N, P, and K in liquid manure. These systems are offered by JOHN DEERE as Manure Sensing System (see ◘ Fig. 4.17) or by SAMSON as NPK Sensor and will find their way into the field in the next few years.

The combination of organic and mineral fertilization, as presented by JOHN DEERE a few years ago with its "Connected Nutrient Management Circle", can contribute to optimal utilization of organic nutrients. Here, the nutrient components are recorded by means of an NIRS sensor during organic fertilization. In a further step, mineral fertilizer is applied on top up to the target values of the application rate map.

4.6.7 Conclusion and Outlook

Based on sensor and satellite data for small-scale determination of the nutrient requirements of plants and the nutrient quantities in the soil, the digital systems presented above create the basis for optimized, field-zone-specific, and thus needs-based nutrient supply.

By means of the sensors, actuators, and automation systems already available for machines, fertilizer application can be optimized, a great deal of fertilizer can be saved, and high yields can still be achieved. The use of pneumatic spreaders, in particular, opens up new possibilities for precise fertilizer application according to high-resolution application maps.

Agriculture can make its contribution to environmental protection and CO_2 reduction by applying less mineral and organic fertilizer. In this context, less fertilizer means lower amounts of ammonia emitted into the air and reduced leaching of nitrates into the groundwater.

◘ Fig. 4.17 System components of JOHN DEERE Manure Sensing [Jos20]

Agronomy Perspective

4.7 Crop Protection: Diverse Solutions Ensure Maximum Efficiency

Stefan Kiefer

Abstract

This section summarizes the mechanical and chemical solutions to manage field crops. The description includes digital components available today and outlook what solutions reach market next.

4.7.1 More Than 50 Years of Modern Plant Protection Technology

The principles of preventive crop protection achieved by the selection of suitable crop rotations, crops, varieties, and cultivation strategies are the basis for the most efficient use of plant protection products. In modern crop cultivation according to "good professional practice", these indirect crop protection measures should be comprehensively used by experienced practitioners. This section deals exclusively with the direct, technical possibilities of modern crop protection [Iva21].

In the late 1980s, mechanical crop protection still played a certain role in weed control. The innovative products of chemical crop protection with ever better and more diverse active ingredients and high levels of efficacy increasingly made it possible to largely do without mechanical methods. The consequences of excessive use of chemical methods, which are visible today, hardly played a role at that time.

The desire to reduce the use of chemical crop protection has been on the rise since the mid-1990s. However, early research and development efforts regarding the more efficient use of crop protection products, such as direct injection or image recognition, never translated into marketable products.

From today's perspective, it can be seen that many factors are interacting positively that lead us to expect the breakthrough of new technologies into widespread practice in the coming years. Above all, this will lead to a decrease in the use of pesticides in practice, as desired by politicians and society (see ▶ Sect. 1.6).

4.7.2 How Did the Plant Protection Technology Commonly Used Today Come About?

4.7.2.1 Farm Organization as a Driver for Machine Development

At the time of low producer prices in the late 1990s the goals in crop production were clear: few, profitable crops with intensive crop management in tight crop rotations required high outputs per hours and per day. Initially, it was important to develop trailed implements with hopper volumes up to 6000 l and working widths up to 36 m for large crop farms. Medium-sized crop farms asked for large-volume mounted implements up to 1800 l and 27-m working widths. The progressive structural change demanded new series of machinery from industry, which were also to reach ever new dimensions in terms of their durability.

4.7.2.2 New Systems for More Precision and Comfort

In addition to the factors of volume and working width, since the beginning of the 2000s, it has been possible to improve additional factors in order to increase performance. Different systems spread in the market that were important for the extension of working hours and the increase of hectare output:

- Ultrasonic systems for automatic boom guidance to increase speeds and use large working widths in hilly terrain (DistanceControl)
- GPS-supported automatic section control (SectionControl) to avoid double treatments at the headland and to relieve the driver's workload
- Boom and spray cone lighting systems to ensure work quality during the day and at night

In the 2010s, these systems were perfected further and further: section control is now possible at the level of a single nozzle (50 cm). New, highly dynamic boom guidance systems (Horsch Boom Control, Hardi AutoTerrain, Amazone ContourControl) control the position of the nozzles not only in height but also in driving direction (Amazone SwingStop).

It is foreseeable today that the current technological level of the leading suppliers will spread dynamically in the market. The uniform distribution of the necessary quantities of active ingredients across field zones is possible very precisely with the technology available today. All the technologies available today form a very good basis for increasing variability within field zones in the future.

4.7.2.3 Avoidance of Negative Environmental Impacts

The fact that very fine droplets can drift outside the target areas led to the introduction of low-drift injector nozzles in agriculture in the mid-1990s. In a very good cooperation of chemical industry, official consulting, and AgMachinery industry, it was possible to establish these environmentally beneficial nozzles on the market—also internationally. In many European countries, the general recommendation for the use of this nozzle technology has become a comprehensive set of regulations accepted by practitioners.

Then, during the 2000s, the issue of Point Entries to surface waters, e.g., during sprayer filling came into focus. In a combination of farmer demand, regulations, recommendations, and investment subsidies, automated filling and cleaning systems are now commonplace in the market. Today, it is easy to accurately fill the required quantities and to clean the field sprayers in the field.

The automated processes of filling, application, and cleaning are an ideal basis for comprehensive automation of complex processes in crop protection equipment.

4.7.2.4 First Systems for Field-Zone-Specific Crop Protection

The desire for field-zone-specific application of crop protection products was already researched in the first phase of precision farming and demanded by practitioners. It has always been possible to change the application rates for the entire working width. Concepts for the application of fungicides and growth regulators came from individual players in the market [LVL18]. However, the knowledge about the interrelationships of all influencing factors (including soil, crop density, varieties, weather) in the field zones was not sufficient to attain conclusive concepts outside of research approaches. Until today, uniform forecasting models for the entire field are the first choice [Isi21] and variable application of active substances on individual field zones is the exception.

Direct injection systems that can switch individual active ingredients on and off within the field or can vary the quantity [Rav21] have so far remained a market niche or the subject of research projects [PRN+18]. In this topic, too, the question of a positive balance between the effort and the benefit of small-scale assessment and application remains unanswered until today.

As early as the mid-1990s, research was concerned with the detection of weeds and

Agronomy Perspective

◘ Fig. 4.18 Use of Amazone AmaSpot system in Kazakhstan

the idea of small-scale shutdown of nozzles. First systems for the detection of green plants on unvegetated soil ("green on brown") have been offered in the market since the beginning of the 2000s (Trimble Weed Seeker, Rometron WeedIT, Amazone AmaSpot). These systems only allow the application of total herbicides (e.g., glyphosate) (◘ Fig. 4.18).

After several years of market introduction and maturation, systems have become established in a few regions of the world. In the largest market for spot-spraying systems, Australia, there is a good match between the agricultural structure, the climatic conditions, and high acceptance for new technologies [Cro21a]. There, the balance between high effort in terms of technology and costs on the one side and the benefits on the other side is good.

More complex systems that are capable of detecting weeds in growing crops ("green in green") have been developed in promising research projects [Agr21], but have never reached market maturity. In the coming years, more advanced camera technology, image processing, and nozzle technology will be able to achieve the required speed and precision and, most importantly, distinguish between crops and weeds reliably.

4.7.3 Which Progress can be Expected in the Next 5 Years?

4.7.3.1 Increased Research on Preventive Crop Protection

In addition to focusing on technological developments, the impact of standard field cultivation practices must be considered. Many chemical crop protection applications emerged in the economically positive, close crop rotations of the last 20 years. If, in the future, more attention or encouragement are again given to crop rotation spacing that makes sense from a crop cultivation perspective, there will be immediate reductions in the effort required for chemical crop protection.

Furthermore, a variety of methods are being tested that can have an indirect influence on the intensity of crop protection. Besides intermediate crops and undersowing, there is also potential in the cultivation of mixed crops or companion plants [Eip21a]. There is also increasing interest in the use of biological control methods such as the use of ichneumon wasps against the European corn borer.

4.7.3.2 New Focus on Mechanical Weed Control

A short-term effect is the renaissance of mechanical weed control. Following some innovative years in the niche of professional organic farming, some larger equipment manufacturers are now developing systems further and making them widely available (the Lemken company with Steketee, the Amazone company with Schmotzer). In general, mechanical systems are much more demanding in terms of application timing, adjustment, and assessment of work quality than classical crop protection equipment. The positive effect of a reduction in chemical crop protection is accompanied by more intensive tillage with corresponding negative consequences for biodiversity on and in the soil, humus content, and greater risk of erosion. The additional passes with small working widths (6 to 12 m with hoeing equipment vs. 27 to 36 m with field sprayers) also lead to additional impairments of the soil (gas exchange, water infiltration).

Modern Harrow systems offer solutions for more effective operation of the individual tines as well as remotely controlled, finely metered adjustment of the harrow intensity (Treffler, APV). As row-independent and relatively inexpensive implements, harrows are easy to use in existing processes. However, the quality of work is highly dependent on soil, crop, and weather conditions. Grasses, in particular, are difficult to control.

Hoeing implements must be adapted more specifically to the individual crop and the soil due to the reference to a specific row regarding the arrangement of the hoeing blades and the selection of additional implements. Various ridging blades or finger wheels also provide certain options for weed control in the row. Current models have the option for automatic control of the implements. Slide frames can guide implements to within a few centimeters of the crop, independent of the tractor. They get their orientation from camera systems that recognize the rows of crops, or from additional RTK-GPS receivers on the implements. Section Control for clean connections at the headland are also being offered. Depending on the requirements for precision, crop, and growth stage, working speeds of up to 12 km/h are possible. If, in the coming years, working widths above the 3 to 9 m common today are offered, enormous hectare capacities will be possible (◘ Fig. 4.19).

Combining hoeing technology with the band application of herbicides is a classical practice from the 1980s. Ideal hoeing timing (dry, warm) does not always match that of optimal herbicide application (moist, cool). New conventional methods with a solo hoe and band spraying using the boom of a classic field sprayer guided with high precision are a new alternative [Agr19] (◘ Fig. 4.20).

In addition to pure weed control, today's hoes can be equipped with special devices for undersowing or fertilizer application. Undersowing is also capable of suppressing weeds and reducing the frequency of use of hoeing implements.

4.7.3.3 Industry Priorities are Changing

The offer of technical solutions must always be seen in the interaction with the offers of the chemical industry and the interest of crop cultivation experts. Accord-

◘ **Fig. 4.19** Modern hoeing technology with camera guidance

◘ **Fig. 4.20** Band spraying with the field sprayer and AmaSelect Row

ing to the "chicken-and-egg principle", the AgMachinery industry has asked for many years who makes the rules and where the active ingredient can be varied. At the same time, crop farmers are asking why they still cannot apply the measure variably on every square meter.

In addition, the chemical industry recognizes that their business models must change from volume business to solution offerings (e.g., Climate Cooperation by Bayer, xarvio by BASF). Investments in digital systems are now being made with an eye on the big turnovers in world markets, and are therefore moving faster.

As there are many positive signs, the AgMachinery industry is also investing in the development of new intelligent systems.

4.7.3.4 Looking at Individual Field Zones Becomes Possible

Today, we are able to distribute the active ingredients perfectly across the field. Currently, however, the system is still oriented to the tramline systems. The logical next step is to go further into precision and vary the management according to yield potential, microclimate, or soil conditions. While many algorithms have been tested and are known in fertilization, in crop protection it is much more difficult to evaluate the many different influencing factors. The farmers' decisions are based on variety, growth stage, and experience. A comprehensive assessment of small-scale differences in larger farming units is still almost impossible today.

The use of current satellite images alone is not sufficient as a basis for decision-making. When more of the influencing factors mentioned above are available digitally, expert systems can generate high-resolution application maps, for example for fungicides. The zones can vary greatly in size depending on the terrain structure and require the possibility of small-scale variation of the application rates.

4.7.3.5 New Technological Solutions

Currently, several technical solutions can be identified that will contribute to more efficient crop protection in the coming years:
- Common crop protection implements are becoming smarter
  - Variable application rates within the working width enable precise response to difference in field zones (pulse width modulation technology).
  - Camera technology for weed detection and online spot application in crops ("green in green") will be available on the market. Based on threshold values, the application will be determined on "every square meter" and high savings will be achieved (SmartSprayer by BASF and Bosch e.g., available via Amazone UX5201).
  - Direct injection systems of crop protection products will make it possible to dispense one or more active ingredients independent of each other.
- Use of drones for the field-zone-specific assessment of plant populations
  - Modern camera systems are capable of generating application recommendations from high-resolution RTK-accurate images (Dronewerkers [Dro21]).
  - AI systems will use these images to generate maps for small-scale spot application against weeds or for the precise control of plant diseases.
  - Common standard field sprayers with high-quality operator terminals are able to process these maps accurately.
- Robotic systems for high-precision crop protection
  - The first robotic systems are able to carry out specific crop protection on the smallest field zones (<10 cm^2). Generally, this involves applications for weed control. Chemical systems for online spot application (e.g., Ecorobotix [Eco21]) hit the smallest weeds and largely spare the crop plants.
  - Systems for precise mechanical weed control within the row are mainly used in high-value vegetable or sugar beet crops in organic farming (e.g., Farming Revolution [Far21a] or Farmdroid [Far21b]).
- Drones for direct application of active ingredients
  - This form of precision crop protection has so far been used mainly in Asia (e.g., by the company XAG [Xag21]) or in individual research

projects in viniculture (Diwakopter Experimental Field [Diw21]). In the EU, application with air vehicles is generally prohibited to a large extent.
- In regions that still practice crop protection by airplane (Ukraine, Russia, Kazakhstan), large drones could take over this work in the future.

Currently it looks like all systems will reach different market segments and slowly establish themselves. Whereas up to now we have seen very uniform systems on the market, which differ more or less only in size, we will soon have many different solutions on the market. The factors influencing this development are very diverse:
- In high-quality vegetable crops, the already well-known trend to eliminate as many active ingredients as possible from the cultivation is intensifying.
- In many Western European regions, the population's discomfort with chemical crop protection is growing. Farmers are reacting, out of self-interest if nothing else, and policymakers are intensively promoting and demanding reduction measures.
- The large crop farming regions—especially outside the EU—are benefiting from the trend toward developing technologies for the reduction of crop protection products. The economies of scale in large farms are enormous and new, efficiency-boosting solutions are being adopted primarily by farms with solid financial means, without the need for government subsidies.

4.7.4 What are the Long-Term Prospects 2025–2030?

Modern crop protection has evolved continuously over a long period of time from the 1970s. The basic principle of whole-field application of active ingredients has become the almost exclusive method since the 1990s. Technical progress has enabled enormous development steps in the perfection of the implements.

Currently, many new and also well-known approaches are in dynamic development and the socio-political desire for change (see ▶ Sect. 1.6) increases the willingness to invest. Whether technologies and methods will only remain a marginal phenomenon or whether disruptive systems will displace the processes common today is not yet foreseeable.

4.7.4.1 Progress in Sensor Technology and AI

The key to the breakthrough of "digital" plant protection methods are crop measurement methods and Artificial Intelligence systems. In addition, the technical and financial effort for assessing the smallest field zones is primarily an issue: Are satellites capable of detecting many things or do cameras and sensors mounted on drones and machines have to answer the questions of the crop farmers?

We can already see very clearly the possibilities of targeted weed control—here the main question is which effort makes sense for the specific detection of individual weed species and how precise the application may be down to the individual plant. The question of effort and benefit is decided mainly on the basis of the regional characteristics of the crops, the weather conditions, and the structure of the farm. In vegetable cultivation, for example, high-precision systems will certainly do the job in order to grow the crop as completely free of herbicides as possible. The "cash crops" in the large agricultural regions, on the other hand, will certainly continue to work with conventional implement systems for a long time to come, where high-quality spot spraying systems are justified if the yield per hectare is high.

The situation is considerably more complex and less clear with regard to the application of fungicides and insecticides. The prevailing basic principle of prophylactic application according to forecast models is still very important today. As a result of easy accessibility of soil, weather, and satellite data via open cloud systems, it may be possible to resolve these forecast models at increasingly small scales. The biggest obstacle here are the complex biological relationships and the look at the weather development in the following weeks. For the AgMachinery industry, the way forward is clear: The possibility to vary the application rates and products on the smallest possible farming units will be much sought after.

Indirect or direct measurement of plant diseases or insect infestation is currently still the subject of research. The first market-relevant solutions are becoming visible (e.g., Phenoinspect [Phe21]).

Ultimately, for the time being, the question remains for which crops and crop rotations these systems are financially advantageous or whether traditional or new crop cultivation methods such as crop rotation, breeding, and companion plant cultivation will not make the systems so stable that "chemical" intervention in the biological processes will become superfluous.

In addition to the technical and biological contexts, the economic situation in agriculture is a decisive factor: Should producer prices remain high on a long-term average (what is to be expected with increasing global demand), the desire for maximum yield and high quality will be paramount and demand many technologies.

4.7.4.2 Crop Protection Technology—Incremental Versus Disruptive

The obvious development of classical agricultural technology lies in the perfection of the common systems—the vision is clear: The main tank contains only clear water, 5–10 individual active ingredients from reusable containers that never need to be rinsed are dispensed with square meter precision and are supplemented with an individual formulation according to the leaf condition and the crop. A "sensor" is capable of detecting both weeds and plant diseases as well as the nutritional needs of the plants. All this is integrated into a machine that is fully automated and relieves the driver's workload almost completely. At this point in time, it is impossible to foresee when this vision will be realized.

Regarding the methods used in crop protection, there are alternative approaches that will potentially also play a role:

- Thermal systems such as scorching have long been known and only have limited innovation potential.
- Electrical systems for chemical-free killing of weed (Cropzone [Cro21b]).
- UV light irradiation for chemical-free suppression of fungal diseases (Sagarobotics [Sag21]).
- Laser systems for weed control (Laserzentrum Hannover [Eip21b]).

All methods are very specific to one task. Development in the vegetable cultivation niche will maybe make it possible to establish them. For the development to be disruptive, there is a lack of efficiency, widespread application, or even an acceptable cost–benefit ratio.

4.7.5 Conclusion: Future Crop Protection will be More Specific and Diverse

Considering all technical possibilities, market opportunities and social requirements, many different methods and systems will be available on the market in the future, in contrast to the past.

For farmers, this diversity offers the enormous advantage that they can choose the best system and are not left with a rigid system without any design options. The challenge certainly is also how to disseminate professional expertise—which system is the right one, and if the investments are justifiable.

Industry will face the biggest challenge: The risk of bad investments in the further development of AgMachinery is increasing and in the medium term, only strong, internationally operating companies will be able to afford the development costs. Whether the digital business models with AI systems and cloud networks will be profitable in agriculture is not yet foreseeable. At the same time, however, the growing niches also offer specialist providers the potential to find a market.

As costs shift from chemistry to technology, sales of the agrochemical industry, in particular, will come under pressure. The heavy investment in research and development will be impacted.

The greatest opportunities for environmentally sound crop protection will come from the further development of indirect, prophylactic crop cultivation processes and methods. The general relaxation of the overall system through wide-spaced crop rotations and alternative crops could ultimately lead to greater reduction than technological developments. The remaining chemical crop protection will then be applied in a very targeted manner. The foreseeable declining revenues will ultimately have to be absorbed by society.

4.8 Weather and Irrigation

Gottfried Pessl

Abstract

This section provides an overview about weather measures and their relevance for farming, how the data is monitored, transferred and processed, as well how farmers use the weather forecast services. The section includes global examples across crop production systems as well as different channels (e.g., FMIS) and features (e.g., crop management and irrigation).

4.8.1 Introduction

Agriculture and weather are inseparably linked. This is because weather is the single most important production factor in open field agriculture and heavily impacts a crops success or failure. Over the last 20 years the weather has become more extreme with heavy rains combined with hail and heavy wind often followed by droughts have become common even in central Europe. In 2016 alone, hail caused more than $23 billion in crop damage across the U.S. alone [Nws16]. The most important message is that more disruptive weather events have been happening, and it is very likely it will continue to happen at an increased frequency over the coming decades.

Today's growers and farming practices have reached a point where simply using the instincts and gut feeling are no longer a viable option to determine how much water and other inputs are needed to accommodate plant needs and to maximize the genetic potential of the crop. From a farming perspective greater risk awareness is a key element for precision farming as higher probability of increased weather volatility is a fact we have to deal with. Based on all available information at this time, further adaptation and preparation for greater risks is a reasonable course of action which will lead to mitigation, together with risk assessment tools (see ▶ Sect. 4.4 und ◘ Fig. 4.21).

The most important climatic factors which should be monitored are: temperature, relative humidity, rainfall intensity, soil moisture, Evapotranspiration, wind speed and direction, hail, frost, excessive heat, soil temperature, and solar radiation. Understanding these data sets helps

◘ Fig. 4.21 Irrigation in Australia

the grower to find the best windows for work planning, spraying, harvesting as well as the other myriad of operations required during the growing season (for a detailed description of possible sensors to measure that see [Met21]).

Climate change and the relevant changes in water supply for the crop are additional challenges for future farming practices. Increased regulations and restrictions for water resources have brought about the need for fast, easy and efficient tracking of both water and environmental conditions within each field. By utilizing technology to monitor micro-climate and soil moisture on site, growers now have easy and effective solutions to improve irrigation competency. Efficient use of water in irrigation can not only improve yields, but also save water and energy, improve crop quality and shelf life, and reduce nutrient leaching. The ability to measure, monitor and manage root zone soil moisture with easy to use, reliable technology is crucial to maximize yield and quality. Numerous studies clearly demonstrate that moisture management using soil moisture probes can increase yield and, in some cases, lower input costs for corn and soybean crop. Irrigation planning should be calculated based on crop water uptake, evapotranspiration, infiltration rate of the water to have good drainage and to avoid long time water saturation of the soil. New digital tools have come to the market in recent decades and are helping farmers worldwide to use their resources (seed, water, fertilizer, diesel) better and at the same time in being more resilient to farming risks (drought, water excess, frost, heat stress, insect damages, fungal infections etc.), making more informed decisions and having last but not least a better bottom line.

4.8.2 Current Status and Tool Kits Availability for Mitigation of Risks for Farmers

Weather tracking on the farms is still not common practice. In the last 20 to 30 years many European countries and the USA have made available to farmers regional weather networks financed and run by the government or regional advisory service for free or a subsidized cost. These regional services often have a resolution of 30–50 km diameter which is suitable for a region, but not for single farms or fields and give a general understanding of the weather conditions (◘ Fig. 4.22).

Weather forecasts services from government or private sectors are often combined with historically measured data to help give a better understanding of the situation. The most prominent companies here are: DWD for Germany, Meteo France,

Agronomy Perspective

Fig. 4.22 Monitoring soy in Brazil

ZAMG Austria to name a few public services and DTN, IBM Weather, Climacell, Meteoblue to name of few private suppliers. Farmers and their advisors do understand how these tools can help them to effectively find and act on specific problems. It is however expected that a big paradigm shift is happening in the next years to come which is due to the changes in technology and sensors (lower cost and good quality), distribution of LPWAN [Wik21a] technology which drives the Internet of Things (IOT), uptake of decision support models (DSS), and the integration and development of Artificial Intelligence (AI) (see ▶ Sect. 3.5). From here the farmers/advisors can create timely and robust operations plans, identifying which fields need to be scouted, how often, and any other key risks. The update frequency of data is also essential (needs to be close to real time) in managing crop production systems and finding specific problems in time for farmers to control them before real damage can happen.

4.8.3 Key Components and Technologies Presently Available

What can farmers do to mitigate the impact of the weather to their daily management practice? How about climate change and the related changes in the risk? What are the technologies available today to help farmers to better control weather risks? How about new services, such as Parametric Insurances, bank ratings on crop as collaterals and real estate/land value on base of long-term precise climate of fields to understand the weather risk situation? How can ROI be improved?

New technologies like LoRaWAN, NB-IoT [Wik21b] are fundamentally changing the way how farmers can get accurate weather and field data in real time. The mobile or cell phone has become the primary device for receiving and displaying data, as close to real time as possible enabling the farmer or manager to intervene without delay.

■ Fig. 4.23 Monitoring tulips in Holland

Companies like Pessl Instruments, DAVIS, Spectrum, and many more are offering practicable solutions for the farming industry. The future will be, that farmers have low-cost sensors on each field which are logging data all year round combining this with highest precision weather and disease risk forecast, together with satellite and yield maps, drone data, tracking information from vehicles (location, working speed), and AI will give them real time insights when it is needed, all in a smart way (■ Fig. 4.23).

The integration of various sensors for different places are vital for the acceptance and large-scale uptake in the farming communities. Farmers, like managers in other industries, are changing practices only if they are being forced (by law) or if they see a fast ROI and convenience gain from the uptake. A good example of this can be seen with GPS guidance systems. As most farmers are spending much more time in the field and not in the office, the smartphone is the tool where all the data needs to be concentrated. In order for this to be a practical solution, many industry partners need to work together in an ecosystem, as farmers are not willing to use dozens of apps which solve individual problems. Weather information is being used in many different applications so the companies offering services in the field need to open their data streams seamlessly to different stakeholders like John Deere, CLAAS 365Farmnet, xarvio and many other providers via smart API's (see ▶ Sect. 4.2). Here the data exchange needs to be secure with all privacy rules applied but open enough not to be the bottleneck of a technology breakthrough.

An Example of a Real Time Application A farmer shares a weather station with his neighbor, both have the fieldclimate app from Pessl Instruments with integrated local and weather station adjusted weather forecast from Meteoblue, Spray weather forecast and the disease model risk forecast for his sugar beet grown on that field. Sunday afternoon the farmer is planning his work for Monday morning and looks at both measured and forecasted data and decides to plan a spray application against Cercosopora Beticola on that rented field 25 km away from his farm. He plans to start at 3:00 on Monday morning where wind is low and DeltaTvalues are perfect for spray penetration into the Canopy. 18:00 Sunday evening he looks at all the from his weather station and local forecast, and despite some local thunderstorm in the area, he decides to plan his application. At 2:00 in the morning he receives a warning message from his weather station that this planned spray will be impossible as a thunderstorm changed direction and 25 mm of rain has fallen so field access will not be possible. With this information he saved an unnecessary trip to his field which costs money and time and he could sleep until 6:00 and changed his work planning to 20:00 on Monday allowing the field to dry up and the weather to improve to allow field access. This is a simple example of how technology can improve life quality and at the same time avoid costly errors.

More applications available are: the smart integration of real time weather data with partner solutions from xarvio, Climate, Pessl Instruments. Many others are generating smart and accurate alerts that tell which fields are at risk for specific insects and diseases, as well as the growth stages for optimal herbicide applications. Accurate weather and crop growth stage models help prioritize fields, ensuring that producers and advisors visit each field during key growth stages when problems are likely to occur. These alerts are based on low-cost Hyper-local Weather Station networks which provide critical insights into an individual farm and literally by field basis. Weather-based risk models can also predict disease susceptibility and combat foliar disease by identifying the ideal timing for post-emergent herbicide or fungicide applications. Degree-day-based Phenology Models combined with remote automatic insect traps and crop cameras can help farmers and advisors anticipate insect flights in an automated way. These electronic devices combined with hyper-local weather stations that measure local rainfall, temperature, etc., further combined with smart traps, satellite imagery, and human observations detect insects present in fields and report the data wirelessly.

Agriculture and water use is tightly linked together, and rational use of irrigation water is absolutely critical for the growing population, as 80% of the freshwater is used in agriculture. Therefore, monitoring soil moisture and applying the correct amount to the field has become a key factor. Last but not least, plant protection using decision support systems (DSS) to understand better which field or which crop needs spraying is a key element for future sustainability of the agriculture industry. Combining weather data with satellite imagery like Sentinel [Wik21c] and/or Landsat [Wik21d] helps to identify growth stages automatically and provides crucial information for insurance companies (Parametric Insurance on Abiotic stress) to offer tailored insurance solutions to farmers (◘ Fig. 4.24).

4.8.4 Future Development

Super low cost but professional Weather Stations combined with AI based virtual weather stations will be the backbone of the new era of weather driven decision

◘ Fig. 4.24 All technology in the palm of the hand

making. These devices like the Pessl Instruments product called LoRAIN are deployed in an ultra-small grid between 20 to 100 ha to gather crop and field condition data, providing farmers in real time with the most accurate and dependable weather information like rainfall, soil moisture, soil temperature, and leaf wetness. This will be combined with tremendously detailed field-level forecasts of sub-acre conditions, and insights for specific areas and geographic regions. Customized alerts can also be set up for specific weather or field conditions to further reduce risk and support strategic field operations. The ultra-fine grid field-level weather stations data will help to calibrate other remote sensors (i.e., rain radar data and satellite imagery) to provide reliable information for field accessibility with full or semi-autonomous machinery. The real time data can be part of the national crop health monitoring of a country. The automatic archives generated of local weather are combined with historical yield and this will become climate information layers for specific locations and crop zones to foster future decision making. All this improves planning and scheduling for future decision making and will mitigate risks based on climate change. The small grid, automatic weather information system is being used also by animal farmers to avoid heat stress for cows, swine, and chickens on the pastureland and barn. Last but not least the high grid rainfall and temperature data can be used to forecast local floods caused by small rivers, local frost risks on small roads and on bridges to save lives and improve quality and protect environment. The COVID-19 crisis has shown us how fundamentally important food security is, and farmers can ensure the food supply and that they protect the environment at the same with smart technology.

4.9 Harvest Sensing and Sensor Data Management

Sebastian Blank and Ignatz Wendling

Abstract

The section provides an overview of sensing approaches and applications in the harvesting domain covering both onboard and external data sources. Furthermore, methodologies for ensuring proper data quality and consistency are introduced.

4.9.1 Introduction

After decades of being a domain driven by mechanical innovation, the field of harvesting technology has changed gears. In the last decades, electronics and sensing technology has taken a more dominating role in delivering process and performance improvements of machines. This is no longer limited to the harvesting equipment itself but should also (and maybe even more so) be seen through the agronomic lens. The role of electronics and sensing technology has evolved from "just" the last step of crop production in the crop growth cycle to becoming a valuable contributor of observations and insights that can be of great benefit to the effectiveness and profitability of the overall plant production process.

As a logical consequence of the large variety of crop characteristics that have to be harvested, a diverse and far-reaching range of specialized vehicle forms exist. They range from self-propelled or drawn harvesters for potatoes, beets, vegetables, and cereals to forage and specialized cultures such as fruit, nuts, and viniculture applications just to name a few [ZAA15]. As this spans an area too wide and too diverse to be discussed here, we will focus on two common vehicle forms: combine harvest-

Agronomy Perspective

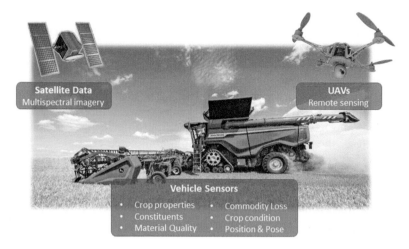

Fig. 4.25 Section content overview

ers and Self-propelled Forage Harvesters (SPFH).

In the following paragraphs these two machine forms will serve as models to exemplify the impacts and implications of sensing technology. Firstly, vehicle-based and remote sensing systems will be introduced. Here, the focus will be on the measurement principles that are commonly applied and the sensed parameters of interest. In the subsequent two sections, applications that make use of the gathered data will be discussed. The emphasis here will be on data quality management and automation solutions. To conclude the section, a brief outlook towards future trends around sensing systems and their application in the harvesting technology domain will be provided (◘ Fig. 4.25).

4.9.2 Vehicle-Based Sensing Systems

In this first subsection the focus will be on the sensing systems that are directly mounted to the harvesting equipment and sequentially moved across the field with progressing harvest.

Among the most common sensor inputs for harvesting equipment today is the **yield sensing system**. Here the basic idea is to determine the amount of crop harvested over time. This information can then be combined with other measurement systems to produce a geospatial representation of the sensed data commonly referred to as yield maps. In other applications, the mass signal is directly used to adjust one or more vehicle settings to optimize the harvesting process regarding different metrics. This use of mass and yield data will be discussed in greater detail later in this section.

There is a large variety of common sensing principles in use [BBF17]. One of the most popular principles used throughout commercial applications is force sensitive elements in the path of the cleaned (free) crop inside the machine. Typically, either piezoelectric elements or strain gages are combined with a mechanical design to either take advantage of gravity ("scale") or crop inertia (the deflection plate). This results in a relatively low-resolution measurement with a significant susceptibility to the machine 3D pose or dynamics as well as environmental conditions. Overall, this method delivers reasonable quality and robustness for most documentation tasks with respect to the harvest conditions. The limiting factor here is that as the sensing happens after the crop has entered the machine and has at least been partially processed (e.g., threshed and cleaned). Its spatial resolution is limited to the width of the front-end equipment (i.e., harvesting header) and the distance the machines travel between two consecutive measurements. As machine sizes are increasing, there is a desire to maximize productivity, and thus ground speed within the application limits and safe operating conditions resolution will decrease further due to larger distances traveled per sample time [AD16]. This presents an increasing limitation for use in future agronomic solutions. Another issue is sensing bias due to changing crop and external conditions.

The most common way to manually perform calibrations of the sensing system is to compare the onboard measurement with an external (certified) weight reference. For this, typically a certain amount of crop is harvested and then unloaded on a grain cart equipped with a scale system or the grain cart is weighed on a stationary scale. Entering this ground truth measurement back into the onboard system allows it to adjust the calibration parameters in a way that reduces measurement tolerances around a given working point [Dem13]. This must be repeated every time there is a significant change in conditions to ensure an overall result improvement. In more academic applications, local sampling techniques with plot combines or hand harvesting and weighing can be found. Those methods are typically time and labor intense and thus not sustainable for larger areas that are typically found in modern farming application. One solution that targets a middle ground between increased precision and application feasibility is vehicle-based references such as load cells in the grain tank of a combine that allow automation of the recalibration and thus facilitate continuous updates without the time commitment needed otherwise.

Another way to improve the measurement precision is to supplement these measurements with those of sensing systems based on different measurement principles. Commonly used examples of these complementary sensors are optical (camera, LIDAR) or electromagnetic (radar) systems. These sensors can either be vehicle-based facing towards the crop ahead of the machine or remote to the vehicle.

A second group of vehicle-based sensors which have received an increasing amount of attention in the last years is **crop constituent sensors**. Today's industry standard is near-infrared (NIR) technology although other applications in the mid-infrared or ultraviolet band are also known. Here the crop is exposed to a broadband source. The characteristic reflection indicating the content across different bands is measured by a sensor. As the measurement is not absolute but an indirect measurement, a library of proprietary reference curves generated by lab trials is used to determine the readings of interest.

This group of sensors measures constituents like protein, starch, fiber, sugar, oil content, as well as general properties such as moisture or dry matter. These measurements primarily serve two applications: agronomic considerations and machine setting automation. An example for the latter group is the automated length of cut adjustment to ensure uniform forage quality for a SPFH [Noa18]. There are also other applications in nutrient management such as slurry (nitrogen, phosphorus, ammonium, potassium). Moreover, active research investigates the area of direct constituent measurements via microwave technology that eliminates the need for the reference curves. However, no commercially viable solution has been developed yet.

A related but different group of crop-facing sensing applications is commonly referred to as **grain quality sensors**. The name may be slightly misleading as the parameter of interest is not the crop itself but rather the way it has been processed inside the harvesting equipment. In combine harvesters, cameras in the visual spectrum are used to generate images at specific locations of the machine (such as the clean grain elevator or the tailings) to determine free and damaged grain (threshing quality) as well as the material other than grain (cleaning quality). The images are either fed to the machine operator for interpretation or are run through state-of-the-art image recognition systems to automatically determine the crop state as described above. In the latter case, these processed inputs can also be used for settings automation to ensure continuous adjustment of threshing and cleaning related machine settings. In SPFH applications closed-loop length of cut control are also commercially available. For effective image-recognition statistics (e.g., optical flow) or AI-based (e.g., clustering, Deep Neural Networks) classification techniques are typically applied.

The next group of sensors that have been considered a standard in combine harvesters for decades now are crop loss sensors. Both the separation (threshing) and cleaning have natural trade-offs between process productivity, quality, and commodity losses given a finite separation and cleaning bandwidth. This means that a certain level of grain loss is inevitable, and it requires the skill and experience of the machine operator to balance the tradeoffs through changing crop and field characteristics. Similar to the fundamental concept introduced for yield sensors earlier, a common sensing principle employed for loss sensing is the use of force sensitive surfaces realized via piezoelectric elements or capacitive films located in the areas of the machine which are of interest (end of the separator, area behind the cleaning shoe) [Klü09].

The sensing elements transfer the impact characteristics of grain kernels as well as non-grain material (e.g., stems, husks, soil components) into an electrical signal that then needs to be processed to extract the true kernel loss. It has become evident that this approach works well for grains with a high kernel to non-grain material ratio such as corn or soybeans but is challenging where grain to non-grain kinetic impact energy and other characteristics become too similar. This makes rapeseed and other light and small kernels challenging to accurately detect. A slightly different approach is pursued by some equipment manufacturers where the focus is on increasing the sensed area in the material flow that is relatively small for the previous sensors. This allows for a more representative sample across a wider range of conditions. One commercial example of this are acoustic sensors.

For research and development efforts, more effort intensive solutions are available that become cost prohibitive for widespread applications but are used for occasional spot checks. The most common ground truthing approach are loss pans. They possess a defined surface area that the machine travels over and catch the material before it hits the ground. The content of the pans is then manually cleaned, and any residual grain is weighed or counted to determine loss per area for the sample location. In recent years, mobile apps have been developed to support this step by using image recognition of the kernels in the pan and thus receive a kernel loss count per area [Klü09].

The last area to be discussed in the section of harvester-based sensors is forward-looking field and crop condition sensing by means of visual images captured by cameras. Although a wide variety of possible configurations exists, the most common forms include monocular or stereo camera systems. While a monocular camera just detects texture and color differences, a stereo setup adds the ability to directly measure distances via triangulation (i.e., determining relative distances of the same object between the two camera frames with a known baseline). As image processing is typically required to be done in real-time to be of relevance for automation tasks, sophistical image processing infrastructure such as FPGAs, GPUs, and other processor types that lend themselves to massive-parallel processing can be found on harvesters today. This setup enables pre-cut crop yield estimation, cut height automation, cutting edge detection, header reel carryover, and obstacle detection. In addition, or complementary to the described visual systems, other sensors such as LIDAR, Time of Flight cameras, or imaging radars can also be found on occasion.

4.9.3 Remote Sensing Support

In contrast to the previous section, the paragraphs below do not focus on sensors that are mounted in or on the harvester itself but originate from remote sources. As they are designed for sensing purposes only, the ratio of cost and weight that can be dedicated to sensing is very different from the previous case. In current applications two main classes of sensor carriers can be categorized: Unmanned Aerial Vehicles (UAV) or Earth Observation Satellites (EOS).

Both serve the common purpose of acquiring multispectral and radar imagery but with very different payload, cost, and distance ratios. Although these images are used for different agronomic purposes through the plant growth cycle, they can also be of significant interest as an input in harvesting equipment automation and efficiency improvement. This is because it offers a glimpse of the crop and field properties pre-harvest, while most vehicle-based sensors are "facing backwards" in time from a crop processing perspective. Therefore, this external knowledge facilitates proactive approaches while vehicle sensed inputs are typically limited to reactive means [Ste15, LP17].

As today multispectral images are readily available on demand for almost all of the surface on earth, the differentiator becomes the way insights are extracted from these data-dense sources. This is reflected by a large body of academic and applied research dedicated to this complex field. Example applications include attempts to estimate crop yield, canopy height, moisture, and constituent content. There are further combinations of images at different stages of plant growth with a physiological plant growth model along with weather data and other inputs [MBB+09]. It has been shown that this dual-path approach overcomes the shortcoming of using imagery alone, especially in areas where only infrequent updates are available due to cloud coverage etc. [BBF17].

4.9.4 Data Quality Management (Post Correction)

So far, the discussion has focused on the sensing methods themselves. To provide a more complete overview of this topic, possible uses of the sensed data should also be discussed. There is a wide variety of potential causes of measurement error and bias as well as (temporary) sensor failure. To deal with these shortcomings, a diverse range of approaches to overcome or mitigate the potential sources has been developed. To further improve sensor data, this section considers a domain that has received particular interest since the early days of precision farming technology: yield maps.

The most common (and also the crudest) approach is direct geospatial interpolation. This is sometimes also referred to as smoothing or blending depending on the context. Here, geospatially adjacent points are averaged using different schemes to remove artifacts like pass-to-pass bias (also known as "striping") or headland sawtooth pattern originating from uncorrected relative temporal delays in the sensing data relative to the true crop location in the field. Even though naive interpolation methods create a less "suspicious" looking result, the true source of concern remains relatively unaddressed as no real attempts to systematically correct sources for imprecision are included [WT13].

To overcome this issue, a group of statistics-based approaches will be discussed next. The common theme for this class of methods is that inherent relationships in the noisy or biased data are used. Moreover, the fact that they originate from different observations (ideally even different observers such as multiple machines in the same field) is exploited and combined with statistical theorems. This allows for a more systematical elimination of sources of error without the need to explicitly model them in the causal sense. In one embodiment, a Generalized Additive Model approach [HT90] is used and based on the assumption that sample locations in the field are randomly attributed with error. Assuming that multiple harvesters provided the measurements, the resulting yield distributions should follow a similar distribution pattern. Further assuming a normal distribution, one could, for example, try to reduce the distance between expectations and variance while considering relationships with the relative location of origin and their relationship with each other. This results in a minimized general and local error and most likely correct solution given the imperfect observations.

Finally, there is another approach that can be found in the domain of yield map correction and other measured characteristics on harvesting equipment: sensor fusion. There are two main classes of fusion approaches—heterogeneous and homogenous fusion which refer to the nature of the sensing inputs that are used. In homogenous fusion, a method to combine multiple sources observing the same aspect of interest is applied in a way that minimizes the resulting error typically below that of the best single input. Thus, the whole of observations become greater than the sum of its parts.

The second class of heterogeneous sensor fusion seeks to overlay sensor inputs of different domains using a mathematical model to infer the parameter of interest that is either unobservable or the observations are not sufficiently reliable. For our example of yield maps, this could be the combination of machine yield information with UAV based yield estimates pre-harvest or the combination of proxies for yield (such as load levels or volume flow in different parts of the machine to compute a secondary yield estimation. This can then also be combined with the direct measurement via homogenous fusion approaches, see [PMA+16]).

4.9.5 Automation

To stay with the theme of downstream uses of sensor data, this section is dedicated to automation applications in harvesting machines. Automation has received a lot of attention in the past years as it is seen as a key enabler for future efficiency gains and the natural progression along the path to semi or fully autonomous machines [LHG+20]. At present, there is a long journey lined with tough challenges ahead as there are still many open questions until Off-highway Autonomy can transition to mainstream application. However, considerable advances in automation have been made in recent years and adoption rates by practitioners are a testament to the efficiency and comfort gains enabled by technology.

There are a variety of subsystems within the harvesting machine that now have automated counterparts as an alternative to manual operation. One of the most common examples is ground speed in harvest automation. Here a sensing system input is used to adjust the machine velocity to the biomass amount either currently going through the system or ahead of the system with respect to engine loads, threshing, or cleaning limitations. The most straight-forward approach is to use the engine load or mass sensing system. The general disadvantage of this is that it is a purely reactive approach which measures the inputs to the control system originating late in the processing chain (several seconds after the crop entered the machine). Therefore, in case of load spikes the system will likely react too late and overload the machine resulting in undesirable effects. To combat this, a more conservative load target is typically used which leaves significant performance headroom.

Thus, more advanced systems use look-ahead sensing via camera systems, UAV, or satellite based estimated biomass maps to allow for proactive load management. Another subsystem example is the terrain adjustment of the crop processing system with respect to the field slope. Here the information from an inertial measurement unit is used to adjust either mechanical actuators that level the grain processing system relative to gravity or purposefully bias functional settings such as cleaning fan speed to avoid increase of uphill losses.

The chopping system is the major driver of forage quality. Therefore, the SPFH automation system has focused on the knife adjustment. Here a setup of inductive sensors is used to perform closed-loop control of the shear bar position (and thus the cutting force). Simultaneously, the sensor can be employed to monitor the condition of the knife edge (sharpness) and automate decisions on sharpening cycles for optimal cut and thus forage quality [MTG15].

In addition to these subsystem examples, there are also a variety of functional settings automation approaches. They are used to automate the adjustment of all major parameters related to threshing and cleaning in a combine harvester. In their most simple form, they use lookup tables with predefined setting combinations. As this approach is not very flexible in reacting to the large spectrum of conditions and crop combinations that exist, more advanced concepts have been introduced. Many are based on either co-simulation (model of relevant components) or an approach based in the wider AI family such as classical expert systems and others. With technical advances and more readily available sophisticated embedded processing ability (deep) Neural Network-based approaches have become more popular but have not been included in full commercial applications yet.

4.9.6 Outlook: Ubiquitous Sensing and the Autonomy Challenge

In its mission to feed and clothe a growing global population with decreasing amounts of arable land due to urbanization and climate change, agriculture as a whole (and harvesting technology in particular) has significant challenges ahead.

Some of the trends that have evolved of the last decades will continue to gain momentum. One of them being the increase in number, diversity, and precision of sensors. On the one hand, this can be seen as the foundation to an increase in automation and an enabler to begin the journey towards (semi) autonomous machines to address the need to improve outcomes and overcome the shortage of skilled labor (see book Sects. 3.4 and 3.5 for more on these topics). On the other hand, the sensing needs attributed to the steps inside of a crop production system will become more and more blurred as the information generated is increasingly looked at in a more holistic way. Thus, data will be gathered year-round to be used for agronomic optimization as well as food safety and traceability efforts. The purpose of sensing in the harvest step might then be focused on determining which crops to grow and how to improve yields in the next year. As stated, there are tremendous challenges ahead but also tremendous opportunities to produce more with less, to break established paradigms, and pave new ways. With all this ahead, the next years will certainly be an exciting journey and bring numerous new innovations to the domain of harvesting technology.

4.10 Direct Agricultural Marketing and the Importance of Software: It's not Possible Without Digitalization! Which Software Solutions Help to Digitalize the Administrative Work Steps of Direct Agricultural Marketing?

Sebastian Terlunen

Abstract

Regional is the new organic. It is a megatrend and an opportunity for farmers to take on more self-responsibility through direct agricultural marketing. However, direct agricultural marketing also means increased marketing and administrative effort. What usually starts with a simple farm store rapidly evolves into a complex marketing business with multiple sales channels. Without digitalization, this opportunity quickly becomes a losing proposition. This section describes the trend that direct agricultural mar-keting has experienced in recent years, the administrative work steps that are necessary for direct agricultural marketing via multiple sales channels, and how these steps can be digitalized.

4.10.1 Direct Agricultural Marketing—Blessing or Curse?

D2C—Direct to Consumer or direct marketing is a macroeconomic trend that has been receiving great attention not only since the COVID-19 pandemic. Rather, the societal need for traceability of food seems to lead more direct agricultural marketers to completely or partially forgo the middlemen in their marketing strategy—and this is also economically profitable for them [BK20]. Thus, the consumption and marketing of regional agricultural products are subject to a sustainability movement.

According to the German Federal Ministry of Food and Agriculture, 73% of consumers prefer regional products [Bel19]. On the other hand, the survey "The farm of the future" by the agricultural news portal Agrarheute revealed that 69% of the farmers surveyed consider the direct marketing of regional products to be the future. In contrast, only 35% of the farmers surveyed consider organic products to be the future [Agr16]. The results of this survey are supported by the consumers' purchasing behavior: "Regionality has surpassed organic. Regional products account for at least 20% of the weekly shopping baskets. Organic products, on the other hand, only account for around 10%" [Lan17].

According to [WM20], the general term direct marketing describes direct sales from producer to end consumer. In the application context of agriculture, direct agricultural marketing describes the direct sales of agricultural products, mostly with a regional focus, to the consumer. Regional agricultural products include food, such as meat and sausage products, dairy products, eggs, fruit, and vegetables, but also non-food products such as (fireplace) wood, natural cosmetics, and natural fertilizers.

However, the entry into direct agricultural marketing must be conceived and planned well. It is clear that newly emerging tasks and work steps cannot be done "on the side". New workflows must be established on the respective farm and must be structured optimally (see ▶ Sect. 4.1). New workloads arise, in particular, in the areas of order acceptance and processing, production and inventory planning, quality assurance, picking and delivery, as well as accounting and employee management. Customer acquisition and support also require the use of additional manpower because marketing directly from the farm also requires doing marketing for the farm and its products. To achieve long-term economic success, appropriate digital administrative structures must be created and paper-based management must be abandoned. Only in this way can the additional workload be managed in a time- and cost-efficient manner.

The fact that direct agricultural marketing is a financially interesting alternative to established agriculture results from two aspects. On the one hand, trade levels (e.g., refiners, wholesalers, and retailers) can be bypassed. On the other hand, higher trade markups (usually between 27 and 42%) can be achieved [Bel19] (see ◘ Table 4.3). Direct agricultural marketing thus represents an economically attractive marketing alternative. It is particularly interesting for small and medium-sized farms, most of which can provide the necessary labor resources to produce high-quality food products.

4.10.2 The Evolution of Direct Agricultural Marketing

Historically, a high density of direct agricultural marketers can be found in metropolitan areas and especially in metropolitan areas with increased purchasing power. However, according to a study by the Department of Agricultural Economy and Rural Development, a more rural location does not preclude successful direct marketing [WSM18]. Especially marketing channels such as delivery service, online trade, market stalls, as well as marketing to resellers (e.g., food retail, gastronomy) enable direct agricultural marketers in rural lo-

◘ Table 4.3 Share of traded products in direct marketing and their trading margins [GHH+16, Key16, Kur18, WSM18]

Product	Germany (%)	Austria (%)	Markup (%)
Meat and sausage products	26	27	40–50
Fruit	25	15	60–76
Vegetables	21	7	60–76
Dairy products	11	16	25–76
Eggs	7	15	40–50
Cereal products	6	14	28.5–50
Other products (e.g., wine, onions, wood)	4	6	33–50

Agronomy Perspective

Table 4.4 Sales channels of direct agricultural marketers (Germany) [Kre16, NW16]

Sales Category	Sales Channel	2016 (%)	Change from 2010 (%)
Direct sales	Farm store	80	1
Direct sales	Vending truck	36	4
Direct sales	Branch system	36	3
Direct sales	Private home delivery service	30	1
Direct sales	Own restaurant	26	1
Direct sales	Farmers' markets	24	4
Direct sales	Online shop	16	3
Indirect sales	Regional gastronomy	64	8
Indirect sales	Food retail	54	8
Indirect sales	Other direct marketers	26	1

cations to tap a commercially viable source of income.

Both the developments in the established sales channels of direct agricultural marketers from 2010 to 2016 shown in Table 4.4 and the purchase channels of agricultural products shown in Table 4.5 underscore the current development in the marketing channels of direct agricultural marketers. A change from the administratively simple marketing channel "Farm store" to administratively more complex marketing channels can be observed. Especially the B2B marketing channels "Regional gastronomy" and "Food retail" as well as the B2C marketing channels "Online store" and "Vending truck" are to be mentioned here [Pri18].

However, all marketing channels except the marketing channel of a farm store generate additional administrative effort (e.g., picking, preparation of delivery bills and invoices, accounting). Only with suitable software support can the increased administrative workload be handled cost-efficiently and economically viable direct marketing can be achieved. In the following section, we will therefore show which administrative work steps are part of professional direct agricultural marketing and which supporting software solutions are currently available on the market.

Table 4.5 Purchase channels of agricultural products [Sta17]

Purchase Channel	Percentage of Respondents (multiple answers possible)
Supermarket	63
Discounter	43
Specialty store (e.g., bakery, butcher, delicatessen)	38
Farmers' market	36
Consumer market	33
Farm store	22
Organic food store	18
Online shop	5

4.10.3 Current Software Solutions for the Digitalization of Administrative Work Steps in Direct Agricultural Marketing

Direct agricultural marketing goes far beyond the operation of a simple farm store. Today, professional direct agricultural marketing means managing complex administrative work steps: from taking orders via online and offline channels, inventory management with batch tracking, picking, and route optimization to invoicing for private and commercial customers (see ◘ Fig. 4.26).

In detail, professional direct marketing means selling agricultural products via different sales channels (farm sales, vending truck, food retail, online store) and using different delivery methods (self-collection, possibly with advance ordering, personal delivery by the farmer, or postal delivery) as well as enabling payment via different payment modalities (cash payment, SEPA direct debit, bank transfer, credit card, PayPal, or other payment service providers) [LOO+20].

The current practical approach in direct agricultural marketing is to support administration either in a paper-based manner or with functionally separate software solutions. For example, individual software solutions exist for recording and invoicing customer orders or for managing inventory.

However, the use of functionally separate software solutions leads to data and information breaks between the individual process steps. The result is a lot of manual, repetitive rework and low data quality due to inconsistencies. Even a few customers thus already generate a high level of administrative effort.

As far as the digitalization of administrative work steps is concerned, four groups of functional providers of software solutions (POS systems, sales software, order management software, and process software) can be identified. These groups of functional providers can be differentiated according to the two software characteristics automation (from low to high) and support for marketing processes (from isolated to holistic). In ◘ Fig. 4.27, the four groups of providers as well as representatives of these groups are listed.

The functional scope of the POS systems currently available on the market focuses on the electronic documentation of local sales, in some cases with a software interface to advance VAT returns. Compared to the functional scope outlined above, which is necessary for the most complete digital support possible for administrative work steps, POS systems cover only a very small range of functions—primarily the documentation of cash sales.

The sales software available on the market enables a customer to perform the main sales tasks (incl. management of customer orders, product management, or billing)

◘ **Fig. 4.26** Possible administrative work steps in direct agricultural marketing

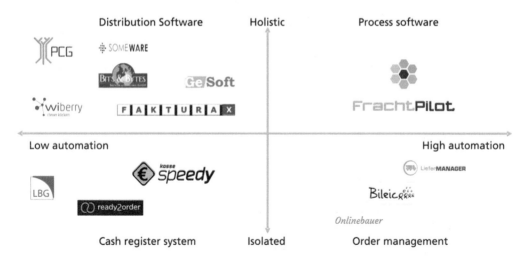

Fig. 4.27 Market overview of different groups of providers of software solutions for the digitalization of administrative work steps in direct agricultural marketing

with the support of software. All software solutions on the market only offer a low degree of automation. Data entry of information relevant for the respective sales processes often has to be done manually. Functions for automated entry of the necessary information are often missing altogether. For example, with current sales software, customer orders often have to be entered manually and cannot be recorded automatically in real time via a web store.

The software providers for order management software that are currently on the market mainly offer digital solutions for the administrative work steps involved in marketing and delivery. For example, the software provider LieferMANAGER offers automated recording and management of customer orders from a web store as well as automated billing for these. However, all providers from the order management software category have in common that digitalization of the administrative work steps for inventory planning and management is missing.

Software solutions offering a high degree of automation and a comprehensive range of functions for marketing and distribution processes are classified in the process software group. Here it is desirable that as high a proportion as possible of the administrative work steps described in Fig. 4.26 is covered by the respective software solution. So far, only one provider can be found in this category of competitors: FrachtPilot from FlexFleet Solutions GmbH. Its unique selling point is the complete digitalization and automation of all administrative work steps described.

4.10.4 Future Developments of Direct Agricultural Marketing

The previous sections have shown that consumers are increasingly demanding more regional food products. In addition to the desire for high food quality, and taking into account a more transparent food value chain, other key reasons include:
- Fairer producer prices without intermediary trading stages
- Climate-friendly, more sustainable shopping through shorter transport routes and less packaging material (see also ▶ Sect. 4.3), as well as
- The desire to economically support regional farmers.

Furthermore, it has been demonstrated in the previous sections that in addition to the mostly established farm store, professional direct marketing includes especially new marketing channels such as delivery service, online trade, market stalls, and marketing via resellers. However, the new marketing channels entail an increased complexity of the necessary administrative work steps, which can only be managed in a time- and cost-efficient manner by a professional, comprehensive software solution.

Digital marketing via an online store, in particular, appears to be an increasingly important form of marketing, considering the generally continuing e-commerce trend [Rol20]. However, the long-term success of an online store is determined to a large extent by a broad product range and targeted marketing [LOO+20]. Fulfilling these two characteristics continuously in a manner that satisfies the consumer usually presents two major challenges for the respective direct marketer. As a consequence, many online stores of direct agricultural marketers do not generate the desired response or profitability (see ▶ Sect. 4.11).

To address these two challenges, a new digital marketing concept is currently gaining ground: the concept of direct marketing platforms [Sch20]. The aim of direct marketing platforms is to offer as comprehensive a range of food products as possible, comparable to what a supermarket offers and consisting of regional products, via a web store. Targeted marketing is used here to achieve long-term consumer loyalty. In order to be able to offer a comprehensive range of food products, direct marketing platforms bundle different direct agricultural marketers with mostly different products. The usual assortment includes both fresh and dry goods and can range from fruit and vegetables, dairy products, fish and meat to fruit spreads, coffee, pasta, and spices. In this context, direct marketing platforms usually take over the recording of customer orders and payment processing in addition to targeted marketing [BK20]. The other administrative work steps presented in ▶ Sect. 4.10.3 must be performed by the respective direct agricultural marketer or by a corresponding service provider. Examples of operators of direct marketing platforms are the companies PIELERS, Bauerntüte, or Wochenmarkt24 [Dic19].

If the marketing concept of direct marketing platforms takes hold, future software solutions must not only offer support for the administrative work steps outlined in ▶ Sect. 4.10.3, but also connection options to direct marketing platforms via interfaces.

In conclusion, it remains to be noted that, following the general trend towards increased digitalization in agriculture, increasing digitalization in direct agricultural marketing can also be observed [Tae16]. The simultaneous use of different online and offline marketing channels, in particular, increases the probability of economic success. The consequence is steadily increasing complexity of the administrative work steps. Managing this complexity in a time- and cost-efficient manner requires a comprehensive software solution with a high degree of automation.

4.11 Challenges and Success Factors on the Way to Digital Agricultural Direct Marketing: "We Do Not Need a Homepage"

Heike Zeller and Martin Herchenbach

Abstract

This section considers digitally supported direct marketing from a grower and local food producer perspective. The status and challenges of direct marketing as well as success factors are discussed. Examples on bread, cheese, wool and general regional products are shared. At the end of the section a short reflection on COVID-19 impact on direct marketing completes the discussion.

4.11.1 Introduction

The external conditions for profitable production and marketing in agriculture have changed sustainably in recent years, which puts farmers under increasing pressure to develop through farm-specific adaptation strategies [HT14]. One way out can be the direct marketing of products from the farm. Direct marketing describes the direct sale of agricultural products to consumers and resellers as well as to bulk consumers and the gastronomy industry [Mün20]. On the one hand, digitalization is creating more and more opportunities to reach and retain customers (See ▶ Sect. 4.10), but on the other hand, many farms are also successful without digital tools. In this context, the following questions arise: How can direct marketing be successfully implemented in practice and what role do digital tools play in this?

The dynamism with which new business models are developing in direct marketing requires an up-to-date, practical view of the situation, which is why the following article is based, among other sources, on an expert interview with Heike Zeller, who has a broad view of the practice of direct marketing as a consultant for regional marketing strategies at aHEU, as moderator, and as keynote speaker.

4.11.2 Status Quo of Direct Marketing

4.11.2.1 Analog and Digital Direct Marketing in Practice

Analog and digital direct marketing essentially differ in the marketing work and the type of contact between people. One can imagine the saleswoman in the hustle and bustle of the Hamburg fish market calling out offers to other people—in contrast to the social media manager who advertises a daily offer in her Instagram story from her smartphone. Completely different activities, both are direct sales. ◘ Tab. 4.4 in ▶ Sect. 4.10 shows other possible sales channels for direct marketing. Countless forms of marketing make use of a mix of analog and digital media. For example, when a baker adjusts his opening hours according to a high visitor frequency at his neighbors on Google or when a city-famous butcher photographs his notice board and puts it in his WhatsApp status. Social media, online stores, and merchandise management systems, as well as the evaluation of social media process data to optimize the online presence play a central role in a digitally set-up direct marketing. On Instagram, for example, products are advertised via stories and sold via private message or the classic online store. The transition between analog and digital direct marketers is fluid.

4.11.2.2 Drivers of Demand for Direct Marketing

Trust and transparency play an important role in the demand for direct marketing [BK20, WVL+15]. Both factors are mutually dependent: On the one hand, trust is built up if everything is transparent. On the other hand, transparency would not be necessary if the trust is given. When customers develop a feeling for what happens on a farm day by day, for example via social media, they develop a stronger connection to agriculture. For the marketer, therefore, it is always about creating a relationship with the origin of the product. Demand is driven by sustainability, heritage, and product quality on the one hand [BK20, WVL+15], and on the other hand by the rural idyll on the farms. Customers want to buy where it is still good, where it is still real, where the animals are treated well, can be outside and still have names and are not numbers. This relationship to the land, to the origin of the food and to this idyll, which is seen as lost, makes up the added value for the direct marketer customer. The relationship to the origin of the product is a key demand driver for direct-marketed products. One challenge is to make this visible and tangible using analog and digital tools.

4.11.3 Challenges of Digital Direct Marketing

4.11.3.1 Company Individual Usefulness

Which digital solutions are useful must be weighed up and determined on an individual company basis. If the cost of creating and maintaining a homepage exceeds the expected benefit that the customer expects from the internet presence, it does not make sufficient sense. One experience was the following: during a consultation with a farmstead cheese dairy, the customer listed his marketing needs to me: "We need a new logo, product labels, delivery bills, stamps, vehicle lettering, roll-up". When I asked him about a homepage, he replied: "No, we don't need a homepage". In his eyes, a company presence on the internet was not necessary for his marketing. Digitalization must not be the purpose and needs to be a means to an end. It is therefore very important to weigh up for which farm or branch of business the changeover makes sense.

4.11.3.2 Costs and Workload

The use of digital solutions such as online stores or enterprise resource planning systems (for examples, see ▶ Sect. 4.10) incurs various costs that can inhibit acceptance. Practical observation often shows that farmers are very price-sensitive regarding software, although software or development costs play a minor role compared to the level of investment in AgMachinery. Especially for online stores on small farms, the additional workload is an inhibiting factor for implementation, as the additional sales are low compared to the additional effort.

4.11.3.3 Online Presence

Many marketers ask themselves what they can or should show from the farm and what is better not to show (this question is also discussed in the research project "Social Media in Bavarian Agriculture" [HSWT21]). In this context, the correct contextualization of the online presentation is extremely important to avoid misunderstandings by the consumer. If the picture of a dead calf appears in the timeline on social media without context or explanation, the risk of misunderstanding is much higher than if something like this occurs during an analog farm tour and can be explained and caught by the farmer's family. Online, this is more problematic because consumers often cannot properly assess the husbandry conditions. For example, on barn webcams, some concerned consumers cannot distinguish what are normal situations in a barn and which ones need an intervention. Consumer ignorance is also due to the gap between image and reality created by decades of food advertising [Zel21]. Advertising images, which have become increasingly cheesy and unrealistic, have shaped the image of agriculture in the minds—and hearts!—of consumers. Today, consumers demand in reality what they were promised in advertising [Zel20]. Digital communication therefore needs to contextualize the content to catch up with today's consumers.

4.11.4 Success Factors of Digital Direct Marketing

4.11.4.1 Location

Digital success models often emerge where the need for alternatives to analog solutions in distribution drives digitalization. This is particularly the case when a farm cannot open a farm store due to its location or cannot supply a nearby town, and thus must develop other distribution channels. Horbacher Mühle is an example of a mill dealing with a remote location through direct online marketing with an online store.

Case Study: Classic direct marketing with online store

Horbacher Mühle produces and markets mill products. After most of its traditional customers had given up, the mill faced closure in 2004. With the generation change, the miller family around Johannes Dobelke focused on direct marketing of baking mixes with a special grinding process. To make the mill "findable" given its remote location, an internet presence with an initial online store was established in 2008. The online store has been developed over the years and the ordering processes have been optimized by an expert for online stores in the team. Today, the online store carries a wide range of mill and bakery products for domestic and international markets and is the central sales channel for Horbacher Mühle [Dob20, Hor20], see ◘ Fig. 4.28.

◘ Fig. 4.28 Online shop Horbacher Mühle [Hor20]

4.11.4.2 Generation Change

The generational change is another possible success factor, as the younger generation often brings digital open-mindedness and/or knowledge from other training and education, which can then steer business development in a new direction. Another positive factor is that older people are increasingly using digital means of communication and thus also for marketing, such as WhatsApp.

4.11.4.3 Small Business Agility

Another opportunity lies in the agility of small companies, which often work on a small scale and can therefore implement customer wishes more individually, for example via messenger marketing or social media. This offers niche products with a special target group, such as fine meat from ostriches, Iberico pigs or Wagyu cattle. One successful example of this is "Paula & Konsorten" for the marketing of wool.

> **Case Study: Crowdfunding project for the marketing of wool**
>
> Paula & Konsorten is a crowdfunding project to market the wool of wandering sheep in the Swabian Alb. Here, a very successful crowdfunding (145,698 € financed by 1285 supporters until the end of 2020, co-funding by Krombacher Naturstarter [Sta20]) ran through targeted and skillful addressing of "do it yourself" communities via social media, in which the supporters could, among other things, also co-determine the color shades of the wool. Otherwise, the approach would have been very time-consuming and lengthy. The wool project shows that sufficient customers can be found for a niche product via the internet [Sta20].

4.11.4.4 Use of Software Solutions

For farms that struggle with digitalization, there are current software solutions that aim to meet the needs of smaller companies so that they do not need their own IT department (see ▶ Sect. 4.10.4). These solutions are particularly attractive to practitioners if they are easy to use, adaptable to individual farms and comparatively inexpensive. They should offer interfaces to any existing IT systems of the producers. Assistance in setting up the systems and personal support play an important role. Such digital software solutions can facilitate the entry into digitization.

4.11.5 COVID-19 and the Digitalization

Observation in consulting practice shows that farmers increasingly encountered digital solutions through the COVID-19 crisis, for example with video conferencing systems. The crisis had a particularly positive effect on marketers who supply directly to end customers. Sales to the hospitality industry were down because of the closures. Alternative marketing concepts have been developed with the help of newly established initiatives. Delivery services have become more important in general [BK20]. People have become more accustomed to ordering online and having goods delivered, including from direct marketers. Picnic is an example of a food delivery service that also deliver regional products.

> **Case Study: Delivery service for regional food**
>
> Picnic is an online-only supermarket, without a stationary store. Customers order products such as fruit, vegetables, meat, and dairy products via a special app. The offer includes regional products from farmers and bakers. Consumers can use the app to determine the route of the delivery vehicle and receive the name and a photo of the driver via it—Picnic refers to this as the "milkman principle." To successfully implement this principle, employees are hired specifically according to personality. The goods are delivered free of charge [Pic20].

Further, the demand for recreation and experience offers has increased in the COVID-19 crisis, which holds development opportunities for the farmer. The linkage of real experience, digital co-experience and various sales and information channels (cross-channel approaches) offer a real opportunity for customer acquisition and retention in this regard. For example, when a mountain cheese discovered in a vacation in the Alps can be ordered home through an online store. All in all, COVID-19 has had a rather positive influence on digitalization in direct marketing.

Agronomy Perspective

4.11.6 Conclusion

Successful direct marketing does not have to be digital, but depending on the conditions, digital solutions can provide added value. Digitalization should not be the purpose itself. Digitalization should facilitate, enable, be a means to an end, and offer added value to producers. This added value, can be represented by improved relationships between consumers and producers and better accessibility to the customer, especially for remote farms. The farm or branch of operation for which digital solutions make sense, must be looked at on a farm-by-farm basis, although the factors described give a rough framework of understanding. Despite all the digital solutions, direct marketing will always remain analog at its core, because it will probably never be possible to send a beet, a salad, or a cheese by data transmission.

Acknowledgments Martin Herchenbach wrote this article during his time as consultant with AFC Consulting Group AG and now works at Industrieverband Agrar e. V. (IVA).

4.12 Digitalization in the Food Industry

Fabio Ziemßen

Abstract

Technologically advanced startups are challenging established food retailers in the still young e-food segment. New infrastructures, processes, and business models are emerging. All players along the food value chain benefit from this development, which is evolving into a network with the customer at the center.

4.12.1 How Digitalization is Changing Food Sales

In the past, everything was more manageable and easier to plan: The value chain for food from the producer to the consumer was straightforward. The food industry followed a logically structured value chain from the field to the plate. Stationary food retailers were at the end of this chain. As intermediaries between consumption and production and processing, retailers took over distribution, refinement, communication, and price balancing. Accordingly, the dependencies on and the need for the established food retail trade increased over the years.

Digitalization in the food sector is changing this situation fundamentally. For all participants in the value chain, the topic of digitalization is on the agenda, with different motivations and goals. The first step is usually about internal digitalization, the creation of more transparency, the optimization of processes and the economic situation, as well as about faster and more transparent communication with all stakeholders. Technological innovations allow demand to be analyzed more precisely and further development to be predicted through Predictive Analytics (see ▶ Sect. 4.13). The collection and evaluation of data forms the basis for building new cycles and systems with the help of Block Chain technology.

Throughout the entire food industry, there is an expectation not only to communicate better with existing direct business partners in the future, but also to gain more information and insights overall about the value chain—preferably in real time—in order to adapt faster and better to customer wishes. Silos are being broken down, inclusion is the order of the day.

4.12.2 More Transparency: Close to the Customer and Yet Far Away

The value chain is now centered around the customer. But direct access to the customer remains predominantly the preserve of food retailers. This development is bringing about fundamental changes, which in turn have enormous implications and offer opportunities for all participants of the Ag-Food industry—if they move away from the image of a "chain" and start thinking more in terms of a "network".

An important driver of this change are new infrastructure solutions. These include, for example, automation for recording orders, machine suggestion systems, robotics for more efficient picking processes, algorithms for optimized route planning (see ▶ Sect. 4.10), as well as Artificial Intelligence for customer interfaces and product range optimization. Sensor technologies help reduce food waste and speed up the process of detecting cases of fraud.

Established food retailers must face up to these innovations—and are being challenged at various points by technologically strong new competitors. Startups that are more IT and communication companies than retailers are vying for customers who have high expectations regarding quality, trust, and digital services. Customers are increasingly using digital applications as tools for tracking their dietary behavior, are constantly searching for inspiration and information to order something, and then communicate their consumption to their own "peers" or "followers" on social media channels. Startups specifically target this new customer behavior. "You are what you eat": Consumption is becoming a statement of one's own values. This resonates especially with younger customers, who define their sense of belonging through an outwardly communicating consumption lifestyle. Startups attract this customer group via their activities on social media and via influencers (e.g., Instagram and TikTok).

Digitalization is changing customer communication. But it has a much deeper impact. New models, infrastructures, and processes are currently emerging that are relevant for the entire food industry.

4.12.3 "Direct to Consumer": Not New, but Different

For agriculture, the model "Direct to Consumer" is nothing new. After all, there have always been weekly markets. The restrictions imposed by COVID-19 accelerated the development of digital farm stores where the customers could order boxes of fruit and vegetables and have them delivered, even during lockdown. Specialty grocers, such as those in the beverage industry, also shifted even more to online sales and offline delivery services.

Smaller brands, in particular, have managed to bypass retailers and build up customer relevance. Brands such as "The Rainforest Company" (superfood), "Koro" (drugstore), "Foodspring" (nutrition for athletes), or "JustSpices" (spice blends) are experiencing a strong response and customer demand online.

One example of exemplary digital customer loyalty is the company Kale&Me, which specializes in juice cleanses and involves its customers in its "juice cleanse system" via subscription models and cleanses. But established companies like Mondelez and Rittersport also use the online sales channel to market "special editions" and build up customer loyalty. However, their online proportion is negligible compared to their offline sales. Here, the first verticalization is taking place through the acquisition of individual online shops. In 2020, the Oetker Group took over the online delivery service Flaschenpost for several hundred million euros, according to media reports, and the Swiss Nestlé Group acquired a majority stake in the British cooking box provider "Mindful Chefs".

Ultimately, the "Direct to Consumer" solutions presented here are also examples from the "e-food" segment. The term electronic food ("e-food") is used to describe the purchase of food via digital sales channels. In marketing practice and scientific market research, terms other than e-food are also in use, such as online grocery retailing (OGR), online food retailing, online grocery shopping, and electronic grocery shopping (EGS).

Farmers are also benefiting greatly from the increasing digital buying behavior of consumers. On the one hand, a visit to a farm is part of an authentic shopping experience; on the other hand, the digital presence of a farm store is increasingly accepted and desired. Different meta-services aggregate demand and bring customers and farmers together on their platforms. The company "Bauernbox" in the Münsterland region and the company "Bauerntüte" in Cologne, which market regional products directly from the producers, are already well established. The "Höfegemeinschaft-Pommern" goes one step further: Startups such as "LunchVegaz" or "Tlaxcalli" have set up shop locally in Rothenklempenow so that they can offer their products not only on the farm but also online. This creates an ecosystem consisting of growers, producers, and marketing within a small area and offers maximum transparency and traceability.

4.12.4 E-Food Startups: New Models, Processes, and Infrastructure

Various players are currently positioning themselves in the e-food segment: traditional retailers with multichannel approaches and pure online companies, the "pure players". In between, strategic collaborations are also becoming established, such as that of the Dutch company Picnic with the Edeka Group and that of Amazon with Tegut.

Startups in the e-food market define themselves by their technological edge and their proximity to customers. Some just aim to link supply and demand better. Others bring together participants in the value chain to form a network. Examples of this include "crowdfarming" (buying together from a farmer) or "crowdbutching" (buying a cow together and using all slaughtered parts). Startups help match customers with suppliers.

Until now, food retailers had a strong function as quality assurers and curators. Customers could trust the tested product ranges. This is precisely where the startups are now positioning themselves. For example, they are now acting as retailers by offering shopping options from partner retailers via shopping apps such as "Instacart". Some companies work directly with manufacturers—but only if the appropriate digital interfaces are in place. Other startups have opened digital marketplaces and subscription models, for example for leftover goods (Motatos), dishes (Hellofresh), and specialties (Gourmondo).

4.12.5 New Distribution Channels: "Tiny Stores, Dark Stores, Ghost Stores"

In addition to the new business models, new infrastructures are emerging. The need for contactless shopping in the age of COVID-19 is accelerating the development of new "customer touchpoints". These include "Tiny Stores", which do not require staff and are filled by producers in the evening. Customers order online and can pick up their goods or food directly from a locker in the Tiny Store. With HelloFreshGo, HelloFresh also offers a POS solution in the form of automated vending machines for fresh convenience products.

Furthermore, supermarkets are increasingly being converted into distribution centers, so-called "Dark Stores". In larger cities, "Ghost Stores" are springing up. With these mini distribution centers, e-food retailers promise lightning-fast delivery within 10 min to any location in the city. The company Gorillas was able to convince investors with this approach and raise 44 million euros in 2020.

4.12.6 Agile Approach: Learning from the Start-ups and Joining in

The agile, step-by-step approach is the startups' big advantage; e.g., the fact that they build their IT infrastructure from scratch. This allows them to incorporate purchasing data and changing user behavior into their business decisions almost in real time. They can also optimize purchase suggestions and personalized user interfaces at high speed and further improve the customer experience. Shopping and delivery are becoming faster, easier, and more compelling for customers.

Traditional food retailers must keep pace with this development. They must upgrade both technically and in terms of the digital presentation of their products because online customers are increasingly less interested in detailed product features and more in the "story", the "shopping experience" that is conveyed. With intuitive user interfaces (User Experience), the pure players convey confidence in their own food expertise and present the customer with a complete brand world. The great art lies in transferring this online experience offline—for example, with an "unboxing experience" when the customer receives goods or through pop-up stores as brand experience locations for better customer loyalty.

4.12.7 The Digital Path to "B2B2C"

For all participants in the food industry, the digital transformation not only improves performance and merges processes with those of other participants, but also creates new business opportunities. These include, for example, collaboration on overarching platform solutions that flexibly adapt to changing customer needs.

In addition, there is an opportunity for traditional B2B companies to make the leap to B2B2C. If it fits their strategic orientation, they can address new customer groups online in the future and do business with them. In this context, it is advisable to work in partnership with the e-food startup scene, for example as part of corporate venturing activities. For this, however, companies do not only have to provide the financial and technical capacities; a new way of thinking and an agile approach are also necessary.

Yesterday's value chain may seem more solid, more manageable, but the impact of digitalization has already broken up the chain in many places and replaced it with a complex network. This requires a quick rethink, because for all digital models, in the end the winner is the one who has data sovereignty and access to the customer and offers customer-centric solutions in distribution and communication in addition to performance marketing.

Agronomy Perspective

4.13 Artificial Intelligence and Sustainable Crop Planning: Better Planning and Less Waste Through Digital Optimization

Wolf C. Goertz

Abstract

Organic vegetable production has very high production costs. Thus, farmers suffer greatly from price and market fluctuation, especially when it comes to overproduction (prices become too low) and out-of-stock situations (prices become high but there is nothing left to offer). As a result, it may take several years before a new organic farm can operate profitably. These problems deter farmers who are willing to convert to sustainable production. This essay is based on the hypothesis that precise recommendations for farmers what vegetable and variety to grow can facilitate a conversion to sustainable production. Precise recommendations are now conceivable, thanks to the increasing development of learning algorithms (artificial intelligence). To make artificial intelligence usable for small- and medium-sized farms, we need to decide which recommendation and information may make sense for farmers to share and let calculate through intelligent algorithms. These questions are addressed in this article.

4.13.1 Artificial Intelligence—Mystery or Helpful Tool

Artificial Intelligence, is one of the hot topics in the IT industry (see ▶ Sect. 3.5). Usually, it goes hand in hand with seemingly all-encompassing terms such as "algorithm" and "Big Data". Behind these terms some entrepreneurs of traditional industries may assume something very big, mysterious and sometimes frightening at the same time. But, especially since our generation is in a position to help shape this important development right now, it is now the right time to demystify [GRU19]. However, in many areas there is still a lack of ideas on how to make use of AI. But it already helps us on a daily base, mostly unnoticed, from daily shopping to analysis in quantum physics and the fight against COVID-19, and as well in the sector of large-scale agriculture.

AI basically means computer-aided processing of data, the results of which become more and more accurate over time. To make it more tangible: an ordinary e-mail spam filter is an example for AI already [KOS19]. The intelligent filter checks the e-mails that you moved to the trash or spam folder, finds similarities on its own and learns to classify them. The software does not know what spam is at all, but it makes suggestions. If the result is not correct and an e-mail has to be taken out of the spam folder again by the user, the software adjusts the algorithm slightly. In this way, the program is tested again and again and "learns" to recognize patterns and regularities. This experience leads to ever better results, recommendations and predictions [ENG18].

This kind of machine learning is heavily turning into a key technology in IT. In agriculture, AI already plays a role, for example, when it comes to "smart farming" such as the digital farm management system other authors of this anthology impressively demonstrate (see ▶ Sect. 3.5.5). The analysis of soil samples, plant and weather data, for example, can help to find the right dosage of fertilizer. In this sense, the project "ArtIFARM – Artificial Intelligence in Farming" of the University of Applied Sciences in Stralsund pursues the goal of using smart farming approaches to make more effective use of operating resources and at the same time improve the CO_2 balances of agricultural enterprises [HOC20].

As will be shown in the next section, a switch to sustainable production methods is increasingly desired by society and politics and is—technically—possible. If it is true that AI can make a meaningful contribution in almost all business areas, it should also be possible to create a case of application for organic vegetable growing, which, as a result, may make it easier for farmers to convert to organic production. The question discussed in this essay is therefore: How may AI help to make vegetable growing more efficient?

4.13.2 Why Don't You just Go Sustainable?

Like all parts of agriculture in Germany, vegetable production in Lower Saxony faces the challenge of producing safe and healthy food with the least possible negative impact on the environment. Many cultivation techniques of regenerative and ecological production can conserve or even process ecological resources such as biodiversity or soil structure [TWM+14]. In its sustainability strategy, the German government has therefore set the goal of increasing the proportion of agricultural land with organic farming to 20% by 2030 [BUN18]. However, the current share of 9.1% in Germany (ca. 4% in Lower Saxony) is still far from this target [BME20]. This is remarkable, as the market for organic food in Germany has been growing for years [AMI19]. Why is this share still so low?

In organic horticulture, this small share is characterized by many small family-owned farms. Due to the high costs and the lack of willingness to innovate, small and middle cultivating farms are often helpless in the face of digitalization. Those farms will only have a chance to keep up with the large farms if they join forces and work closely together. Moreover, in the food value chains in Europe, many food products are thrown away and thus produced for nothing [IPC19]. Especially in the case of perishable products such as fresh organic vegetables, producers prefer to produce too much rather than not being able to deliver [KRA12], which in case of sustainable produced vegetables is extremely cost-intensive. This maladministration leads to a situation, where many organic farmers have no chance to anticipate the real market need and the prices paid by the market players once the vegetables are ready for harvesting. It also is one of the main entry hurdles for new producers [BB03]. The expansion of areas of existing organic farms is also limited by the lack of market information. Many farm managers say that they would like to grow more or different, e.g., cultivate rare and old vegetable varieties for the regional market, but do not venture into new areas because of the uncertainty in sales. Thus, we shall sharpen our research question: How can AI help farmers to anticipate sales and decide which vegetable to grow and in which quantity?

4.13.3 Thinking Vegetable Production and AI Together

Farmers constantly have to make decisions, especially about which crops they grow and how much of them. It is crucial that in the end there is demand for the products and that they can therefore achieve a good price. But is this even possible to support by AI?

At least in online trade and reselling, AI is already helping to increase profit. There, learning algorithms analyze factors such as overall market development, market prices, market potential, and even weather and sales history to make suggestions on which products shall be ordered by the retailer and what reselling prices he/she can possibly achieve. Following the AI suggestions, according to a McKinsey study from 2018, "a reduction of up to 65% in lost sales due to out-of-stock situations and a reduction of between 20–50% in stock levels are possible" [GLÄ18].

A feasible example that points in a similar direction is the "BAReS" app for farmers in Bavaria. It offers a promising way to support small and medium-sized farms. There, regional purchase and sales prices for market crops are to be recorded by farmers and made available to users in real time. This increases the "bargaining power" of the individual farmer. The app is intended to give farmers a knowledge advantage over retailers. Based on the prices reported by the farmers, the app calculates a comparison price at the district level. This is coordinated with the delivery quantity, transport costs and current price development at the futures exchange market. The prerequisite for the creation of the regional comparison price is that enough farmers enter the prices they got for their crop and fruit. In order to be able to record enough prices, the principle of give and take applies to the app. Members must enter at least 12 prices per year in the app themselves.

This app can only show the actual price ranges, so that the farmer has a stronger background for bargaining. This is a downstream process, helping after the processes planning, growing, and harvesting.

The Vegetable-Cloud project recently initiated by the Chamber of Agriculture of Lower Saxony and the start-up Foodsupply from Osnabrück follows the idea to use market data like that to be used in a framework to predict a forecast of the possible prices of certain products. Thus, the aim is to help the farmers in their decision before growing and harvesting. Theoretically, exactly the advantages of the BAReS app could be transferred. However, the BAReS app is an exception and operates only locally. Moreover, the sector "trade" and the sector "agricultural primary production" differ immensely in terms of their degree of digitalization. The basis of price forecasting is the access to data on relevant factors on a larger scale.

4.13.3.1 Data

The decisive factor is that "data [rarely] already exist in such a form that learning algorithms can be applied to them without difficulty. On the contrary, it is usually a very complex process to prepare data in such a way that they can be used as material for machine learning processes" [ES18]. Thus, it becomes a central question of how this can be addressed accurately. As a rule, a distinction is made between primary and secondary data, with primary data only being directly generated by the farmer and secondary data already available from other sources (such as magazines, market reports, credit agencies, trade directories, catalogues, price lists, advertising material, and internet research) [BIC10]. Primary data offer the advantage of being geared to solving decision-making problems, but their collection involves greater effort than the use of secondary data. Therefore, the development of suitable systems for primary data acquisition is becoming increasingly important. This is the task of the Vegetable-Cloud project. Among other things, the following questions are discussed in this project: 1) Which data of which parameters must be available for a suitable learning algorithm, 2) how can they be collected, 3) how can the results be used to guide action?

4.13.3.2 Factors

First of all, it should be noted that possible learning algorithms are not a classification like in the spam-filter example (spam or non-spam), but much more of a regression. Similar to a multiple regression analysis, it is important to recognize a connection between multiple factors based on their values. This should help to make predictions. It can be assumed that a horticultural business can increase its profit and strengthen its market position if a precise assessment is made of the time at which sowing is promising based on future prices. It can also be assumed that digital platforms can even strengthen regional markets.

For this purpose, besides the above-mentioned factors such as regional and international price and exchange development, the duration of cultivation (time from sowing to harvest), which is partly field-dependent, and the current cultivation plans of the regional co-suppliers must be known for each vegetable crop in order to avoid parallelism and thus, again, overproduction.

In addition, the software needs information on influencing weather data on an international scale and should be able to estimate when which foreign crops could come onto the market and at what price.

Also, information about the overall market development in Germany and worldwide, as well as global nutritional trends (increased demand for organic products in China could possibly lead to lower cheap imports) are important. Further quantity information and data on overproduction and shortages are necessary. In order to better assess consumer behavior, data on changes in regional income is important. If a smart farm is set up on the farm, data on soil and plants could provide the AI with further valuable information. It will be necessary to differentiate these data for individual vegetable crops in order to make precise and differentiated predictions. The demand of the downstream parts of the supply chain should, if possible, be recorded historically and up-to-date. Thus, cooperation with new and established trading companies must also be sought.

4.13.3.3 Sharing is Caring

In related areas of the industry, such as trade in corn, wheat or sugar, data on trade is already being used by AI-systems. Different exchange markets such as Matif or CBot provide information on price trends. In addition, farmers use information services such as the commodity letter from "KS-Agrar" or the information service from Kaack-Terminhandel to keep up to date with market developments and consult them for future crop planning. These open data can also be used as indicators for crop planning. In the vegetable cultivation sector, quarterly market reports from the Federal Agency for Agriculture and Food can be used. However, there is a lack of digitalized data that can be used, especially for organic farming.

New data platforms must be created for a more dynamic exchange between farmers themselves and between buyers/traders and farmers. Start-ups such as Foodsupply, House of Crops or Cropspot are already demonstrating this to some extent. In the foreground for the collection of this primary data, especially prices and quantities, parallel efficiency-enhancing services must be developed to support farmers in optimizing their business processes and in selling their products. Network-oriented platforms aimed at cultivation planning in the regional context of the farms must be further expanded. Only with the central data exchange of small and medium-sized farms can potential market developments for the future be identified more quickly.

4.13.3.4 How are the Results Communicated?

Using AI/machine learning, the data thus obtained can be evaluated more and more precisely over time. Prices, competition and demand potentials can be generated and communicated as estimates and action-guiding recommendations on demand. This strengthens the autonomy of the builders, prevents monopolies and oligopolies (as they receive a range of advice and can still make their own decisions) and gag contracts (as they are not dependent on one buyer). In the future, companies will have to work together according to the principle of swarm intelligence. The decisive factors are the sales volumes, the price achieved, the actual desired price, and then a should-is comparison. This gives swarm intelligence the ability to recognize that many others are planning to grow a particular fruit. In this way, the AI can recognize at which point farmers themselves, but also the others, have farmed according to their ideas and thus in turn free scope for exploratory cultivation vegetable and crops that have not been in scope before.

4.13.4 Conclusion

This section discusses the potential that AI has for the expansion of sustainable vegetable growing. Precise, computer-supported cultivation planning is not only possible for large trading companies, but also in a joint network of small and medium-sized organic farms. Prerequisites are openness of the farms to share true data of the own farm and their own estimations anonymously. In order for farmers to do this, it is important to offer an easily accessible interface and to provide them with well-prepared data so that they can see their own benefit directly and in real time for their own farm.

This would make it easier for conventional horticultural enterprises to convert to sustainable production, as supply and demand can be presented immediately in a logical manner. With greater sales security, it will also be possible to cultivate rare crops. This will also contribute to even greater regional bio-diversity and independence from international supply chains. In addition, surplus (under-ploughing) and profit losses due to shortfalls will be minimized.

References

[AD16]   Auernhammer, H., and M. Demmel. 2016. State of the Art and Future Requirements. In *Precision Agriculture Technology for Crop Farming*, 299–346. Boca Raton: CRC Press.

[AEF18] AEF Online. 2020. ▶ www.aef-online.org. Accessed 18 Aug 2020.

[Agr16] Agrarheute. 2016. *Betrieb der Zukunft: Hier liegt das größte Potenzial*. ▶ https://www.agrarheute.com/management/betriebsfuehrung/betrieb-zukunft-liegt-groesste-potenzial-524403. Accessed 5 Mar 2021.

[Agr19] Agritechnica. 2019. ▶ https://www.agritechnica.com/de/innovation-award-agritechnica/gold-und-silber-2019. Accessed 24 Feb 2021.

[Agr21] Agritechnica. 2021. *Uni Hohenheim*. ▶ https://agrar.uni-hohenheim.de/organisation/projekt/intelligenter-optischer-sensor-fuer-den-teilflaechenspezifischen-herbizideinsatz-im-online-verfahren-h-sensor. Accessed 24 Feb 2021.

[Ama20a] AMAZONE. 2020a. *AMAZONE Argus Twin*. ▶ https://amazone.de/de-de/produkte-digitale-loesungen/digitale-loesung/terminals-hardware/sensorik/argus-duengetechnik-84490. Accessed 13 Mar 2022.

[Ama20b] AMAZONE. 2020b. *AMAZONE mySpreadApp*. ▶ https://amazone.de/de-de/produkte-digitale-loesungen/digitale-loesung/software/agapps/myspreader-app-84568. Accessed 13 Mar 2022.

[AMI19] AMI. 2019. *Agrarmarkt Informationsgesellschaft mbH: Öko-Landbau – Marktdrends*. ▶ https://www.ami-informiert.de/.

[ATL20] ATLAS. 2020. ▶ https://www.atlas-h2020.eu/. Accessed 18 Aug 2020.

[BUN18] Bundesregierung. 2018. Zukunftsstrategie ökologischer Landbau – Impulse für mehr Nachhaltigkeit in Deutschland. *Herausgegeben vom Bundesministerium für Ernährung und Landwirtschaft*. ▶ https://www.bmel.de/SharedDocs/Downloads/DE/Broschueren/ZukunftsstrategieOekologischerLandbau2019.pdf?__blob=publicationFile&v=4.

[BB03]   Bokelmann, Wolfgang, and Bettina König. 2003. *Hinderungsgründe für die Umstellung von Wein-, Obst- und Gartenbaubetrieben (Gemüsebaubetrieben) auf ökologische Wirtschaftsweisen in verschiedenen Regionen Deutschlands und Möglichkeiten ihrer Minderung*. Bonn: Bundesanstalt für Landwirtschaft und Ernährung (BLE), Geschäftsstelle Bundesprogramm Ökologischer Landbau. ▶ https://orgprints.org/4784/.

[BBF17] Balafoutis, A. T., B. Beck, and S. Fountas. 2017. Smart Farming Technology – Description Taxonomy and Economic Impact. In *Precision Agriculture: Technology and Economic Perspectives*, 21–77. Cham: Springer International.

[Bel19] Bundesministerium für Ernährung und Landwirtschaft [German Federal Ministry of Food and Agriculture]. 2019. Deutschland, wie es isst: *Der BMEL-Ernährungsreport*.

[BHR16] Bovensiepen, Gerd, Ralf Hombach, and Stefanie Raimund. 2016. *Quo vadis, agricola? PricewaterhouseCoopers AG*. ▶ http://www.pwc.de/de/handel-und-konsumguter/assets/smart-farming-studie-2016.pdf. Accessed 13 Mar 2022.

[BIC10] Bichler, Klaus, et al. 2010. *Beschaffungs- und Lagerwirtschaft: Praxisorientierte Darstellung der Grundlagen. Technologien und Verfahren*. Wiesbaden: Gabler. 9783834919478.

[Bit20] Bitkom Research. 2020. *Digitalisierung in der Landwirtschaft*. ▶ http://www.bitkom.org/Bitkom/Publikationen/Digitalisierung-in-der-Landwirtschaft.html. Accessed 19 Jul 2020.

[BK20] Böhm, Michael, and Christine Krämer. 2020. Innodirekt – Neue und innovative Formen der Direktvermarktung landwirtschaftlicher Produkte [Innodirekt – New and innovative forms of direct marketing of agricultural products], *BMEL*. ▶ www.orgprints.org/37311/. Accessed 28 Dec 2020.

[BLE21] Bundesanstalt für Landwirtschaft und Ernährung (BLE) [German Federal Agency for Agriculture and Food]. 2021. *Machbarkeitsstudie zu staatlichen, digitalen Datenplattformen für die Landwirtschaft*. ▶ https://www.ble.de/DE/Projektfoerderung/Foerderungen-Auftraege/Digitalisierung/Machbarkeitsstudie/Machbarkeitsstudie.html. Accessed 1 Feb 2021.

[BME20] BMEL. 2020. *Strukturdaten zum Ökologischen Landbau*. Pressemitteilung Nr. 125/2020. ▶ https://www.bmel.de/SharedDocs/Pressemitteilungen/DE/2020/125-strukturdaten-oekolandbau.html;jsessionid=80A1F8F8A34F8FC984FC-C140B66EF33A.intranet922.

[BMW21a] Bundesministerium für Wirtschaft und Energie [German Federal Ministry for Economic Affairs and Energy]. 2021a. *Agri-Gaia*. ▶ https://www.bmwi.de/Redaktion/DE/Artikel/Digitale-Welt/GAIA-X-Use-Cases/agri-gaia.html. Accessed 1 Feb 2021.

[BMW21b] Bundesministerium für Wirtschaft und Energie [German Federal Ministry for Economic Affairs and Energy]. 2021b. *GAIA-X Eine vernetzte Datenstruktur für ein europäisches digitales Ökosystem*. ▶ https://www.bmwi.de/Redaktion/DE/Dossier/gaia-x.html. Accessed 1 Feb 2021.

[Bos20] Robert Bosch GmbH. 2020. *NEVONEX*. ▶ https://www.nevonex.com. Accessed 29 Dec 2020.

[BT21] Bernardi, Traphoener. 2021. *Artificial intelligence*. See Chapter 3.7.

[CCI20] CCI. 2020. ▶ https://www.cc-isobus.com. Accessed 18 Aug 2020.

[Cro21a] Croplands. 2021. ▶ https://www.croplands.com.au/. Accessed 24 Feb 2021.

[Cro21b] Crop.Zone. 2021. ▶ https://crop.zone/de/. Accessed 24 Feb 2021.

[Dar19] Darr, M. 2019. High Definition Yield Maps for Precision Ag Decision Support. In *VDI Berichte 2361, 77th International Conference on Agricultural Engineering*, 327–334. Hannover: VDI Wissensforum GmbH.

[Dem13] Demmel, M. 2013. Site-Specific Recording of Yield. Precision in Crop Farming, 314–329. Dordrecht: Springer.

[Dic19] Dicks, Henning. 2019. *Wo sich Direktmarkter tummeln*. ▶ https://f3.de/wo-sich-direkt-vermarkter-tummeln. Accessed 5 Mar 2021.

[Diw21] Diwakopter. 2021. ▶ https://www.diwakopter.de/. Accessed 24 Feb 2021.

[Dob20] Dobelke, Johannes. 2020. *Digital direct marketing at Horbacher Mühle, expert interview by telephone, conducted by Martin Herchenbach*.

[Dro21] Dronewerkers. 2021. ▶ https://www.dronewerkers.nl/english/. Accessed 24 Feb 2021.

[EC20a] European Commission. 2020a. *Agricultural Interoperability and Analysis System*. ▶ https://cordis.europa.eu/project/id/857125. Accessed 13 Mar 2022.

[EC20b] European Commission. 2020b. *Expert Workshop on a Common European Agricultural Data Space*. ▶ https://ec.europa.eu/eip/agriculture/en/event/expert-workshop-common-european-agricultural-data. Accessed 13 Mar 2022.

[Eco21] Ecorobotix. 2021. ▶ https://www.ecorobotix.com/de/. Accessed 24 Feb 2021.

[Eip21a] EIP. 2021a. ▶ https://www.eip-nds.de/anbau-von-raps-mit-begleitpflanzen-im-anbausystem-einzelkornsaat-und-weiter-reihe.html. Accessed on 24 Feb 2021.

[Eip21b] EIP. 2021b. ▶ https://www.eip-nds.de/LURUU.html. Accessed 24 Feb 2021.

[ENG18] Engemann, Christoph, and Andreas Sudmann, Eds. 2018. *Machine Learning. Medien, Infrastrukturen und Technologien der Künstlichen Intelligenz*. Bielefeld: transcript.

[ES18] Engemann, Christoph, and Andreas Sudmann, Eds. 2018. *Machine Learning. Medien, Infrastrukturen und Technologien der Künstlichen Intelligenz*. Bielefeld: transcript.

[Fa+18] Faust, Eberhard, et al. 2018. *Heatwaves, Drought and Forest Fires in Europe: Billions of Dollars in Losses for Agricultural Sector. ▶ https://www.munichre.com/topics-online/en/climate-change-and-natural-disasters/climate-change/heatwaves-and-drought-in-europe.html.

[Far21a] Farming Revolution. 2021. ▶ https://www.farming-revolution.com/. Accessed 24 Feb 2021.

[Far21b] Farmdroid. 2021. ▶ https://farmdroid.dk/de/willkommen/. Accessed 24 Feb 2021.

[GRU19] Grunwald, Armin. 2019. Künstliche Intelligenz – Revolution oder Hype? *Physik in unserer Zeit* 50 (5): 211. ▶ https://doi.org/10.1002/piuz.201970502.

[GHH+16] Gremmer, Pia, Corinna Hempel, Ulrich Hamm, and Claudia Busch. 2016. *Zielkonflikt beim Lebensmitteleinkauf: Konventionell regional, ökologisch regional oder ökologisch aus entfernteren Regionen?*.

[GLÄ18] Gläß, Rainer. 2018. *Künstliche Intelligenz im Handel 2 – Anwendungen: Effizienz erhöhen und Kunden gewinnen*. Wiesbaden: Springer Fachmedien. ▶ https://doi.org/10.1007/978-3-658-23926-8.

[Gra17] Graf, Alexander. 2017. *Wie groß wird der Lebensmittelhandel online bis 2020*. ▶ https://www.kassenzone.de/2017/01/14/wie-gross-wird-der-lebensmittelhandel-online-bis-2020. Accessed 5 Mar 2021.

[GSE18] Gandorfer, Markus, Sebastian Schleicher, and Klaus Erdle. 2018. Barriers to Adoption of Smart Farming Technologies in Germany. In *Paper from the proceedings of the 14th International Conference on Precision Agriculture Montreal/Quebec Canada*.

[HOC20] Gläß, Rainer. 2018. *Künstliche Intelligenz im Handel 2 – Anwendungen: Effizienz erhöhen und Kunden gewinnen*. Wiesbaden: Springer Fachmedien. ▶ https://doi.org/10.1007/978-3-658-23926-8.

[Hor20] Horbacher Mühle Production- and Trade GmbH. 2020. ▶ www.horbacher-muehle.de. Accessed 28 Dec 2020.

[HSWT21] Social Media in der bayerischen Landwirtschaft [Social Media in Bavarian Agriculture]. 2021. *Research project at the University of Applied Sciences Weihenstephan-Triesdorf*. ▶ https://forschung.hswt.de/forschungsprojekt/1553-social-media-in-der-bayerischen-landwirtschaft. Accessed 4 Jan 2021.

[HT14] Heise, Heinke, and Ludwig Theuvsen. 2014. Erfolgsfaktoren in der Landwirtschaft: Status Quo und Bedeutung der IT für die Wirtschaftlichkeit der Betriebe [Success Factors in Agriculture: Status Quo and Importance of IT for Farm Profitability, 2014]. In *IT Standards in the Agricultural and Food Industry – Focus: Risk and Crisis Management*, Eds. M. Clasen, M. Hamer, S. Lehnert, B. Petersen, and B. Theuvsen. Bonn: Gesellschaft für Informatik e. V.

[HT90] Hastie, T. J., and R. J. Tibshirani. 1990. *Generalized Additive Models*. Chapman & Hall/CRC Monographs on Statistics and Applied Probability, 1st ed. ISBN 978-0412343902.

[iGr20] iGreen. 2020. ▶ http://www.igreen-projekt.de. Accessed 18 Aug 2020.

[ILV20] ILVO. 2020. ▶ www.ilvo.vlaanderen.be. Accessed 18 Aug 2020.

[IPC19] IPCC. 2019. Summary for Policymakers. In *Climate Change and Land: An IPCC Special Report on Climate Change, Desertification, Land Degradation, Sustainable Land Management, Food Security, and Greenhouse Gas Fluxes In Terrestrial Ecosystems*, Eds. P. R. Shukla, J. Skea, E. Calvo Buendia, V. Masson-Delmotte, H.- O. Pörtner, D. C. Roberts, P. Zhai, R. Slade, S. Connors, R. van Diemen, M. Ferrat, E. Haughey, S. Luz, S. Neogi, M. Pathak, J. Petzold, J. Portugal Pereira, P. Vyas, E. Huntley, K. Kissick, M. Belkacemi, and J. Malley (In press).

[Isi21] ISIP. 2021. ▶ https://www.isip.de/isip/servlet/isip-de. Accessed 24 Feb 2021.

[Iva21] IVA. 2021. ▶ https://www.iva.de/iva-magazin/schule-wissen/was-bedeutet-integrierter-pflanzenschutz. Accessed 24 Feb 2021.

[Joi20] JoinData. 2020. ▶ www.join-data.nl. Accessed 18 Aug 2020.

[Jos20] JOSKIN. 2020. *YouTube JOSKIN Manure Sensing*. ▶ https://www.youtube.com/watch?v=aWLGKjmBMXo. Accessed 19 Jul 2020.

[KER19] Kersting, Kristian, et al. 2019. *Wie Maschinen lernen: Künstliche Intelligenz verständlich erklärt*. Wiesbaden: Springer. ▶ https://doi.org/10.1007/978-3-658-26763-6.

[Key16] KeyQuest Marktforschung GmbH. 2016. *Landwirte-Befragung zu Direktvermarktung*.

[Kos19] Kossen, Jannik et al. 2019. Wie Maschinen lernen! in: Kersting, Kristian et al. 2019. *Wie Maschinenlernen: Künstliche Intelligenz verständlich erklärt*, 3–10. Wiesbaden: Springer. ▶ https://doi.org/10.1007/978-3-658-26763-6.

[Klü09] Klüßendorf-Feiffer, A. 2009. *Druscheignung als zentrale Führungsgröße im Erntemanagement*. Berlin: Humboldt-Universität zu Berlin – Landwirtschaftlich-Gärtnerische Fakultät.

[KRA12] Kranert, Martin, et al. 2012. *Ermittlung der weggeworfenen Lebensmittelmengen und Vorschläge zur Verminderung der Wegwerfrate bei Lebensmitteln in Deutschland*. Universität Stuttgart. ▶ http://www.bmel.de/SharedDocs/Downloads/Ernaehrung/WvL/Studie_Lebensmittelabfaelle_Langfassung.pdf?__blob=publicationFile.

[Kre16] Krenn, Katharina. 2016. *Direktvermarktung – Dahin geht der Trend*. ▶ https://www.agrarheute.com/management/betriebsfuehrung/direktvermarktung-dahin-geht-trend-521310. Accessed 5 Mar 2021.

[Kur18] Kuratorium für Technik und Bauwesen in der Landwirtschaft e. V. 2018. *Direktvermarktung landwirtschaftlicher Erzeugnisse.*

[Lan17] Klüßendorf-Feiffer, A. 2009. *Druscheignung als zentrale Führungsgröße im Erntemanagement.* Berlin: Humboldt-Universität zu Berlin – Landwirtschaftlich-Gärtnerische Fakultät.

[LHG+20] Lowenberg-DeBoer, J., I. Y. Huang, V. Grigoriadis, and S. Blackmore. 2020. Economics of robots and automation in field crop production. *Precision Agriculture* 21:278–299. ▶ https://doi.org/10.1007/s11119-019-09667-5.

[LOO+20] Lehr, Thomas, Laura Oppermann, Thomas Osterburg, and Markus Schubert. 2020. *Online-Marktplätze für regionale Lebensmittel in Sachsen.*

[LP17] Lind, K. M., and S. M. Pedersen. 2017. Perspectives of Precision Agriculture in a Broader Policy Context. In *Precision Agriculture: Technology and Economic Perspectives*, 262. Cham: Springer International.

[LVL18] Leithold, Peer, Thomas Volk, and Hermann Leithold. 2018. *26-7 – Teilflächenspezifische Wachstumsreglerapplikation – Ergebnisse von 10 Jahren OFR Versuche.* ▶ https://www.openagrar.de/receive/openagrar_mods_00042058. Accessed 24 Feb 2021.

[MBB+09] Migdall, S., H. Bach, J. Bobert, M. Wehrhan, and W. Mauser. 2009. Inversion of a canopy reflectance model using hyperspectral imagery for monitoring wheat growth and estimating yield. *Precision Agriculture* 10:508. ▶ https://doi.org/10.1007/s11119-009-9104-6.

[Met21] Metos Sensors. 2021. ▶ metos.at/sensors/. Accessed 17 May 2021.

[Mi21] Migende, Jörg. 2021. *Agribusiness from products to solutions*, see Section 4.1.

[MR20] Munich Re. 2020. *aiSure: Insure AI – Guarantee the performance of your Artificial Intelligence systems.* ▶ https://www.munichre.com/en/solutions/for-industry-clients/insure-ai.html.

[MTG15] Münch, P., J. Teichmann, and A. Günther. 2015. *Operator Assistance System for Forage Harvesters, VDI-MEG Agricultural Engineering*, 445–452. Hannover.

[Mün20] Susanne V. 2020. *Münchhausen, Vom Hof auf den Teller? Hemmfaktoren und Handlungsansätze für landwirtschaftliche Direktvermarktung [From farm to Fork? Inhibiting factors and approaches to action for agricultural direct marketing]*, 2015, Expert talk on local supply in the countryside. Berlin: Bündnis 90/Die Grünen German Parliamentary Group.

[Noa18] Noack, P. O. 2018. *Precision Farming – Smart Farming – Digital Farming.* Heidelberg: Wichmann.

[NW16] Nefzger, Nicole, and Christina Well. 2016. *Kreativität gefragt – Innovative Wege der Direktvermarktung in Zeiten des Internets.*

[Nws16] National Weather Service. 2016. ▶ http://www.nws.noaa.gov/om/hazstats/sum16.pdf. Accessed 19 Mar 2021.

[Pa+18] Panagos, Panos, et al. 2018. *Cost of agricultural productivity loss due to soil erosion in the European Union: From direct cost evaluation approaches to the use of macroeconomic models.* ▶ https://onlinelibrary.wiley.com/doi/full/ ▶ https://doi.org/10.1002/ldr.2879.

[Phe21] Phenoinspect. ▶ http://www.phenoinspect. de/. Accessed 24 Feb 2021.

[Pic20] Picnic GmbH. 2020. ▶ https://www.picnic. app/de/uber-picnic. Accessed 28 Dec 2020.

[PMA+16] Pantazi, X. E., D. Moshou, T. Alexandridis, R. L. Whetton, and A. M. Mouazen. 2016. Wheat yield prediction using machine learning and advanced sensing techniques. *Computers and Electronics in Agriculture* 121:57–65. ▶ https://doi.org/10.1016/j.compag.2015.11.018.

[PRN+18] Pohl, Jan-Philip, Dirk Rautmann, Henning Nordmeyer, and Dieter von Hörsten. 2018. *Teilflächenspezifische Applikation durch Direkteinspeisung – Mehr Präzision und weniger Mitteleinsatz.* ▶ https://www.openagrar.de/receive/openagrar_mods_00042100. Accessed 24 Feb 2021.

[Pri18] PricewaterhouseCoopers. 2018. *Online-Lebensmittelhandel vor dem Durchbruch in Deutschland.*

[pro20] profi. 2020. *Neue Schnittstellen bei John Deere.* ▶ https://www.profi.de/aktuell/neuheiten/neue-schnittstellen-bei-john-deere-12133462.html. Accessed 13 Mar 2022.

[Rau20] RAUCH. 2020. *RAUCH EMC2.* ▶ https:// rauch.de/duengerstreuer/axis-h-30-2-emc-w.html. Accessed 13 Mar 2022.

[Rav21] Raven. 2021. ▶ https://ravenprecision.com/products/application-controls/sidekick-pro-direct-injection. Accessed 24 Feb 2021.

[Rol20] Rolff, Marten. 2020. *Klein, fix, digital.* ▶ https://www.sueddeutsche.de/stil/lebensmittel-klein-fix-digital-1.4856047. Accessed 5 Mar 2021.

[RV20] Roberson-Vogel, Lisa-Marie. 2020. *Autonomes Fahren: Wie weit ist die Technoloie?* ▶ https://www.blog-ergo.de/selbstfahrendes-auto/.

[Sag21] Sagarobotics. 2021. ▶ https://sagarobotics. com/. Accessed 24 Feb 2021.

[SB+21] Späth, Jan-Hinrich, Roland Barth, and Christian Astor. 2021. *Deep dive autonomous: Safe surround sensing.* See Section 2.7.

[Sch20] Schweikert, Andreas. 2020. *Digitale Hofläden – 5 Beispiele für den kontaktlosen Einkauf von Lebensmitteln.* ▶ https://www.bitkom.org/

Themen/Digitale-Hoflaeden-in-Zeiten-von-Corona. Accessed 5 Mar 2021.

[SL20] Stuchtey/de Liedekerke. 2020. *Regenerating Europe's soils: Making the economics work.* ▶ https://www.systemiq.earth/wp-content/uploads/2020/01/RegeneratingEuropessoilsFINAL.pdf.

[Sta17] Statista. 2017. *Regionale Lebensmittel in Deutschland.*

[Sta20] Startnext Crowdfunding. 2020. *A Wool Project from Our Walking Sheep Farm.* ▶ https://www.startnext.com/paulaswolle. Accessed 13 Mar 2022.

[Ste15] Stehr, N. J. 2015. Drones: The Newest Technology for Precision Agriculture. *Natural Sciences Education* 44 (1): 89–91. ▶ https://doi.org/10.4195/nse2015.04.0772.

[Ste19] STENON. 2019. *Räumliche Verteilung von Nmin im Boden bei organischer/mineralischer Düngung.* Potsdam.

[Sto11] Stöcklin, V. 2011. Trends und Innovationen in der Düngetechnik. In *Landtechnik für Profis, Effiziente Technik für Düngung und Pflanzenschutz.* Hasbergen: VDI Wisssensforum GmbH.

[Sto19] Stöcklin, V. 2019. Development of Future Machine Concepts for the Needs Based Fertilisation of Individual Plants. In *LAND.TECHNIK AgEng*, 447–454. Hannover: VDI-MEG.

[Ta21] Taube, Friedhelm. 2021. *From ecological intensification to hybrid agriculture?*, see Section 6.3.

[Tae16] Taenzer, Miriam. 2016. *Digitalisierung in der Landwirtschaft.*

[Thu11] Thullner, C. 2011. *Methode zur Berechnung und Optimierung der Mineraldüngerverteilung im Bereich eines rechtwinkligen Vorgewendes*, Gingen/Fils.

[TWM+14] Tuck, Sean L., Camilla Winqvist, Flávia Mota, Johan Ahnström, Lindsay A. Turnbull, and Janne Bengtsson 2014. Land-Use Intensity and the Effects of Organic Farming on Biodiversity: A Hierarchical Meta-Analysis. *Journal of Applied Ecology. John Wiley & Sons Ltd: British Ecological Society* 746–755. ▶ https://doi.org/10.1111/1365-2664.12219.

[UBA20] Umweltbundesamt. 2020. *Bodenerosion durch Wasser – Eine unterschätzte Gefahr?* ▶ https://www.umweltbundesamt.de/themen/boden-landwirtschaft/bodenbelastungen/erosion#bodenerosion-durch-wasser-eine-unterschatzte-gefahr.

[Usm20] Usmani, Fahad. 2020. *A Short Guide to Expected Monetary Value (EMV).* ▶ https://pmstudycircle.com/2015/01/a-short-guide-to-expected-monetary-value-emv/.

[Wik21a] Wikipedia. 2021a. ▶ https://en.wikipedia.org/wiki/LPWAN. Accessed 19 Mar 2021.

[Wik21b] Wikipedia. 2021b. ▶ https://en.wikipedia.org/wiki/Narrowband_IoT. Accessed 19 Mar 2021.

[Wik21c] Wikipedia. 2021c. ▶ https://en.wikipedia.org/wiki/Copernicus_Programme#Sentinel_missions. Accessed 19 Mar 2021.

[Wik21d] Wikipedia. 2021d. ▶ https://en.wikipedia.org/wiki/Landsat_program. Accessed 19 Mar 2021.

[WM20] Wirthgen, Bernd, and Oswin Maurer. 2020. *Direktvermarktung. Verarbeitung, Absatz, Rentabilität, Recht.* Stuttgart (Hohenheim): Ulmer. ISBN: 3800142074.

[WSM18] Wille, Stefan Clemens, Achim Spiller, and Marie Meyer-Höfer. 2018. *Lage, Lage, Lage? Welche Rolle spielt der Standort für die landwirtschaftliche Direktvermarktung?.*

[WT13] Whelan, B., and J. Taylor. 2013. *Precision Agriculture for Grain Production Systems.* Collingwood Australia: CSIRO.

[WVL+15] Wiesmann, Janina, Luisa Vogt, Wolf Lorleberg, and Marcus Mergenthaler. 2015. Erfolgsfaktoren und Schwachstellen der Vermarktung regionaler Erzeugnisse [Success factors and weak points of marketing regional products]. *Research Reports of the Department of Agricultural Economics Soest* 35.

[Xag21] XAG. 2021. ▶ https://www.xa.com/en. Accessed 24 Feb 2021.

[ZAA15] Zujevs, A., V. Andrejs, and P. Ahrendt. 2015. Trends in Robotic Sensor Technologies for Fruit Harvesting: 2010–2015. *Procedia Computer Science* 77:227–233. ▶ https://doi.org/10.1016/j.procs.2015.12.378.

[Zel20] Zeller, Heike. 2020. *Milchkanne versus Melkroboter – Wunsch und Wirklichkeit in der Werbung für Lebensmittel. Oder: Wie geht zeitgemäße Werbung für Lebensmittel? [Milk can versus milking robot - desire and reality in food advertising. Or: How does contemporary advertising for food work?].* Acatech German Academy of Science and Engineering. ▶ https://www.acatech.de/allgemein/milchkanne-versus-melkroboter-wunsch-und-wirklichkeit-in-der-werbung-fuer-lebensmittel-oder-wie-geht-zeitgemaesse-werbung-fuer-lebensmittel/. Accessed 13 Mar 2022.

[Zel21] Zeller, Heike. 2021. Anreiz und Anspruch. Ländliches in der Lebensmittelbranche – Ein Streifzug durch die Praxis regionaler Vermarktung [Incentive and aspiration. Rural in the food industry – A foray into the practice of regional marketing]. In *Gutes Leben auf dem Land? Imaginationen und Projektionen vom 18. Jahrhundert bis zur Gegenwart*, Eds. Werner Nell and Marc Weiland. Bielefeld.

Farming System Perspective

Tom Green, Emmanuelle Gourdain, Géraldine Hirschy, Mehdi Sine, Martin Geyer, Norbert Laun, Manuela Zude-Sasse, Dominik Durner, Christian Koch, Noura Rhemouga, Julian Schill, Christian Bitter and Jan Reinier de Jong

Contents

5.1　Arrival of Digital Ag at Scale: The Farming Perspective – 280
5.1.1　Background and Context – 280
5.1.2　What's New? – 281
5.1.3　How Come? – 281
5.1.4　So What? The Prize – 282
5.1.5　Who Cares? – 284
5.1.6　Where's the Catch? – 285
5.1.7　Outlook – 286

5.2　The Digital Revolution, a Performance Accelerator from a French Perspective: The Issues and a Panorama of Possibilities for French Cereal Crops – 286
5.2.1　Introduction – 286
5.2.2　Observe – 288
5.2.3　Record – 289
5.2.4　Analyze and Decide – 291
5.2.5　Act – 292
5.2.6　Impact of Digital Agriculture on the Multi-Performance of Farms – 293
5.2.7　Conclusion – 294

5.3　Digital Transformation of Vegetable Production – 295
5.3.1　Seedling Cultivation and Planting – 296
5.3.2　Fertilization – 297
5.3.3　Plant Protection and Irrigation – 297

© The Author(s), under exclusive license to Springer-Verlag GmbH, DE, part of Springer Nature 2022
J. Dörr and M. Nachtmann (eds.), *Handbook Digital Farming*,
https://doi.org/10.1007/978-3-662-64378-5_5

5.3.4	Climate – 298
5.3.5	Weed Control – 298
5.3.6	Harvest – 298
5.3.7	Processing – 300
5.3.8	Outlook – 302

5.4 Digital Transformation of Fruit Production – 302
5.4.1	Challenges in the Supply Chain of Fresh Fruit – 302
5.4.2	Irrigation – 302
5.4.3	Crop Load Management – 304
5.4.4	Fruit Quality Post-Harvest – 306
5.4.5	Conclusions – 306

5.5 Digital Transformation in the Wine Business – 307
5.5.1	Innovation Versus Tradition – 307
5.5.2	The Wine Value Chain – 308
5.5.3	The Potential of the Vineyard – 308
5.5.4	Wine as the Role Model for an Authentic and Sustainable Agricultural Product – 309
5.5.5	From the Wine Value Chain to an Operational Network – 310
5.5.6	Challenges in Communication – 310
5.5.7	Digital Transformation in the Wine Business – 311
5.5.8	Conclusion – 315

5.6 Transparency of Animal Welfare Through Digitalization: A Dairy Farming Example of "Hofgut Neumühle" – 315
5.6.1	Dairy Farming in Germany – 316
5.6.2	Legal Basis – 318
5.6.3	Automation in Dairy Farming – 318
5.6.4	Outlook – 323

5.7 German Farmers Perspective on Digitalization – 323
5.7.1	Cross Farm Comparison – 323
5.7.2	Field-Specific Open Data – 327
5.7.3	Conclusion – 329

5.8 A Farm Case Study from the Netherlands – 329
5.8.1 The Arable Farm of the Family De Jong – 330
5.8.2 Precision Agriculture – 332
5.8.3 Future – 333

References – 334

5.1 Arrival of Digital Ag at Scale: The Farming Perspective

Tom Green

Abstract

Digital Ag is strongly influencing the practice of farmers. This section discusses the influence on various farming systems and relates Digital Ag to production factors. It provides an overview of the suite of digital tools that can be used by stakeholders and discusses what is needed to bring Digital Ag to scale.

5.1.1 Background and Context

The past six decades of modern field scale agriculture have been characterized by tremendous innovations in mechanization, plant protection, nutrition, and genetics. These innovations have supported and enabled substantial increases in absolute production (total tons produced) in response to steadily increasing global demand (◘ Fig. 5.1), increases in harvested yield (tons/ha) (◘ Fig. 5.2), and in overall agricultural productivity (◘ Fig. 5.3). The consequence of all this, for consumers in developed countries, is that the cost of food as a proportion of household expenditure has consistently fallen over the period. In the UK, in 1960, food represented over 25% of household expenditure, in 2020 it is less than 10% (source: UK ONS). Additionally, during this period, the range of foods available to consumers has increased dramatically, and independent quality standards regarding the environment, agricultural production systems and supply chains have also been established. In summary, the ag industry has responded to demand from consumers—who now benefit from greater choice, reliability of supply, and competitive pricing than ever before.

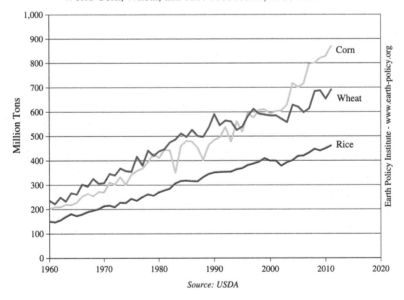

◘ Fig. 5.1 Total global ag production 1960–2011

Farming System Perspective

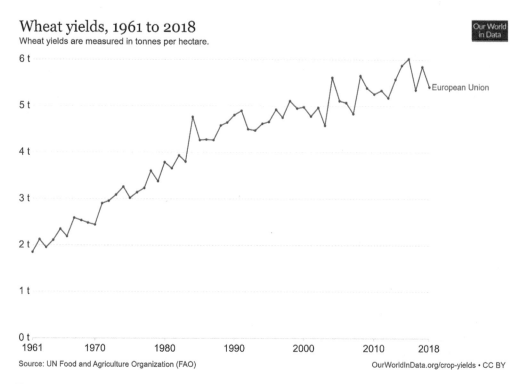

Fig. 5.2 Increasing agricultural yield 1961–2018

5.1.2 What's New?

Digital Agriculture is not new—many of the benefits and performance improvements referred to above have already been made possible by digital technologies. The most obvious of these is perhaps the Global Positioning System (GPS) and other satellite positioning and navigation systems which enabled the first combine yield mapping in the 1980s, and subsequent innovations including variable rate application (VRA) which offers input efficiency gains, and controlled traffic farming (CTF) which has offered reduced soil compaction, both of which have now been widely adopted. What is new is the prospect of these technologies and many more digital solutions becoming available, in theory, **at scale**, as the title of this section suggests. That is a game changer: Digital Ag available at scale offers a paradigm shift in our industry's ability to meet the challenges of the future, both in food demand, and how that food is produced, processed and delivered to consumers. I insert the words **in theory** because there remain many impediments to successful deployment of Digital Ag at scale, these are discussed in ▶ Sect. 5.1.6.

5.1.3 How Come?

The main drivers for the arrival of Digital Ag *at scale* are common to many other industries which are also benefiting from and being reshaped by the digital revolution, these include:

- unlimited availability of cloud storage
- unlimited availability of data processing capacity
- widespread application of machine learning (ML) and artificial intelligence (AI) in product development

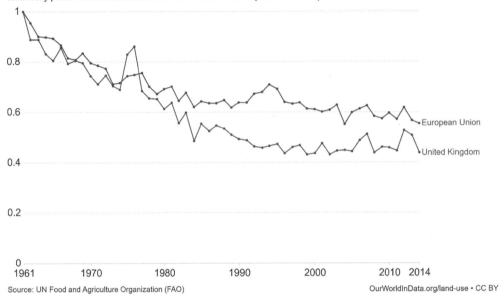

Fig. 5.3 Increasing Agricultural Productivity 1961–2014

- widespread connectivity opportunities including Satellite (GPS) Radio transmission, NB-IoT and LoRaWAN, enabling real-time remote sensing
- improved manufacturing of hardware devices including circuit boards
- widespread availability of Open APIs which enable access to, and sharing of proprietary software and web-based tools

As each of these elements have become more widely available so has their cost of adoption come down, often by orders of magnitude. Additionally, these features have encouraged and enabled synergistic combination of complementary technology disciplines—this has also led to improved user experience and value, and progressive reduction in cost of deployment.

5.1.4 So What? The Prize

The prize offered by successful exploitation of Digital Ag at scale is of global significance and is capable of becoming materially impactful not only for farmers and farm businesses, but also for the future well-being and flourishing of our planet, humankind, and all of the biodiversity with which we co-habit.

Digital Ag offers the opportunity for farmers to farm **to biological potential with nature**. That means understanding and being able to measure and monitor:

- crop phenology, physiology, and pathology
- soils
- livestock and its interaction with the farming system and environment
- weather and climate

- carbon footprint of different crops and growing and post-harvest handling systems

Digital Ag tools enable decisions to be supported by objective data sets, modeling and analysis that go far beyond the capacity of what the farm manager can observe. This, combined with the farmer's experience and judgment, ensures that decisions of What? Where? and How? to grow can be optimized. What crop(s) and rotation are most suited to this location, to the soils, topography, and climate? And with all such decisions capable of being made on a regional, local, field, and sub-field scale. These huge digital data sets, largely gathered remotely, when combined with knowledge of plant growth models, enable the farmer to make every judgment and decision against the background of the theoretical biological potential for that precise location and crop. This is a new dimension from which to view agriculture, allowing the farmer to methodically harness all the natural benefits offered by nature, deploying technologies to dramatically enhance performance.

The distribution of average crop yield is typically depicted as a classic "bell-curve" (◘ Fig. 5.4). If this curve was depicting the UK wheat crop, then the peak of the curve would be approximately 8 tons per hectare, with the flanks representing extremes of poor and good performance with "High Performers" delivering 10-12t/ha and "Low Performers" 4-6t/ha. The opportunity and prize offered by Digital Ag is to reposition and reshape this curve:
- eliminating crop failure
- reducing sub-optimal performance
- increasing proportion of high yielding crops and those which get close to or meet biological potential

This "Prize" is not just about improving yield, productivity, and financial performance—it also enables farmers to farm more sustainably. The same tools which are deployed to enhance yield and decision making around cropping plans and growing systems also enable the farmer to monitor performance in relation to the environment. Furthermore, it is often the insight offered by digital tools that demonstrates

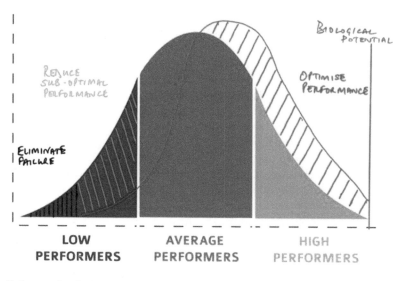

◘ **Fig. 5.4** Bell-curve distribution of agricultural yield

how farming more sustainably can also be the key to increased crop yield and improved long-term financial performance. Sustainable farming is a "journey" not a "destination"; it is a process of continuous improvement which is enabled by monitoring and measurement of key environment metrics both above and below the ground—these include many indicators of biodiversity; soil, water and air quality; and potentially a life-cycle "true cost" analysis of all production. Digital Ag at scale makes this possible.

Global agriculture is responsible for substantial greenhouse gas emissions (GHG), and thus is correctly identified as a major cause of climate change (see also ▶ Sect. 1.4); what is discussed less frequently is the role played by crops and farming systems to remove CO_2 from the atmosphere and, furthermore, for a proportion of the carbon absorbed by plants as they grow to become sequestered (Carbon Sequestration) in the soil (see ▶ Sect. 4.5). This narrative of agriculture becoming a major part of the solution to climate change, referred to as Regenerative Agriculture, sees farmers seeking to change their farming systems in ways that demonstrate year on year improved sustainability and carbon footprint. The demonstration of carbon capture by this means offers the prospect of an entirely new income stream for farmers as they sell carbon credits in the future. In the USA, Indigo Ag is paying farmers approximately $15/ton for carbon capture. This whole subject area is enabled by Digital Ag which, by monitoring, measurement, and data-driven objective evidence, not only informs change of practice on the farm, but also describes impacts and outcomes.

The suite of digital tools which a farmer may wish to deploy includes:

Real-time and historic analysis, at regional, local, field or sub-field level (potentially down to cm accuracy) of:

− cropping, cultivations, inputs, and outputs (yield)
− soil health, including soil organic matter content and carbon, water percolation rates, biodiversity, and nutrient status
− weather and climate conditions
− above and below ground physical measures capturing crop canopy and rootzone conditions
− presence and prevalence of indicator species and biodiversity

These measures will inform predictive models including:
− crop growth
− crop disease susceptibility
− harvest maturity, quality parameters, and yield

Choices of management practice including:
− crop choice and rotation
− variable rate cultivation, seeding, fertilizing, crop protection and irrigation

Most digital data collection is accomplished remotely and automatically by devices which are capable of deployment below ground in the rootzone, in or above the crop. Devices may be fixed, handheld, machine mounted or, for aerial imaging, mounted on drones, aeroairplanes, or satellites. Satellite imaging is the best example of "scale" where global coverage is available at increasingly high image resolutions—this offers the prospect of very low cost per unit area analysis.

5.1.5 Who Cares?

Good news for Digital Ag at scale is that there is strong demand for it. This comes from the following key constituencies:
− Farmers
− Consumers
− Brands
− NGOs and Government regulators

Farmers want Digital Ag tools because they want to produce more (crops and profit), produce better, and do so sustainably. Digital tools enable them to balance these four goals. Digital tools are not only capable of decision support and decision making, but also critically, they provide the auditable evidence base and justification for all decisions. This will grant access to the premium consumer and brand markets described below.

Consumers want digital tools because, increasingly, they want to see evidence that the food that they are buying for their families is healthy and safe and has been procured via a sustainable supply chain. This goes beyond commonly available "traceability" and seeks to give deeper insight into provenance, ingredients, life cycle analysis and carbon footprint. Consumers are increasingly aware of and concerned about issues of pollution and the environmental impact of the supply chains that they depend on; Digital Ag will enable premium supply chains to demonstrate positive performance over a range of such measures and will inform robust independent consumer assurance standards.

Brands respond to consumer trends, and hence their procurement, production and distribution practices all need to evidence delivery of the consumer demands described above. Leading Brands and retailers will seek to be "ahead of the curve," trying to shape their supply chains and supplier (farmer) relationships to exceed consumer requirements.

NGOs and Government regulators are at the forefront of policy creation in this area. Democratically elected governments are strongly guided by consumer trends and demands. There is growing evidence, with widespread adoption of "Net-zero Carbon" targets being adopted by governments, that Digital Ag will be required to play a pivotal role in national food, ag and environmental policies in the future (see ▶ Sect. 1.6).

5.1.6 Where's the Catch?

The good news is that today global farming systems have unprecedented access to transformational new technology which is capable of dramatic economic, environmental, and social impact (see Chapt. 3). The bad news is that technology companies routinely fail to deliver the benefits that they promise, and farm adoption of new tools and practices is fragmented and slow. Too often, individual technology companies are inhibited by their "siloed" approach to R&D (see ▶ Sect. 2.8). While this high level of focus has benefits, especially in early stage companies and novel technology developments, the downside is frequent failure to embrace complementary technologies which, when deployed together, result in outcomes greater than the constituent parts would imply.

Technology adoption on farm is inhibited by many factors including:
- farmers' limited financial resources, often a consequence of overall farm cash generation, and other more pressing calls on cash than new tech investment
- limited human resources, especially lacking expertise in technology, and over stretched management
- lack of "interoperability" between technologies, and failure to deploy complementary, synergistic new technologies. This results in poor return on both time and financial investment, consequently undermining confidence. Lots of farmers have had bad experiences with underperforming or failed technology.
- nervousness of system change especially when there is only one harvest each year
- financial and reputational commitment to previous years' technology investments

This mixture of inhibitors both on farm and inside technology companies results, too often, in technology deployment, which is sub-scale, cautious, and lacking in "buy-in" from all stakeholders. This threatens to impair or destroy the paradigm shift in environmental sustainability, farm productivity, and consumer assurance that is offered by the Digital Ag revolution. Successful addressing of these issues will result in accelerated and de-risked technology development and deployment. Failure to do so risks a loss of confidence in many areas that offer such value, and hence the cost and frustration associated with years of delay.

5.1.7 Outlook

A concentrated demonstration of Digital Ag can be seen in vertical farms (VF) where the entire growing system is contained in a digitally controlled environment. While there are challenges around financial returns from VF, evidence of transformed productivity and environmental sustainability is impressive. The Digital Ag revolution of the 2020's will see many of the tools used for micro-management in VF, (remote sensing, machine learning, crop specific algorithms and plant by plant agronomy), also deployed on an open field scale. The combination of this micro-management with similarly novel macro-management and insight from satellite derived monitoring of crops, fields and entire ecosystems is of sufficient power and potential to deserve the description revolution.

Where the deployment challenges referred to above (in ▶ Sect. 5.1.6) can be overcome, this revolution will hasten and de-risk R&D and create new value and premiums through enhanced performance and transparent insight. This is a new dimension which offers previously unimaginable scrutiny and control to farmers, consumers, brands, and regulators.

We can be optimistic that Digital Ag can not only respond to the demands of growing population and changing diets, but do so also in a manner that increasingly protects and enhances our planet and each of the ecosystems that farmers interact with. The challenge that remains is the gap between the theoretical arrival of Digital Ag at scale and the reality in practice on the farm. My sense is that there is still a way to go; the deployment challenge should be uppermost in the minds of technology companies.

5.2 The Digital Revolution, a Performance Accelerator from a French Perspective: The Issues and a Panorama of Possibilities for French Cereal Crops

Emmanuelle Gourdain, Géraldine Hirschy and Mehdi Sine

Abstract

This section follows the classical crop management process starting with field observation, analysis and decision, and action. Specific digital product examples combined with scientific publications show how this crop management process can be digitally supported today. The focus is on French cereal farmers, including selected examples from other field crops.

5.2.1 Introduction

Agricultural activity has become a source of big data, through the use of sensors, software and telecommunications networks. By endowing itself with new skills in data science, and through advances in modeling, R&D must transform these data into knowledge to render them useful, usable and being used by farmers in their decision-making processes or directly by agricultural equipment, which is becoming increasingly automated and accurate.

Decision support systems (DSSs) are one of the means being developed to meet some of the needs of producers. These tools are increasingly based on the coupling of modeling and sensor data. Crowdsourcing, the Internet of Things (IoT) and artificial intelligence, including machine learning in particular, are providing new opportunities, which are beginning to be explored. Nevertheless, these tools must increasingly be integrated into the logistics of software platforms, which requires a real effort to ensure interoperability, particularly with agricultural equipment, and, in the future, with robots. The deployment of these technologies is quite contrasted in France and their appropriation by farmers depends heavily on the service provided and the gains made.

In this section, we use the definition of Digital Agriculture initially proposed by [BH16]: an agriculture that uses Information and Communications Technology (ICT), that is, data acquisition (satellites, sensors, connected objects, smartphones, etc.), transfer and storage (3G/4G cover, low-speed terrestrial or satellite networks, clouds) technologies, and embarked or remote processing technologies (supercalculators accessible via communications networks). These technologies can be used at all scales of agricultural production and of the agroecosystem, at the level of the individual farm, in outreach and advisory services, or at larger scales, such as an agricultural region or the upstream–downstream value chain.

Economically, the AgTech sector is still largely dominated by the United States and Asia/India, which account for 90% of the funds raised in this sector over the last five years (155 billion Euros). Europe struggles to reach about 10%, while France, the leading agricultural power in Europe, lags behind at 2%. However, this does not necessarily mean that France is not dynamic, and in 2017, almost 200 start-ups operating in the agricultural domain were listed in France [SGP17]. French initiatives such as La Ferme Digitale, an association founded in 2016, aims to combine forces in the French environment and to ensure that its voice is heard by the French government and the European Commission. The forty or so start-ups belonging to this association have raised 270 million Euros to date, 60% of this sum in 2019. Other recent initiatives have demonstrated the dynamic nature of the French AgTech sector, including the AgDataHub initiative, which raised 3.7 million Euros in 2020 with the aim of providing French and European agriculture with a shared, sovereign, technological infrastructure of consent, data storage and exchange.

Since 2016, Arvalis-Institut du végétal has been accompanying this movement by dedicating three of its experimental farms to the application of the most recent technologies, for their evaluation, and to facilitate the digital transition in agriculture. These field-testing facilities, known as Digifermes®, have since been extended to other sectors of the network of technical institutes, and now comprise 13 experimental farms [BP17]. Digifermes® are characterized by the application of this approach to the entire cycle, while ensuring that the needs of the farmers are always at the heart of the system, particularly as concerns working comfort, the triple performance (economic, environmental, and social) of the system and control over processes. In this cycle, various different technologies and methods are used and must be mastered: sensors for observation and measurement, IoT and associated telecommunications networks, and the cloud for data transmission and recording, data platforms, and modeling for data analysis and processing, Decision Support Systems (DSS) to help the producers to make decisions and, finally, precision agricultural equipment or robots to ensure precise, efficient interventions. Other farm networks are being developed with the same goal, serving as a "living laboratory," a laboratory of open innovation. These networks include the Fermes

Leader (IN VIVO) and Fermes Numériques (Brittany Chamber of Agriculture).

The growth of the digital agriculture is continuing, but it is still rather weakly developed in the French territories. In 2019, a survey of 500 farm managers conducted by the BVA group on behalf of groupama [BVA19] constituting a representative sample of French professional farmers (field crops, livestock, mixed farming and livestock, fruit, vegetables, and flowers), showed that although two-thirds of farmers use at least one digital technology, their equipment rate is still modest. The first is the GPS-equipped tractor (31% of farmers equipped), followed by surveillance cameras (20%). 18% are equipped with connected weather stations and 18% use satellite imagery services. Alarms by GSM, heat detection sensors, connected electric fences, calving sensors, connected flowmeters or water probes, drones, RFID chips, connected traps, and robots are still of relatively marginal use.

In the next sections, we will paint a wide panorama of the technologies used in the framework of this digital agriculture, applied to cereal production in France and illustrated with real examples and elements extracted from scientific publications. We will also try to highlight several prospective elements of the work of Arvalis and its partners to help cereal producers.

5.2.2 Observe

5.2.2.1 Sensors and Phenotyping

The applications of phenotyping sensors have progressed considerably in recent years, particularly through the work of UMT CAPTE at Avignon and the teams of INRAE and Arvalis [SBT+18]. In aerial phenotyping, these sensors, carried by new vectors, such as drones, phenomobiles (autonomous vehicles for high-throughput phenotyping), the ALPHI arch (phenotyping trailer), or gantries mounted on heavy machinery, are used in production in several large R&D programs (including the PHENOME program in particular). High-resolution RGB cameras, LIDAR, infrared thermal cameras, multispectral cameras and radiometers can be used to obtain data for new variables of interest, with very broad possible uses. The possible applications of this technology include the monitoring of leaf development, the counting of plants or organs, biomass evaluation, varietal characterization coupled with genomic studies, the control of irrigation according to evaluations of water stress and of nitrogen applications according to nitrogen status and the chlorophyll content of the plants. These applications have been made possible by new capacities for processing the data generated by the sensors. Artificial vision techniques based on deep learning (see ▶ Sect. 3.5) are currently being tested and are very promising but must be integrated into complete data processing chains and automated for the management of data from cycle to cycle. In addition to aerial or proximal phenotyping, combinations of imaging techniques with chemometrics tools for soils also appear promising for studies of the development of root systems [EFB+18]. Hyperspectral imaging in the near infrared is proving an effective tool for the automatic identification of roots in soil samples, overcoming the need for manual sorting, thereby saving researchers a considerable amount of time.

5.2.2.2 Iot

The first applications of the Internet of things (IoT) date back to the end of the 1990s, but IoT has really taken off in the last decade [SDL17]. The applications of IoT in this domain involve collecting data from connected things or machines, processing them and then acting according to the results of the analysis. IoT is particularly relevant in the agricultural domain, because it provides real-time access, with unprecedented precision and in a simplified form, to information about the state of crops and their growing environment. The application of IoT technology to weather stations, a technology that is already largely mature, has led to a large decrease in costs and a new boom. By providing data every 15 min that can be consulted in real time, connected weather stations make it possible to perform reasoned interventions and to optimize the organization of activities.

Other more innovative technologies are being designed to facilitate the monitoring of crop development. One notable example is the IOF2020 project on wheat (Internet of Food and Farm R&D program, Horizon 2020). A complete processing chain has thus been developed, incorporating a wheat crop model developed by ARVALIS (the CHN model) and observed data obtained from proximal sensors and by remote sensing, within a dynamic correction procedure. The prototype was implemented in 2019 on a collection of agricultural plots, to demonstrate the benefits and limitations of directing nitrogen applications with this approach, and to evaluate its economic, technical, social and environmental performances. Based on the preliminary results obtained, the robustness of the service could be improved by assimilating other observations acquired by sensors in the cropping system model, such as plant phenology and nitrogen absorption data. Indeed, an error in the estimation of nitrogen absorption by the model can lead to erroneous decisions being made [STD+20].

5.2.3 Record

5.2.3.1 Farm Management Information Systems

Farm management information systems have a long history of development, dating back to the mid-1980s in France. They initially focused solely on the financial management of the farm, but have since become a veritable information system, incorporating highly diverse data acquired on farms. A recent literature review [FCS+15] identified the principal functions of these tools and the services they provide to producers. The authors analyzed 141 international software suites, including 10 in France. Over the last few years, these software suites have become veritable systems of information and knowledge management for the producers using them in France. They include mesparcelles from APCA, Geofolia from Isagri and Smag-Farmer from Smag. In 2017, 80% of farm advisers assessed that farmers are using these solutions on a regular basis along the crop season [Lac18]. Increases in the number of functions and connections with tools, including agricultural equipment in particular, with standards such as ISOBUS and mapping, have made it possible to connect DSS to precision materials. This is the case for farm management systems, such as FARMSTAR®, which can be integrated into farm management information systems and allows the downloading of maps for modulating fertilizer applications at within-plot scale. IoT, web services and application programming interfaces (API) are also making it possible to extend the range of function of these tools beyond their standard functionalities. Nevertheless, a lack of interoperability remains one of the principal brakes on the adoption of these tools, due to the need for multiple data entries between tools (see ▶ Sect. 3.4).

5.2.3.2 Interoperability and Data Platforms

Conscious of the lack of communication between data acquired on the farm, by equipment or in the various software suites used, Arvalis has worked on a project management dashboard project [Lau17]. The objective was to display, on a single interface all the data, aggregated and combined together so as to provide a panoramic vision of the information available at the level of a single farm. This preliminary work demonstrated the need for standardized interfaces for the various tools and software suites used by producers, facilitating the development of applications for exchanging data. These interfaces and API are beginning to emerge in the domain of agriculture through the API AGRO platform [SHE15]. This platform aims to bring together and make available databases or API for all the animal and plant production sectors. The company recently evolved to become AgDataHub, with the objective of developing a sovereign, shared technological infrastructure of consent, storage and exchanges of agricultural data. It has become clear that access to agricultural data, and the valorization of these data, are major issues. A white paper produced by the agricultural technical institutes illustrated this issue with several concrete examples based on agricultural data, and formulated proposals to facilitate this access [BSG+16]. Furthermore, most agricultural data generally lie outside the legal framework relating to data, and instead enter the field of contractual relationships. Greater attention must therefore be paid to the possibility of situations in which there is an imbalance in the power relationships in contracts [Tom19]. For this reason, several collective initiatives for protecting exchanges of agricultural data have emerged in recent years. These initiatives include the EU Code of Conduct on Agricultural Data Sharing by Contractual Agreement put forward by Copa-Cogeca [EU18] and the DATA-AGRI charter [Cha01]. This charter arose from an initiative of the French agricultural unions (la Fédération Nationale des Syndicats d'Exploitants Agricoles—FNSEA, and les Jeunes Agriculteurs, JA) and aims to lay down guidelines concerning the ownership, sharing and use of agricultural data. In accordance with a proposal relating to transparency, the MULTIPASS project [Lau18] aims to establish an interoperable environment for the management of consent and to demonstrate to the agricultural market the potential utility and feasibility of such a solution, through several real and specific cases of use. Such a consent management environment would make it possible to protect exchanges of data and to provide farmers with answers concerning the control of information and the use made of it. The project allowed to develop a router to manage consents, the router being integrated for production by AgDataHub.

5.2.3.3 Blockchain Services

Although producers need to keep control over their data, sharing them, particularly for the purposes of food product traceability, remains a necessity [IP15]. The creation of links with consumers, ensuring the traceability or certification of production specifications, and the contractualization of exchanges from the producer to the consumer, without the generation of addition costs, are among the major concerns of farmers. Blockchain technologies, particularly when associated with IoT, can help to provide greater transparency and efficiency in agriculture, from the management of the data from farms to their supply chain. For example, the French start-up Connecting Food offers a solution based on the use of a blockchain to trace and certify food products, in real time and in a non-falsifiable fashion. The producers and other actors of the agrofood industry can, thus, create added value for consumers by displaying, with total transparency, the differentiation of their products. Arvalis, together with other partners, including the writers of agricultural software and the telecommunications operator Orange, is also testing this technology in the framework of the MULTIPASS project [Lau18], and is applying it to the collection of consent from farmers for the generation of data flows between applications without a trusted third party.

5.2.4 Analyze and Decide

5.2.4.1 Modeling and Data Science

Two major approaches can be distinguished for the modeling of processes: mechanistic approaches and statistical approaches. Classically, when modeling biological processes, a data acquisition step is required, which has, until now, been based on experimentation, often over several years. It is now possible to access new sources of data acquired for reasons generally far removed from the questions raised during modeling. This paradigm shift has led to the creation of a new discipline, data science. Data science aims to produce useful information through the sorting and automatic analysis of big data, principally digital in nature, emanating from data sources of various complexities, connected to various extents [Cle01]. Within Arvalis, a study implementing several of these methods was performed on three cases of use for the prediction of epidemiological phenomena on cereals [GPC+18]. This study demonstrated that the modeling of biological phenomena by the application of data mining techniques to data sets originating from a collaborative network—the Vigicultures® portal for the biological surveillance of agricultural areas, in this case—can generate interesting results for predictive modeling, but less useful results for explanatory modeling. The statistical modeling of biological phenomena comes up against one major difficulty, that of climatic variables frequently being correlated and mostly acting through interactions. It therefore appears that the choice of variables requires more careful consideration upstream, to decrease the number of variables considered, restricting the choice to the most pertinent. Finally, these methods have been little used to date by agronomists, who prefer more mechanistic models, the output of which they can easily understand and interpret. This is the case for the CHN cropping system model [SBL+18], and for other cropping system models, such as STICS and CIRIUS, which can be used to break down the functioning of the plant into functional compartments—the soil, the plant and the atmosphere—and then to model all the processes within a given compartment, together with the dynamic flows of carbon (C in the model acronym), water (H) and nitrogen (N) between compartments. The implementation of the CHN model, validated for wheat, durum wheat and maize, paves the way for new DSS for managing irrigation and nitrogen fertilizer applications, particularly since the advent of sensors that can use data assimilation methods to adjust the model in real time and propose assistance at the finest possible level for a plot (Digipilote, SmartAgriHubs project). New approaches coupling statistical and mechanistic models and data from experts are also used in the framework of wheat yield prediction and have given highly satisfactory results.

5.2.4.2 Decision Support Systems (DSS)

When managing a farm, the producer must take many decisions to meet long-term objectives (strategic decisions) and concerning short-term technical actions during the cropping season (tactical decisions). Arvalis has, for many years, been involved in the development of a large number of models and DSS on several subjects and species: crop protection, fertilizer applications, irrigation, quality, storage, and environment. Some were designed with technological partners, such as Airbus Defense and Space (FARMSTAR®) and Météo France (TAMÉO®), others were designed exclusively by Arvalis (Prévi-LIS®, Irré-LIS®, etc.), and some have been commercialized by other entities but include Arvalis models (xarvio® from BASF, Optiprotech® from l'APCA, etc.).

FARMSTAR® provides guidance concerning fertilizer application at the within-plot level, based on a combination of satellite data and agronomic models from Arvalis. It is currently used on 700,000 ha (wheat, barley, and rapeseed) by around 18,000 farmers. Thanks to remote sensing technologies (satellites, planes, and drones), leaf area index and the chlorophyll content of plants can be estimated by modeling. These biophysical data serve as the input data for the fertilizer guidance model which suggests the dose of nitrogen to be applied, particularly for the final application [SCN18]. The authors estimate that the actual model could be improved by a more dynamic approach, in which interactivity with farmers is increased by integrating the amounts of nitrogen already applied during the growing season. This should soon be possible with the CHN model and further improvements may be possible with other sensors in the field automatically providing data concerning plant cover. This example provides an illustration of how the coupling of a model with sensors, which has already been successfully explored in the past, remains the clear way forward for further increasing the precision of DSS.

DSS are just one of a number of elements available to support producers in their decision-making processes. Decision making in a complex environment requires methods shedding light on the nature and role of knowledge, past experience, the processes of perception and inference, and the decision-maker's appreciation of the situation. With the advent of new technologies for decision-making by farmers, the relationship between farmers and their advisors is undergoing major changes. Advisors have to acquire new skills (including a certain mastery of technology), and are increasingly becoming facilitators for farmers, and vectors of learning and comparison. Their technical dimension (agronomic knowledge and expertise) is extending into a technological dimension (modeling and computing), but also into social and behavioral aspects.

5.2.4.3 Artificial Intelligence

According to Research and Markets, the AI market in agriculture was estimated at nearly 519 million in 2017 and should grow by more than 22.5% to reach 2.6 billion by 2025. Still little developed in the agricultural field, the prospects are nevertheless interesting, particularly through recognition (plants, diseases, etc.) or predictive models. Arvalis has recently implemented this type of approach for image analysis [SBT+18] and the recognition of plant organs. The application of the convolutional neuronal network (CNN) deep learning method to wheat ear recognition on images gave results consistent with estimates of ear density obtained by eye and with a detection algorithm, with an error rate of 21 ears/m^2. This work will pave the way for a large field of application, once the methods have been stabilized and the processing tools have been industrialized and placed on a cloud system that can be interrogated directly by new measurement tools, such as drones, smartphones, portable imagers, and imagers carried by AgMachinery.

5.2.5 Act

5.2.5.1 Precision Agricultural Equipment

Agricultural equipment is increasingly automated in arable farms growing field crops, particularly for the guidance of machinery and for spraying [DMC17]. GPS-assisted guidance emerged in the first decade of this century. It optimizes the passages of the tractor within plots, limiting missed areas and preventing the repeated treatment of areas during a given application. The combination of GPS and RTK can have a precision of the order of a centimeter and can be used to prevent the same area being treated twice during tillage, spreading and sowing operations (retreatment rates may reach 2 to 12%, depending on the type of intervention). In recent years, there have been a growing number of innovations in the domain of spraying (spray heads) and, particularly, in precision weeding. The association of agricultural equipment with sensors makes it possible to localize the weeds, either during the intervention itself or during a second passage. With the advent of image analysis and advances in sensor technology, other companies are now offering prototypes capable of recognizing weeds, to improve the targeting of interventions. For mechanical weeding operations, the cultivator may be guided with the aid of a camera, or by autoguidance of the tractor. This second type of guidance is used particularly by seed producers and in organic agriculture. Tests performed by Arvalis have shown that the mechanical weeding of cereal crops at an interval of 15 cm is possible, provided that the RTK autoguidance is perfectly parameterized and the tools are well-centered. Agricultural equipment and the digital services associated with it are important levers for the agroecological transition, and the two approaches should certainly not be seen as in opposition [BH17]. The use of these technologies should increase efficiency (better yields, with a smaller impact on the environment), make it possible to perform more precise interventions (the right dose at the right moment), and to improve the integration of environmental factors (closing flows, use and preservation of biodiversity).

5.2.5.2 Robotics

Agricultural robotics took its first steps almost 30 years ago in the domain of animal husbandry, with the development of milking machines, but is now enjoying a new boom. The limitation of input use (phytosanitary products, fertilizer, seeds, water, etc.), while maintaining high levels of production, is one important issue. Labor-saving, given the lack of attractiveness of agricultural careers and the need to maintain competitiveness, is another. The progress made in robotics, in terms of autonomy, image recognition, visual perception of the environment, and geolocalization, is opening up new perspectives, particularly in the domain of chemical and mechanical weeding. The first robots for mechanical weeding were designed for market gardening, which has a large added value. The Oz robot produced by Naïo can hoe between rows of vegetables but, given its width and working speed, the work rate of this robot is low (it can cover just under 0.1 ha/h for a crop sown with an 80 cm in-terrow). Tests performed by Arvalis on Digiferme® for maize, showed that several passages were required to decrease weed density [DMC17]. The robot produced by Suisse Ecorobotix was tested on Digiferme® on beet. With its artificial vision system and its robotic arm, it can spray weeds directly. This weeding robot is powered by a solar panel and is, therefore, autonomous in terms of energy, which is a major asset, but limits its workrate. In France, with a view to dynamizing its market and participating in the ecological transition in agriculture, the agricultural equipment industry launched the Robagri association in October 2017. This large consortium, associating public and private actors with actors from other sectors, has the objective of designing the robots of tomorrow [Ber18].

5.2.6 Impact of Digital Agriculture on the Multi-Performance of Farms

As we have seen, sensors and IoT, precision equipment, software, DSS and robotics constitute a set of resources that can be mobilized to attain the objectives of sustainable agricultural production. Nevertheless, the real impact of digital solutions on the agro-ecological multi-performance of farms is too often only partially measured, thus slowing down their deployment. The impact of digital agriculture should be analyzed in terms of its consequences for the performance of farms: economic and productive performances and respect for the environment in its various dimensions, together with social acceptability, for the farmers themselves and for society [BH16]. The pertinence of these digital solutions is evaluated with field systems deployed notably within Digifermes®, but also at the regional scale and involving farmers, through the SYPPRE (Systèmes de Production Performants et Respectueux de l'Environnement—[TCP+15]) network, for example. In these systems, multiple criteria are evaluated to obtain the results, and the use of software is indispensable, to describe the cropping systems and to calculate performance indicators by the SYSTERRE® tool. It can be used for the description and multi-criteria evaluation of the cropping systems of farms aiming to achieve a high triple performance in very different production contexts. Using the information provided on equipment, labor, crops, and plot layout, the farmer can plan interventions on each plot, estimate prices, and make measurements and observations to use in decision making and soil analysis as needed. It is then possible to calculate the indicators of technical, economic and cropping practice sustainability. The indicators calculated by SYSTERRE® can be used at the scale of the plot or the farm, over the course of a growing season or a rotation [CCG+12]. Mechanisms of interoperability have been developed, to simplify the work of the users, by making it possible for this tool to communicate directly with other plot management software. As pointed out in the introduction, the impact of digital agriculture on the triple performance of farms is not limited to measurements of the effect of introducing a particular technology on yield or on the economic success of farms. This impact is much deeper and is likely to revolutionize the way in which agricultural products are produced and to activate multiple levers in highly diverse contexts subject to increasingly unstable hazards that are particularly difficult to predict.

5.2.7 Conclusion

The digital transition in agriculture is a major question for society. It encompasses much more than a simple question of new production tools. Distribution, sales, alimentation and consumption, the entire chain of action and the actors involved, from field to plate, require repositioning in the face of new collaborative practices, the generation of massive amounts of agricultural data and the demands of consumers, who already use digital devices as tools for information and mobilization.

As a means of control, the technical institutes, such as Arvalis for arable crops, are placing themselves at the junction of all this knowledge, thereby enabling farmers to keep control over their link to this exponentially growing universe of data. The intrusion of tools designed to acquire a greater knowledge of our behavior can generate suspicion, sometimes even outright rejection. What are the barriers to the adoption of digital technologies for farmers and their technical advisors? The Observatory on the Uses of Digital Agriculture conducted several surveys between 2017 and 2020 on the use of digital technologies. These studies make it possible to identify the main obstacles to adoption and the expectations of farmers and technicians in various sectors. In arable crops (2018 study), the first obstacle expressed by the technicians questioned is the lack of time (25%), followed by the cost of equipment (18%), the fact that it is not sufficiently adapted (16%), the lack of training (14%) and the lack of visibility on the offer (6%). For farmers, the first brake would be the cost (23%), followed by a lack of visibility on supply (21%), a lack of training (20%) and a low interoperability between tools (12%). Whether it is an opportunity or a threat, technicians nevertheless see digital as a profound change in their profession and believe they have an important role to play in the democratization of digital agriculture on farms. However, they do not feel sufficiently trained (68%). Training and support are therefore their first expectations. In his article, [Maz17] exposes the current obstacles to the deployment of digital innovations among French farmers, mentioning the issue of their reliability, poor interoperability between tools, distance from the field and the fear of losing decision-making autonomy. The question of the acceptability of DSSs by the agricultural profession, but also by society, is all the more important as it is sometimes difficult to obtain information on the technical and scientific knowledge they incorporate, as well as the philosophy on which their design is based [MBT09]. They are therefore often referred to as "black box" tools. This opacity is also increased with the advent of predictive models based on artificial intelligence methods, and in particular machine learning approaches. Artificial intelligence methods are indeed an important aid in analyzing a large amount of data that would be impossible for the human mind to grasp as such [LM20]. They are very useful for making predictions. However, they show only correlations between variables and not causal relationships [ZH19], which could be a problem for answering research questions and acceptance by farmers.

There is still a long way to go, and many hurdles need to be cleared before the advent of digital agriculture on field crops in France. These locks can only be lifted by a clear and shared policy among stakeholders on data management, as well as digital solutions whose economic, environmental, and societal gains are clearly identified and quantified. The main challenge is therefore to create the efficient ecosystem between the actors of the agricultural world and AgTech in order to take up this challenge and make digital agriculture a reality, i.e., a sustainable and resilient agricultural model in the face of the many challenges of tomorrow's agriculture.

5.3 Digital Transformation of Vegetable Production

Martin Geyer and Norbert Laun

Abstract

Compared to agriculture, the global range of different types of vegetables is extremely variable. Root, hypocotyl, sprout, leaf and fruit vegetables require different crop conditions, harvesting methods and post-harvest technologies. In addition to outdoor production, many vegetable crops (seedlings, herbs, fruit vegetables, salads) are produced seasonally in unheated polytunnels or all year round in heated greenhouses. Depending on the technology used, the level of automation is correspondingly high.

A distinction must be made between fully mechanized crops with single harvest, such as washing carrots, tomatoes for industry or spinach, and crops with selective hand harvest, such as tomato, pepper, white asparagus, cauliflower for the fresh market or pickles. The labor and the cost of harvesting add up to 50% or more of the total production costs. In addition, it is becoming increasingly difficult to find suitable seasonal workers. Therefore, great efforts are being made to automate such work.

Vegetable crops like tomato, bell-pepper or cucumber are grown protected in unheated plastic film tunnels in the south and in heated greenhouses in northern areas. Climate control in greenhouses is a highly complex process which is mainly carried out with the help of sensors, actuators and control computers. Temperature, relative humidity, CO_2 concentration, lighting, water, and fertilizer must be kept in an optimal range to optimize growth and yield.

For most vegetable crops, a high degree of automation in cultivation, fertilization, crop protection, and irrigation is a prerequisite for successful crop management.

5.3.1 Seedling Cultivation and Planting

Crops like salad and many cabbages are transplanted mechanically. The seedlings are produced in greenhouses and the cultivation time is reduced.

With the help of fully automatic sowing lines, peat pots are pressed or trays filled with substrate and the seeds are deposited with pinpoint accuracy. After watering, the propagation boxes are placed in germination rooms at optimal temperatures for several days before they are cultivated in the greenhouse for several weeks until the planting date. Picking robots and robots for re-planting of missing parts are used to achieve completely filled propagation boxes [Vis20a].

During the cultivation of vegetable plants, barcode- or QR-code-supported labeling is usually used for the respective batches, which enables documentation of the production process from sowing to delivery. For outdoor planting, RTK-supported guidance systems (see ▶ Sect. 3.2) are being used with a rapidly increasing tendency, in order to ensure precise arrangement of beds and rows and to increase the efficiency of planting (◘ Fig. 5.5).

This is mainly done semi-mechanically, i.e., removal and insertion into a transfer mechanism is done manually by workers but the actual planting is done automatically. Fully automatic machines are also available, which are technically complicated, therefore relatively expensive and still susceptible to faults.

Important automated techniques supporting fertilization, plant protection, and weed control during the cultivation of vegetable crops are partly established, partly in development and will be further described in the following.

◘ Fig. 5.5 12-m wide transplanter for soil press pots (Photo: Geyer)

5.3.2 Fertilization

In addition to complex calculation procedures that require considerable computing support, site-specific fertilization methods are currently being developed and tested for determining fertilization requirements. These are often based on sensor technology developed in arable farming, but need to be refined and adapted for the wide variety of crops and varieties in vegetable growing. The high demands on the quality and marketability of the individual plant require an adapted and small-scale controllable fertilization technology. Options for this are bedding spreaders with weighing technology, already available on the market [Rau20], but also pneumatic spreaders. In conjunction with appropriate sensors, it is expected that site-specific fertilization of vegetable crops with reduced fertilizer application (mainly nitrogen) can achieve the economically necessary high harvest rates. Furthermore, these techniques using appropriate GIS support can exclude free sub-areas from fertilization, e.g., the tramlines in bedding systems or transport and irrigation lanes, and thus offer potential for further savings in fertilization of open field vegetables. During production in the greenhouse in soilless cultivation systems, the plants are continuously supplied with all necessary nutrients via a nutrient solution. The concentration is automatically adjusted via the measured conductivity of the nutrient solution. The control is carried out together with the regulation of the temperature (ventilation, heating), the CO_2 supply and the irrigation with the help of appropriate climate computers (e.g., [Ram20]).

5.3.3 Plant Protection and Irrigation

Plant protection in the open field is carried out with systems described in ▶ Sect. 4.7. In the greenhouse, spraying robots are also used, which drive independently on the heating pipes laid as guide rails between the crop rows (e.g., [Ste20]). Due to the high level of technology, further procedures can also be implemented in the sense of an integrated plant protection concept. For automated greenhouse management, for example, self-propelled devices are available that maneuver between the rows and are equipped with various spectral cameras. For any conspicuous feature detected, a yield check is carried out on tomatoes, for example, and, if necessary, UV light treatment is used to combat fungal diseases [Eco20].

A high standard of daily updated information is necessary for the selection of suitable plant protection measures. The reasons are, in addition to the great diversity of vegetable species, the highly complex approval situation, the wide range of different requirements and their sometimes rapid and extensive change. Database systems for the approval situation are available for this purpose. A digital compilation with further relevant information (e.g., maximum residue limits) is currently being processed [Hor20] as well as the availability of the requirements on the tractor for controlling the sprayer. Prognosis models for plant protection are partly offered by manufacturers of weather stations. The models developed in cooperation of the national plant protection services are available almost nationwide on the website of ISIP [Isi20]. The available models (e.g., for the evaluation of the infection risk in case of downy mildew on onions or for the development of different vegetable flies) focus on highly important and economically significant pests in the great variety of crops. Accordingly, significant gaps remain here.

For the future, further developments in an improved recording and modeling of the crop-specific microclimate can be expected through increasing computer performance and faster data availability.

The same applies to irrigation control, where currently precipitation resolutions of

1000×1000 m can already be made available by interpolating precipitation values (e.g., [Wet20]). These data can be used for irrigation control in field crops via climatic water balances [PKM09]. Alternatively, point measurements via soil moisture sensors can be used.

5.3.4 Climate

The temperature control possibilities in the greenhouse area are very diverse due to heating, ventilation, and shading possibilities. They allow a highly intensive and productive plant production, using a lot of different computer-aided control possibilities.

In the open field, the possibilities of influencing the climate are less, but still lead to astonishing extensions of the harvesting season, mainly due to considerable premature effects through the use of film and fleece. In addition, digitally available climate data has led to a considerable optimization of produce quality and harvesting processes. For example, the harvest of asparagus begins more than 3 weeks earlier by using film and tunnel systems than with unprotected crops. In addition, different film covering systems to control the temperature guarantee a staggered harvest. The temperature management of the crops (film turning from black to white side, tunnel covering) can be done either by own app assisted measurements (e.g., [Dee20]) or by online available data from representative sites [Wet20]. A modeling of temperature data, which allows a site-specific prognosis with own covering systems, is currently being validated and should be available in the foreseeable future.

5.3.5 Weed Control

The restriction of the use of herbicides for single crops or the ecological production requires alternatives to chemical weed control. At present, row-guided systems are increasingly used. Their effectiveness is based on a reduced distance between the tractor-drawn hoe and the row of plants by GPS or camera support, larger working widths and driving speeds. Working between single plants in a row can be supported by camera or sensor guidance (e.g., [Dul20], [Gar20]). Alternatively, all individual plants could be located via GPS as a basis for hoeing all free areas. Actually, there are already different, partly autonomous hoeing robots on the market, which are successfully used, for example, in planted crops such as lettuce or cabbage (e.g., [Kre20]). This hoeing machinery is able to improve the effectiveness and operational reliability enormously. The future challenge will be autonomous hoeing robots that can distinguish and eliminate weeds from the crop at the seedling stage, for example in carrots [e.g., Nai20]. At present, however, the legal situation allows self-propelled machines only in fenced areas, which still limits their use.

5.3.6 Harvest

A distinction must be made between fully mechanized crops with single harvest, such as washing carrots, tomatoes for industry, or spinach, and crops with selective hand harvest, such as tomato, pepper, bleached asparagus, cauliflower for the fresh market or pickles. The labor and the cost of harvesting add up to 50% or more of the total production costs. In addition, it is becoming increasingly difficult to find suitable seasonal workers. Therefore, great efforts are being made to automate such work under protected conditions as well as in the field. The following difficulties have to be considered:

— Compared to industrial products, vegetables are very variable in shape, size and color and therefore have to be handled specifically and individually

- Most horticultural products have high mechanical sensitivity to pressure and impact loads
- Arrangement of the products in space requires new and highly accurate detection, gripping, separating, depositing, transport mechanisms and logistics
- Plants grow in the field or in the greenhouse; the technical equipment must move toward the plants. The harvesting unit therefore requires chassis, drive and control
- For agronomic and economic reasons, the cultivation systems can only be adapted to technical processes to a limited extent
- Strong temperature dependency of the growth or ripening development of vegetables requires a high efficiency of the harvesting systems
- From an agronomic point of view, a high harvesting rate close to 100% must be aimed for
- Equipment must be suitable for use in all weather conditions
- Short harvesting periods lead to low time utilization
- Only small numbers of units are required resp. built
- Procedures must be nevertheless inexpensive and easy to use
- Security issues must be clarified

The mechanization of harvesting of greenhouse cucumbers and peppers has already been intensively researched, but the performance and detection rate are not yet satisfactory [SWH+18]. The sweet yellow pepper harvester Sweeper [Swe20] moves autonomously between the rows, recognizes the fruits, a robot arm grips and separates the fruits thermally and transfers them to a conveyor [HHV+02], [BBB+20].

Intensive studies are underway for the selective harvesting of iceberg lettuce [BHC+19], broccoli [BBB16], pickled cucumbers, and asparagus open land.

The major technical challenges in harvesting pickled cucumbers, for example, are the visual recognition of the fruits lying on the ground between leaves, stems and mulch film [FMS+18] and the subsequent gripping and separation (Fig. 5.6).

Fig. 5.6 Selective pickling cucumber harvest prototype "CATCH" [FRA18] (Photo: ATB)

Fig. 5.7 Single-row harvesting aid with electric drive (Photo: Geyer)

Compared to the human hand, the number of degrees of freedom of robot arms is very limited, which makes it difficult to grip quickly and safely in the correct orientation [Fra18].

Another labor-intensive crop is white asparagus. It is grown on ridges under opaque film to prevent the white asparagus tips from discoloring. Electrically driven single-row semi-automatic harvesting aids are increasingly used (e.g., [Eng20]) (◘ Fig. 5.7).

The asparagus cutter is thus only responsible for digging up, cutting and depositing the asparagus spears; all ancillary work, picking up the film and putting it down again and transporting the asparagus spears, is done by the machine [Gey18]. The next stage of development is fully automatic selective harvesting. The manufacturer Cerescon uses electrical conductivity to detect asparagus spears in the ridge. Electricity is introduced into the side of the ridge and fingers on the surface of the ridge locate the asparagus spears before they reach the surface [Cer20]. A special pricking and cutting mechanism sinks into the ground, separates the asparagus spear, lifts it up and puts it down on its side. Afterward, the soil is smoothed and the film is put down again. The extent to which this procedure successful depends on the machinery costs, performance, and maintenance, because sandy soil is extremely abrasive, and the losses that will be incurred if "blindly" pricking into the soil will injure other spears.

5.3.7 Processing

In comparison with agriculture, processing into a marketable product usually takes place within horticultural companies, which entails an immense need for technical equipment and logistics for cleaning, sorting, packaging, and logistics.

Using the examples of carrots and white asparagus, the technical status of preparation to the ready-for-sale product will be described.

After harvesting, the boxes with the asparagus spears are transported to the farm as quickly as possible. Asparagus cutters apply adhesive labels with bar or QR code to the boxes, giving details of the person, the field and the row, so that the cutters can subsequently be remunerated according to the quantity harvested and the quality of the cut asparagus. Reaching the farm, the boxes are scanned and weighed, pre-washed, and cooled in cold water to prevent the asparagus tips from turning red [BGZ08]. The chilled asparagus is cut to length and washed with the aid of computerized washing and sorting machines, and sorted using optical methods according to up to 20 sorting criteria, for example length, diameter, color, curvature, and hollowness (e.g., [Neu20], [Hmf20]). For this purpose, the spears are turned under the camera and photographed up to 20 times. Up to 40,000 spears per hour can be processed with such machines with approx. 8—12 workers. Placing the spears from the boxes onto the conveyor belt is done by hand and limits the output. The computer stores all the data, so that it is possible to relate the quality and quantity of harvested asparagus to the person cutting it, the field and the row, and to weigh up, for example, whether a field should be taken out of production early or whether the covering film should be turned from white to black. Subsequently, the asparagus spears are put back into clean plastic crates, stored in the cold store until distribution or until bundling and packaging.

Washed carrots are either processed directly after harvesting or stored refrigerated in big bins for several months in autumn. After washing in drum washing machines and polishing/peeling with rotating brushes (e.g., [Wym20]), they are sorted according to length, diameter, and defects. Increasingly, state-of-the-art computer-aided sorting, weighing, and packaging machines are being used in processing (◘ Fig. 5.8).

In the computer-aided optical sorting for defects, the carrots are individually guided past cameras and photographed from all sides according to the asparagus spears and deflected in free fall by air

◘ Fig. 5.8 Preparation of washing carrots with packing machine and multihead weigher in the background (Photo: Geyer)

pressure into various chambers. Depending on carrot diameter, the performance is between 2.5 and 8 t/h (e.g., [Vis20b]). In this way, four to six workers can be saved. For carrots, but also for all other vegetables, which are offered in uniform weight units, weighing devices are standard, which compile the optimum target weight from several load cells under process control (e.g., [New20]).

5.3.8 Outlook

The biggest problem in vegetable production for the future is the lack of available labor suitable for the heavy work. This means that labor-intensive crops will either have to be mechanized or they will migrate to low-wage countries. The most time-consuming tasks are mechanical weed control and selective harvesting. There is also a lack of decision support systems and forecasting models for the large number of different vegetable crops, which provide vegetable growers with short-term assistance in solving problems quickly via apps or other media. A big challenge for the development of automated technology is the high diversity of crops and growing conditions. It is expected that the degree of automation is related to the possible savings esp. for manual labor and the size of the market. For vegetable production we see a development of small, but worldwide markets for specialized technology.

5.4 Digital Transformation of Fruit Production

Manuela Zude-Sasse

Abstract
Digital Farming is requested in fruit production to achieve the fruit bearing capacity of the plants, while avoiding to waste resources.

This section describes the potential and the challenges of combined application of sensors, ecophysiological models, and actuators along the fruit supply chain.

5.4.1 Challenges in the Supply Chain of Fresh Fruit

Sustainable fruit production has been approached based on ecophysiological knowledge and detailed experience in cultivation practice. However, a huge source of errors is still the individual properties and responses of each plant. The resulting ranges of plant variables lead to individual requests considering the input resources for a certain amount of yield. Consequently, the plant variables need to be known for avoiding management errors that lead to waste of resources.

Digital twins of orchards or individual fruit trees are reflecting the plant status in digital format. They are constructed by means of in-situ plant sensor data, which have been introduced over the last decade [ZFG+16]. From a fruit grower's perspective "farming with sensors is much easier," since knowledge on the crop in real time assists precise management decisions. To meet this goal, sensors should collect data automated along the supply chain in the production processes and post-harvest. Crucial processes that can benefit from the precision horticulture approach are
- irrigation
- crop load management
- fruit quality keeping in post-harvest

5.4.2 Irrigation

Irrigation is a relatively expensive and resource-intensive production measure, but it must be used for fruit crops to enable or improve yield, yield security, and product quality. For example, in Brandenburg, 85%

of the orchard crops are irrigable and similar percentages are found in other regions of temperate and subtropical climate. Water management practice has developed continuously in recent years and, in addition to the water supply during dry periods, irrigation fulfils more and more additional functions such as fertigation. The controlled supply of irrigation water and nutrients can make a significant contribution to environmental protection, as it optimizes mineral conversion in the soil and reduces nutrient leaching. Furthermore, water is demanded for risk minimization as a frost protection measure and for crop climate control during longer heat periods. The latter becoming more crucial due to global warming.

Several soil variables, most pronounceable its water holding capacity, affect the water supply to the root system. Due to its Pleistocene and post-Pleistocene origin, many soils in the temperate climate exhibit a high degree of variability in even one orchard. Consequently, individual fruit trees face varying soil and water supply conditions. Existing soil maps are too coarsely resolved to characterise orchards and to evaluate them with regard to the necessary irrigation management.

Correspondingly, the sensors used in irrigation practice up to now record weather data and spatial variability of soil, which is an important step to more precise irrigation. However, it means that the water supply to the plant can only be determined indirectly—by the soil water supply. Actually, with modern fertigation systems and the tree's ability to adapt its phenotype to the varying conditions, the influence of the soil is decreasing. Adaptation of individual plants to the growth factors capture the rooting depth, regulation of osmotic potential of the roots, and daily course of stomatal conductance to avoid water loss while still enabling photosynthesis. Recent approaches in the context of precision horticulture addressed the soil and root interaction based on spatial information of the soil's water holding capacity and root responses considering rooting depth and root's osmotic water potential [TGZ20]. Deeper roots were described in sandy soil showing low water holding capacity and penetration resistance. Such plant-individual adaptation enables the water uptake from an increased volume of soil.

Consequently, the most precise information can be gained directly from the plant. The effective, varying water requirements during plant and, particularly, fruit development was hardly considered in practice so far. Main approaches have been undertaken by means of defining crop coefficients, gas exchange measurements, and dendrometers, which enabled the development of comprehensive physiological models [MS19], [BMB+19]. The challenge so far is the lack of automated plant sensors, which would support their application in the field with the necessary temporal resolution. Consequently, a reason for the lack of more precise irrigation in orchards was the limited availability of plant sensors and its implementation. However, emerging methods in remote sensing are closing this gap at present [Kin17], [HRL+20]).

Currently, the integration of plant sensors is undertaken to determine the effective water demand of the different fruit species, cultivar/rootstock combinations and individual trees, which has a high economic and ecological potential. A plant variable with high importance for calculating the water needs of a plant is its leaf area and temperature. Both plant variables can be estimated by means of remote and proximal sensing methods (◘ Table 5.1). Such empirical data can serve as input in existing irrigation models. Most recently, the calculated water demand has been provided to the farmer by means of mobile apps [Kin17]. The automated control of irrigation based on plant sensor data used in ecophysiological models is, however, forecasted for the near future.

5.4.3 Crop Load Management

Crop Load Management (CLM) aims to adjust the optimum fruit number per tree, since in the majority of orchards the flower set exceeds the capacity of the trees to bear the developing fruit. The production of apples needs to find a compromise between high crop load and the tree's ability to provide carbon for fruit growth and maintenance.

"The higher the crop load, the smaller the fruit" is a general rule. The fruit size directly affects the economic success of farmers. Reaching minimum fruit size is necessary for marketing according to trade standards, although recently these standards allow also smaller fruit size for specific markets, e.g., "kids fruit." In addition, fruit size is the main quality parameter, e.g., for stone fruit, for the buying decision by consumers and results in high market prices. Yielding fewer, but big fruit may also lead to disadvantages, since the yield given in kg per tree or ton per ha is decreasing with the reduced number of fruits per tree. Additionally, the big fruit may have a reduced storability. Without appropriate CLM, indeed the fruit sizes achieved in practice are frequently not sufficient to be marketed in accordance with the trade standard or too big with a risk for physiological disorders in storage and reduced yield. Consequently, in each orchard, the aim is always to find the optimum in terms of the quality and yield. The optimum fruit number per tree depends on the economically desirable fruit size at point of sale, the daily fruit carbon demand to achieve this fruit size compromised by the growth capacity of each individual tree.

Fruit reduction can be achieved through thinning measures. Early thinning of flowers or fruitlets improves fruit quality, since no carbon budget is lost for the subsequently removed fruit. Furthermore, flower thinning can also reduce alternate bearing in susceptible crops. In practice, thinning is carried out by means of pruning to reduce flower buds per branch, mechanical or chemical thinning of flowers or fruitlets, and often corrective hand thinning after fruit drop. Generally, all these treatments can be applied according to the concept of variable rate application.

Why should we apply variable rate CLM in orchards? The growth capacity of fruit trees is represented by the tree's leaf area. Therefore, the leaf area, determining the carbon supply, limits the generative

Table 5.1 Examples of remote and proximal sensing of leaf area and temperature differences in fruit trees for supporting the irrigation management

Crop	Plant variable	Sensor	Reference
Apple	Leaf area	LiDAR laser scanner	[TPF+19]
Citrus	Leaf reflectance	Spectrophotometry	[DVS11]
Plum	Leaf temperature	Thermal Imaging	[KBG+17]
Vineyard	Temperature, reflectance	Thermal imaging, photogrammetry	[MAC06]
Vineyard	Leaf area	LiDAR	[FPO14]
Vineyard	Leaf temperature	Thermal camera	[RBN+19]
Vineyard	Leaf area	LiDAR	[SEA+18]
Mango	Leaf area	video	[WPJ+18]

and vegetative growths directly. Furthermore, the partitioning of carbon to the fruit or vegetative organs results in variable allometric factors for the fruit, leaves, stem, branches, and roots. Nevertheless, the carbon supply depends mainly on the leaf area, global radiation (better: light interception), and temperature. These climate variables show similar values in the entire orchard, while the leaf area of individual trees may be highly variable within one orchard. Consequently, the fruit bearing capacity of individual trees varies accordingly within the orchard. Field uniform CLM, therefore, always results in crop load above or below the fruit bearing capacity of individual trees in the orchard. Variable Rate Application of the thinning intensity could avoid such errors. Each tree should be treated based on its fruit bearing capacity. A prerequisite for this adaptive approach is the availability of leaf area data of each individual tree by the application of proximal and remote sensing. Many proximal and remote sensing methods have been developed in the last two decades. Inexpensive ranging sensors, based on ultrasonic measurements have been employed to measure the plant height and identify small trees and gaps. Photogrammetry was applied with various cameras to derive the leaf area from RGB or NDVI data. 3D point clouds obtained by means of structure-from-motion approaches or stereo vision provide advanced methods to get a better insight into the canopies. Light Detection and Ranging (LiDAR) sensors were employed to measure the leaf area per tree (Fig. 5.9). Depth cameras, providing RGB and depth information, have a high potential to measure the leaf area with even inexpensive devices. However, the approach is challenging due to the targeted early CLM treatment, already during bloom, when the leaves are hardly expanded. Here, the use of allometric growth models is requested. It would be highly valuable, if the precision horticulture community bridges the sensor development and agronomic modeling as has been shown in few examples already [PLT20], [TFZ22].

Fig. 5.9 3D point cloud of sweet cherry tree (leaf area = 17.4 m^2)

As an alternative for the analysis of the leaf area per tree, it may be beneficial to know the number of flowers and remove excessive flowering with enhanced thinning intensity. Adjusting the thinning intensity according to the number of flowers is an indirect method, because the fruit bearing capacity determined by the leaf area (density) and resulting light interception is not taken into account. Consequently, it cannot be expected that the trees are managed at their optimum. The advantage of setting the thinning intensity according to the flowering intensity is provided by the easy implementation of the approach. Simply the sensor, e.g., photogrammetry [DTM18] or thermal cameras [WUW18], [BCA+20], but no physiological growth model is requested.

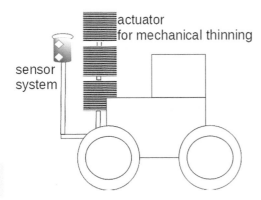

Fig. 5.10 Schematic drawing of georeferenced flower detection and actuator for mechanical flower thinning for adjusted thinning intensity according to the actual number of flowers per tree

By means of non-destructive sensing, the flower density can be counted, and in the same process, the Variable Rate Tinning application can be carried out (Fig. 5.10). The information of the flowering density can be displayed to the driver, who adjusts the thinning intensity of the actuator. Also, a machine-to-machine communication can be enabled, supporting an automated thinning treatment. This approach already provides a reasonable solution to work more precisely based on the in-situ obtained plant data. Here, mainly ergonomic studies are needed to support the usability of the system.

5.4.4 Fruit Quality Post-Harvest

The analysis of fruit quality after harvest has been approached since more than two decades. The majority of the sensors recently introduced in practice are based on spectroscopy in the electromagnetic wavelength ranges from visible to near infrared. In the visible range, the fruit pigments can be estimated in a non-destructive way even with hand-held devices [WBZ+20]. The data analysis is approached by means of indices, mostly developed in remote sensing, and multivariate methods. The near infrared wavelength range provides information on internal browning, water content, percentage of dry matter, soluble solids content, and in the wavelength range up to 1700 nm even sugars and organic acids. Further quality variables, e.g., vitamin C content, were approached, but so far, no robust calibration was published confirming the results with an independent test set capturing fruit from different orchard or season. Meanwhile, hyperspectral imaging is gaining importance in the scientific approaches. Recently released systems allow fast acquisition of the data. Furthermore, fluorescence and Raman Spectroscopy [QKC+19] are discussed to provide information on more specific molecules, but the perturbating effects of quenching by the fruit matrix is an unsolved challenge so far.

The fruit flesh firmness is an interesting quality parameter considering marketing and the monitoring of fruit in shelf life. Spectroscopy has been employed for this task, but with no success. At present, it is assumed that firmness can only be analyzed non-destructively, if the absorption and scattering properties of the samples are recorded independently [LVS+20]. This was approached with time- and spatially resolved spectroscopy. So far, the methods lack commercialization. However, the quality traits analyzed by means of visualization of NIR spectroscopy are meanwhile considered for shelf-life prediction in physiological models, which is exactly what is needed for more precise management in post-harvest [NDL19].

5.4.5 Conclusions

Tools of information and communication technology—such as satellites, drones, autonomous platforms, wireless networks, data management techniques—exist for all scales from fruit to orchard level to support the data acquisition by means of remote and proximal sensors directly in the production and post-harvest. On the other hand, the translation of sensor data into information on the crop and knowledge of the process is still challenging. Early stage stress detection could allow the growers to cost effectively manage the stress and prevent the adverse effects of that stress on yield and profit. The integration of the newly available plant data into agronomic models is the present challenge for reaching the next step of automation.

5.5 Digital Transformation in the Wine Business

Dominik Durner

Abstract

Wine has been a part of human culture for thousands of years. During this long history, wine has always been associated with high culture, tradition, origin, and dignified social settings. As an enduring cultural symbol of prosperity, the societal role of wine has changed in the last few decades from an important source of nutrition to a cultural complement to food and conviviality compatible with a desirable lifestyle. While changing its role, wine is still synonymous with art, beautiful regions, and indulgence. Wine is produced in small—often family-owned—businesses. Even today, winemakers all over the planet are overseeing the complete production cycle, from grape growing, harvesting and processing of the grapes, to fermenting and producing the wine, and finally selling the final product to customers. Wine is manufactured and delivered "from farm to fork" (or as wineries might say "from grape to glass") as it was done thousands of years ago.

The global area under grapevines is approximately 7.4 million hectares (18 million acres) and the world production of grapes was 77.8 million tons in 2019 [Int20]. Nearly, 60% of the harvested grapes are used for winemaking [Int20]. The rest is processed to grape juice, consumed as table grapes or raisins, or utilized for phytochemical and pharmacological needs. Considering the high prices for a bottle of wine, grapes bear a high potential for added value. Therefore, quality parameters in viticulture and enology are regarded crucial; sometimes even more than parameters affecting the production quantity. The global wine market has witnessed a steady growth over the years and was valued at 355 billion US dollars in 2018 [For20]. It is projected to grow 21% by 2023 to over 429 billion US dollars [For20].

5.5.1 Innovation Versus Tradition

Viticulture and wine production are seen as traditional trades. The tradition of grape growing and winemaking is vastly protected by those in the trade. The geographic origin of the grape, the grape variety, the vintage and processing details, such as barrel aging, are strictly controlled. These parameters are the main quality parameters for wine and, at the same time, important criterions for retailers and customers to ensure authenticity of the products.

The commodities of the modern world, however, have long made their way into wine production. In fact, viticulture is often the first mover among other sectors when it comes to innovations. From grafted grapevines to selected yeast cultures used for fermentation, from mechanical harvesters to dynamic crossflow filters, from NDVI-based vineyard sampling protocols to SEO-based wine marketing approaches, the wine sector has always been an innovative sector. In fact, [Dre13] revealed a high level of innovation activity throughout all areas in the wine value chain. He showed that a high innovation activity does not correlate with the size of the winery, the strategic orientation or the management tools applied in wineries. Accordingly, the innovation power of the wine sector seems to be detached from economy of scales.

An explanation for the high innovation power may lie in the fact that the wine sector is highly fragmented by means of the vast number of products and producers. There are more than one million wine producers worldwide operating and competing in more or less saturated markets. No single firm accounts for more than 1% of global retail sales [Rob11]. In France, there are 232,900 wine producers and the top 10 brands control only 4% of the market [Rob11]. The high number of actors in the wine sector, mature markets in countries where grapes are grown, and the strict regulations controlling many aspects in grape growing and winemaking, including planting, irrigation, classification, and labeling, create a stimulating environment for innovations.

The winemaking process lies in the hands of many. Each hand creates a part of the value of wine. However, the wine value chain itself is not—or at least not heavily—broken up into different trades. Unlike cereals and other crops, which are cultivated and harvested by farmers, sold to agricultural trading companies and processed by the food industry, grape growing and winemaking is still considered as one trade. Especially in Europe, where 75% of the global wine production and wine consumption take place [Rob11], many wineries still control grape growing, winemaking and the sales of their products to customers. Hence, the wine value chain can be regarded as more or less undisrupted.

5.5.2 The Wine Value Chain

In order to understand, initiate and successfully implement innovations in the wine sector, it is necessary to study the wine value chain. A value chain begins with the supplier and ends with the customer, thus emphasizing the relationships between a company, its suppliers, and customers. The core parts of the wine value chain are comprised of grape growing (viticulture), winemaking (enology), and wine sales (◘ Fig. 5.11). Other segments such as grape breeding, wine analysis and wine marketing are associated parts of the wine value chain. The operational links between grape growing, winemaking and wine sales lie mostly in the hand of one company. However, a closer look reveals that this is not always the case. The three main business models in the wine sector may actually be defined by whether and how far the wine value chain is disrupted. According to the operational structure throughout the wine value chain, the three main business models in the wine sector are:

1. Wineries with own vineyards: grape growing, winemaking, and wine sales are controlled by the winery. The wine value chain is not disrupted
2. Cooperative wineries that unite grape growers and jointly operate a winery: grape growing is controlled by the cooperative members; winemaking and wine sales are controlled by the winery.
3. Winemaking companies that purchase grapes from private grape growers: grape growing is separated from winemaking and wine sales.

5.5.3 The Potential of the Vineyard

Grape growing is considered as an integral component of the wine value chain. Since they control themselves, wineries determine and execute their own measures in the vineyards (often regarded as the philosophy of the winery) for quality management purposes, and they utilize viticulture practices and information from the vineyards for fur-

(1) Winery: A grape grower makes own wine and trades intermediate products and/or sells finished products to customers or retailers.

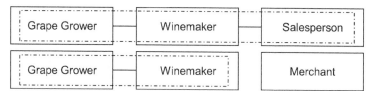

(2) Cooperative winery: Several grape growers are united in a cooperative, jointly operate a winery and trade intermediate products and/or sell finished products to customers or retailers.

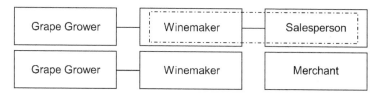

(3) Winemaking company: A grape grower sells the harvest to a winemaking company. The winemaking company trades intermediate products and/or sells finished products to customers or retailers.

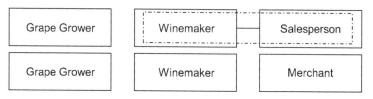

◘ **Fig. 5.11** Wine value chain (own illustration inspired by [Gon16] and [Gon17]). The solid lines between boxes indicate connected divisions within one entity. The dashed lines indicate that the operational flow is in one entity

ther processing and marketing. Of course, this is feasible for other (disrupted) value chains, too. The food industry and retailers have learned to set comprehensive quality standards for their agricultural providers and, of course, they use these standards for processing, communication, and marketing as well. Due to the high protection standards and the limited possibilities of interventions in winemaking, the vineyard itself sets the standard for the quality of wine. And customers appreciate this. They know that the vineyard, the growing site, the soil, the grape variety, and the actions taken by the grape grower are key features for quality, authenticity, and sustainability of the final product.

5.5.4 Wine as the Role Model for an Authentic and Sustainable Agricultural Product

People all over the planet are becoming aware of what environmental responsibility means. Wineries are able to transport relevant information about environmental and sustainability concepts from the vineyard to the customer directly with their products. This advantage needs to be expanded for wine and other products from agricultural origin. An undisrupted value chain, reaching from the field to sales, allows innovations to be envisioned from the top-down, as many famous innovation gurus suggest, from demand to supply: from the customer to grape growing, from the market to the vineyard. At first glance, the top-down innovation approach might be in conflict with statements that resource dependency should guide strategic innovation management [Tou10]. However, as long as grape growing and winemaking are seen as two sides of the same coin and as long as wineries regard their vineyards and their grapevines as key resources and quality drivers, innovations should be thought, pushed, and implemented back and forth the entire value chain of wine.

5.5.5 From the Wine Value Chain to an Operational Network

The illustration of the wine value chain in ◘ Fig. 5.11 should not disguise the fact that wineries can only function within a broad network of stakeholders. An extended wine value chain involves grape nurseries, experts and advisors for pest control, marketing and PR agencies, and providers for all kinds of technical equipment and IT solutions. Accordingly, the wine value chain can be seen rather as a wine value network (◘ Fig. 5.12).

The basic idea of the wine value network is to build bridges, not walls. Wineries cultivate vineyards, produce wine, and sell their products. The vinification process is seasonal and most operation steps are carried out discontinuously. Although it usually lies in the hands of one company, it should not be regarded as a closed job. Winemakers rely on suppliers and services from numerous providers. To run a winery successfully, winemakers deal not only with other businesses and small companies, such as wine laboratories, but also with large enterprises, such as agrichemical providers and machinery producers. Besides processing, wineries sell their products directly to customers and/or to wholesalers and retailers.

5.5.6 Challenges in Communication

To maintain good contact with customers, wholesalers, retailers, wine laboratories, agrichemical providers, machinery producers, and eventually also to staff, it is a challenge

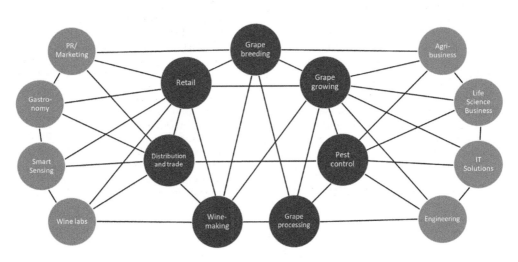

◘ **Fig. 5.12** Extended wine value chain illustrated as an embedded network with stake-holders (own illustration). The lines indicate connections for communication/data exchange

for wineries to manage their relationships. Similar to what is observed in general, wineries shift their communication channels from regular mail to email, from circular letters to electronic newsletters, from phone calls to text messages, to maintain contact with customers, stakeholders and colleagues. Considering the high degree of specialization and the small size of most wineries, B2B relations are often asymmetric and, thus, challenging for winemakers due to major differences in composition and size of the involved companies. However, a good collaboration across companies is crucial for innovation [BBB20]. Well-organized wine innovation clusters could help to strengthen the bonds and to support the communication between stakeholders.

Besides private stakeholders, wineries need to record specific data for administration authorities. The accounting obligation includes recordkeeping in cultivated vineyard acreage, use of pesticides and fertilizers, yield of harvested grapes, produced wine volumes, etc. B2A relations are often time-consuming and not related to the production efficiency and the commercial success of a winery. Logbooks for planting grapevines, pest control, grape harvesting, winemaking activities, and wine bottling and storage are more and more converted from paper to digital. Many software companies have specialized to provide digital recordkeeping systems for wineries. Several countries legally require recording and reporting of data and, therefore, it is important for wineries to use officially authorized digital recordkeeping systems.

5.5.7 Digital Transformation in the Wine Business

Wineries manage many different processes: strategic, operational, and financial; and various processes related to staff, production, marketing, and sales. Digital technologies are slowly replacing paper-based administration, are utilized for communication purposes, and are used to control production-related processes. So far, digital technologies are considered as a help to the operational processes in a winery. Still, wineries continue to operate in analog, as digital solutions have entered into only some divisions inside the winery. The potential of digitized information, which lies in the easy transfer, the strategic use and interpretation of data [Was16], is neither exploited nor fully recognized in the wine sector. Obstacles to overcome include the lack of technical sensors that would allow for noninvasive and automatic acquisition of data throughout the vinification process, missing links between data entry/acquisition and data processing, and the absence of uniform standards for computational networking applications. It needs to be mentioned that digital transformation itself is not the objective for a winery. Rather, improved assistance for internal processes in the winery and improved support for external processes between the winery and different stakeholders should be the goal.

The need of process control in vineyards and wineries, and the requirement of a close coordination between grape growing and winemaking processing, has initiated multiple digital transformation processes along the value chain from vineyard to wine sales (◘ Fig. 5.13).

Some examples for innovative digital technologies, which have entered the wine sector recently, are described in the following:

5.5.7.1 Smart Vineyard Management

In a time when wineries are increasing in size [Deu21] and skilled labor is short [FME21], grape growers are searching for remote techniques ensuring reliable yields and quality in fruit. Smart vineyard management solutions comprise of databases and geographic maps or navigation

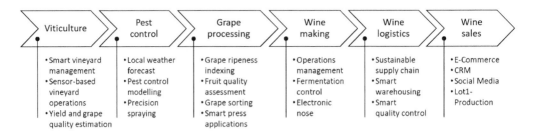

Fig. 5.13 Digital transformation processes along the value chain from vineyard to wine sales (own illustration)

systems with high resolution allowing to identify single rows or even single grapevines. A high spatial accuracy provided by geo-spatial positioning systems combined with functional landscape characterization makes it feasible for workers and machines to conduct assisted or autonomous driving and guided or automated vineyard work operations [THK03]. Cloud-based web applications allow to control single work steps carried out by machines and workers in precise relation to place and time. Additionally, databases, which are usually operated by private providers, can be fed with quantitative and qualitative data facilitating a new form of vineyard management enhancing the work efficiency of wineries and providing valuable data for grape processing, communication, and marketing.

5.5.7.2 Sensor-Assisted Vineyard Operations

While smart vineyard management is regarded as the digital infrastructure, additional sensor-assisted technologies are needed to enable so-called precision viticulture. Soil-adapted tillage [CBT06], risk-related pest control [Mic17] and vigor-related canopy management [XFS+19] are thought to contribute to an improved plant physiology, better fruit quality and a sustainable environment. Rather than following standard procedures, all needs-oriented vineyard operations are following the principle "as much as necessary, and as little as possible." Since they are both highly valuable and susceptible to a wide range of biotic and abiotic stressors, grapevines and grapes need to be closely monitored by their growers. Important aspects for grape growers are accurate local weather forecasts (temperature, humidity and rain; see ▶ Sect. 4.8) for the timing of sprays; specific recommendations for pest control, irrigation, and vine nutrition; and an accurate estimation of yield and grape ripeness to allow for the planning of harvest and grape processing. Mathematical models for pest control are based on local weather forecasts derived from a dense network of weather stations operated by state authorities. This open-access information is highly interesting for every grape grower. The websites are easily accessible by PC or Smartphone and the information can be freely integrated in the operational decisions of a grape grower.

Also, private providers strive to implement weather stations directly in vineyards. Sensor-assisted monitoring of meteorological data at multiple points in vineyards ought to provide precise recommendations for pest control, irrigation needs, and nutrition for optimal health and vigor of grapevines and grapes. Modern meteorological sensors should be protected against external forces, such as machinery interference, and equipped with signal transmission to send information to a relay node. From there, the collected data are transferred to the base station and eventually to the cloud computing platform by Wi-Fi for processing and generating site-specific advice.

Wireless sensor networks are greatly beneficial for those with large vineyards and widely distributed acreage [KNP+14] as micro-climates often cause changes in weather conditions over small distances [SFM+20].

Irrespective of the sensor type and its intended use, the positioning of noninvasive sensors in relation to the objects to be measured is a crucial factor (◘ Fig. 5.14). Remote sensing begins with the decision "satellite, aerial, or close-to-ground sensing." Indeed, spatial resolution, accuracy and 3D imaging become more precise when a sensor is closer to the object [STD09]. Also, in theory, data transmission is simpler when distances to overcome are closer. However, satellite and aerial sensing cover large areas and are in the focus of research and development because of their broad application possibilities.

Unmanned aerial vehicles (UAVs), commonly referred to as drones, are discussed to provide real-time images that enable early detection of plant diseases and provide information about the nutritional and water status of grapevines. Most promising are drones which are paired with sensors to compute the NDVI in order to compare vegetation data and recognize grapevine health issues. In recent years, the commercial use of drones has become possible [QZM+19]. With technology now available for use and at reasonable costs, drone solutions are designed for applications in viticulture. Drones are flexible devices due to their size, maneuverability, and ability to access and view areas that are difficult for humans to access [MHP14], such as steep slope vineyards. Drones provide advantages for precision viticulture.

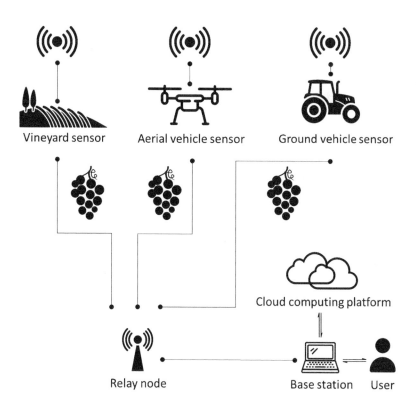

◘ **Fig. 5.14** System framework with three remote sensing approaches in vineyards (own illus-tration)

5.5.7.3 Sensor-Assisted Winemaking Operations

Grape processing and winemaking processes are comprised of a large number of process possibilities eventually influencing the quality of the final products. Yet, winemaking is considered as a traditional trade. It is not based on specific formulas, and the use of processing aids is limited by state or federal laws protecting wine. At grape reception, winemakers must thoroughly investigate their harvest and determine adequate processing steps with regard to the production goals and the available process possibilities. In general, winemaking aims to extract valuable substance from grapes and transform it into wine. Being a traditional trade should not lead to losing sight of the fact that winemakers are attracted to new methods aiming for better process reliability, product safety, and wine quality. Data originating from vineyards, as well as data acquired during the winemaking process, is crucial.

Grape sorting, de-stemming, and berry maceration are processes carried out before pressing, which is the important step in winemaking where the juice and all of the valuable components are extracted from the grapes with the aid of a wine press. The aim of modern wine presses is to gently increase the pressure and to use as minimal movement of the grapes inside the press as possible. At the same time, winemakers desire maximum yield and maximum quality. Innovations in wine presses include sensors that monitor the obtained volume of juice, temperature increase indicating unwanted friction, and quality parameters such as pH to determine the end point of pressing. Future innovations could include vintage- and varietal-related data from vineyards to tailor the control parameters by using artificial intelligence (AI) to obtain the desired juice.

Fermentation is carried out after juice clarification. Control over fermentation has several advantages, in terms of tank use optimization, controlling the characteristics of the wine, and control over energy expenses for the regulation of temperature [Sab09]. Online fermentation monitoring is promising because it is much more accurate and has a higher temporal resolution than manual measurements. Eventually, it will allow the implementation of new control strategies such as PID feedback controlling for precise dissipation of fermentation heat [SDW+19] or the addition of yeast nutrients in a fed-batch approach [FH18].

Automation and supporting technologies in pressing, fermentation, and other processes during winemaking will guarantee robust analyses. Better data reliability—as well as a higher level of freedom for winemakers to focus on core tasks—is seen as beneficial for efficiency, process security, and product quality. Indeed, the discontinuity of the winemaking process and small business sizes must be seen as impediments to sophisticated and expensive automation technologies. However, contractor companies or machinery ring associations might be the solution providing specific technologies for certain process steps in wineries.

5.5.7.4 Blockchain to Ensure Authentic Wines

The inevitable need for digital transformation may be explained by an increasing demand for manufacturing traceability and product authenticity and the necessity of pesticide compliance and water compliance for a sustainable environment. Regardless of the perspective taken, today's challenges call for an integrated approach that recognizes the interdependencies between single processing steps, trade, and consumers. Digital transformation in the wine business has advanced to an extent which is comparable to that existing in the private sphere. Single process steps in grape growing and winemaking, which are seen as isolated campaigns defined by seasonality, need to be merged—data-wise. How this works has been shown in the early 1990s when the first blockchain solutions were introduced. Back then, blockchain was nothing more than an online system for recording transactions, available to those who are party to the transactions. It remained obscure until a decade ago when Satoshi Nakamoto used blockchain to create the digital cryptocurrency Bitcoin [Nak08]. No doubt, there is also great potential for the use of blockchain in the wine business. Increasing visibility and transparency across all areas of the process from grape growing, winemaking, strategic sourcing, procurement and supplier quality to marketing and wine sales, blockchain can allow for strategic advances. By scaling growing data accuracy, product quality and track-and-traceability, wine producers could be able to enhance product quality, better hit trade deals and ultimately sell more.

The wine business has always been a trade that is set in its own ways. However, with technology such as blockchain playing a more influential role, the winery of the future might look and operate differently. As blockchain technologies mature, they could allow grape growers and winemakers to clear some hurdles that are basically justified in a lack of information exchange throughout the complex operational networks of wine production. As a result, more efficient operations that use data sharing and collaboration among the complex networks will be created in the future and set as a new norm across the wine business.

5.5.8 Conclusion

The digital transformation of wine business is reaching from vineyard to cellar to end-consumer marketing. The mentioned examples confirm, digitalization can improve products and processes. This provides opportunities for better production, efficiency, sustainability, and differentiation.

The historic role of early adopters of innovations in agriculture provides an opportunity beyond the vineyard to crop farming, e.g., enrich products with data.

5.6 Transparency of Animal Welfare Through Digitalization: A Dairy Farming Example of "Hofgut Neumühle"

Christian Koch and Noura Rhemouga

Abstract

In the millennia of domestication of farm animals, the focus was on securing our own food base in a decentralized dairy farming system. Owning and living with animals was the basis for settling down and increasing prosperity. Only through precise visual observation and interpretation of the animals' signals was it possible to make the right decisions for effective livestock management and thus to provide the animals with the best possible care and use their capital for a long time. As the number of farms worldwide has been steadily declining over the past decades, but the global demand for animal-based food is increasing dramatically, agricultural dairy farming will inevitably lead to larger farms and centralized production. According to estimates, the global demand for animal-based foods will double by 2050 [RNH+19].

Due to limited natural resources (such as land and water), more sustainable and efficient ways of dairy farming need to be found and developed in order to meet the future demand for animal-based food. Currently, digitalization technologies (computers, sensors, cloud computing, machine learning (ML), artificial intelligence (AI)) and their meaningful use for more sustainable and efficient production are being deployed and further developed in all industries, which can increase and improve economies of scale and efficiency [WGV+17].

The possibilities and pioneering effects that digitalization will have in the future in the field of dairy farming are clearly illustrated in the following section with the help of a few examples. The further discussion concentrates on German dairy farming in particular.

5.6.1 Dairy Farming in Germany

The German dairy industry is the most important branch of the German agricultural and food industry and occupies a leading position within the EU. In May 2020, the Federal Statistical Office of Germany counted around 58,351 dairy farms with just under 3.97 million dairy cows in Germany, producing around 32 billion kilograms of milk annually [Gen20].

Looking at the last ten years, it is estimated that 37% of dairy farms have given up production. The number of dairy cows has decreased by about 6% in the same period. The herd size per farm has grown about 33%. A simultaneous increase in milk yield per cow and year results in a 44% increase in the amount of milk produced per farm (see ◘ Table 5.2).

The German dairy industry is thus affected by a strong structural change, which is characterized by decreasing farm numbers and increasing milk production.

5.6.1.1 Strategy 2030 of the German Dairy Sector

The German dairy sector faces many challenges. Volatile markets, increasing regulations with a high intensity of international

◘ Table 5.2 Structural change of German Dairy Farms, Source: [Gen20a], [Gen20b], [Mil20a], own calculations

Year	Number of kept animals	Number of dairy farms	Decrease in farms per year in %	Number of cows per keeper (rounded)	Average milk yield (kg per cow per year)	Milk production in kg per farm
2010	4,181,679	91,550	(-6.04)	46	7,080	323,389
2011	4,190,103	87,162	-4.79	48	7,240	348,046
2012	4,190,485	82,865	-4.93	51	7,323	370,324
2013	4,267,611	79,537	-4.02	54	7,343	393,994
2014	4,295,680	76,469	-3.86	56	7,541	423,619
2015	4,284,639	73,255	-4.20	58	7,628	446,157
2016	4,217,700	69,174	-5.57	61	7,746	472,292
2017	4,199,010	65,782	-4.90	64	7,763	495,529
2018	4,100,863	62,813	-4.51	65	8,063	526,147
2019	4,011,674	59,925	-4.60	67	8,200	547,542
2020	3,921,410	57,322	-4.34	68	8,400*	574,645*
Change 2010–2020 in %	−6.22	-37.39	-4.57 per year	+33.23	+15.71	+43.72

competition, and growing criticism from society about existing animal welfare standards and environmental impacts of the industry are just a few examples. Milk production on farms is therefore characterized by a constant and increasing structural change.

The "Strategy 2030 of the German Dairy Sector" developed in 2019 is intended to actively meet these challenges and includes measures and solution approaches that are supposed to strengthen the German Dairy Sector sustainably [Mil20b]. In this respect, the digitalization of dairy farming offers opportunities to develop more efficient and sustainable solutions and to increase value creation.

In the following, two of the nine main topics of the Strategy 2030 are addressed: digitalization and sustainability. The explanations are intended to show which opportunities digitalization can offer in the area of animal welfare. In this context, the QM-Milch sustainability tool is designed to provide figures, data, and facts as part of the Sector Strategy 2030. Using the example of the Teaching and Research Institute for Animal Farming, Hofgut Neumühle (▶ www.hofgut-neumuehle.de) in Münchweiler an der Alsenz, currently existing digital information for the assessment of animal welfare criteria is presented as an example and its useful application is described.

5.6.1.2 Digitalization

The economic success of a dairy farm depends on many factors. Above all, the health status of the entire herd has a considerable influence on the milk yield. Dairy cows that are healthy and feel well perform well. Therefore, animal welfare plays an essential role for every dairy farmer.

Here, more and more automatic and digital systems support the farmer in managing the farm. Today's digital information ranges from an individual animal view to a complete picture of the dairy farm. Digitalization thus changes processes and offers opportunities for economic efficiency while simultaneously increasing animal welfare.

5.6.1.3 Sustainability

The debate on sustainability is becoming increasingly prevalent around the world. The German dairy industry, and in particular milk production, is also in the focus. Rising consumer expectations, demands form the food retail sector at national and international level, as well as politics, call for more sustainable milk production along the entire value chain. In this process, the pillars of economy, ecology, social, and animal welfare are taken into account. Above all, the topics of animal welfare and animal protection in agriculture are increasingly being discussed and consumer acceptance of agricultural livestock farming is declining. This is due to a change in the human-animal relationship in society. The species-appropriate farming and the needs of farm animals are increasingly viewed in ethical and moral terms [MBS20].

Therefore, the topic of animal welfare is given a special role in the Sector Strategy 2030. The aim here is to achieve the highest possible coverage of dairy farms with the objective of a uniform approach throughout the country. The QM-Milch sustainability tool accompanies the topic of animal welfare on a scientific based level [Mil20b].

5.6.1.4 QM-Milch Dairy Sustainability Tool

The QM-Milch Sustainability Tool has been developed in several stages by science and practice since 2015. For about three years, the elaborated system was tested by 34 dairies and a milk producers' association. It includes essential criteria in the areas of economy, ecology, social aspects, and animal welfare and serves as a status quo survey to highlight the strengths and weaknesses of milk producers in the respective areas and to support the dairies and their production companies in the sustainable further development of milk production [LCJ+20]. A total of 34 dairies successfully participated in the pilot phase and further developed the module in all areas. In developing the module, it was important to ensure that it meets scientific requirements on the one hand and can be used by farmers and dairies with reasonable effort on the other hand [LCJ+20].

In the following, the defined animal welfare criteria of the milk sustainability tool are discussed. Based on these criteria, the example of Hofgut Neumühle will be used to consider which of these criteria can already be mapped today using existing digital information from sensors and herd management systems that are used. ◘ Table 5.3 shows possible animal welfare criteria that could be used to calculate a potential animal welfare index in the future. Taking the example of the Neumühle farm, it can be seen that more than 60% of the defined animal welfare criteria can already be mapped by digital systems. Especially in the area of "good health," there already exists an almost complete digital recording of information.

5.6.2 Legal Basis

Due to various legal foundations (Animal Welfare Act, Animal Health Act, Veterinary Home Pharmacy Ordinance, Livestock Traffic Ordinance,), the animal keeper is obliged, among other things, to document information on animal health permanently and at all times in order to be able to maintain and care for the animals in the best possible way. For example, §1 of the Animal Protection Act states: "The purpose of this Act is to protect the life and well-being of animals out of man's responsibility for them as fellow creatures. No one shall cause pain, suffering or harm to an animal without reasonable cause." Since in agricultural animal farming the daily income is earned directly with the animals or their products, it is a very own interest of every animal keeper to establish the health and well-being, which he is also legally obliged to do, on a high level on a permanent basis. In order to comply with these legal obligations in view of the increasing number of farms and simultaneously less time and personnel per animal, it will be essential in the future to collect data and information on animal health and welfare by means of digital techniques, to evaluate it, to present it in a way that is easy for the farmer to understand (data mining), and to document it permanently.

5.6.3 Automation in Dairy Farming

The work in dairy farming is becoming increasingly automated. Automatic Milking Systems (AMS) in particular have experienced rapid development here. In 2016, about 7,800 AMS were in use in 5,500 companies in Germany. Two thirds of all dairy farmers decide to buy a milking robot when they buy a new machine. According to the German Board of Trustees of Technology and Construction in Agriculture (KTBL), the use of an AMS results in a seven percent higher milk yield, with a significant increase in labor productivity at the same time. In addition to AMS, sensor-based retrieval feeding or automatic feeding systems are also widely used. Here and there, robots are even used to clean the treads and to supply the basic fodder.

5.6.3.1 Digitalization

The dynamic development of various sensors for documenting behavior, activity, health, location systems, etc., has increased dramatically in recent years, resulting in more and more people using these

Farming System Perspective

Table 5.3 Dairy Sustainability Tool 2.0—Animal welfare indicators Source: [CJL+20]

	Animal welfare indicators	Digital capture*
	Good husbandry	
1	Type of cubicle and cubicle floor covering & Management of lying areas	-
2	Ratio: Resting areas to dairy cows	-
3	Special Needs: Existence of areas for sick cows and calving pens & Management of calving pens	-
4	Cow comfort: Existence of cow brushes & Facilities for thermo-regulation	✓
5	Care of resting areas for new-born calves	-
	Good feeding	
6	Ratio: Feeding areas to dairy cows	-
7	Access to water: Number of troughs & Inspection and cleaning of drinking troughs	-
8	Calf rearing management: Supply of colostrum and water	✓
	Good health	
9	Metabolism profile of dairy cows: Frequency of analysis & Share of dairy cows with a fat-protein-ratio > 1,5 or < 1 for the last year	✓
10	Share of dairy cows with max 100.000 somatic cells/ml at average for the last 12 months	✓
11	Strategy for the use of antibiotics for mastitis	✓
12	Control of joint injuries	✓
13	Carrying out of cow-individual lameness control	✓
14	Frequency of prophylactic hoof care of the herd	✓
15	Calf mortality for the last year	✓
16	Dairy cow mortality rate for the last year	✓
17	Share of difficult calving	✓
18	Way of disbudding calves	?
19	Herd care by external experts	✓
	Species-appropriate behaviour	
20	Freedom of movement for the dairy cows (husbandry system)	✓
21	Access to outdoor climate stimuli for lactating cows, dry cows and young cattle	-

*using the example of the Teaching and Research Institute for livestock husbandry, Hofgut Neumühle technologies at work and also in everyday life to obtain information or to work more efficiently. This trend has also found its way into agriculture and agricultural production. Due to the very rapid development of the "Internet of Things," "Cloud Computing," ML or AI, the development of so-called Smart Farming has been strongly pushed and is thus already being usefully applied in practical agriculture in various areas (animal farming, agriculture, grassland, environment, etc.) [SVW+16]. In addition to location-based data an information (e.g., localization, temperature, humidity), Smart Farming also includes information on current events and changes, e.g., in the barn, which is supplemented by real-time information [WSG14]. This real-time information is needed especially to detect very short-term changes (e.g., in animal health) in order to make timely management decisions (e.g., in herd management of dairy cows). These applications require a complex interplay of the most diverse functions in the field of computing, ML, and AI in order to be able to comprehensively describe and apply the concept of Smart Farming along a functioning management system [WSG14]. Figure 5.15 shows the cyber-physical management cycle of Smart Farming.

For the sensible practical application of Smart Farming, it is essential to permanently collect all relevant data and store it for evaluation. Only then is it possible, through the adequate application of various mechanisms and algorithms, to make

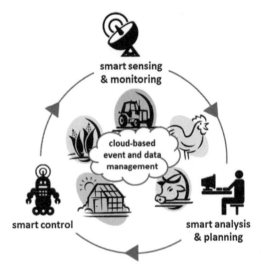

Fig. 5.15 Cyber-physical management of Smart Farming [WSG14]

appropriate information available to the farmer, on which bases basis management decisions can be made. You can see which steps are necessary for this in Fig. 5.16.

Thus, it is fundamentally important that all relevant data is collected securely and stored permanently through secure data transfer (e.g., cloud). The collected data must promptly undergo a process-oriented transformation and then be made available for a corresponding data analysis, with the help of which the data can be evaluated, e.g., through complex algorithms. After this analysis, it is important to prepare the data in such a way that the end user is able to interpret the data correctly and derive appropriate management measures with very lit-

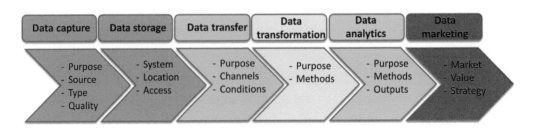

Fig. 5.16 Necessary data chain for the practical application of Smart Farming [CML14]

tle effort. Currently, efforts are being made to recommend or suggest possible management measures to farmers and dairy farmers already through the analysis of the data.

5.6.3.2 Practical Application of Smart Farming

The aim is to document existing and recordable information completely and digitally, starting with the birth of each animal until the animal leaves the farm. This includes, for example, information on the animals' weight, diseases, treatments, localization, performance, age, chewing duration, activity, behavior, feed intake, oestrus, just to name a few. ◻ Table 5.4 shows some information on important data for the use of Smart Farming as an example.

In order to be able to evaluate the collected data in a meaningful way and use it for practical applications, it must be possible to automatically evaluate this permanently stored data, which can currently still lead to problems due to different data and file formats from different manufacturers. The aim is to formulate simple and easily applicable decision-making aids or management recommendations for the farmer from very large amounts of data through "data mining." In order to be able to implement this in the future in a practical and profitable way for farmers, further targeted scientific findings and their implementation in practice are of outstanding importance in order to permanently collect automated data, e.g., on the topic of animal welfare (◻ Table 5.3). It is therefore possible, for example, to collect and permanently store all data and information from the birth of a calf to the departure of the dairy cow from the herd, partially automatically. At Hofgut Neumühle, the birth weight of each calf is documented, the Colostrum quality as well as colostrum supply is checked, measured and stored in the herd management program HERDEplus (dsp-agrosoft). Furthermore, all information and measures of each animal, e.g., diseases and treatments, fertility data (e.g., oestrus, insemination, etc.), daily milk yield, milk contents, feed intake, chewing duration and behavior, animal behavior, walking behavior, lameness score, etc. are determined and stored individually for each animal. Since a lot of data are collected by different and often incompatible systems, there is currently still a lack of comprehensive interfaces for all systems in order to generate adapted and optimized Smart Farming solutions or applications for dairy farming. In the future, these problems should be solved and further developed in additional scientific research approaches and research alliances with industry. A major challenge at present is to permanently store the multitude of complex

◻ **Table 5.4** Important data to be collected for the application of Smart Farming technologies

Animal category	Data
Calf	Birth weight, genotype, weight development, colostrum intake, feeder intake, concentrate intake, water intake, nutrient intake, health data, diseases, treatments, activity, feeder behavior, number of visits to feeder, body temperature, calf appearance imaging features, body condition, etc.
Cattle	Weight development, body condition, diseases, treatments, chewing duration, chewing activity, activity, body condition, water intake, heat detection, insemination index, lameness score, etc.
Dairy cow	First calving age, weight development, milk yield, milk constituents, lifetime performance, lifetime performance, rumination duration, rumination activity, localization, heat detection, activity, diseases treatments, lameness score, etc.

data mentioned in a standardized system in such a way that defined data or data sets can be made available to specific users in a timely, automated manner and in the required data structure at all times. Only then it is possible to evaluate data in real time and make it available to the farmer so that it can be transferred promptly into management measures or sick animals can be recognized and treated much earlier. Through the intelligent and sensible application of Smart Farming technologies, it will be possible for the first time to collect objective data, also with regard to animal welfare criteria in dairy farming, from calves to young cattle cows, and to be able to compare them between farms. How such a data structure as well as a sensible data networking between different systems can look like is exemplified by the herd management program "HERDEplus" of the company dsp-agrosoft, which is used at Hofgut Neumühle (◘ Fig. 5.17).

Thus, with the help of HERDEplus, it is possible to collect and store a variety of important data permanently, automatically and digitally, even with very different data structures. Through appropriate read-out procedures, it is possible to generate user-specific data without any problems and make it available to various end users from business, public authorities, or research. In this regard, it is important that ownership and data protection regulations are respected and, if necessary, must be clearly regulated by the state. In addition, with HERDEplus it is possible to automatically read data into the herd management program also from external sources, such as veterinarians, hoof trimmer, the dairy, accounting, breeding values, or also from authorities etc., whereby important information can be generated directly for the farmer to constantly receive information about the current production in real time. Only through timely evaluations it is possible to initiate and take correct and important management decisions and measures.

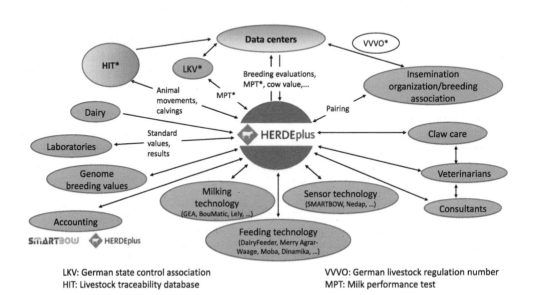

◘ Fig. 5.17 Example of data networking (dsp-agrosoft)

5.6.4 Outlook

The trend toward increasingly larger dairy herds described in Sect. 5.6.1.2 highlights the need for digital solutions in dairy farming. The main goal of digital developments is to make the processes more efficient and thus more sustainable. Across the entire milk value chain, the topics of transparency, traceability and the optimization of processes are in particular focus [Lan20].

The entire industry is still facing a number of challenges, such as the reduction in interface problems, increasing data security requirements and the possibility of ensuring secure and up-to-date data transfer between the value creation levels. Only through increasing integration, the targeted goals can be achieved and costs along the value chain can be sustainably reduced [Lan20]. In conclusion, this means that digital transformation is directly linked to the success and sustainability of the dairy sector.

5.7 German Farmers Perspective on Digitalization

Julian Schill and Christian Bitter

Abstract

The digitalization of the agricultural industry is mainly driven by machinery manufacturers, agrochemical companies, suppliers, and technology providers. In 2016, companies in the US invested $1.8 billion in the development of digital solutions for farmers [Ber19]. The range of digital products and services offered covers nearly all the tasks of a farmer. The spectrum ranges from digital marketplaces for input goods, to solutions for planning and executing field activities, to marketing services for agricultural products [Bet20].

Farmers are confronted with this wide range of products and services. However, they must decide for their business whether and which digitalization measures are suitable for them, improve their products and/or process efficiency. This section is not intended to document the rapid development of digital products and services. Rather, it aims to take a closer look at the digitalization of agriculture from the perspective of farmers. For this purpose, the following section is divided into two sections.

The first section deals with the specific role that digitalization plays for farmers. This section is based on five interviews with farmers from different regions of Germany. It focuses on the goals they are pursuing with digitalization and the challenges they are trying to meet with digital products and services. In this section we also examine the impact of digitalization and how farmers can quantify it.

The second section focuses on agricultural data, which forms the backbone of digitalization. We want to give an overview of agronomic data that are freely accessible to farmers. In contrast to the digitalization driven by companies, we want to show what possibilities exist for farmers to digitize themselves based on this freely accessible data.

The section concludes with practical recommendations for the future development of agricultural digitalization.

5.7.1 Cross Farm Comparison

After a short introduction of the interviewed farmers, this section will provide insights gained from the interviews.

5.7.1.1 Introduction of the Interviewed Farmers

Five farmers were interviewed in the context of this section. In selecting the interview partners, attention was paid to covering a wide range of viewpoints. The interviewees come from different regions of Germany and differ in the size and orientation of their farms. What all five farmers have in common is that they not only perform their daily tasks on their farms but are also involved in other projects. These include activities as a test farm for AgMachinery manufacturers, participation in research projects and pilot projects, or the self-marketing of their own products.

- Dominik Bellaire is the manager of the Schmiedhof in Neupotz, in Rhineland-Palatinate. He runs the family business in the third generation. On the farm with its 200 hectares, besides dairy production, cereals, corn, sugar beets, carrots, and tobacco are cultivated.
- Sven Borchert is the manager of the Betriebsgemeinschaft GbR Groß Germersleben. The farm is located within the Magdeburger Börde and covers an area of 1,750 hectares. Cereals, rape, potatoes, sugar beets, and corn are cultivated. A biogas plant is also operated, which produces green energy.
- Michael Freiherr von Gemmingen is a managing director of the Kraichgauer Güterverwaltungen. It consists of a total of three managing directors and six shareholders. Together they run the farm with 1,400 hectares of arable land in Kraichgau, Baden-Württemberg. They have specialized in the cultivation of corn.
- Manfred Hurtz is the manager of a farm in Nideggen, in North Rhine-Westphalia. He cultivates 100 hectares of his own arable land and also offers various contract work for other farmers (200 hectares of threshing, 80 hectares of fertilization and plant protection, 100 hectares of sowing).
- Broder Preuß-Driessen is the manager of the Herzogliche Gutsverwaltung Grünholz in Thumby, Schleswig–Holstein. On an area of about 1,500 hectares between the Baltic Sea and the Schlei, he cultivates wheat, winter rape, winter barley, energy beets, and silage corn.

All five farmers share the common perception that digitalization plays a supporting role within agriculture. Digitalization supports farmers in three ways. Firstly, agricultural processes can be simplified and automated. Secondly, digitalization has a supportive function in helping farmers make decisions. In addition, we have identified various functions of digitalization that support farmers in achieving their personal goals.

5.7.1.2 Creating Close Relationships with End Customers

For Dominik Bellaire, establishing a relationship with the end customer is an important concern in the orientation of his farm. His intention is to "show the consumer how modern agriculture and sustainability can work hand in hand." In particular, he is interested in bringing his profession closer to his customers and showing them why, for example, it is necessary to apply crop protection products. To achieve his goal, digitalization can support the farmer, for example by collecting and processing agronomic data. Already today, Dominik Bellaire can present a complete documentation for his products. This documentation starts with the sowing, includes all plant protection applications, the harvest, and ends with the time of handing over to the distributor. However, to generate real added value, it is necessary at this point to integrate the following stages in the supply chain. There is also the question of how the farmer can create financial value from this data. As things stand today, it is not yet possible to assign a real price to this data, which could generate additional income for the farmer. Not least because of this, it is currently easier for Dominik Bellaire to generate added value through special products such as tobacco. This is also the case because they have a higher price range than cereals.

5.7.1.3 Making Sustainability Measures More Efficient

Sustainability and biodiversity within the agricultural industry are of great importance to Sven Borchert. He is convinced that much more needs to be done in this area, especially within his industry. However, he does not see politics as being responsible. Rather, he believes that farmers have a responsibility. For him, the focus is on the efficient design and economic viability of existing sustainability measures. According to Sven Borchert, digitalization can make a significant contribution to increasing the efficiency of sustainability measures. Starting with the planning of the numerous measures, he sees digitalization as a decision-making aid. On the one hand, suitable areas can be identified, and on the other, various measures can be evaluated according to their potential sustainability effect. It would also be conceivable to suggest the scope of the sustainability measures, considering the possible field yield. Thus, digitalization must be able to determine optimal locations and assign the appropriate measure to them. The next step is to implement the planned sustainability measures. To do this, it is necessary that the measures are automatically considered in the various fieldwork activities without causing additional work for the farmer. Finally, the various sustainability measures must be documented and exchanged with the public authorities. This would make it easier for Sven Borchert to confirm that, for example, he fulfills greening requirements or agri-environmental programs. In short, he believes that digitalization must be able to make sustainability measures more efficient in the future and integrate them seamlessly into the farm's processes.

5.7.1.4 Differentiation Despite Farm Size

Michael Freiherr von Gemmingen and his fellow managing directors are exploring opportunities to further differentiate themselves within the existing product portfolio. The Kraichgauer Güterverwaltungen are specialized in the cultivation of grain. In the past, attempts were made to convert small areas to organic farming. However, it is not their aim to convert the entire farm to organic farming. Their aim is rather a product-specific differentiation. The role of digitalization is to be seen in two different ways. Firstly, the farm is dependent on generating profits in classical cultivation on high-yield areas. Digital products such as historical yield maps can be used as a decision-making aid for the selection of land.

Secondly, Michael Freiherr von Gemmingen needs to find new partners for product-specific differentiation using new developments within digital agriculture. This includes the development and preservation of strategic partnerships. In this context it is necessary to ensure the know-how of the farmer. In contrast to industrial production, which takes place behind closed walls, farmers' fields are easily accessible. Activities and processes in the field can be seen by others and easily copied. If other farmers copy his new product, this can lead to a differentiation strategy no longer being economically viable. However, if the differentiation is not based on a single product, but on a related strategic partnership, this strategy is more difficult for others to copy. This is especially true when non-standardized machines are required for cultivation or processing, which involves high investments. Here, Michael Freiherr von Gemmingen and his partners benefit equally from the cooperation.

5.7.1.5 Importance of Digitalization for Contractors

Manfred Hurtz notes that many farmers who demand his contract work are still quite closed to digitalization. Many of his customers are farms with 40–70 hectares of arable land, which do not have a strong digital infrastructure. In contrast, Manfred Hurtz can offer a complete digital equipment. For some of his customers, however, it is irrelevant when choosing a contractor whether the contractor offers the digital equipment or not. When in doubt, they choose the contractor who makes the cheaper offer.

In the meantime, however, there are signs that the willingness to use digital technologies is increasing. He is of the opinion that resource-saving applications are slowly moving into the focus of many small farms. For the future, Manfred Hurtz sees further potential for his digital services in fertilizer and crop protection application. On the one hand, farms are required to provide complete documentation, which is driven by ever more far-reaching regulations. On the other hand, the quantities of fertilizers and pesticides are reduced and restricted by law. His digital products enable resources to be used more efficiently while at the same time providing complete documentation.

For Manfred Hurtz to actually use the full potential of his digital equipment for the contract work on offer, it is necessary that his customers recognize this benefit for themselves in the future. In order to achieve this, he is trying to sensitize smaller agricultural businesses in particular to the topic of digitalization in general. At the same time, he shows them the advantages of digitalization in farm management and documentation. He currently achieves access to his customers best against the background of strict fertilizer regulations and comprehensive documentation obligations.

5.7.1.6 Support the Experienced Farmers, not Replacing Them

For Broder Preuß-Driessen his own experience and the experience of his employees is the most important factor for the agricultural success. He and his employees can rely on their agricultural know-how, which has grown over the years. This includes field-specific knowledge and the assessment of the weather. For Broder Preuß-Driessen his own know-how is the benchmark against which he evaluates digital products. If digital products and services cannot stand up to this evaluation, they are not considered. Broder Preuß-Driessen sees the potential of digitalization in the fact that he has access to additional information for decision-making. However, the decision is still up to him as a farmer and not a digital decision model. For him, it is essential "to be out in the field, to take the spade in his hand and look at what is going on in the field."

5.7.1.7 Interim Conclusion—What Contribution Does Digitalization Provide

To answer this question, it is necessary to clarify in advance how farmers measure the effects of digitalization activities. None of the farmers interviewed measure the explicit impact of individual digitalization activities on their farm. However, the financial impact of digitalization is reflected in the contribution margin calculation (see ▶ Sect. 2.4). If the components of the contribution margin accounting are considered separately, the following influences of digitalization can be identified.

In terms of variable costs, all the farmers interviewed agreed that the introduction of digital products and services is associated with savings. For example, Michael Freiherr von Gemmingen was able to achieve savings of around 20€/ha by using GPS control and variable application maps. Manfred Hurtz was able to reduce the amount of nitrate applied by 30 kg/ha through the same measures. Furthermore, the interviews showed that digitalization reduces the workload of the plant manager. Standardized processes can be executed automatically, and data can be exchanged more easily between machines, farm management, and public authorities. This positive effect can be attributed to the area of labor costs. The problem here is that the farm manager's workload is rarely included in the contribution margin calculation. Therefore, the influence of digitalization is difficult to quantify at this point. The interviews also demonstrated that the plant managers do not notice the time savings for existing tasks, as additional regulations, new tasks and administrative processes represent additional expenditure. Nevertheless, in the end, the positive effects of digitalization dominate.

On the revenue side, digital products and services can also help farmers to increase their income. For example, yield forecasting models can be used to help farmers make decisions in case they sell their crops in advance.

From the point of view of fixed costs, digitalization offers the farmers interviewed no potential for savings. On the contrary, digitalization of a farm involves high investments, which at this point have a negative impact on fixed costs.

5.7.2 Field-Specific Open Data

Farming even more field zone specific requires availability of accessible, digitized data in good quality (see ▶ Sect. 3.6.). In the second part of this section, we look at exemplary open data offerings that farmers can use. To give an overview of these freely accessible offers, we start with information about the geographical location of one field and try to describe it as best as possible.

5.7.2.1 Geographical Mapping

Equipped with the information about the geographical location of the field (longitude and latitude), we can already get a first picture with the help of publicly available data. Thanks to the free availability of services such as Google Maps or Google Earth, we have the possibility to determine and narrow down the location of our field. Thus, individual fields of the same cultivation can usually be delimited from others purely visually. Furthermore, public satellite data imagery, such as Sentinel-2 or Landsat-8, allows tracking the change of a piece of landscape over time and can serve to better map fields. In addition, publicly available and anonymized information from the official Real Estate Cadastral Information System (ALKIS) can support geographic mapping.

5.7.2.2 Historical and Actual Field Development

In addition to geographic coverage of the field, satellite data can be an effective tool for assessing the historical and actual development. Subsets of these data can be used to estimate vegetation activity or the water balance. This may require integration of additional data such as historical weather or information on regional and local soil conditions. The German Weather Service offers access to numerous publicly available weather and climate data such as air temperature, air pressure, precipitation, and derived variables measured at weather stations. These are available in various temporal resolutions (hourly, daily) in real time, as well as historically. Location-specific weather forecast data are often only available in low spatial resolution (e.g., regional) and time-limited form (1–14 days). Often, information is provided via a graphical user interface (web page) in textual or summarized graphical form, such as the agricultural weather ISABEL. This representation is primarily suitable for selective evaluation of a few fields. Some commercial weather data providers, such as Open Weather Map, also allow users to programmatically retrieve weather data on a small scale for non-commercial use. Here, it must be noted that the quality of the data offered can vary greatly between providers, depending on the type of weather parameter retrieved, the desired forecast horizon, and may well be region-specific.

5.7.2.3 Soil Quality and Condition

To visualize the soil quality and texture of our field, information from sources such as Soilgrids or state specific BÜK50/200/1000 can be used. These thematic maps provide information on local phenomena such as soil erosion, hydrology, or general terrain shape. For example, the geoportal of the federal state of Rhineland-Palatinate offers maps for characterizing the soil and site typing with regard to water or soil nutrient balance.

5.7.2.4 Growth Stage and Type of Arable Crop

Another important parameter for assessing the field is the growth stage. Known in Germany as BBCH stage, it is largely derived from the influencing variables of crop and variety characteristics, as well as the location-specific weather (e.g., temperature). This information is available from the public sector, such as ISIP, which provides a statement on the age of the crop by specifying the field location. Besides the growth stage, remote sensing data are also used for the identification of the type of cultivated field crops. With the help of these data, statistical statements about historical or currently cultivated arable crops can be derived. Such statements exist partly in predefined, publicly freely available offerings such as CORINE land cover (CLC). CLC provides information on the historical cover of a standardized 5 ha plot. Although the available information is neither species-specific nor applicable in all cases, it allows, together with aggregated information such as statistical yearbooks, knowledge of regional and local practices that can be validated using optical sensor data (satellite imagery) to provide a sufficient seasonal picture of a field.

5.7.2.5 Timing and Type of Agronomic Measures

Even if a concrete operational measure (time, character), such as sowing, plant protection or harvesting remains hidden from the external eye, the artificial satellite eye may offer a remedy here. Satellite images are already successfully used for the documentation of commodity flows such as oil transporters at sea. Accordingly, imaging techniques can be used for temporal containment or exact determination of measures such as harvesting, sowing, or fertilizing.

5.7.2.6 Purchase and Sales Prices

If costs for production factors such as fertilizers or the sales price for agricultural products need to be determined, regional reference values provided by e.g., chambers of agriculture of individual federal states such as North Rhine-Westphalia can be useful. In addition to reference values, online providers of agricultural inputs such as Amazon provide product catalogs including product master data, recommended uses, and prices for research.

5.7.2.7 Interim Conclusion

In general, it can be stated that through initiatives such as open data and open government, a variety of different types of data and thus data for characterizing agricultural land are available. As noted, the characteristics, quality and quantity of the respective data varies, sometimes strongly, due to various factors, such as the federal structure of the Federal Republic, or protection of private data. However, the private sector also provides limited information for site characterization and agronomic use. The more IT expertise there is on the part of the end user, the higher the possible degree of use of the available data. Even though the Federal Republic's Open Data Action Plan was already adopted in 2014, it must be noted that there is still a lot of potential for open data in uniform provision of documented, up-to-date, agriculturally relevant data. It is precisely the partly different handling of the topic in the individual federal states that makes a uniform consideration of the potentials more difficult.

5.7.3 Conclusion

Learnings for Future Digitalization For agricultural enterprises, the potential of digitalization currently lies in the optimization of their existing processes. Resources can be conserved, and savings made. All interviewees were able to confirm that digital products and services have a supporting function for them. For the future development of digitalization, it is important to be aware of this function. Digitalization is not able to replace the farmers, but it can support them in their daily tasks. Similarly, the role of the farmer as an entrepreneur cannot be questioned. The farmer remains the decision maker on his or her farm.

The digitalization of farms will enable all agronomic data to be collected and processes to be made measurable and comparable. The process of data collection currently represents additional efforts for the farmer. However, this effort must be compensated by an added value that justifies the additional tasks. Currently, the added value corresponds to the savings achieved by optimizing existing processes. The trend toward more sustainability and biodiversity in agriculture is generally acknowledged. This is driven on the one hand by stronger regulation by lawmakers and on the other hand by a broad public debate. Based on these developments, new opportunities for digitalization are opening. The aim must be to use digitalization to introduce, maintain, and document sustainability and biodiversity measures without having a negative impact on farmers' yields.

Digitalization in agriculture must not only be capable of enabling new business models and sustainability and biodiversity measures independently of each other. Rather farmers are interested, that profitability and sustainability goals are pursued jointly. It must be shown that it is possible to generate additional revenues by introducing sustainability and biodiversity measures.

The question of the extent to which farmers can digitize themselves remains open. There is a clear trend toward more freely available data. However, it is questionable whether this potential is being exploited, or can be exploited short term. This is also linked to the question of which qualifications the future farmer must have. It is already apparent today that information technology skills are becoming increasingly important.

5.8 A Farm Case Study from the Netherlands

Jan Reinier de Jong

Abstract

In this section history, development and current digitalization status of the de Jong family farm is described. The historic development is vividly described and provides an example for family farm heritages. Despite the historically solved issues also current sustainability and Digital Farming practices are described. An outlook for potential future on farm digitalization activities closes the section.

5.8.1 The Arable Farm of the Family De Jong

The arable farm of the family de Jong was reclaimed around 1920. Before that it was rough land and not used for agricultural purposes. The first years after reclamation were difficult. The soil quality was poor and the crops suffered from a lot of so-called reclamation diseases which is a combination of several diseases and deficiency disorders. These arose particularly because there was a lack of organic manure in the region and standard artificial fertilizers were used. This resulted in a lack of copper elements. When trace elements were developed an addition of copper sulphate solved this problem.

The first farmer on this site was not successful. Jan Reinier de Jong's great-grandfather heard that the farm was becoming vacant. At that time he had a mixed farm about 50 km from Odoorn and this was actually too small. After some negotiating the farm was leased and the family moved to Odoorn. In those days such a trip was comparable to emigrating abroad.

The new location was then a real traditional mixed farm. All kinds of agriculture were present, combined with cows, chickens, and pigs. In the following years a lot of soil improvement was done with manure from their own animals. Some years later the farm was purchased and became in complete ownership of his great grandfather. The farm was taken over by Jan Reinier's grandmother. She continued the family business with a farmer's son from a village nearby.

The grandparents focused more on the growing of seed potatoes and less on animal farming. First the pigs were removed, then the cows and finally the poultry. Growing seed potatoes was particularly interesting. The light soil was excellent for growing high quality seed potatoes. The grandfather of Jan Reinier was an excellent cultivator of plants. His potatoes were sold via a cooperative and were sold worldwide.

In the early 1970s, Jan Reinier's parents took over the business and from that time on Crop Registration was commenced. This registration was mainly used to counteract problems with the potato cyst nematodes Globodera rostochiensis and Globodera pallida. The potato variety was noted for each field every year, so at the end it was clear what recontamination the previous variety had left in the soil. In this way, thanks to the various resistances, the potato cyst nematode problem could be kept reasonably well under control.

From that time on the farm expanded and the cultivation of seed potatoes remained the basis, but starch potatoes were also introduced to the farm and the acreage of sugar beet and malting barley was expanded. Investments were also made in a new barn for the storage of seed potatoes and mechanization. Crop registration was in its infancy at that time. Suiker Unie's Unitip growing program was the first to use crop data to improve yields. Unitip is the cultivation registration and advice program of Cosun Beet Company—formerly the Suikerunie—in which beet growers register their beet cultivation (Unitip=Suiker Unie Teelt Informatie Programma). Growers who register in Unitip receive crop reports and crop advice. They also receive comparison reports comparing their own growing methods with those of other growers in the region and elsewhere. Unitip was launched in the early 1980s and was one of the first registration programs at the time. In the meantime, much has changed about the program. In the beginning it was a paper questionnaire, now of course it is completely digital, with fully automated reports for insight, benchmarks and advice. This advice is given at field level because historical data is used. The main goal is improvement of the beet cultivation. Approximately 10 years ago, chain transparency and accountability for sustainable production of sugar beets, for buyers of Cosun Beet Company, were added. In 2018, this resulted in Unitip becoming a supply requirement of Cosun. Sustainable sugar production is an important consideration in the sale of sugar. To make sustainability requirements concrete, a large number of customers have created the organization SAI (▶ www.saiplatform.org). Participants in Unitip meet the requirements for the highest sustainability category according to SAI and hereby meet the sustainability requirements of large sugar customers.

With a registration system the customer and the farmer now have the possibility to govern and control the crop on how it performed compared with their colleagues. In this way, beet growers could learn from each other and jointly improve the yield of the crop. The Netherlands is unique in this respect. Farmers do not see each other as competitors and are not afraid to share data and knowledge. The effect is that the entire crop is raised to a higher level.

After finishing his studies in 1996 at the agricultural college Jan Reinier joined the company together with his parents. Jan Reinier continued the business which at that time had an area of 63 ha. Consumption potatoes were grown on a limited scale and if these were delivered together with a crop registration, a small bonus was paid. Due to this sale procedure the purchasing party achieved more insight into the use of crop protection products and fertilizers. Later, this became a food safety certificate, followed by the EurepGAP certificate, an initiative of a number of European supermarkets (e.g., Tesco, Delhaize, Albert Heijn, Laurus and Schuitema). In 2007, the name EurepGAP was changed to GLOBALGAP, due to the internationalization (globalization) of the trade in food products. Nowadays, crops grown for direct consumption must have this GLOBALGAP Certificate.

Crop registration is also required for a food supply chain project with malting barley. For this project, all actions during the cultivation of malting barley were registered. An additional benefit for the producer in this project is that he obtains a better understanding of his cost price. The cost price is easy to calculate by means of the crop registration. That information is useful and valuable for a farmer, it allows you to focus on costs. There is a good overview of all the activities that have taken place in the crops in the past period.

At present the farm area is 120 hectares. In 2021, the following crops will be grown on the farm: 43 ha of seed potatoes, 19 ha of starch potatoes, 24 ha of sugar beets, 25 ha of malting barley, and 9 ha of biodiversity measures (agricultural nature management). In total 110 ha are owned, the rest is rented. A few hectares are also exchanged with a dairy farmer.

Nowadays, the registration of crop protection products per plot is required by law. Several customers have started yield improvement programs, for which crop registration is the basis.

The weather is an important factor and has a great influence on the cultivation. The weather forecasts are therefore followed very closely and an adequate respond to this is crucial. Weather data is also recorded. Rain gauges are installed on several plots. Over the past four years, experience has been gained with soil moisture and soil temperature sensors. The soil moisture is controlled by irrigation. The data can be viewed via an app and provides real-time data, but also an overview of the past months. Because the soil is of varying quality, it is difficult to assess the real value or benefit of the data. A tour on the field gives a good idea how to interpret the data of the moisture sensor. By means of a soil scan we have a good view of the soil quality. Based on all this data, it is possible to irrigate specifically for each location.

5.8.2 Precision Agriculture

Since 2011, we have been working with a GPS system from Ag-Leader. Initially to enable the tractor to drive in a straight line across the fields. By not having to steer you can better focus on the work. This gives considerable relief to the driver. Two years later, a new CHD sprayer was purchased. In consultation with the dealer, supplier, manufacturer and importer of the various systems the sprayer was modified so that it is controlled by the GPS system. This means that a spraying computer is no longer required, and the sprayer can work even more accurately thanks to the GPS-RTK system. This was the first GPS-controlled sprayer in Europe to use the tractor's GPS system instead of its own GPS antenna. In recent years, Jan Reinier has been working with the OptRx crop sensors from Ag-Leader, which enables him to monitor the crop growth better. The "eye" of the sensor can detect deviations in the crop quicker than the eye of the farmer. This allows you to respond more quickly. Meanwhile, experience has been gained with site-specific dosing of fungicide in potato cultivation. In recent years a lot of data has been collected. But what do you do with this data? This is a question that more arable farmers are asking.

As of 2018 Ger Evenhuis, Alko Tolner and Jan Reinier de Jong are working together with Nicole Bartelds on the precision agriculture project MAXSUS. The MAXSUS project aims to increase the predictability of the harvest and to stimulate crop growth site-specific in order to gain a homogeneous and higher yield within the field. MAXSUS is a contraction of Maximum Sustainable. With all the data points collected from the crop scans, the foundation was laid for this project. However, crop scans alone are not sufficient to gain insight into crop growth. Therefore, the MAXSUS project also scanned the field plots to achieve greater knowledge on the soil quality. Different methods of scanning have been examined. The difference between the scanning methods were mainly the costs of the scan. The measured values correspond reasonably well. The next step is to gather more insight into the yield at a specific location. For this purpose, substantial investments were made in yield sensors on the harvesters of the participating farmers. By means of direct weighing on the harvesting machine, the yield can be determined very site-specifically. This gives a good picture of the yield differences within a field. These differences can vary, from half the average to twice the average of a field. This gives of course a good indication of the potential that still exists for agriculture.

With the collection of extra data, the farmers within the project hope to take the next step. Currently, investing in precision agriculture techniques is still an expensive hobby. Due to the high cost of the equipment, it is not easy to recoup these costs. There is a lot to learn and knowledge has value, but a better financial return is not possible yet. This of course has to do with the relatively low selling prices of agricultural products. And this slows down investments in these techniques which of course increase the price.

Landscape and nature management has become a serious crop (in the sense of source of income) on the farm. Investment in biodiversity was started in 2014. The main reason at the time was to prepare for the new European agricultural policy. Biodiversity could play an important role in this. By becoming a member of an agricultural collective (united in Boeren-Natuur—"FarmandNature") the company could gain experience with this type of cultivation. A start was made with half a hectare of winter feed field for birds. This is a mix of cereals that are not harvested but remains on the field in order to provide feed and shelter for birds during the winter. Nowadays, there is almost 8 ha of these winter feed fields on the farm and in total

1 ha (almost 3 km) of field margins (flower strips). These field margins are sown with different flowers, Perennials and herbs. Some of these strips were created to protect amphibians (esp. for the common spadefoot toad), others to support local birds and insects.

In close cooperation with BASF, an insect monitoring project has taken place from 2018 to 2020. Goal of this project was to assess possible differences in numbers and species diversity of pollinators (wild bees, hoverflies, and butterflies) between two different flower mixtures. The first flower mixture consisted only of native species, the other mixture had also many non-native species. The Dutch Butterfly Foundation conducted the monitoring. There were surprisingly many insects in the field margins despite being right next to the crops.

Jan Reinier, together with the eldest daughter, counted night moths during the summer of 2020. This was part of a national monitoring network of 50 farmers throughout the Netherlands. Also here many butterflies were observed. The data from these two projects are of great importance and will provides greater insight into the status of on-farm biodiversity. Public opinion is negative about agriculture. In particular, the use of pesticides is under pressure. The use of plant protection products would be at the expense of biodiversity, is generally thought. Thanks to monitoring data you can show that biodiversity, and here more specific insects, are not greatly affected in agricultural areas. With scientifically based data, it is easier to create a discussion on the effects of agriculture on biodiversity. But also to come with science-based measures to effectively support on-farm biodiversity.

The soil is the basis of the arable farms. A lot of time and energy is spent on improving the soil. Examples are the drainage of wet areas, breaking up disruptive layers and site-specific irrigation. By far the most important thing on light sandy soil is the organic matter content of the soil. De Jong family does everything possible to increase this. An organic matter balance is regularly made to see how much organic matter is added and removed. It is not easy to equal the input and output. De Jong therefore does the maximum to add extra organic matter. During the harvest of the cereals the straw is chopped and a cover crop is sown after the cereals are harvested. Furthermore, animal manure and other organic fertilizers are applied as much as possible. No nitrogen or phosphate fertilizers are used on the farm. Furthermore, natural grass from the natural reserves nearby is scattered over the fields. This is also a simple way of adding extra organic material. Every 4 years the fields are sampled on organic matter content. With this data it is easy to understand what the effects of all efforts are. This shows how much CO_2 is being stored in the soil. CO_2 storage is becoming increasingly important. In time, there may be compensation for this. Agriculture could play an important role in storage of CO_2 in the soil or in permanent strips or thickets. In time, this could even become part of the CAP but has to be rewarded then of course. Unfortunately, the Dutch manure policy works against soil improvement/organic matter storage. The policy is mainly based on preventing over-fertilization in cattle-density areas. It is not a soil policy, but a livestock policy. The arable sector does not benefit from this.

5.8.3 Future

What will happen in the future is unpredictable. Data are going to play an increasingly important role in agriculture. But certainly, also in economics and politics. Satellite images are making it easier to estimate how world food production is progressing. If yields are lower than expected, global turmoil may arise. In stable countries, this will not be a problem and can be absorbed. But in certain regions, lower food production can lead to serious problems. In extreme cases, it can even lead to war or toppling of a regime. In Western countries, this will mainly have financial consequences.

Large multinationals and food companies can benefit from data. Insight into the supply of raw materials is of great importance to them. Large French fry producers monitor the cultivation and supply of French fry potatoes worldwide. This enables them to anticipate major fluctuations in yields and thus keep costs down. Whether these benefits the primary producer remains to be seen.

In time, blockchain will also play an increasingly important role. This is a system that can be used to record data. This can be, for example, a crop registration, but can also contain appointments, personal messages or other data. The special thing about blockchain is that falsification or adjustment of the recorded data is not possible. With this, a product can be traced back to the plot, making all data available. And the end user has precise insight into what has been done along the chain.

Finally, the question of who owns the data remains. Not everyone thinks the same about this. Some large tractor manufacturers claim that the data collected with their machines belongs to them. Farmers sign the purchase agreement that they relinquish the data the machine collects.

In the Netherlands, the agricultural sector is still fairly much in the hands of farmers. The cooperative plays an important role. Fortunately, cooperatives are developing initiatives and platforms to collect data, but this data remains the property of the grower. Nothing can be done with this data without the prior consent of the grower.

In short, a lot of data are collected and used on Jan Reinier de Jong's farm. Some of these data have been kept for almost 50 years and have direct value in business operations. With the arrival of all sorts of new equipment, sensors and precision agriculture, more and more data are becoming available. The question remains whether this data always has added value. The downside is that obtaining and using this data involves a lot of additional cost. With low prices for agricultural products, it is not easy to make these techniques profitable. Opportunities lie in a better financial distribution throughout the chain. But also, by compensating the green and blue infrastructure services provided by the agricultural sector.

References

[BBB20] Brown, P., N. Bocken, and R. Balkenende. 2020. How do companies collaborate for circular oriented innovation? *Sustainability* 12(4):1–21.

[BBB+20] Boaz A., J. Balendonck, R. Barth, O. Ben-Shahar, Y. Edan, T. Hellström, J. Hemming, P. Kurtser, O. Ringdahl, T. Tielen, and B. van Tuijl. 2020. Development of a sweet pepper harvesting robot. *Journal of Field Robotics*. ▶ https://doi.org/10.1002/rob.21937.

[BBB16] Blok P.M., R. Barth, and W. van den Berg. 2016. Machine vision for a selective broccoli harvesting robot. *IFAC-PapersOnLine* 49(16):66–71. ▶ https://doi.org/10.1016/j.ifacol.2016.10.013.

[BCA+20] Byerlay, Ryan A.E., Charlotte Coates, Amir A. Aliabadi, and Peter G. Kevan. 2020. In situ calibration of an uncooled thermal camera for the accurate quantification of flower and stem surface temperatures. *Thermochimica Acta* 693. ▶ https://doi.org/10.1016/j.tca.2020.178779.

[Ber19] Roland Berger GmbH. 2019. Landwirtschaft 4.0 – Digitalisierung als Chance. ▶ https://www.rolandberger.com/de/Insights/Publications/Landwirtschaft-4.0-Digitalisierung-als-Chance.html. Accessed 13 Mar 2022.

[Bet20] Better Food Ventures. 2020. *Farm Tech Landscape*. ▶ https://betterfoodventures.com/agtech-landscapes/farm-tech-landscape-2020. Accessed 13 Mar 2022.

[BGZ08] Brückner, B., M. Geyer, and J. Ziegler. 2008. *Spargelanbau Grundlagen für eine erfolgreiche Produktion und Vermarktung*, 128. Stuttgart: Ulmer Verlag, (ISBN 978-3-8001-4627-7).

[BHC+19] Birrell S., J. Hughes, J.Y Cai, and F. Iida. 2019. A field-tested robotic harvesting system for iceberg lettuce. *Journal of Field Robotics* 37(2020):225–245. ▶ https://doi.org/10.1002/rob.21888.

[BH16] Bellon-Maurel, V., and C. Huyghe. 2016. L'innovation technologique dans l'agriculture. *Géoéconomie* 3:159–180.

[BH17] Bellon, Maurel V., and C. Huyghe. 2017. Putting agricultural equipment and digital technologies at the cutting edge of agroecology. *OCL* 24(3):D307.

[Ber18] Berducat, M. 2018. *Vers la possibilité de repenser la mécanisation agricole en Grandes Cultures grâce à la robotique?* Conférence PHLOEME Paris.

[BMB+19] Boini, A., L. Manfrini, G. Bortolotti, L. Corelli-Grappadelli, and B. Morandi. 2019. Monitoring fruit daily growth indicates the onset of mild drought stress in apple. *Scientia Horticulturae* 256:108520.

[BP17] Bouttet D., and P. Pierson 2017. *Digifermes : un laboratoire des technologies numériques* (446): 38–39

[BSG+16] Brun F., M. Siné, S. Gallot, B. Lauga, J. Colinet, Cimino., T.P. Haezebrouck, and S. Besnard. 2016. *ACTA - Les Instituts Techniques Agricoles: L'accès aux données pour la Recherche et l'Innovation en Agriculture.* ISBN: 978-2-85794-298-6.

[BVA19] BVA Study. ▸ https://www.bva-group.com/sondages/agriculture-nouvelles-technologies-on/. Accessed 5 July 2021.

[CBT06] Coulouma, G., H. Boizard, G. Trotoux, P. Lagacherie, and G. Richard. 2006. Effect of deep tillage for vineyard establishment on soil structure: A case study in Southern France. *Soil and Tillage Research* 88 (1–2): 132–143.

[CCG+12] Cerrutti, N., G. Chaigne, M. Gayrard, E. Emonet, and A. Chabert. 2012. Description des systèmes d'exploitation de référence. In *Actes du Colloque Pollinov*, 23. France: Poitiers.

[Cer20] Firma Cerescon. ▸ http://www.cerescon.com/NL/home. Accessed 14 July 2020.

[Cha01] Charte sur l'utilisation des données agricoles DATA-AGRI. 2018. ▸ https://www.data-agri.fr/Asset/Charte_Data-Agri-Utilisation%20des%20donn%C3%A9es%20agricoles.pdf.

[CJL+20] Claus, Anna-Sophie, Julia Johns, Tomke Lindena, Hiltrud Nieberg, and Heike Kuhnert. 2020. *"Nachhaltigkeitsmodul", Vortrag im Rahmen des Molkereitreffens in der Pilotphase.* Berlin. Accessed 18 Febr 2021.

[Cle01] Cleveland, W.S. 2001. Data science: An action plan for expanding the technical areas of the field of statistics. *International Statistical Review* 69:21–26.

[CML14] Chen, M., S. Mao, and Y. Liu. 2014. Big Data: A Survey. *Mobile Networks and Applications* 19 (2): 171–209.

[DMC17] Desbourdes, C., P. Métais, and D. Chavassieux. 2017. L'automatisation a le vent en poupe. *Perspectives Agricoles* 446:46–49.

[Dee20] Firma Deepfield-Connect. ▸ www.deepfield-connect.com. Accessed 14 July 2020.

[Deu21] Deutsches Weininstitut. 2021. *Deutscher Wein Statistik.* ▸ www.deutscheweine.de. Accessed Mar 2021.

[Dre13] Dressler. 2013. Innovation management of German wineries: From activity to capacity – An explorative multi-case survey. *Wine Economics and Policy* 2 (1): 19–26.

[DTM18] Dias, P., A. Tabb, and H. Medeiros. 2018. Apple flower detection using deep convolutional networks. *Computers in Industry* 99:17–28. ▸ https://doi.org/10.1016/j.compind.2018.03.010.

[DVS11] Dzikiti, S., S.J. Verreynne, and J. Stuckens, et al. 2011. Seasonal variation in canopy reflectance and its application to determine the water status and water use by citrus trees in the Western Cape. *South Africa. Agr Forest Meteorol.* 151:1035–1044.

[Dul20] Firma Dulks. ▸ www.dulks.de. Accessed 14 July 2020.

[Eco20] Firma Ecoation. ▸ www.ecoation.com. Accessed 14 July 2020.

[Eng20] Firma Engels. ▸ www.engelsmachines.nl. Accessed 14 July 2020.

[EU18] EU Code of conduct on agricultural data sharing by contractual agreement. 2018. ▸ https://www.copa-cogeca.eu/img/user/files/EU%20CODE/EU_Code_2018_web_version.pdf.

[EFB+18] Eylenbosch D., J.A. Fernández Pierna, V. Baeten, and B. Bodson. 2018. *Utilisation de l'imagerie hyperspectrale proche infrarouge combinée aux outils de la chimiométrie dans l'étude de systèmes racinaires.* Conférence PHLOEME Paris.

[FCS+15] Fountas, S., G. Carli, C.G. Sørensen, Z. Tsiropoulos, C. Cavalaris, A. Vatsanidou, … and B. Tisserye. 2015. Farm management information systems: Current situation and future perspectives. *Computers and Electronics in Agriculture* 115:40–50.

[FH18] Frohman, C.A., Mira de Orduña, and R. Heidinger. 2018. The substratostat – an automated near-infrared spectroscopy-based variable-feed system for fed-batch fermentation of grape must. *OENO One* 52:4.

[FME21] Federal Ministry for Economic Affairs and Energy. 2021. *Dossier on skilled professionals for Germany.* ▸ www.bmwi.de. Accessed Mar 2021.

[FMS+18] Fernandez R., H. Montes, J. Surdilovic, D. Surdilovic, P. Gonzales-de-Santos, and M. Armada, et al. 2018. Automatic detection of field-grown cucumbers for robotic harvesting. *IEEE Access* 6: 35512–35527. ▸ https://doi.org/10.1109/ACCESS.2018.2851376.

[For20]   Fortune Business Insights. 2020. *Wine market size and industry analyses by type, flavor, distribution channel and regional forecast.* ▶ www.fortunebusinessinsights.com. Accessed Jan 2021.

[FPO14]   Fuentes, S., C. Poblete-Echeverría, and S. Ortega-Farias, et al. 2014. Automated estimation of leaf area index from grapevine canopies using cover photography, video and computational analysis methods. *Australian Journal Grape Wine R.* 20:465–473.

[Fra18]   Fraunhofer Gesellschaft. ▶ https://www.fraunhofer.de/content/dam/zv/en/press-media/2018/February/ResearchNews/lightweight-robots-harvest-cucumbers.pdf. Accessed 13 Mar 2022.

[Gar20]   Firma Garford. ▶ https://garford.com/de/robocrop-inrow-weeder. Accessed 13 Mar 2022.

[Gen20]   „Haltungen mit Rindern und Rinderbestand für Mai 2020 und November 2020", Genesis Online Datenbank. ▶ https://www.destatis.de/DE/Themen/Branchen-Unternehmen/Landwirtschaft-Forstwirtschaft-Fischerei/Tiere-Tierische-Erzeugung/Tabellen/betriebe-rinder-bestand.html. Accessed 18 Febr 2021.

[Gen20a]   „41311-0001, Gehaltene Tiere: Deutschland, 2009–2020, Rinder 2 Jahre und älter, Milchkühe", Genesis Online Datenbank. ▶ https://www-genesis.destatis.de/genesis/online. Accessed 18 Febr 2021.

[Gen20b]   „41311-0003, Betriebe: Deutschland, 2009–2020, Rinder 2 Jahre und älter, Milchkühe", Genesis Online Datenbank. ▶ https://www-genesis.destatis.de/genesis/online. Accessed 18 Febr 2021.

[Gey18]   Geyer, M. 2018. Mechanisation of white asparagus harvest–overview and perspectives. *Acta Horticulturae (Acta Hort 1223)* 239–249. ▶ https://doi.org/10.17660/ActaHortic.2018.1223.33.

[Gon16]   Goncharuk, A. 2016. The challenges of efficiency and security of international food value chains. *Journal of Applied Management and Investments* 5 (4): 241–249.

[Gon17]   Goncharuk, A. 2017. Wine value chains: Challenges and prospects. *Journal of Applied Management and Investments* 6 (1): 11–27.

[GPC+18]   Gourdain E., F. Piraux, G. Couleaud, G. Grignon, M. Launay, D. Deudon, D. Gaucher, F. Moreau, and X. Le Bris. 2018. *L'apport des datasciences dans la modélisation des maladies des céréales à paille.* Conférence PHLOEME Paris.

[HHV+02]   Van Henten E.J., J. Hemming, B.A.J. Van Tuijl, J.G. Kornet, J. Meuleman, J. Bontsema, and E.A. Van. 2002. Os An autonomous robot for harvesting cucumbers in greenhouses. *Autonomous Robots* 13 (3): 241–258.

[Hmf20]   Firma Hermeler. ▶ https://www.hmf-hermeler.de. Accessed 14 July 2020.

[Hor20]   Firma Hortisem. ▶ www.hortisem.de. Accessed 14 July 2020.

[HRL+20]   Huang, Y.R., Z.H. Ren, D.M. Li, and X. Liu. 2020. Phenotypic techniques and applications in fruit trees: A review. *Plant Methods* 16:107. ▶ https://doi.org/10.1186/s13007-020-00649-7.

[Int20]   International Organization of Vine and Wine. 2020. *State of the world vitivinicultural sector in 2019.* Statistic report. ▶ www.oiv.int. Accessed Jan 2021.

[IP15]   Isaac H., and M. Pouyat. 2015. *Les défis de l'agriculture connectée dans une société numérique*, 106. Rennaissance numérique: Livre blanc.

[Isi20]   Informationssystem Integrierte Pflanzenproduktion e. V. ▶ www.ISIP.de. Accessed 14 July 2020.

[KBG+17]   Käthner, J., A. Ben-Gal, R. Gebbers, A. Peeters, W.B. Herppich, and M. Zude-Sasse M. 2017. Evaluating spatially resolved influence of soil and tree water status on quality of European plum grown in semi-humid climate. *Frontiers in Plant Science* 8:1053. ▶ https://doi.org/10.3389/fpls.2017.01053.

[Kin17]   King, A. 2017. Technology: The future of agriculture. *Nature* 544:S21–S23. ▶ https://doi.org/10.1038/544S21.

[KNP+14]   Koshy, S.-S., Y. Nagaraju, S. Palli, Y. Prasad, and N. Pola. 2014. Wireless sensor network based forewarning models for pests and diseases in agriculture: A case study on groundnut. *International Journal of Advanced Research and Technology* 3:74–82.

[Lac18]   Lachia N. 2018. *Numérique et Conseil en Grandes Cultures.* ▶ http://agrotic.org/observatoire/wp-content/uploads/2018/07/20180130_ObsDossierGC.pdf.

[Lan20]   Greta Langer. 2020. Der digitale Pfad der deutschen Milchwirtschaft – Ein Überblick. ▶ https://www.milchtrends.de/fileadmin/milchtrends/5_Aktuelles/2020-10.pdf. Accessed 18 Febr 2021.

[Lau17]   Lauga B. 2017. *Un tableau de bord pour un pilotage plus efficace* 446:40.

[Lau18]   Lauga B. 2018. *Faire émerger de nouveaux services pour l'agriculteur dans une chaine de confiance gérant les consentements d'accès aux données des exploitations.* Conférence PHLOEME Paris.

[LCJ+20]   Tomke Lindena, Anna Sophie Claus, Julia Johns, Hiltrud Nieberg, and „QM-Nachhaltigkeitsmodul Milch – es geht weiter!". 2020. ▶ https://media.diemayrei.de/57/722557.pdf. Accessed 18 Febr 2021.

[LM20]   Lecoeur, J., and B. Moureaux. 2020. Adaptation aux aléas : Identifier les leviers d'action en intégrant la modélisation et le Big Data. *Perspectives Agricoles* 475:42–43.

[LVS+20] Lu R.F., R. Van Beers, W. Saeys, C.Y. Li, and H.Y. Cen. 2020. Measurement of optical properties of fruits and vegetables: A review. *Postharvest Biology and Technology* 111003.

[Kre20] Firma Kress-Landtechnik. ▶ https://www.kress-landtechnik.eu/de/produkte/robovator.php. Accessed 14 July 2020.

[MAC06] Moller, M., V. Alchanatis, Y. Cohen, et al. 2006. Use of thermal and visible imagery for estimating crop water status of irrigated grapevine. *Journal of Experimental Botany* 58:827–838.

[Maz17] Mazaud, C. 2017. « À chacun son métier », les agriculteurs face à l'of numérique. *Sociologies pratiques* 1 (34): 39–47.

[MBS20] Mehlhose Clara, Gesa Busch, and Achim Spiller. 2020. Tierwohl im Molkereiproduktregal – Neue Herausforderungen für Erzeuger und Molkereien. ▶ https://www.milchtrends.de/fileadmin/milchtrends/5_Aktuelles/2020_04_Tierwohl_Moproregal_final.pdf. Accessed 18 Febr 2021.

[MBT09] Mésséan, A., H. Bernard, and E. Turckheim. 2009. *Concevoir et construire la décision : démarches en agriculture, agroalimentaires et espace rural, Editions Quae*.

[MHP14] Malveaux, C., S.G. Hall, and R. Price. 2014. *Using drones in agriculture: Unmanned aerial systems for agricultural remote sensing applications. Proceedings of the 2014 annual meeting of the American Society of Agricultural and Biological Engineers from July, 13th to July, 16th in Montreal*, Canada, pp.1–10.

[Mic17] Micheloni, C. 2017. Diseases and pests in viticulture. Starting paper of the EIP-AGRI Focus Group. ▶ https://ec.europa.eu. Accessed Dec 2020.

[Mil20a] "Deutsche Milchindustrie in Zahlen 2010–2019", MIV. ▶ https://milchindustrie.de/wp-content/uploads/2020/04/Milchmarkt-in-Zahlen_2010-2019_Homepage_neu.pdf. Accessed 18 Febr 2021.

[Mil20b] "Strategie 2030 der deutschen Milchwirtschaft", MIV. ▶ https://milchindustrie.de/wp-content/uploads/2020/04/Milchmarkt-in-Zahlen_2010-2019_Homepage_neu.pdf. Accessed 18 Febr 2021.

[MS19] Maes, W., and K. Steppe. 2019. Perspectives for remote sensing with unmanned aerial vehicles in precision agriculture. *Trends in Plant Science* 24:152–164. ▶ https://doi.org/10.1016/j.tplants.2018.11.007.

[Nak08] Nakamoto, S. 2008. *Bitcoin: A Peer-to-Peer Electronic Cash System*. ▶ https://bitcoin.org. Accessed Mar 2021.

[NDL19] Nordey, T., F. Davrieux, and M. Léchaudel. 2019. Predictions of fruit shelf life and quality after ripening: Are quality traits measured at harvest reliable indicators? *Postharvest Biology and Technology* 52–60.

[Nai20] Firma Naiture. ▶ https://www.naiture.org/. Accessed 14 July 2020.

[Neu20] Firma Neubauer. ▶ https://www.neubauer-automation.de/. Accessed 14 July 2020.

[New20] Firma Newtec. ▶ www.newtec.com. Accessed 14 July 2020.

[PKM09] Paschold, Peter-Jürgen, Juergen Kleber, and Norbert Mayer. 2009. „Bewässerungssteuerung bei Gemüse im Freiland", Landbauforschung, Sonderheft 328.

[PLT20] Penzel, M., A.N. Lakso, N. Tsoulias, and M. Zude-Sasse. 2020. Carbon consumption of developing fruit and individual tree's fruit bearing capacity of 'RoHo 3615' and 'Pinova' apple. *International AgroPhysics* 34:409–423. ▶ https://doi.org/10.31545/intagr/127540.

[QKC+19] Qin J.W., M.S. Kim, K.L. Chao, S. Dhakal, B.K. Cho, S. Lohumi, C.Y. Mo, Y.K. Peng, and M. Huang. 2019. Advances in Raman spectroscopy and imaging techniques for quality and safety inspection of horticultural products. *Postharvest Biology and Technology* 149:101–117.

[QZM+19] Qi, F., X. Zhu, G. Mang, M. Kadoch, and W. Li. 2019. UAV Network and IoT in the Sky for Future Smart Cities. *IEEE Network* 33:96–101.

[Rau20] Firma Rauch. ▶ www.rauch.de. Accessed 14 July 2020.

[Ram20] Firma Ram. ▶ www.ram-group.com . Accessed 14 July 2020.

[RBN+19] Romero-Trigueros, C., J.M. Bayona Gambín, P.A. Nortes Tortosa, J.J. Alarcón Cabañero, and E.Nicolás Nicolás. 2019. Determination of crop water stress index by infrared thermometry in grapefruit trees irrigated with saline reclaimed water combined with deficit irrigation. *Remote Sensing* 11:757.

[RNH+19] Rojas-Downing, M.M., A.P. Nejadhashemi, T. Harrigan, and S.A. Woznicki. 2017. Climate change and livestock: Impacts, adaptation, and mitigation. *Climate Risk Management* 16:145–163. ▶ https://doi.org/10.1016/j.crm.2017.02.001. Accessed 18 Feb 2021.

[Rob11] Roberto, M. 2011. The changing structure of the global wine industry. *International Business and Economics Research Journal* 2 (9): 1–14.

[Sab09] Sablayrolles, J.-M. 2009. Control of alcoholic fermentation in winemaking: Current situation and prospect. *Food Research International* 42 (4): 418–424.

[SBL+18] Soenen, B., P. Bessard Duparc, M. Laberdesque, J.C. Deswarte, A. Bouthier, F. Laurent, and X. Le Bris. 2018. *Piloter conjointement la fertilisation azotée et l'irrigation par couplage d'observations sol/plante avec le modèle de culture CHN*. Conférence PHLOEME Paris.

[SBT+18] de Solan, B., F. Baret, S. Thomas, S. Madec, A. Comar, K. Beauchêne, and A. Fournier. 2018.

Systèmes de phénotypage haut-débit au champ, méthodes associées et premiers résultats. Conférence PHLOEME Paris 2018.

[SCN18] Soenen B., M. Closset, A. Nonnard, and X. Le Bris. 2018. *Le pilotage de l'azote sur blé dans le service FARMSTAR.* Conférence PHLOEME Paris.

[SDL17] de Solan, B., O. Deudon, and F. Leprince. 2017. *L'internet des objects impacte tout le secteur agricole* (446): 42–45.

[SDW+19] Schwinn, M., D. Durner, M. Wacker, A. Delgado, and U. Fischer. 2019. Impact of fermentation temperature on required heat dissipation, growth and viability of yeast, on sensory characteristics and on the formation of volatiles in Riesling. *Australian Journal of Grape and Wine Research* 25:173–184.

[SEA+18] Sanz, R., J. Llorens, A. Escola, J. Arno, S. Planas, C. Roman, and J.R. Rosell-Polo. 2018. LIDAR and non-LIDAR-based canopy parameters to estimate the leaf area in fruit trees and vineyard. *Agricultural and Forest Meteorology* 260:229–239. ▶ https://doi.org/10.1016/j.agrformet.2018.06.017.

[SFM+20] Santos, J.-A., H. Fraga, A.-C. Malheiro, J. Moutinho-Pereira, L.-T. Dinis, C. Correia, M. Moriondo, L. Leolini, C. Dibari, S. Costafreda-Aumedes, T. Kartschall, C. Menz, D. Molitor, J. Junk, M. Beyer, and Schultz, H.-R. 2020. A review of the potential climate change impacts and adaptation options for European viticulture. *Applied Sciences* 10 (3092): 1–28.

[SGP17] Siné, M., E. Gourdain, and X. Pinochet. 2017. La dynamique des startup agricoles. *Perspectives Agricoles* 446:52–53.

[SHE15] Siné, M., T.P. Haezebrouck, and E. Emonet. 2015. API-AGRO: An Open Data and Open API platform to promote interoperability standards for Farm Services and Ag Web Applications. *AGRÁRINFORMATIKA/Journal of Agricultural Infomatics* 6 (4): 56–64.

[STD09] Sansoni, G., M. Trebeschi, and F. Docchio. 2009. State-of-The-Art and applications of 3D imaging sensors in industry, cultural heritage, medicine, and criminal investigation. *Sensors (Basel)* 9 (1): 568–601.

[STD+20] de Solan B., S. Thomas, G. Deshayes, J. Labrosse, W. Li, P. Piquemal, P. Porrez, D. Bouttet, O. Deudon, S. Jézéquel, P. Braun, F. Aubertin, A. Vanhoye, C. Vivens, K. Velumani, F. Baret, A. Comar, F. Leprince, and M. Siné. 2020. *Modèle de culture et mesure par capteurs : quelle complémentarité pour l'aide à la décision ?* Conférence PHLOEME Paris 2020.

[STE20] Firma Steenks. ▶ www.steenks-service.de. Accessed 14 July 2020.

[SVW+16] Sundmaeker, H., C. Verdouw, S. Wolfert, and L. Pérez Freire. 2016. "Internet of food an farm 2020", Digitising the Industry – Internet of Things Connecting Physical, Digital and Virtual Worlds, 129–151.

[Swe20] Sweet pepper harvesting robot. ▶ www.sweeper-robot.eu/. Accessed 13 Mar 2022.

[SWH+18] Shamshiri, R., C. Weltzien, I. Hameed, I. Yule, T. Grift, S. Balasundram, L. Pitonakova, D. Ahmad, and G. Chowdhary. 2018. Research and development in agricultural robotics: A perspective of digital farming. *International Journal of Agricultural and Biological Engineering* (4): 1–14. ▶ https://www.ijabe.org/index.php/ijabe/article/view/4278/1737.

[TCP+15] Toqué C., S. Cadoux, P. Pierson, B. C. Flenet, F. Angevin, and P. Gate. 2015. *SYPPRE: A project to promote innovations in arable crop production mobilizing farmers and stakeholders and including co-design, ex-ante evaluation and experimentation of multi-service farming systems matching with regional challenges.* 5th International Symposium for Farming Systems Design. Montpellier – France.

[TFZ22] Tsoulias, N., S. Fountas, and M. Zude-Sasse. 2022. *Tree growth modelling by means of LiDAR laser scanner.* Biosystems Engineering.

[TGZ20] Tsoulias, N., R. Gebbers, and M. Zude-Sasse. 2020. Using data on soil ECa, soil water properties, and response of tree root system for spatial water balancing in an apple orchard. *Precision Agriculture* 21:522–548. ▶ https://doi.org/10.1007/s11119-019-09680-8.

[THK03] Trapp, M., G. Hörner, and R. Kubiak. 2003. *Functional landscape characterisation with object-oriented image analysis for a GIS-based local risk assessment. Proceedings of the XII Symposium on Pesticide Chemistry from June 4th to June 6th, 2003 in Piacenza,* 649–655. Italy.

[Tom19] Tomasso L. (2019). *Analyse juridique contractuelle des données de l'agriculture numérique.* ▶ https://numerique.acta.asso.fr/multipass-analyse-juridique-contractuelle-des-donnees-de-lagriculture-numerique/.

[Tou10] Touzard, J.-M. 2010. *Innovation systems and the competition between regional vineyards. Proceedings of the Innovation and Sustainable Development in Agriculture Symposium from June, 28th to July, 1st, 2010 in Montpellier,* 1–13. France.

[TPF+19] Tsoulias, N., D.S. Paraforos, S. Fountas, and M. Zude-Sasse. 2019. Estimating canopy parameters based on the stem position in apple trees using a 2D LiDAR. *Agronomy* 9:740. ▶ https://doi.org/10.3390/agronomy9110740.

[Vis20a] Firma Visser. ▶ www.visser.eu. Accessed 14 July 2020.

[Vis20b] Firma Visar. ▸ www.visar-europe.com. Accessed 14 July 2020.

[Was16] Wassan, J.T. 2016. Big data paradigm for healthcare sector. In *Managing Big Data Integration in the Public Sector*, Eds. A. Aggarwal, 169–186.

[WBZ+20] Walsh, K.B., J. Blasco, M. Zude-Sasse, and Sun Xudong. 2020. Review: Visible-NIR 'point' spectroscopy in postharvest fruit and vegetable assessment: The science behind three decades of commercial use. *Postharvest Biology and Technology* 168. ▸ https://doi.org/10.1016/j.postharvbio.2020.111246.

[Wet20] Agrarmeteorologie Rheinland-Pfalz. ▸ www.wetter.rlp.de. Accessed 13 Mar 2022.

[WGV+17] Wolfert, S., L. Ge, C. Verdouw, and M. J. Bogaardt. 2017. "Big data in smart farming – a review", *Agricultural Systems* 153:69–80. ▸ https://doi.org/10.1016/j.agsy.2017.01.023. Accessed 18 Febr 2021.

[WPJ+18] Wu, D., S. Phinn, K. Johansen, A. Robson, J. Muir, and C. Searle. 2018. Estimating changes in leaf area, leaf area density, and vertical leaf area profile for mango, avocado, and macadamia tree crowns using terrestrial laser scanning. *Remote Sensing* 10:1750. ▸ https://doi.org/10.3390/rs10111750.

[WSG14] Wolfert, J., G. Sørensen, and D. Goense. 2014. "A Future Internet Collaboration Platform for Safe and Healthy Food from Farm to Fork, Global Conference (SRII), 2014 Annual SRII. *IEEE, San Jose, CA, USA* 2014:266–273.

[WUW18] Wang, Z.L., J. Underwood, and K.B. Walsh. 2018. Machine vision assessment of mango orchard flowering. *Computers and Electronics in Agriculture* 51:501–511. ▸ https://doi.org/10.1016/j.compag.2018.06.040.

[Wym20] Firma Wyma. ▸ https://www.wymasolutions.com/. Accessed 14 July 2020.

[XFS+19] Xue, J., Y. Fan, B. Su, and S. Fuentes. 2019. Assessment of canopy vigor information from kiwifruit plants based on a digital surface model from unmanned aerial vehicle imagery. *International Journal of Agricultural and Biological Engineering* 12(1):165–171.

[ZFG+16] Zude-Sasse, M., S. Fountas, T.A. Gemtos, and N. Abu-Khalaf. 2016. Applications of precision agriculture in horticultural crops – A REVIEW. *European Journal of Horticultural Science* 2016(81):78–90. ▸ https://doi.org/10.17660/eJHS.2016/81.2.2.

[ZH19] Zhao, Q., and T. Hastie. 2019. Causal interpretations of black-box models. *Journal of Business & Economic Statistics*, 1–10.

Sustainability Perspective

Keith A. Wheeler, Friedhelm Taube, Klaus Erdle, Markus Frank and Isabel Roth

Contents

6.1 Role of Multi-Stakeholder Organizations in Digital Sustainable Agriculture Transformation – 343
6.1.1 Introduction – 343
6.1.2 Discussion – 345
6.1.3 Conclusion – 348

6.2 Digitalization Towards a More Sustainable AgFood Value Chain – 350
6.2.1 Setting the Stage: Sustainable and Resilient Agriculture – 350
6.2.2 Sustainability Assessment and Management Systems – 351
6.2.3 Case Studies for Effective Interoperability of FMIS and Sustainability Assessment – 352
6.2.4 Value Creation Through New Business Models – 355
6.2.5 Bringing Sustainability to Life—The Role of Digitalization – 357

6.3 From Ecological Intensification to Hybrid Agriculture— The Future Domain of Digital Farming – 357
6.3.1 Introduction – 357
6.3.2 Why Should the Production Narrative be Questioned? – 358
6.3.3 Why is Ecological Intensification the New Paradigm? – 359
6.3.4 Is an Ecological Intensification in Western Europe Justifiable in View of the Global Hunger Problem? – 360
6.3.5 Animal Sourced Foods—Less and Better? – 361
6.3.6 Back to the Roots: Integrated Crop—Livestock Systems Including Grass-Clover Leys? – 362

© The Author(s), under exclusive license to Springer-Verlag GmbH, DE, part of Springer Nature 2022
J. Dörr and M. Nachtmann (eds.), *Handbook Digital Farming*,
https://doi.org/10.1007/978-3-662-64378-5_6

6.3.7	Implementation of Ecological Intensification Through Hybrid Agriculture and Public Goods Boni? – 362	
6.3.8	Conclusion – 364	

6.4 Digitization as Co-Designer for Cropping Systems: Technology Shapes Cropping Systems, Now and in Future – 365

6.4.1	Mechanization and Its Excesses on Farming – 365
6.4.2	Precision and Efficiency, the Evil Twins – 365
6.4.3	Back to Reality – 367
6.4.4	Digitization for Initiating Promising Cropping Concepts – 367
6.4.5	Conclusion – 370

6.5 Fighting Climate Change Through "Carbon Farming": A Future Business Opportunity for Digital Farming? – 371

6.5.1	The Carbon Cycle and Its Interaction with the Nitrogen Cycle – 371
6.5.2	Carbon Sequestration in Agricultural Soils—A Synopsis – 372
6.5.3	Digitization as Key Enabler of Carbon Trading in Agriculture – 377
6.5.4	Conclusion – 377

References – 378

Sustainability Perspective

6.1 Role of Multi-Stakeholder Organizations in Digital Sustainable Agriculture Transformation

Keith A. Wheeler

Abstract

Over the past two decades there has been an increasing awareness for the linkage between food security, sustainable farming standards and practices, and digital agriculture. Several sustainable agriculture organizations and associations play a critical role in a digital sustainable farming transformation to achieve the Sustainable Development Goals (SDGs). Regional and global associations have emerged to serve as an intellectual and practical convener for a range of sustainable farming themes including: (1) developing vision consensus for sustainable agriculture, (2) adopting sustainability metrics and best practices, (3) creating assessments and tools to support agricultural supply chains, (4) engaging multi-stakeholder partnerships to implement voluntary sustainability standards (VSS), (5) and support the open-sourced availability for quality data and platforms to drive innovation in digital sustainable farming. Together these opportunities are transforming sustainable agriculture, with a holistic and integrated perspective, to help achieve agricultural resiliency and food security. Several sustainable agriculture multi-stakeholder organizations will be highlighted in this article, representing regional exemplars engaged with multi-regional audiences.

6.1.1 Introduction

Farming associations have been one of the key drivers for change in agricultural practices and markets over the past century in Europe and North America. These organizations have provided a key networking and capacity development role for adopting new scientifically based practices, the evolution and acceptance of new technologies, and are a catalyst for shaping agricultural policies at the regional, national and international levels.

Agricultural sustainability has been a theme in farming since the mid-twentieth century both in Europe, the UK, and the USA. It has evolved over time in various iterations to promote self-reliance through adapting farming systems based on the stewardship of the land, healthy soil, food quality, food sufficiency, human health, and maintaining healthy farming communities [MAC90]. In the 1960's and 70's, the Green Revolution was catalyzed by the need to address the issue of malnutrition in the developing world. The primary breakthroughs during the Green Revolution were to increase yields through technology enhanced plant breeding combined with increased chemical pest control and fertilizers, and the expansion of large irrigation efforts. These efforts reshaped farming systems worldwide, especially in the supporting agricultural supply chains.

In 1987, the Report of the World Commission on Environment and Development: Our Common Future [WOR87] outlined a "global agenda for change" and proposed long-term environmental strategies for achieving sustainable development. It was this challenge that set a globalized framework for a more formal action on Sustainable Agriculture. The Food and Agriculture Organization of the United Nations (FAO) defined sustainable agricultural development as "the management and conservation of the natural resource base, and the orientation of technological and institutional change in such a manner as to ensure the attainment and continued satisfaction of human needs for present and future generations. Such sustainable development (in the agriculture, forestry and fisheries sectors) conserves land, water, plant and animal genetic resources, is environmentally non-degrading, technically appropriate, economically viable and socially acceptable" [FAO98].

An appeal was made to farming groups, non-governmental organizations, the private sector, educational institutions, and the scientific community to take up the challenge of increasing production to keep up with increasing demand, while retaining the essential ecological integrity of agricultural production systems as core to the sustainable agriculture paradigm. Significant dialog and debate occurred from 1987 through 2000 by many organizations on sustainable agriculture focusing on definition articulation, sustainability metrics, and farm and supply chain-based assessments. Despite the efforts for making sustainability assessments in the food and agriculture sectors accurate and easy to manage, no internationally accepted benchmark emerged that defined a common rubric for what sustainable food production entails.

In the first decade of the 2000's, new multi-sectorial sustainable agricultural organizations emerged and began playing a significant role in refining and implementing sustainable agriculture definitions and voluntary standards for large scale arable agriculture in Europe and North America. The 2012 United Nations Conference on Sustainable Development (UNCSD), also known as Rio+20, established a framework for the Sustainable Development Goals (SDGs) and in 2015, seventeen broad goals and specific targets were ratified by the United Nations [UNI15] (see ▶ Sect. 1.3). A number of the SDGs are directed at sustainable agriculture and the food supply systems to end hunger and all forms of malnutrition by 2030.

During the same period, a number of corporate socially responsible (CSR) sustainability reporting standards were evolving that would influence and shape the agriculture supply chain including: the Global Reporting Initiative (GRI), the United Nations Global Compact, Sustainability Accounting Standards Board (SASB), Rain Forest—UTZ standards, Carbon Disclosure Project (CDP), and the International Organization for Standardization (ISO). In 2014, FAO produced an analysis of 38 different assessment schemes and tools by various organizations for Sustainable Food and Agriculture Systems [FAO14]. Points of view and processes tended to be both bottom-up and top-down as it related to the food and agriculture supply chain.

Addressing sustainability from the farmer perspective has proven to be a daunting task over the past 30 years. Farmers were being urged by their input suppliers (pesticides, fertilizers, seed, machinery, and Digital Farming) to adopt and track these new innovations that will support more sustainable practices and allow the suppliers to report on their contributions to the sustainable agricultural supply chain. Farmers were simultaneously being pulled by the consumer market and its value chain companies to use the best possible practices to grow food that is environmentally and socially responsible for the best price, and track critical supply chain metrics for the companies' sustainability reports.

Many barriers to sustainable practice adoption exist, ranging from: increasing numbers of national, regional and global sustainable agricultural organizations that have differing vision and roles along the food supply chain; voluntary or mandatory standards; a web of metrics for all aspects of complex farming operations; significant amounts of data to acquire, collect and manage; lack of direct linkage to existing farm management systems; lack of farmer training and support; lack of benchmarking on universally accepted best practices; increasing market demands for sustainable certification; low return on investment (ROI), and lack of value chain transparency.

6.1.2 Discussion

In the twenty-first century a number of multi-stakeholder sustainable agricultural organizations have emerged with significant input from farmers through existing grower associations, government agencies, the food and agriculture private sector—both on the supply side and the market side, NGOs, and financial institutions. Serving in a catalytic capacity, these organizations supply capacity development to their constitutes in navigating the complexity of issues influencing the changing landscape of sustainable agriculture by developing and testing new unified methods and scale it in the farming community. Three organizations will be highlighted (see ◘ Table 6.1) with a brief discussion showcasing their leadership and impact in the remainder of this section. These include: Sustainable Agriculture Initiative (SAI), Field to Market (FTM): The Alliance for Sustainable Agriculture, and The Committee on Sustainability Assessment (COSA).

6.1.2.1 The Sustainable Agriculture Initiative (SAI)

The Sustainable Agriculture Initiative (SAI) Platform was launched in 2002 by the European-based food and beverage companies Danone, Nestlé and Unilever as part of a sustainable agriculture corporate social responsibility (CSR) supply chain initiative [PHB12]. Today, the SAI non-profit organization has offices in Brussels Belgium focused on pre-competitive networking, best practices and capacity development for its 100 plus corporate members. The membership, ranging from small farmer organizations to large multi-nationals food and agricultural companies, collaborates to develop tools and recommendations to support local and global sustainable sourcing by accelerating widespread adoption of sustainable agricultural practices for the food value chain.

Through a multi-stakeholder process, SAI's membership defined sustainable agriculture as: "The efficient production of safe, high quality agricultural products, in a way that protects and improves the natural environment, the social and economic conditions of farmers, their employees and local communities, and safeguards the health and welfare of all farmed species" [SAI16].

SAI invited member organizations in commodity-specific working groups to articulate what implementing sustainable agriculture meant for their supply chain and develop principles for best practice. The Crops Working Group is made up of members from across the crops supply chain spectrum. Each working groups principles were translated into practices, tested, and refined in the field for more than four years. In 2009, SAI created the first industry-agreed set of tested and aligned practice guidelines, and validated that a pre-competitive collaborative approach was an effective means for implementing sustainability at the farm level [SAI09; BRA15]. These guidelines were utilized to develop web-based self-administered Farm Sustainability Assessment (FSA) tool [SAI18a] in 19 languages across the supply chain membership.

SAI members are actively engaged in benchmarking the assessment tool with innovative farmers in key supply chains, thus accelerating change and expanding sustainable agriculture globally. Examples of several efforts include the SAIRISI ([SAI19a] rice project in Italy, the Doñana Berry Project [SAI18b] in Spain, and the European Sugar Beet Project [SAI19b]. To date, over 60 standards have been benchmarked to obtain an equivalent FSA level rating with their existing specific commodity supply chain certifications. This allows farmers and suppliers to directly use the Sustainable Agriculture Initiative (SAI) FSA program as a universal standard with their customers, reducing costly duplication of multiple audits and certifications.

Table 6.1 Comparative table for key attributes of organizations highlighted

Key attributes	Field to Market: The Alliance for Sustainable Agriculture (FTM)	Sustainable Agriculture Initiative (SAI)	Committee on Sustainability Assessment (COSA)
Founders	The Keystone Policy Center	Danone, Nestlé and Unilever	UN Conference on Trade and Development (UNCTAD) and the International Institute for Sustainable Development (IISD)
Year founded	2006	2002	2006
Current organizational structure and headquarter location	Not-for-Profit membership organization headquartered in Washington, DC USA	Not-for-Profit membership organization headquartered in Geneva, Switzerland	Not-for-Profit membership organization headquartered in Philadelphia, PA USA
Primary geographic focus	North America	Global	Africa, Latin America, Asia
Primary partners and members	Agriculture and food companies, government agencies, NGOs, Grower groups	Agriculture and food companies, Grower groups, Individuals	Development organizations, Ag companies, Grower groups, NGOs
Multi-stakeholder developed sustainability indicators	Yes	Yes	Yes
Key primary crops	Commodity crops	Commodity and specialty crops, dairy and beef	Food and beverage small scale specialty crops
Web-based assessment tools	**Proprietary** Fieldprint® Calculator Platform	**Proprietary** Farm Sustainability Assessment (FSA) Platform	**Custom** SMART sustainability and resilience indicators and performance Monitoring tools for specific producer organizations

6.1.2.2 Field to Market (FTM): The Alliance for Sustainable Agriculture

In 2006, the Keystone Policy Center, convened a multi-stakeholder meeting across key sectors of the U.S. agricultural supply chain to explore sustainability. FTM emerged as a science and outcome-based association focused on the various commodity crop supply chains. A definitional consensus formed around building **the foundation for continuous improvement to meet the challenge of producing enough food, fiber and fuel for a rapidly growing global population while conserving natural resources and improving the ability of future generations to meet their own needs.**

In order to have a benchmark to gauge improvement for the U.S., Field to Market produced the first national level indicators report examining the environmental components for sustainability. The report analyzed trends in land use, irrigation water use, energy use, greenhouse gas emissions, and soil loss for U.S. corn, cotton, soybeans, and wheat from 1987–2007. Public scientific data was used to evaluate trends at a national scale and peer-review feedback was used to improve the report. The report Environmental Resource Indicators for Measuring Outcomes of On-Farm Agricultural Production in the United States [KEY09] was published in 2009. Subsequent reports were published in 2012 [KEY12] and 2016 [FIE16] and will be updated every five-years going forward. These reports helped identify trends, gaps and areas for future focus to implement resilient sustainable agriculture practices at the farm level across the commodity crop value chain.

FTM launched its first version of a digital farm web-based Fieldprint® calculator [FIE18] in 2009, based on metrics identified in the national reports that are relevant to individual farmers. This tool allows farmers to evaluate their performance geospatially at the field level against national and state averages for specific indicators across 12 commodity crops. The Fieldprint® analysis provides an overall sustainability score, but allows the farmer to drill down field-by-field and see how changing specific field-based practices can improve their overall farm sustainability. It is now being used in over 36 states and several provinces in Canada with a goal for active use on 7.5 million acres in the next few years.

After four years of incremental growth in participation and scope, Field to Market became an independent not-for-profit multi-stakeholder member alliance in 2013. Headquartered in Washington DC, it has 140 plus members from farmer associations, agribusiness, government organizations, food brands and retailers, civil society, and university research institutions. The members have created more than 65 specific farmer based best-practice management projects across the US and Canada to reduce environmental footprints through FTM's Continuous Improvement Accelerator Program. One sustainability acceleration effort is the Fieldprint® Platform, which enables farmers using leading precision agriculture, decision support and farm management software tools to integrate with FTM sustainability metrics in a seamless way. This will drive an unprecedented scaling of sustainable best practice options in the farm planning process.

In another recent project led by two FTM member companies, the Field to Market/SAI Platform Equivalency Module [FIE19] was utilized to analyze the sustainable performance for the wheat supply chain and align the FTM and SAI platforms. This enables farmers utilizing Field to Market's Fieldprint Platform to achieve the Bronze requirements for the SAI Platform's Farm Sustainability Assessment (FSA) and allows farmers to pursue higher levels of assurance by answering additional questions. In partnership with the Control Union these companies developed a solution that can be fully executed across their supply chains.

6.1.2.3 The Committee on Sustainability Assessment (COSA)

The Committee on Sustainability Assessment (COSA), organized in 2005, is a global consortium of UN development institutions that work collaboratively to advance the systematic and science-based measurement for sustainability in agriculture. They discovered that the lack of sustainable agriculture definitions and metrics were hampering stakeholders' ability to evaluate what was working and what was not. COSA provided strategic input for the UN's Sustainable Development in the 21st Century (SD21) project for achieving consensus on a common vision published as **Food and Agriculture: The Future of Sustainability** [GSN12+]. In 2012, COSA became a US based not-for-profit advancing research and training in the field of sustainable agricultural with a focus on supporting small-scale farmers capacity to deliver to global sustainable food systems.

Supported by USAID, SIDA and BelgianAid, COSA initiated a multi-stakeholder process to begin to define scientifically based sustainable agriculture metrics. COSA has focused on creating a master library of indicators designed to be similar across all countries to allow for global supply chain comparison but can be also adapted to allow for local contexts. This supports their work on responsible sourcing and performance tracking for leading food companies. Utilizing down scalable advanced technology platforms, COSA can deliver results from state-of-the-art research to small farmers by bridging sustainability and resilience measurement and metrics. COSA has been able to bring their expertise integrating performance data to improve farmer standards and certifications across diverse areas of impact.

The key drivers for COSA to expand sustainable agriculture globally have been the creation of sustainable business intelligence platforms. With ten plus years of metric development, COSA created Sustainability Intelligence Systems that utilize advanced digital data capture and geo-spatial mapping to support smart sustainable risk management and decision intelligence, enhance traceability and compliance, and link various verification approaches for farmers and supply chain managers.

Given the challenges of climate change, farmers around the globe are needing to build farming systems that are not only sustainable but also resilient to the rapid changes they are encountering. In 2017 COSA, in partnership with the Ford Foundation, published **Simpler Resilience Measurement** [COS17] a framework for **resilience tools and indicators** to begin to address rural farming communities resilience to climate change. COSA supports the development of sustainability and resilience strategies to meet the needs of the private sector and farming communities worldwide.

6.1.3 Conclusion

The three organizations (COSA, SAI and FTM) have grown significantly both in their scope and resulting impacts across the sustainable food and agriculture sectors. Farmer participation is growing, as measured by the number of hectares utilizing the services of these organizations along with hundreds of value chain institutions that are active members. Collectively these organizations are engaging with farmers thru farmer cooperatives, farming associations dealing across the full range of commodity crops, and governmental and non-governmental organizations that directly support farmers.

As we trace the history of sustainable agriculture and the rapid advancement of digital agriculture defined as: **a set of digital and geospatial information technologies that integrates sensors, analytics and automation to monitor, assess and manage soil, climatic and genetic resources at field and landscape scales** [BA20], we see an equally complex critical path development from the late-1980's to the present. The organizations highlighted have made their advances through multi-stakeholder processes to identify key indicators, define standards and evaluate sustainable practices that lend themselves to use geospatial data analytics and modeling tools to quantify outcomes, and provide alternative options for balancing solutions. In addition, these organizations have recruited finance, investment and risk management partners to enhance the investment in digital tools that will support both farmer ROI and the new value chains. These tools offer greater ability for tracing and transparency for the value chains allowing support for the various sustainable and quality certifications that consumers are demanding.

Key commonalities of these exemplar programs include: a clear vision for sustainable agriculture, voluntary sustainability standards (VSS) based on a science-based systems approach that is accurate, transparent, and scalable; support for supply chain innovation and resilience; providing pathways for continuous improvement for all stakeholders; providing capacity development and technical support; and providing access to digital tools for actionable best practice management that are agile, economically viable, socially and culturally congruent, and environmentally compliant.

As digital agriculture and sustainable agriculture evolves over the next decade a number of challenges remain including:
- the need to increase the farmer utilization and participation with digital technologies
- the need to embrace various resilience strategies to mitigate the impacts of climate change
- the need to design and implement smart digital technologies that are cost effective and can be used across farming at all scales
- enabling new sustainability systems and platforms to seamlessly integrate with next generation digital agriculture platforms to increase efficiency for farmer utilization
- collectively support a digital sustainable farming value chain framework with new policies and necessary financial support to insure universal access to this new farming paradigm
- educate consumers as to the value and benefits for digitally enabled food and agriculture systems
- Perhaps the greatest opportunity to accelerate the expansion of Digital Farming will be driven by climate change mitigation and adaptation. The deployment of on-farm carbon sequestration practices and the emergence of robust carbon markets will offer farmers an opportunity to increase their revenue streams in conjunction with their sustainable practices [MB20]. Digital Farming will support the data ingestion, modeling and transformation into fungible carbon assets that will funnel into these emerging markets. The three organizations described in this section are currently engaged with these emerging carbon market initiatives. Through this coupling of sustainable agriculture and Digital Farming society will see the greatest impact on transforming the state of agriculture over the next decade to keep pace with climate change, thus insuring that nine plus billion people have resilient, quality, sustainable, and healthy supplies of food by 2050.

6.2 Digitalization Towards a More Sustainable AgFood Value Chain

Markus Frank

Abstract

This section describes the opportunity to integrate sustainability assessment into farm management. Case studies from different industry segments focus interoperability and value creation. Case studies show how digitalization assures scale-up and brought application of sustainability measures.

6.2.1 Setting the Stage: Sustainable and Resilient Agriculture

Megatrends such as the population growth, increasing urbanization, climate change and the increasing degradation of arable land and biodiversity [Iaa19] put enormous pressure on the agricultural sector to come up with a more resource efficient production of staples as well as specialty products. On the other hand, the resilience of agricultural production is gaining more and more importance as climate change and biodiversity loss will render agricultural production more and more erratic [Tyl13]. Considering resilience is important when planning for the future of agriculture as some social-ecological changes bear the potential to undermine agricultural development [EGP08]: Climate change, declines in insect biomass as well as shifts in pests and diseases all create instabilities that can disrupt the ecosystem services provided by the agricultural landscape, including food production [GPG08].

Sustainability has thus increasingly become an imperative for the AgFood value chains in all geographies of the globe and has advanced to the number one topic on the business agenda of many food and retail companies [LKJ07; Pet09; FG09]. In particular, in the area of food, consumers and governments are sustainability-sensitive forcing companies to add the sustainability dimension to their "modus operandi". Further, sustainability can influence shareholders' decisions and add value to companies' brands.

The paradigm of "sustainable intensification" as coined by the Royal Society of the UK [TRS09] aims at strategies to produce more and better-quality food with less environmental harm or even positive contributions to natural and social capital [PB18]. Sustainable agricultural systems exhibit a number of key attributes:

1. utilizing highly productive crop varieties and livestock breeds
2. avoiding the unnecessary use of external inputs
3. harnessing agroecological processes such as nutrient cycling, biological nitrogen fixation, Allelopathy, predation and parasitism
4. minimizing the use of technologies or practices that have adverse impacts on the environment and human health
5. making productive use of human capital in the form of knowledge and capacity to adapt and innovate social capital to resolve common landscape-scale or system-wide problems (such as water-, pest-, or soil management)
6. minimizing the impacts of system management or externalities such as greenhouse gas emissions, clean water, carbon sequestration, and dispersal of pests, pathogens and weeds [TRS09]

Agricultural systems emphasizing these principles tend to display a number of features that distinguish them from processes and outcomes of conventional systems. First, these systems tend to provide a broad range of ecosystem services [DP04]. They jointly produce food and other goods for farmers and markets, while contributing to a range of basic and regulating ecosystem services, not to forget leisure and tourism opportunities. In their configuration, they capitalize on the synergies and efficiencies that arise from complex ecosystems, social and economic forces [PB18]. Second, these systems are diverse, synergistic and tailored to their particular socio-ecological contexts. As many pathways towards a more sustainable agriculture exist, no single configuration of technologies, inputs and ecological managements comprises a "one-fits-all" solution. On the contrary, a more sustainable agriculture is dependent upon the need to fit these factors to the specific circumstances of different agricultural systems [HM11].

Some conventional thinking about agricultural sustainability has assumed that it implies a net reduction in input use, thus making such systems essentially extensive (requiring more land to produce the same amount of food; [NHD+11]). In particular, the majority of consumers believe that sustainable agricultural production systems require a less intensive use of resources, resulting in a "more natural" production strategy. In this vein, organic or biodynamic farming serves as a poster child for sustainable agriculture [DP04]. The concept of agroecological intensification summarizes strategies that rely on the application of ecological science to the study, design and management of sustainable agriculture [ANH15] in order to replace agricultural inputs as much as possible. However, recent evidence shows that successful agricultural sustainability initiatives and projects arise from shifts in the factors of agricultural production (e.g., from the use of mineral fertilizers to nitrogen-fixing legumes; from pesticides to emphasis on beneficials; from ploughing to zero tillage). Such approaches often sail under the banner of sustainable intensification of agriculture (SI), making better use of existing resources (e.g., land, water, biodiversity, knowledge) and technologies [Iaa19; TRS09; TBH+11] in combination with a stronger emphasis on agroecology (AEI; [ANH15]). In literature, AEI and SI have often been viewed as two pathways to agricultural sustainability that are polar opposites [GBC+10; WS09]. Complexities like this and the synergies and tradeoffs associated to the chosen strategies result in the necessity to employ sustainability assessment and management systems in order to ensure progress on the path towards a more sustainable agriculture.

6.2.2 Sustainability Assessment and Management Systems

For more than 20 years, multi-criteria sustainability assessment schemes have been launched to inform management strategies on the global, national, regional and local scale in order to improve decision making in with respect to sustainability and resilience of agricultural production. The most widely deployed systems comprise web-based tools hosted by multi-stakeholder initiatives. The "Fieldprint Calculator" launched by the Keystone Initiative "Field to Market" (see also ▶ Sect. 6.1). "Field to Market", a broad group of stakeholders—seed companies, agricultural producers, processors, trade and non-governmental organizations—joined in 2006 to establish sustainability standards in the American agricultural supply chains, particularly those of soybean, maize and cotton, and to make them measurable. Farmers can use the "Fieldprint Calculator" as an aid to check their overall sustainability in terms of energy, soil and water consumption, and use and their effect on climate. In doing so, the "Fieldprint Calculator" compares the method and the performance of individual farmers with the average in their region and in their state. Access to the "Fieldprint Calculator" is free and it is very easy to use [KHS+19]. The aim is to increase the awareness of farmers on the issue of sustainable production by comparison with a benchmark.

In a similar vein, the Cool Farm Alliance, is a membership, science-led, not-for-profit UK registered community interest company with the goal "to enable farmers around the globe to make more informed on-farm decisions that reduce their environmental impact". With a focus on greenhouse gases, biodiversity, food loss and waste and water quantity, the Alliance provides the "Cool Farm Tool" (CFT) as a quantified decision support tool that is designed to be used by agricultural practitioners [CFA19]. The members of the CFA—mainly large input manufacturers and food companies as well as organizations—encourage, reward (or even mandate) farmers to use the web-based suite of tools and to implement improvement strategies. These entities leverage the CFT as a standardized global approach to help farmers take steps to reduce greenhouse gas emissions and negative impacts on biodiversity and water from their products and help share best practice across the sector.

Other privately run initiatives include e.g., the Good Growth Plan by Syngenta or the life-cycle assessment-based method AgBalance® by BASF SE [FFV14] and launched in 2011. In contrast to the aforementioned approaches, AgBalance® assesses the performance of an agricultural production system in all three dimensions of sustainability—economy, environment and society. The tool is designed to calculate the life-cycle impact assessment of agricultural systems, applying principles of the Life Cycle Assessment (LCA) framework (defined by the ISO 14040 and 14,044 standards) and the UNEP-SETAC guidelines for social impact assessment [SFV+12].

All these methodologies are accessible through web-based software tools by either agricultural or sustainability practitioners, requiring inputs of which a farmer typically has good knowledge (and no more) and background information mostly provided through links to specific databases, have in common a specific farm-scale, decision-support focus. Due to their use of only readily available farm data, there is considerable scope for their use in global surveys to inform on current practices and potential for mitigation [KHS+19]. The methodologies and programs all aim at informing optimization strategies towards sustainability and resilience of the farming system. Despite this fact, the deployment of the tools has been rather limited, mainly due to the obligation to manually enter numerous data into these websites on a regular basis. The majority of farmers increasingly make use of web-based documentation systems in the form of field diaries or full-fledged farm management information systems (FMIS; [FST+15]) in order to better manage their farm operation. Such systems often contain the vast majority of data needed to run the aforementioned sustainability assessment schemes, typically directly recorded by farm machinery and then stored and processed by the FMIS. If these data could be made directly accessible through interoperability of the respective sustainability assessment tool and the FMIS, a quasi-automated sustainability assessment and conversion into new, valuable farm-management insight would be possible. Farmers would not be required anymore to manually input farm data and re-translate the sustainability information into an improved farm management practice.

6.2.3 Case Studies for Effective Interoperability of FMIS and Sustainability Assessment

A few pioneering success stories are summarized in order to introduce different concepts for the interoperability of a sustainability:

6.2.3.1 AgBalance

AgBalance® is an LCA-based software tool for growers and agricultural advisors, which evaluates cost/profitability and several environmental and social sustainability metrics for arable crop production [FSG+14]. Examples of indicators include climate change, water consumption or ecotoxicity. Farm field records can be manually entered through an intuitive user interface, but also automatically transferred through existing farm management systems. The users can visualize and compare the cradle-to-farm gate impact of their crop production with regional average results of their peers and create scenarios to support in-season management decisions. A concise report can be created after harvest, summarizing key parameters of input use and sustainability results (including LCIA indicators) that support marketing the produce into the AgFood value chain [SFS+15].

6.2.3.2 Cool Farm Alliance

The CFA has focused their activities strongly on increasing the interoperability of the CFT with FMIS. For instance, the CFT has been embedded into the Agricultural sustainability Code of Unilever through the Quickfire software platform in order to calculate the carbon footprint metric for one "typical farm" on a per crop basis and monitor continuous improvement of the farms. CFA has also started to focus on interoperability with other FMIS such as the Agrible platform. The Agrible platform gives users field-level insights to help them make decisions for their ag-operation that are proactive, not reactive. To bring the adoption to the next level, its integration into OpenTEAM (Open Technology Ecosystem for Agricultural Management) has been initiated. Open TEAM comprises a recently founded farmer-driven, interoperable platform to provide farmers around the world with the best possible knowledge to improve soil health [Ope20a].

6.2.3.3 Syngenta

Back in 2011, Syngenta, a leading company in the global agricultural industry, identified that there was a significant overlap of data stored in FMIS and the data required to conduct sustainability assessments. As more and more value chain players, in particular big brand owners such as Walmart and Unilever [FFV14] started to request information on the sustainability of the production of the agricultural goods purchased, Syngenta initiated a pilot project to hard-code the FTM assessment scheme in their Land.db farm management system in North America [KHS+19]. By utilizing Field to Market's sustainability metrics embedded in Syngenta's Land.db, farmers supplying big brands such as Kellog's or General Mills are able to measure their sustainability performance and identify opportunities for continuous improvement. However, as the issues of not-harmonized data formats and the need to reuse information recorded early in the season in retrospect for the sustainability assessment, the partners decided to develop an Application-Programming-Interface (API) between FTM and Land.db in 2018. The motivation for this approach was that both partners should concentrate on their respective core competency. In particular, only the sustainability experts within the FTM organization should be responsible for the maintenance and further development of the sustainability indicators.

The "Good Growth Plan" (GGP) comprises Syngenta's sustainability and strategic plan, which has a strong focus on the sustainable development of farmers in the developing as well as the developed world [Syn15]. To make the calculation of the sustainability indicators more straightforward for the aggregate GGP reporting as well as for individual farm calculations, Syngenta has integrated the CFT API as well as the FTM API with Land.db. This addresses one of the common issues for farmers when adopting agronomic or farm management tools and applications, which is the need to retype many common pieces of farm information into different systems. By integrating individual tools into broader platforms, and also by connecting those platforms, it becomes possible to map data fields from one system to a second and avoid the need for a farmer to retype the information. This reduces errors but more importantly reduces the barriers to use of a tool. If it is straightforward to get a view of the sustainability of one's existing farming practices and then to compare them to some alternative scenarios Farmers are more likely to adopt practices, which are both more sustainable for the broader environment and economically attractive.

Through the work on the GGP, Syngenta formed partnerships with the Open Data Institute (ODI). The agricultural company has made six datasets related to its GGP available, including data on productivity, soil, biodiversity and smallholder reach. The data will be updated yearly to measure the plan's actual performance against its stated commitments [Syn15]. It is collected by external companies as well as Syngenta. The reporting process, its quality control and evidence is independently assured by PriceWaterhouseCooper. One of Syngenta's key motivations for collecting and publishing data in this way was to develop external trust in the GGP, which is essential for strengthening collaboration between private and public stakeholders for global food security. Making the data available has brought about an unprecedented level of transparency for a multinational company such as Syngenta [Luc16].

6.2.3.4 Wrangler

Apparel manufacturer Wrangler's sustainability program "From Burden to Benefit" is designed to create value for each link in the chain, beginning with the cotton farmer [Wra18]. Wrangler has established a software platform "My Farms", which serves as a hub in which data sets from e.g., data clouds from machinery, seed or input providers can be uploaded. In order to benchmark the farm data from the different suppliers, sustainability assessment metrics by FTM were embedded in My Farms, allowing for automatic sustainability assessment and benchmarking [Wra18]. It further will embed the sustainability metrics of the multi-stakeholder initiative led by retail company Walmart, "The Sustainability Consortium" (TSC). As the multi-stakeholder organizations themselves engage the scientific community to build consensus around data-driven sustainability metrics, Wrangler could implement such a data- and science-based approach to supply chain engagement. The information on the sustainability performance of their farmers should "convey respect for farmers' inherent roles as conservationists", while equipping them with additional information on which to base farm-management decisions. Promoting and managing sustainability in this manner should provide a lever to fundamentally build stronger supplier relationships [Wra18] and brand protection as the environmental impacts of cotton production might put the brand equity of Wrangler at risk.

Recent work on data system interoperability has been covered by the AgGateway consortium ([Luc16], see ▶ Sect. 3.4), working over several years to define standards for agriculture machinery and IT system interoperation. This includes e.g., standardized messages for passing crop protection "catalog" entries—i.e., well-defined fields for a product identifier and product properties. Through such messages plus a common dictionary of terms, agricultural information systems are able to share information in an integrated way [Luc16]. Apart from interoperability of software solutions, modern sustainable agriculture requires a common data ecosystem, ideally on a global scale as supply chains become increasingly globalized. Produced and used by diverse stakeholders, from smallholders to multinational conglomerates, a shared global data space could help build the infrastructures that will propel the agricultural industry forward [God16]. One of the key prerequisites will be finding business models that provide incentives for various entities to collect and share data. If these models provide business value directly to the data providers, the covering as well as the quality of the collected data will be higher [God16]). Furthermore, a strong emphasis should be placed on the automation of the data collection and annotation as automatically collected and annotated data tend to be more accurate and precise than data collected by hand [Luc16]. All of the best data sharing efforts have little impact if the data do not get used in a productive way to pave the way for a more sustainable agriculture.

6.2.4 Value Creation Through New Business Models

Measuring sustainability is one thing, to translate this into an added value for the grower is a different one. Thus far, without pressure from multinational food or retail companies, there has been very little intrinsic motivation by the agricultural production level to use sustainability assessment as reporting or even management tool. Unilever comprises a poster child example for such a multinational food company. Brands of Unilever firming under the company's Sustainable Living Plan [Uni19] haven been reported to grow faster compared to the rest of their business: the Sustainable Living Brands delivered over 60% of the company's total growth in 2016, with 18 of these brands in the company's top 40, as opposed to 12 in 2015 [Uni18]. Also for Unilever's pledge of "100% sustainable sourcing" of food ingredients there are two main routes to ensure raw materials count as "sustainably sourced", either by working towards one of the Unilever recognized certification standards (e.g., Rainforest Alliance, Roundtable for Sustainable Palm Oil or Fairtrade) which have been benchmarked against Unilever's Sustainable Agriculture Code. The alternative is self-assessment using the Unilever Sustainable Agriculture Code that applies to all of Unilever's agricultural raw materials and producing regions and is a key tool in helping them to achieve their sustainable sourcing ambitions. A self-assessment, off-the-shelf software tool for the code called QuickfireTM [Uni13] is now commercially available but still today comprises an exercise that is separate from the daily farming operations—despite the fact that it uses a whole lot of farm management information created or deposited in FMIS [God16].

The continued growth of perceivably more sustainable products brands is said to be boosted by continued consumer demand for products with a "strong social or environmental purpose", indicating that there is a potential business case for sustainably produced solutions. While sustainability assessment and implementation of sustainable management practices is often mandated by value chain players such as Unilever in order to help them prioritizing their suppliers and protecting their brand equity, new interesting business models mushroom that make sustainability in the context with digitalization a tangible business opportunity. Most if not all of them, however, comprise small-scale opportunities and not always bear the opportunity to be scaled up. One nice example comprises the project "Lerchenbrot" [Bok20]. Participating contract farmers from the Southern Palatinate region in Germany use the xarvio™ Field Manager software in order to identify less productive areas within their wheat fields, in which they create skylark plots, i.e., undrilled patches that serve as nesting opportunities for the iconic farmland bird. As consumers are required to pay a premium of 10 € Cent per loaf of bread, there is a win–win for both the farmers as well as for the mill and the bakery selling the bread [Pie21].

An excellent case study for value creation through the interoperability of a sustainability assessment tool with a certification scheme has been recently reported for malting barley in Ireland [Bas20]. In this collaboration, Boortmalt, the leading barley malt provider for the brewing and distilling industries, teamed up with the agricultural solutions company BASF SE and the Sustainable Agriculture Initiative Platform, a global pre-competitive not-for-profit organization aiming to implement sustainability in agricultural value chains (see ▶ Sect. 6.1). The common goal of this collaboration is to promote sustainable barley production in Ireland. Boortmalt has implemented web-based sustainability assessment tool AgBalance® with its barley growers in Ireland. The assessment software uses data from the farm's current farm operation and compares them with a benchmark of other farms in the region and shows where farmers can improve their operations towards a reduced environmental footprint and at the same time increase the efficiency and productivity on the farm. The digital tool, therefore, allows farmers to take a holistic view of their farm operations based on a broad set of sustainability criteria [FFV14]. The software can assess environmental impacts such as carbon emissions, nitrate leaching, water consumption, soil health, biodiversity potential as well as the farm economics. The implementation of the software tool should enable Boortmalt's agronomists to assess the sustainability of the production of their more than 800 malting barley suppliers and to have an inform dialog with them about the continuous improvement of their operations, thus helping Boortmalt to reach their 2030 target of "to source 100 per cent sustainable barley in Europe" (BASF 2020). In addition, the integration of SAI Platform's Farm Sustainability Assessment (FSA) module into AgBalance® will help Irish barley growers supplying Boortmalt to verify the sustainability contributions of their farm operations more quickly and easily. In light of this, the trilateral collaboration provides a good example for "certification 2.0", i.e., the merger of quantitative sustainability assessment with certification schemes, on the one hand allowing for a certification based on direct on-farm data and on the other hand providing farmers with scenarios how to drive their production towards a higher sustainability performance.

Approaches like this one provide a good blueprint for the seamless flow of information between agricultural management, decision support and sustainability assessment tools. As the data flows, it informs and is informed by the functions of each element, building knowledge throughout. Knowledge, informed by context specific research and available in real time should pave the way towards more sustainable farm management practices. Moreover, digitization used to enable interoperability of FMIS with sustainability metrics can help to make the shift from the "era of certification" to the "era of analytics", allowing for a specific interpretation of farm management data to reach a certificate or to document cross-compliance with environmental or social standards. Taken together, through approaches like these, there is a good chance that sustainability will become mainstream throughout most relevant AgFood value chains.

6.2.5 Bringing Sustainability to Life—The Role of Digitalization

Assessing the sustainability and resilience of an agricultural production system provides the basis for a continuous improvement of intensive arable cropping in the context of all three dimensions of sustainability, i.e., economy, environment and society. Digitization can become a key enabler of sustainable and agro-ecological intensification through 1.) allowing for the interoperability of sustainability assessment systems and the on-farm data infrastructure, first and foremost the FMIS, and 2.) providing the basis for new business models for farmers, which are based upon data exchange and transparency. As shown above, using tailor-made IT tools, sustainability data can be converted into value-adding information for the farmer and the entire value chain. Further, in order to scale up sustainable practices in agriculture in at least the more developed agricultural production systems of Europe, digitization plays a key role. By seamlessly interfacing data sets from farm machinery and operations to a data hub system such as a FMIS, where this information can be matched with background information from specific databases, the translation of sustainability information into the logic of sustainability indicators and ultimately certification criteria can be fully automated. Through interfacing of on-farm production data deposited in electronic field documentation systems with the software carrying out the sustainability assessment, acceptance by farmers will be substantially improved and at the same time, data quality will strongly benefit. In sum, interoperability of FMIS and sustainability assessment can comprise a key enabler of a more sustainable agriculture. However, to date activities of substantial scale to increase value creation through sustainability are still rather the exception than the rule.

6.3 From Ecological Intensification to Hybrid Agriculture—The Future Domain of Digital Farming

Friedhelm Taube

Abstract
The future of Northwest-European Agriculture is facing a wide range of challenges including significant contributions to worlds food supply on the one hand, but also new leadership in developing strategies towards climate change adaptation and mitigation, reduced Eutrophication of watersheds and a reversal of biodiversity loss. Synergy effects have to be identified including new food policies which are in line with the carrying capacity of ecosystems. Merging benefits from integrated and organic land use systems towards hybrid agriculture is a promising option for those synergy effects and digitalization offers the tools bringing hybrid agriculture into practice.

6.3.1 Introduction

The current debate about the future direction and intensification level of the agricultural sector in Northwest Europe is shaped by two different schools. On the one hand are the representatives of sustainable intensification (SI), who associate intensification with intensification for increased yields and earnings. This follows the paradigm from the 1980s and 1990s, according to which the Western European countries are expected to make a considerable and increasing contribution to global food security. This view, however, ignores the fact that since the 1960s on the global scale, significant yield increases have taken place in the developing countries (green revolution). This understanding of industry-related European agricultural research and the agricultural associations of SI was comparatively unproblematic as long as environmental problems were not the central focus of the debate. However, at least since the 1990s environmental problems have become of concern. Since then, we have not only been discussing elevated nitrate concentrations in the groundwater and biodiversity loss in agricultural landscapes, but also challenges that arise from climate change. In essence, this resulted in the second school, which initially focused on organic farming as an alternative to high-intensity systems. However, by now we have seen that, although organic farming is superior regarding essential ecosystem services beyond the production of agricultural products, it has a considerable yield gap to conventional high intensity systems. On average yields under organic farming yields are 19 to 25% lower compared with those under conventional farming, but this yield gap varies depending on management, crop rotation and crops [SR17]. In northern German arable farming regions, yield gaps can be well over 50% [BTV+20]. In the public debate, however, representatives of this school are requesting that organic farming is expanded beyond the currently set target of 20% of the agricultural area for Germany. According to Europe's Farm to Fork strategy (see ▶ Sect. 1.6), at least 25% of the EU's agricultural land shall be under organic farming by 2030. In its recent report, the Scientific Advisory Council for Agricultural and Food Policy and Consumer Health Protection, WBAE) has stressed the concept "Organic farming and more". What does this mean and how can such an approach be further developed into a "hybrid agriculture" concept, i.e., an approach where the benefits of organic and conventional farming are mixed in an extended crop rotation cylce? Thus, what might be the frame for Digital Farming in the future? This will be revealed in this section.

6.3.2 Why Should the Production Narrative be Questioned?

After the Second World War, over a period of almost 50 years, a linear yield increase was achieved in Germany in all agricultural crops. The increase in both plant and animal production was realized through technical progress and a corresponding implementation in the agricultural practice—an achievement worthy of recognition. Since almost 20 years, however, yields of the main arable crops (with the exception of sugar beets) have stagnated in Germany. While yields are high, year-to-year variations are considerable, with coefficients of variation frequently over 10%. Although a breeding progress is still documented for all crops, in practice this progress is only realized for sugar beet. This means, that the yield gap, i.e., the yield gap between possible yield (best practice conditions) and actually realized yield, which over decades had decreased to a level of around 20%, has for the last 20 years increased again. This is a completely new situation in a sector shaped by technical innovations. Three central components are obvious as causes: 1. Management mistakes on the farms, caused by crop rotations that are too close and related to this 2. problems and restrictions (e.g., approval and ecotoxicity) in the use of chemical pesticides and finally 3. Climate change effects: climatic changes in Northwestern Europe over the last 40 years have led to an extension of the vegetation period by 10–14 days [TVK+20], increasingly periods of drought stress, especially in early summer, in connection with an increasing number of days with maximum temperatures above 30 °C. This causes on the one hand a premature phenological development and, on the other hand, an increased sensitivity to drought and heat stress, and consequently the above-mentioned year-to-year fluctuations in yield. While this particularly affects dry locations, high-yielding locations in the north are increasingly also affected by drought and heat stresses. In economic terms this means that the high input of production resources, which are geared towards maximum yields, is actually only rewarded with maximum yields in 20–30% of the years. As a conclusion, the following proposition can be derived: the paradigm of sustainable yield increases will become too expensive in the face of climate change.

6.3.3 Why is Ecological Intensification the New Paradigm?

The above-mentioned yield stagnation, with considerable inter-annual fluctuations, are causing additional surpluses of nutrients. Consequently, Germany has not met the normatively anchored environmental goals in the agricultural sector for the last 20 years. Water protection (EU Nitrate Directive, EU Water Framework Directive, EU Marine Strategy Directive) as well as air pollution control (EU NERC Directive) and the protection of biodiversity (e.g., share of High Nature Value areas, EU-Flora Fauna Habitat Directive and EU- Birds Directive), concerning the binding commitments to the EU are not kept. This has resulted in pilot inquiries as a preliminary stage of infringement proceedings and infringement proceedings themselves. Since at the same time the goals of the Sustainability Strategy of Germany are not being fulfilled at the national level (e.g., national maximum nitrogen surplus of +80 kg N ha^{-1} by 2010, and recently set to +70 kg N ha^{-1} by 2030), one can talk of a sustainable policy failure.

In view of the clear evidence of the social costs arising from this environmental pollution, it is indisputable in the science community that a transformation of the agriculture sector towards Ecological Intensification (e.g., [Tit14]) is urgently required. All federal government's scientific advisory boards are pushing for this. Is ecological intensification the same as organic farming? It can be in individual cases, but in principle, it is not, as shown in ◘ Fig. 6.1. The graph shows the use of resources on the x-axis and the yield on the y-axis. Position "a" characterizes the current common situation of agriculture in Germany: a high yield level linked to high environmental costs, for example in the form of nitrogen surpluses per hectare. Ecological intensification means to maintain a high yield level,

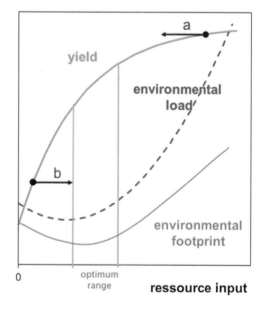

◘ **Fig. 6.1** The principle of ecological intensification derived from the yield function (yield) and the environmental load function (environmental load) per hectare. The optimum range derived from this shows the lowest environmental footprint per product unit

but with a reduction in the use of resources (e.g., nitrogen fertilization) until the ecological footprint of production (expressed as ecological burden per unit of yield, e.g., nitrogen surplus per ton of wheat) is located in a favorable area. In individual cases, the theoretical optimum is determined by the ratio of the entire benefit to society to the entire costs to society. As the overall societal costs of the environmental pollution caused by reactive nitrogen compounds are more difficult to quantify than the benefits (revenue per ton of wheat), a social debate and a political consensus will be needed to narrow down this new optimum. According to the European Nitrogen Assessment Report [SHE+11], an optimum N fertilization rate in the order of 30% below the economic optimum has been derived for wheat in northern Germany, taking into account both, the yield function and the social costs. The result is a drastic drop in the nitrogen footprint, with only very moderate reductions in yield. In contrast, position "b" on the production function indicates the point, which can be found in many developing and emerging countries. According to the Global Yield Gap Atlas, the yield gap for many areas of Sub-Saharan Africa is in the order of 70–80%. Even a small use of profit-increasing resources can trigger high increases in production and profit in these countries. If the focus were really on securing world food supply, then the highly developed countries need to develop international economic, development and trade policies for developing and emerging countries far beyond the dimension previously agreed. This would ensure that production takes place in these countries, and the people would have the prospect of staying instead of getting into the boats.

6.3.4 Is an Ecological Intensification in Western Europe Justifiable in View of the Global Hunger Problem?

For many years, the FAO has been pointing out that world hunger is not a problem of quantity, but rather of poverty. According to FAO estimates, this problem will remain until the expected world population has reached its maximum in around 30 years. In addition, FAO projections up to 2050 and beyond indicate that in less than 30 years, overweight and obesity will be quantitatively a greater problem globally than hunger and malnutrition. "Greedy or needy" is the title of an article by [RBS+17], which convincingly demonstrates that diet and consumption patterns in highly developed countries, and increasingly also in emerging and developing countries, are the drivers of scarcity, not the actual needs in terms of a balanced diet. The facts are: 1. the consumption of animal source foods (ASF) per inhabitant in Germany is twice as high as the recommendations of the German Nutrition Society (DGE), that 2. around a third of the food produced is not consumed and that 3. for this ASF a reduced tax rate is applied, although the production of ASF has the highest ecological footprints [WBAE20]. These facts highlight that a transformation of German and Western European agriculture towards ecological intensification is required. The difference in the current land consumption for the generation of ASF and the requirements according to the DGE recommendations indicates that in Germany, an area as large as 3–4 million hectares could either be used

Sustainability Perspective

for the additional cultivation of e.g., grain for bread and other foods of plant origin [Woi07; MCS+14], or could be dedicated to e.g., nature conservation or other purposes. If only an additional available grain area of 2 million ha in Germany were assumed, with an average yield of 7 tons ha^{-1}, this would result in around 15 million additional tons of grain annually—a real contribution to global nutrition with a favorable ecological footprint.

6.3.5 Animal Sourced Foods— Less and Better?

In its most recently published nutrition report, the [WBAE20] identified political instruments to promote the transformation of the agricultural and food sector towards ecological intensification. For the agricultural sector, the recording and documentation of the ecological footprint, beginning with the "carbon footprint" (for the function of climate protection), plays a central role for assigning social costs (environment, health) due to incorrect production and malnutrition to the underlying causes and the prices for food. In terms of perspective, the CO_2 certificate trade or the CO_2 tax are the instruments that can show these costs approximately and yet acceptably. Besides, for the function of water and biodiversity protection the nitrogen footprint can be used as an indicator of the ecological footprint. With a corresponding orientation of the agricultural sector to the primary reduction of the ecological footprint at high yield levels, at least two year cultivated forage ley systems based on grass-clover mixtures which provide climate protection (soil carbon sequestration), nitrogen fixation (reduction of GHG emissions) and biodiversity (multispecies mix) have advantages over current exclusively annual production systems. The required transformations in land use are particularly pronounced for milk production systems in Germany, because a significant proportion occurs on peatland sites, which are burdened with high greenhouse gas emissions, and which in turn impact on the carbon footprint of milk. Data from intensively managed grassland on lowland peatlands in Northern Germany show that these areas currently emit more than 15 tons of CO_2-C-equivalents per hectare and year [PRK+16]. Agriculturally used peatlands make up around 900.000 ha of around 7% of the agricultural land in Germany, but are responsible for around 35–50% of the GHG emissions from agricultural land use, depending on the source. There is therefore no question that these locations (especially in northern Germany with around 70% of Germany's peatland) must be transformed towards a use, which is geared towards protecting climate and biodiversity [ARP+17]. While this transformation has occurred in Eastern Germany since the reunification (Paludi Cultures etc.), the peatlands in the west are largely characterized by intensive dairy farming. This means that in the short future a transformation from a very conservatively estimated 400–500.000 ha of peatland to higher water levels, and land use geared towards nature and climate protection is pending. However, these would be framework conditions that oppose efficient milk production, rather than alternative extensive- usage concepts. "Less and better" milk production in this context does not mean, that the reduced milk production on peat soils cannot be at least partially offset by an eco-efficient expansion elsewhere. Rather, the problem of "peat soils" appears as a blueprint for basic approaches to solutions for area-based animal husbandry as a whole and for the re-integration of arable farming and animal husbandry.

6.3.6 Back to the Roots: Integrated Crop—Livestock Systems Including Grass-Clover Leys?

The yield stagnation in almost all arable crops, shown above, is largely due to the fact that too few crops result in too little functional diversity in our fields. In current "crop rotations" (rapeseed and grain here and maize in long-term self-rotation there) the historically grown function of crop rotations, to produce food and feed eco-efficient, i.e., via the inclusion of at least a minimum of self-regulation mechanisms, is switched off and instead replaced by external energy supplies (e.g., fertilizers and pesticides, digitalization). A return to the benefits of integrated systems of agriculture and animal husbandry and correspondingly wider crop rotations with a high level of diversity of crops and the resulting functional diversity seems to be necessary, today more than ever. What is the lack of functional diversity in arable farming today? There is no combination of annual and perennial cultures (function: carbon-sequestration and weed suppression), the inclusion of species with the highest root length density (function: water protection) and the inclusion of efficient legumes (function: N fixation) for the domestic protein supply. These functions could only be met by the use of at least a two-year-old clover grass system (ley system), as current studies from our research farm confirm: dry matter yields without any mineral N fertilization of 10 tons ha^{-1} year^{-1} with high energy and crude protein contents, nitrogen fixation capacities of up to 300 kg N ha^{-1} year^{-1}, massive suppression of invasive species through frequent use, negligible nitrate discharges via leaching (< 10 kg N ha^{-1}) due to root length densities of over 100 km per m^2 and year [LMK+20], provision of food habitats for insects visiting flowers through multispecies mixes [LRK+20], no necessary use of pesticides. In short: everything that is currently missing in our agricultural landscapes can largely be provided by such ley systems with certain additional elements (flower strips, etc.)—they are the golden bullet towards resilient agricultural systems in connection with milk production, especially in maritime climate areas. If these multispecies mixtures consisting of grasses, legumes and herbs are also grazed, the ecological footprint—due to the reduced energy requirement and the further increased energy density in the feed at the concentrate feed level (> 7.2 MJ NEL kg^{-1} DM)—decreases further, both in terms of nitrogen—as well as the CO$_2$ footprint [RLV+]. We have carried out this work on "eco-efficient pasture milk production" under the framework conditions of organic farming (▶ www.lindhof.uni-kiel.de) and show that the milk output in energy corrected milk (ECM) ha^{-1} main forage area is over 80% of the output of the top 10% of the conventional farms, and significantly above the average of comparable conventional farms in Northern Germany. Ecological intensification with a very high level of milk yields is thus possible also in the frame of organic farming as long as ley systems are involved.

6.3.7 Implementation of Ecological Intensification Through Hybrid Agriculture and Public Goods Boni?

The dilemma with regard to stabilizing yields at a high level in a crop rotation begins with the subsequent crops following the grass-clover leys, i.e., the connection between forage production and cash crop production under the framework conditions of organic farming. In regions with significantly more than 100 mm drainage/year, a summer culture must follow the grass-clover to allow the nitrogen residues from the stubble and roots of the ley to flow directly into the yield formation of the subsequent crop, and to minimize N losses via leaching in the following winter period. In the first crop following grass clover, the N transfer of over 120 kg N ha^{-1} ensures high yields, of, i.e., oats without additional fertilization and crop protection of about 6 tons ha^{-1}. Problematic are the following classic crops, such as winter wheat (WW) or rapeseed, in which a yield gap of up to over 50% can occur [BTV+20]. Such a yield gap questions the expansion of organic agriculture beyond a certain level. So why not develop a hybrid system that combines ecological and conventional elements to ensure high yields and environmental performance in equal measure? A rotation with 6 crops, each consisting of 50% cultivated area according to organic farming standards (e.g., 2 × clover grass (CG) + oats) and 50% according to conventional standards (WW, winter rape (WR), WW) would, with additional integration of i.e., 10–15% of special "biodiversity areas" [OPE20b], guarantee under the conditions of Northern Germany a large part of what is currently discussed in the political area (EU Green Deal; EU F2F; national insect protection program, etc.). Yield losses in the first part of the crop rotation should be almost compensated by crop rotation-related additional yields of the cash crops in the second part of the crop rotation. There is a need for further research to confirm this and we are working on the corresponding numbers. The relevance of this approach arises in Northern Germany, because the added value of milk production, which in the future should not occur on peat soils, should remain in the country as long as eco-efficient production is maintained. The line of argument is thus: cure the deficits of specialized arable farming with some additional forage and milk production ensuring multifunctional benefits! Virtual mixed farms would be an approach to implementation, whereby neighboring cash crop and forage farms would implement hybrid approaches with joint crop rotations. If such a transformation is to succeed, then the central question arises as to how incentives can be created for the actors on the farms to take this direction? The use of the common good premium model of the German Association for Landscape Management (DVL, [DVL20]) is a supreme tool. This instrument first assesses the "best practice" implementation of good professional practice with high environmental standards in agriculture (e.g., optimal N, P farm balances, see [NDT17]) and the additional benefits of various elements of agricultural production and corresponding additional services via a points model for ecosystem services beyond agricultural production. Based on this expert model the effect of different hybrid options for a 100 ha model farm can be assessed and compared with conventional systems (◘ Fig. 6.2).

In this figure the abbreviations on the X-axis indicate the different uses of the agricultural land (not the crop rotations) of the model farms (DGL: permanent grassland, MA: maize, AG: arable grass, TR: triticale, WR: winter rape, WW: winter wheat, KG: Clover-grass, HA: oats, BS: Flower strip). The abbreviations in the legend as well as calculations are according to [DVL20]. ◘ Figure 6.2 describes the points depending on different setups:

— specialized forage system (see pillar 1: permanent grassland, high proportions of maize, low proportions of grain)
— market crops (see pillar 2: WR, WW, WW)
— various hybrid options (see pillar 3: conventional: WR, WW, WW, ecological: KG, KG, HA; pillar 4: in addition to HA 10% BS; pillar 5: instead of a WW pro rata MA)

Depending on the chosen setup the farmer achieves additional public good bonus points. The result shows an additional

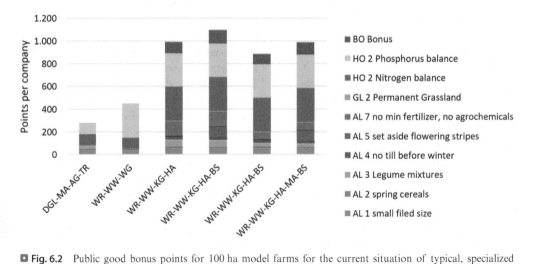

Fig. 6.2 Public good bonus points for 100 ha model farms for the current situation of typical, specialized farms in forage and market crops (pillars 1 + 2) as well as for various hybrid options (pillars 3–6)

quantifiable benefit of up to 700 evaluation points (1000 instead of ~300) for the hybrid options compared to the specialized actual starting options. Currently, in the political discourse on the implementation of the DVL points model in the GAP eco-schemes, a fee of € 30–50 per point is being discussed after 2020, which would correspond to an additional benefit of € 210–350 per hectare in the model farm calculations for the hybrid approaches. In view of the urgency of achieving climate protection and biodiversity goals, there are many arguments in favor of developing and optimizing hybrid approaches that go beyond the existing Extensification in the various regions of Germany. Digital Farming solutions (e.g., spot spray technologies) will play a significant role in this transformation process.

6.3.8 Conclusion

In current agricultural research, technological options of precision agriculture, digitization and artificial intelligence on the tactical level are representing the main stream to fulfil the given goals of global food security on the one hand, and environmental goods and animal welfare on the other hand. This offers, however, only solutions in the existing structures of specialized systems. It seems questionable whether the given goals can be achieved in this way. Rather, the overarching options at the strategic level should—as discussed above—be given greater attention. The re-integration of animal husbandry and arable farming within the framework of ecological intensification and the formulation of hybrid approaches are such key elements of a comprehensive transformation of agriculture in the upcoming 30 years. This includes primarily answering the questions in connection with the transformation of the food system, namely how much animal husbandry, where distributed and how kept, which equally achieves the protection goals and ensures a social consensus that allows agricultural practitioners to make appropriate investments in the future.

Acknowledgements I would like to thank Dr. Iris Vogeler-Cronin, Dr. Thorsten Reinsch, Dr. Ralf Loges, Dr. Arne Poyda, Dr. Carsten S. Malisch and Christof Kluß for valuable contributions to this manuscript.

6.4 Digitization as Co-Designer for Cropping Systems: Technology Shapes Cropping Systems, Now and in Future

Klaus Erdle

Abstract

Crop production changed fundamentally in the past decades. Farm inputs as well as technology allowed cropping systems to be specialized and scaled up to large entities. Even crop varieties were fitted to easy to manage crop stands. While technology and digital tools fueled this reinforcement towards market related efficiency, nature is catching up with pest resistances, soil degradation, and biodiversity loss affecting ecosystem services. However, the same technology being blamed to be a major factor for the negative effects can be used to reshape our way of farming for a more integrated and sustainable way of crop production.

6.4.1 Mechanization and Its Excesses on Farming

Mechanization is one of the major factors which shaped the way of farming during the last decades. The introduction of tractors as a multi-purpose machine led to a series of developments and a broad range of implements and self-propelled machinery. As the number of employees in agriculture decreased in the more industrialized countries and market conditions led to a growing demand on efficiency, farming structures followed this development and formed cropping systems suitable for maximum productivity and control:

- extensive mechanization for low labor input
- increasing weight and working widths of machinery for high spatial efficiency
- highly precise seeding, fertilizing and spraying systems to avoid unproductive subareas
- geometric field structures for a most undisturbed crop stands and the use of machinery
- less crop varieties and adapted crop phenotypes for a highly specialized production
- low variety of soil and crop management methods for a best utilization rate of technology

6.4.2 Precision and Efficiency, the Evil Twins

The mentioned developments led to the actual situation of frail farming systems. Representatively, regions of Mediterranean climate combined with adequate soil characteristics for winter wheat and rape seed production. For decades, the crop production in such regions followed the market's demand by highly specialized farming systems. Short successions of the same crops and low intensity in soil management were compensated by highly precise nutrient inputs and chemical crop protection. Technology allowed to run these systems most efficient and, in turn, technology was adapted to the farming systems by size and function. Subsequently, cropping systems and technology changed the structure of landscapes merging fields to bigger entities which—in consequence—allowed an even more efficient use of even more specialized technology grown in weight and size. The precise application of seeds and inputs resulted in 100% crop-coverage and fields with nearly homogeneous growth conditions for an optimized field use efficiency. Economy tells us, machine use efficiency grows with the size of field units and uniform crop stands. Even breeding was adapted to produce crops of homogeneous appearance best to be managed and harvested by respective technology. Market quality standards asked for uniform crops of constant quality to be processed in the downstream sector. The larger the field and their structure the better AgMachinery of the actual broad working widths can be used and the sooner they pay off. Machines do not have to be moved several times between different fields with high efforts. The more geometric or rectangular the fields structure, the less overlapping and the higher the machine performance. Managing fields of small scale takes a lot more resources of time and unwanted soil disturbance by increased passage frequency. Eventually, fields covered densely with only one single crop and variety are managed more easily because inputs and cultivation is equal for the whole area and machinery can be specialized therefore.

While reading the latter sentences, the effect of a mutual reinforcement of technology and cropping system can be gathered easily. However, we do not only see this effect in agriculture alone. Nearly all economic sectors of the more industrialized regions of this world follow the upwards directed spiral of efficiency. Highly specialized production lines for mining minerals, the production of automobiles, furniture, cloths, electronic devices, etc. can be described similarly strongly relying on technological support. Why should it have been different for the agricultural branch considering an even decreasing number of available labor and increasing global demand for food?

In contrast to most of the sectors mentioned before, agriculture, in particular crop production, is strongly connected to environment which can only be controlled limited. Staying with the example of winter wheat and rape seed rotations in the highly productive regions, agriculture was able to push these limits by ever more sophisticated technology and adaption of their cropping system. In ▶ Sect. 1.2 the challenges of farming practice are described extensively. Specifically, upcoming pests due to narrow crop rotations were successfully avoided by breading tolerant varieties and the development of plant protection methods. Decreasing soil nutrient content was overcome by external nutrient supply. Complex landscape structure was not seen as relevant for a successful cropping system. Eventually, nature caught up by an insidious strengthening of ecological complexity:

– soilborne pathogens hardly to be managed
– resistances of insect pests and weeds to chemical actives
– environmental impacts by nutrient losses
– decreased macro and micro biodiversity
– decreased soil health
– increased soil erosion

Farmers suffer crop failures due to massive weed invasion. Chemical actives are not able to control pests anymore and substitutes are not to be seen in near future. Biodiversity is drawn down by intensive crop management, the loss of landscape features and management diversity. It becomes clear now that agriculture cannot be compared directly to any other industrial process. The interdependencies between crop production and environmental factors leads to effects which hardly can be limited to a specific location or production section. Almost all crop management methods do have impacts affecting crucial ecosystem processes and subsequently society.

It seems that the desire for an ever more precise and efficient production led to the point where methods used in the past decades struggle to control the actual challenges. Even more, well known methods appear to be of harm. Some would even say "nature takes its toll".

Precision and efficiency fueled by increased mechanization and sophisticated technology can at first glance be considered as the cause of all evil.

6.4.3 Back to Reality

Looking at regions where mechanization is not yet adapted accordingly, farming systems seem still to be more various, individual and less specialized. Often it is assumed that in such regions crop production systems would be more sustainable especially in the subject of ecology. However, looking at Southeast Asia as an example where rice cropping is done on small patches, with low tech solutions and by small holders, similar problems can be found. Rice in such regions is cropped on the very same field for even decades and the use of fertilizer, water, and plant protection chemicals is done by low educated farmers. This leads to very similar effects of ground water shortage, loss of soil quality and the eutrophication of water bodies by chemicals, thus, leading to a strong decrease in biodiversity. Similar to that, in Sub Sahara regions traditional farming practices are intensified as arable land decreases with increasing population. And we see, with better education and mechanization, situation improves in such regions.

Technology and the run for efficiency and precise application enabled us to reduce inputs in conventional agriculture during the last 30 years (see ◘ Fig. 6.3). For sure, there are negative consequences of the intensive way the actual crop production is done. At the same time, spraying systems work most precisely to reduce the amount of active agents applied, so do fertilizer spreaders. And this while increasing yield per area and independent of the decreasing number of employees available in the agricultural branch.

Nevertheless, we experience the ecological limitations of our systems and have to react on that like stated in ▶ Sect. 1.4. A majority of these effects are due to a still limited knowledge about the interactions of soil, plant, and atmosphere. Even more, if additional effects of agricultural measures interfere with natural processes. These processes are highly complex and hard to be overlooked by farmers who have limited resources of methods. To reduce negative effects on ecosystems and at the same time produce food and commodities for an international market we have to find ways to understand the interactions between farming practice and environment. If this knowledge is available, farmers will decide upon it as to adapt processes to reduce negative impacts and even increase benefits by a most individual and adapted farming system. Technology, previously demonized, will play a crucial role in this approach.

6.4.4 Digitization for Initiating Promising Cropping Concepts

6.4.4.1 Data, Information, Decision

To change our current cropping systems to more sustainable processes we have to find ways to improve the way of decision making in crop management. Decisions are founded on information. Information should always be based on relevant and sufficient data of high quality. This conclusion is not new even in strongly traditional agriculture. Independent of the way of farming—if organic or conventional—cropping systems rest on a few principles to be sustainable: adapt measures to best fit crops to soil conditions and to climate with a tailored supply of water, nutrients, and the avoidance of pests. Farmers with a deep knowledge about their fields with soil and regional conditions are able to make reasonable decisions to run sustainable plant production. In the past, this knowledge stems from long-lasting experience and passed on knowledge of generations. Today, this information is not to be gathered by a single farmer anymore due to growing farm sizes and management tasks as well as a growing complexity of the challenges

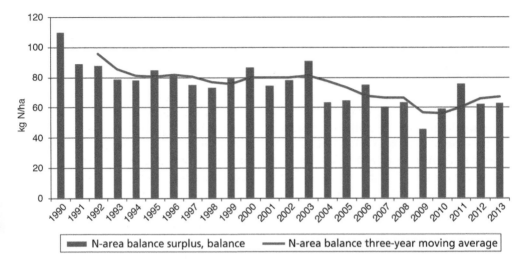

Fig. 6.3 Yearly (blue) and moving average (red) of the nitrogen balance of the German agricultural area. (Source: DLG Sustainability Report 2016)

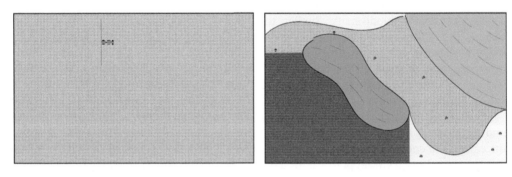

Fig. 6.4 Unstructured field (left) enabling high efficiency due to homogeneous crop stand and conditions for broad working widths. Highly structured field (right) adapted to local soil and landscape being managed by swarm organized robots

ahead. Decreasing numbers of available labor in agriculture, increasing farm sizes, new market requirements and pests as well as climate change ask for a higher level of information to run cropping systems successfully, thus, sustainable.

It was called precision farming when farmers for the first time used sensors for improved data recording in the field to finally gather better information for crop production. This started already three decades ago. Sensors were developed to detect the actual crop status for improved fertilizer application considering spatial differences in the field. A big step to individualize farming measures to actual crop needs. In the beginning, systems worked on the basis of a direct connection between cause and effect: poor crop status means higher fertilizer application. Considering the number of factors influencing our cropping systems, the decision actually cannot be that easy. It needs more detailed and integrated observation of our ecosystem we are working in and we are working with. Here, digitalization can help to analyze interactions, quickly process data into tangible information and ultimately provide a sound decision-making aid for farmers.

To create new cropping systems, it needs suitable crops which allow for a more sustainable management. Our current main crops were adapted not only to pest appearance and yield or quality standards but also to the technology used for their management. Crop pattern and architecture was considered in breeding to meet the limitations of machinery for crop management. Breeding has to speed up to meet the challenges not only of climate change or new pests but also to offer a variety of crops and varieties to enable farmers the adaption of their cropping system to local conditions. Digitalization will also help breeders to precisely detect genotypes and phenotypes in rapid throughput and analyze data for the breed to follow.

6.4.4.2 Size Does Matter

With time ever more sophisticated technology was developed to increase the database for improved information. Satellite data cover soil and crop characteristics for improved fertilizer or water management. Forecasting models provide information on precipitation or the spread of pests through yield forecasts and CO_2 emission and sequestration. But still, data are not interconnected sufficiently to consider potential interactions or offer straightforward information to farmers (see ▶ Sect. 3.4). Artificial Intelligence (AI) (see ▶ Sect. 3.5) promises to be capable to detect complex interactions—by considering and analyzing huge databases—and finally offer sound information for decisions made by farmers in the end. Eventually, this only leads to an even more optimized and efficient crop production. But let us think a step forward: interactions in nature do not only show gaps of inefficiency. There are chances to use natural interactions actively to improve crop fitness, thus, less efforts for crop management.

With the right data acquisition and processing we may get information about privileged plant (crop) communities and even better rotations. Taking advantages out of symbioses between plant and soil borne organisms. Improved understanding of pest development and spreading as well as their antagonists. A better knowledge about the existence of these interactions and the use of their potentials can lead to improved and locally adapted cropping systems. A broad information on such level may lead to a regional organization of cropping systems or—in contrast—to a small scale, patch-based crop production (see ▶ Sect. 6.3). Extensive fields will be analyzed to be subdivided into patches of specific soil and micro climate conditions. Crops can be found that are adapted to these conditions and develop with less effort and higher fitness against pests, climatic influences or better nutrient efficiency. Eventually, this will break with traditional ever-growing entities down to individually managed, small and unstructured patches of different crops. A picture hardly any farmer today is willing to imagine. Even if data tell us that small, unshapely patches of different crops oriented on soil, microclimate and plant community could be favorable for crop production: How should such small areas be managed economically and the use of machines be accomplished?

Small patches need small machinery. The competition for ever larger and heavier AgMachinery with ever increasing working widths contrasts with the actual development of small, intelligent and swarm organized machines. Concepts and already on the field working autonomous machinery shows potentials of very different cropping systems in future. These implements are able to manage areas nearly independent of field size and shape and therefore could support the introduction of intercropping, mixed cropping or patch based cropping systems. With working widths of less than one meter up to a few meters, weights far below 1000 kg and intelligent decision support systems, these new implements may be able to optimize crop stand architecture in a field considering soil parameters, landscape structures, topography, and microclimate. So, technology can follow the needs of the cropping system rather the other way around. If technology can be most efficient independent of field size and shape, farm size must not be the figurehead of farm prosperity anymore. This allows farmers in small structured agricultural regions to have better future prospects as well (◘ Fig. 6.4).

6.4.4.3 Mastering Complexity

Improving cropping systems in that way definitely leads to heavily increasing complexity of production processes with which farmers would struggle in management.

Intelligent technology can support farmers when weather, soil, and crop demands offer only minor time slots to cover crucial measures. If the number of crops in new crop rotations increases (see ► Sect. 6.3), farmers will need support to meet the diverse demands of individual crops and their effects on each other and the environment. Even the design of successful crop rotations in complex unshapely field patterns may be possible by using digitalization for scenario evaluation and strategy adaption. Robotics already seed and treat crops autonomously working 24/7 to get the work done in the short time frames available. By identifying individual plants, crop management focuses only on the crop in question and not on the field as an area itself, which leads to individual treatment and only covers actual needs rather than homogeneously managing the area. Early detection of pests is supported by improved predictive models which better estimate the probability of a pest spreading, and spectral sensors (see ► Sect. 3.2) which detect the earliest stages of infections—far before the human eyes can see them.

The assessment of the threat to our crops will not be limited to pests, but will also include the independent detection and identification of weeds in the field. Digitalization will enable machines to not only detect weeds in general, but also to distinguish between "good" and "bad" weeds and individual plants. Ultimately, only weeds that definitely affect crops or yield quality will be eliminated with slender tools or treated with precisely guided spraying systems.

Considering such developments, biodiversity will improve even in intensively cropped agricultural regions. Patches of different crops can be integrated in a heterogeneous landscape much easier. The patches offer green stepping stones for organisms and harmless weeds may be left in the field as food source for pollinators and beneficial insects. Keeping spinning this thought, an increased beneficial biodiversity may lead to pest management hand in hand with nature. Less insect pests through less uniform crop stands and a higher number of beneficial insects lead to decreased pest management inputs and labor resources for crop management.

If agricultural machines become smaller and work within swarms, the sum of the operational investments will not decrease in reality. However, the performance of future technology will be judged not only by the size of the area it has worked in a unit of time, but more by the success and ultimate benefit of a highly complex farming system in which it operates.

6.4.5 Conclusion

Complexity increases in agriculture by changing natural and political frameworks as well as the changing structures of farm organization. In the past, technology strongly influenced decisions on how to plan and implement crop production measures. With time, production processes got highly efficient by ever more productive machinery even overtaking natural regulatory processes. Without natural interactions being considered, crop production systems seem to lose their balance and affect resources of any kind.

We are at the threshold where our farming systems need to be reconsidered to avoid strong impacts on the ecosystem on which we depend. Technology has enabled us to advance the unification of our production systems and ultimately cause them to struggle. But technology is also capable of breaking the cycle of increasing size and homogeneity. By gathering and using knowledge about our ecosystem and the effects of agricultural practices, future farming systems can be adapted to natural processes and even benefit from interactions within the systems. This will lead to: a) an increase of crop types per crop rotation, b) temporal extension of crop rotations and thus to a higher diversity of crop and farming systems.

Technology with an increase in digitized tools will make this possible. Both the intelligent use of knowledge and the adaptation of plant production to natural conditions can make agriculture more sustainable and a respected economic sector in society.

6.5 Fighting Climate Change Through "Carbon Farming": A Future Business Opportunity for Digital Farming?

Markus Frank and Isabel Roth

Abstract

As governments and multi-stakeholder initiatives more and more discuss the need to develop "carbon farming" strategies to ameliorate the effects of anthropogenic carbon dioxide emissions on global change, there is increasing focus on the question how agricultural land management can be used to mitigate their effects. "Regenerative agricultural practices" are widely considered a key component of the attempts to increase carbon sequestration through agricultural land use. To date, however, the scientific evidence to substantiate this claim is rather slim. Despite this fact, an informal market for the trading of carbon certificates in agriculture is emerging, and numerous stakeholders from the AgFood value chain share very high hopes for a significant new business opportunity in "carbon farming". Digitization has the potential to become the key enabler to scale up of this informal market.

6.5.1 The Carbon Cycle and Its Interaction with the Nitrogen Cycle

CO_2 in the atmosphere is critical for life on earth in two ways. First, CO_2 comprises one of the key greenhouse gases (GHG) in the atmosphere, thus providing a temperature that supports life on earth. Second, photosynthetic organisms use CO_2 in the presence of light to produce biomass, which eventually becomes the basic food source for all microbes, animals, and humans. The burning of fossil fuels as well as changes in land use emit approximately ten billion tons of carbon (C) annually as CO_2 into the atmosphere. These activities are the main drivers of the increase in CO_2 concentration observed since the start of the Industrial Revolution [Chu13]. Approx. 40% ($\pm 14\%$) of this human-generated CO_2 remains in the atmosphere [JHR+05], contributing to a further increase in the temperature on earth. Carbon cycles through different pools on the earth. The size and residence time of C in each of these pools varies; the largest and slowest pool of C resides in sediment and rocks, which is followed in size by the ocean, land, and atmosphere, respectively. Burning of fossil fuels, thus, transfers C from the largest pool (which is also the slowest to accumulate and store C) into the atmosphere, elevating the CO_2 levels and contributing to global warming [Chu13]. CO_2 fluxes between the atmosphere and biological ecosystems are primarily controlled by uptake through plant photosynthesis and releases via respiration, decomposition, and combustion of organic matter. Living biomass, dead organic matter, and soils are the biggest terrestrial carbon stocks, and the increase in these carbon stocks over time represent a net removal of CO_2 from the atmosphere and decreases in total carbon stocks represent a net emission of CO_2 [IPC14].

Other biogeochemical cycles, such as the Nitrogen (N) cycle are tightly interlinked with the C cycle. Human activities have also approx. doubled the quantity of N cycling between terrestrial ecosystems and the atmosphere. Globally, human activities convert N_2 to reactive forms using industrial fixation in the manufacturing of mineral fertilizers and in the planting of N fixing crops. Predominantly agricultural activities have nearly doubled nitrous oxide fluxes from land to the atmosphere through mainly fertilization, cattle, and feedlots [Chur13]. In addition, the flux of ammonia has more than tripled since pre-industrial times because of agricultural activities, animal husbandry being the single largest global source ammonia. Although the amount of reactive N doubled between 1860 and the 1990s [GAE+03], continues to increase, and is mostly deposited on land, N is still a limiting nutrient for many land ecosystems. In some areas N deposition stimulates land C uptake and storage, and this additional uptake may help offset global warming [CBB+09]. However, net fertilizer production and direct field emissions together with emissions from animal husbandry and feedlots comprise the key factors through which intensive agriculture contributes to global warming.

6.5.2 Carbon Sequestration in Agricultural Soils—A Synopsis

Carbon sequestration is the process of capturing carbon dioxide (CO_2) from the atmosphere, storing it, and preventing it from being re-released [IPC14]. Thus, Carbon sequestration describes a blend of man-made as well as natural processes that capture CO_2 as part of either industrial or agricultural and land-use activities. As mentioned above, there are two distinct pathways for carbon emissions and removals: i) land use changes (e.g., cropland converted to grassland) and ii) management practices in an area with no change in main land use, i.e., areas that have not undergone any land use conversion for a period of at least 20 years (as a default period). The amount of carbon stored in and emitted or removed from cropland depends on crop type, management practices, and soil and climate variables [Toe16]. Annual crops (cereals, vegetables) are harvested each year, so there is no long-term storage of carbon in biomass. However, perennial woody vegetation in orchards, vineyards, and agroforestry systems can significantly store carbon in long-lived biomass, the amount depending on species type and cultivar, density, growth rates, and harvesting and pruning practices.

Carbon stocks in soils can be substantial and changes in stocks can occur in conjunction with soil properties and management practices, including crop type and rotation, tillage, drainage, residue management (e.g., burning residues, using them for animal feed or fuel or leaving them in situ) and adding organic matter to the soil (such as manure, sewage sludge or compost [Toe16]). The system of "regenerative agriculture" combines many of these practices,

Sustainability Perspective

such as reduced tillage, cover cropping, and crop rotation [GHA+21]. The carbon sequestration rate of regenerative agriculture has been estimated at 0.6 t/(ha*yr) [SLK+15].

6.5.2.1 Key Strategies for Carbon Sequestration in Annual Cropping Systems

[Lal14] describes the key strategies for carbon sequestration in annual cropping systems:
- Cover Cropping, i.e., planting crops for erosion control, weed suppression, nitrogen fixation, and other benefits rather than as a marketable commodity. Providing nitrogen to crops with N-fixing cover crops, intercrops, undersown crops, and similar strategies comprises a partial alternative to synthetic fertilizers.
- Crop rotation, i.e., alternating crops, usually from different families, rather than growing the same crop season after season. Crop rotations that include a period in perennial grassland for grazing or hay are especially beneficial on carbon sequestration as they tend to build up more soil organic carbon.
- Mulching and residue retention, i.e., leaving crop residues such as stalks and stubble at the soil surface rather than burning or tilling them into the soil. This not only improves soil organic matter but also sequesters more carbon than burning or tilling the residues. Mulching with biomass has similar benefits.
- Organic Amendments, i.e., using manure and compost (as well as other more controversial amendments such as zeolits or biosolids) to improve soil organic matter and build soil carbon.
- Reduced tillage, i.e., besides conventional no-till, a diversity of reduced tillage systems exists for large- and small-scale operations. For example, strip tillage as an alternative to tilling the entire field. Full inversion tillage using a moldboard plow prior to cultivation is widely accepted as the one of the key causes for the loss in soil C [PWG11].

Conversion to these practices can mean reduced yields for three to five years during the transition. After this period, yields may even slightly increase [HCD+14]. The IPCC (2014) rates crop rotations and cover crops to have a potential medium global impact, being easily adoptable by farmers, and being ready for a wide implementation within 5 to 10 years. Tillage and crop residue retention are considered a high potential global impact, easy adoption by farmers worldwide, and ready to be implemented. Strategies to create polycultures or to integrate perennials in annual cropping systems are rated to provide excellent opportunities to increase the carbon sequestration potential of the landscape [IPC14]. Well-designed polycultures maximize the land use efficiency, thereby indirectly improving the carbon footprint on a hectare basis. Integrated perennial/annual systems typically sequester far more carbon than any other type of farming, however these land-use systems are not always compatible with an intensified and efficient production system [Toe16]. Improving soil organic matter, cover cropping, and mulching are also important climate change adaptation strategies [Lal14]. Crop residues as well as cover crops tend to trap carbon dioxide in the soil. A wider crop rotation has proved effective to support the buildup of humus. Moreover, biomass added to soil in the form of manure and composts further supports the generation of soil organic carbon (SOC). Crops release 10–40% of the carbon compounds they synthesize as root exudates, which get metabolized by microbes, immobilizing carbon and increasing SOC [Chu13].

6.5.2.2 Scientific Evidence for Carbon Sequestration Through Regenerative Practices

Tillage and Residue Retention Effects

Conservation tillage to increase the C content of C-depleted agricultural soils is accepted as a key strategy for stabilizing global atmospheric CO_2 concentrations over the next fifty years [Lal14], [BOV07]. However, many studies have shown that tillage affects the distribution of soil organic carbon (SOC) within the soil profile more than its net accumulation [Lal07]. For instance, studies where the soil was sampled to depths at or just below the plow layer have shown a significant accumulation of SOC at 15 to 20 cm below the soil surface for conservation tillage compared with no-till [AE08]. A common observation across all of these analyses is a high variability in the annual rate of SOC sequestration. There are multiple parameters contributing to the variation in annual rates of C accumulation, most of which are related to the spatial and temporal heterogeneity of soil, plant, hydrologic, and atmospheric processes that drive the accumulation and distribution of SOC in agricultural fields, both laterally and vertically. Despite this uncertainty, the general agreement is that reduced tillage might be beneficial to C sequestration, and there needs to be an awareness of the differences that are induced in the soil profile by the changes in tillage practices.

Crop Rotation and Cover Cropping

Enhancing the complexity of the crop rotations has the potential to enhance SOC sequestration in agricultural soils either alone or in combination with conservation tillage practices. Multifunctional cropping rotations that include forage legumes, small grains, and organic amendments from animal manure or compost are more complex and biologically diverse than simple two- or three- year combinations of small grains with corn and soybeans. Implementation of integrated, extended cropping rotations have been shown to increase SOC e.g., [Tea07]. The impacts of this type of agricultural management are especially evident for labile forms of SOC and have been reported to increase biologically available forms of SOC [FOG+07]. Organic systems have been shown to have more microbial biomass C, greater microbial community diversity, and higher microbial activity than conventional for a variety of grain, vegetable, and fruit production systems [EGM+07]. The more highly diverse microbial communities haven been shown to transform C from organic residues into biomass at a lower energy cost [FOG+07], thus resulting in higher retention efficiency of microbial biomass C within organic systems.

Organic Fertilizer Use

Application of organic nitrogen (N) fertilizer is believed to increase SOC sequestration by increasing crop residue inputs to the soil. This belief is supported by numerous studies that relate increases in crop aboveground residue biomass with organic fertilizer application, which results in greater inputs of aboveground residue C to the soil [PWG11]. For instance, [RCL+09] report that corn residue C:N declined and decomposition rate significantly increased with N fertilization. Decomposition rates of crop residue and SOC have been shown to increase with the application of fertilizer nitrogen [HAC07]. Further studies in the literature document that both organic and mineral N fertilization stimulates soil C sequestration in some agroecosystems [GLE+96] but not in others [RCL+09].

The Limits of Carbon Sequestration in Agricultural Soils

The limits of carbon sequestration of the agricultural soil are given by the carbon saturation capacity. "Carbon saturation capacity" is defined as the maximum amount of C that can be sequestered by a soil under specific climatic and management conditions [SCP+02]. Temporal changes in SOC content vary because of complex interactions among different factors, including climate, baseline soil C levels, and agricultural management. An increase in temperature is likely to affect SOC [BLB+05] by influencing soil organic matter decomposition and mineralization rates [CDH+08] as well as soil and root respiration [JMC+05; HAC07], assessed tillage effects and crop sequence effects on SOC dynamics on long-term trials and found that all of the tillage and cropping treatment lost SOC compared to initial SOC, whereas conservation tillage and no tillage lost the least. [SBK+09] observed that SOC declined under conventional management practices relative to baseline, but they did not observe an increase in SOC under conservation practices such as no tillage and planting of cover crops relative to baseline levels. The conservation practices appeared to have only prevented SOC losses compared to conventional management practices rather than facilitating SOC sequestration. The largest SOC losses relative to baseline was observed for virgin grassland sites, which has more than twice as much SOC at the start of the experiments than the agricultural sites. Other studies have also observed that greater SOC losses with time relative to baseline SOC level are associated with higher initial SOC contents [BLB+05]. A possible mechanism to explain this paradoxical relationship is the concept of soil C saturation capacity.

6.5.2.3 "Carbon Farming"— A (Future) Business Opportunity for European Farmers?

Intensive agriculture in Europe has come more and more under pressure as societies expect the adoption of more sustainable production practices by farmers. As sustainability has increasingly become an imperative for the AgFood value chains in Europe, the issue of the reduction of greenhouse emissions has advanced to a high priority topic for most food and retail companies (e.g., [FG09]).

Carbon Certificate Trading in the European Union

In the negotiations and debates leading up to the Copenhagen Accord, there has been growing emphasis on carbon credits for agriculture and the inclusion of soil carbon sequestration into the Clean Development Mechanism (CDM) and other mechanisms including REDD. Soil carbon sequestration has so far been explicitly excluded from the CDM under the Kyoto Protocol, because of major uncertainties in measuring and verifying the permanence of soil carbon stores. But there is now a major push, by agribusiness, the FAO and some governments to change this. A possible way forward might be the compensation through an increase of SOC and the trade of carbon certificates from agriculture to other industries. The EU-Emission-Trading-System (EU ETS) is "the world's first international emissions trading system" [EC18] aiming on the reduction of GHG emissions and creating an international trading system for CO_2 emissions. However, carbon certificates from agriculture or soils have not been introduced to the ETS yet, though this is currently discussed in the context of the implementation of the Farm to Fork Strategy as part of the Green Deal [EC19]. Despite this fact, currently various agri-businesses are lining up to create an informal market for carbon certificates from agriculture and to put secondary standards in place. The number of businesses investigating the potential of trading carbon certificates for carbon sequestration in agricultural soils has been steadily increasing. Substantial funds have been raised by new pioneering organizations (e.g., $850 M by Indigo—agriculture [Bui21]). This raises the question what these players expect as far as the development of this future market is concerned.

Status Quo of Carbon Trading Initiatives in Agriculture

Thus far, certification of SOC buildup through agricultural management practices has been managed on a voluntary and private basis: Companies generating and selling certificates offer programs to growers, mostly focusing on growers cultivating cropland. There are various individual methodologies and systems in place on how to certify and reward carbon sequestration. Some standards, such as for example the "verified carbon standard" provided by Verra [Ver21] and used for example by Indigo agriculture, an internationally active certifying company, are more commonly known than others. Despite different approaches to methodologies, programs by different providers have some principles in common:

- The increase of SOC in comparison to a baseline is certified and the grower is rewarded for an improvement compared to former actions. Therefore, additionality is the key requirement of the certifiability of SOC in soils.
- Certain measures, which must be additional to former management strategies, have to be applied in order to increase SOC in the soil. The portfolio of the accepted measures, however, might vary between programs.
- The buildup of SOC is slow and reversible. Therefore, improved measures, as described earlier in this article, must be implemented for a minimum term. To ensure that, the farmer is not rewarded with the entire share at the beginning of the program but receives the total compensation after finishing the individually defined amount of years depending on the program.

Private standards implemented by multinational food companies and agribusinesses as well as political initiatives such as the "Green Deal" or "Farm to Fork Strategy" of the European Union currently create a "climate" which encourages more and more commercial organizations to bet on the further development and expansion of the market for carbon certificates from agriculture. Accordingly, numerous market entrants such as Indigo agriculture, Truterra, Positerra, Carbocert, and others strongly invest into the business model. At present, the different players have their own strategies and certification approaches in place, without a clear move towards standardization or consolidation. Stakeholders from these different organizations state, however, that scaling must happen fast to allow the different players to capture sufficient market share to justify their substantial investments.

To date, the different organizations pursue different approaches to carbon sequestration, certification, and reward. Besides the increase of SOC, e.g., the use of biochar as input for soils is considered an approach towards increasing carbon sequestration of soils. The use of biochar can be an irreversible sequestration and has positive side effects such as improving water holding capacity (e.g., [LWM+20]) but is not yet certifiable in the programs of the beforementioned stakeholders. In accordance with the "additionality" criterion, the organizations actively recruit farmers who are willing to convert to regenerative practices as already practicing ones are out of scope as far as carbon sequestration in their soils is concerned. All organizations shoot for a long-term commitment by participating farmers as the benefits of regenerative agricultural practices will only materialize after years of adoption. Considering the agronomic strategies pursued by the different organizations, there is a large body of diversity as increasing SOC is a complex and multifactorial task, requiring locally and regionally adapted strategies [GHA+21]. They further use different approaches to measuring the buildup of soil organic matter compared to the baseline at start of the adoption of regenerative practices. The most accurate but also most expensive option is based on soil sampling and includes a minimum of two soil analyses before implementing the program and after 3–5 years, depending on the program. Another option is the use of models, a more cost-effective option providing less accuracy and safety depending on model and area. Sometimes, combinations of soil sampling and modeling are used.

6.5.3 Digitization as Key Enabler of Carbon Trading in Agriculture

The task of measuring and/or modeling SOC buildup in soils as a result of regenerative practices and the subsequent certification of the effect is complex and requires the collection and processing of numerous data from the farm management information system [FCS+15], dedicated databases, and downstream operations. Furthermore, the identification of suitable areas in the agricultural landscape to convert to regenerative practices and the best management practice in order to combine efficient and productivity-oriented strategies with the carbon sequestration, e.g., satellite-based soil and biomass maps, can be of great help [GHA+21]. Further, the collected and processed data needs to be managed and secured to make certificates safe and at the same time to make the entire certification process scalable. The collected data can be used e.g., to automatically determine baseline emissions and help to implement individual strategies for regenerative practices anticipated to increase the SOC accumulation in the soil. Benchmarking is made possible only through digital tools. Moreover, modeling of the plant/soil system is data intensive (e.g., [BWP+20]) and cannot be done realistically without the support of digital tools. The documentation around the implementation of regenerative practices and their impact on SOC need to be simplified and automated to the extent possible in order not to overwhelm the farm operation and the players involved in the downstream certification processes. Lastly, digitization such as block chain technology can create transparency and credibility to the entire process, which is the key prerequisite to allow for the transition from a rather informal to an established and predictable marketplace in the nearest future.

6.5.4 Conclusion

Taken together, as outlined in the other sections of this book, Digital Farming strategies can not only support farms in using their resources more efficiently (e.g., [FCS+15]) but also to become the catalyst of new business models and to create transparency and credibility in a complex value chain approach such as certification of "carbon farming". Even though there is no scientific consensus on the effectiveness of regenerative practices on SOC accumulation and thus on the mitigation of climate change, there is high likelihood that multi stakeholder and governmental initiatives will put the topic on their agenda. As a result, building conventions and standards in the accounting for "carbon farming" practices could happen fast, e.g., under the roof of the ETS. Against this background, there is a realistic chance that "carbon farming" as a new business opportunity for European farmers will come of age rather sooner than later.

References

[AE08] Angers, D. A., and Eriksen-Hamel, N.S. 2008. Full-inversion tillage and organic carbon distribution in soil profiles: A meta-analysis. *Soil Science Society of America Journal* 72:1370–1374. ▶ https://doi.org/10.2136/sssaj2007.0342.

[ANH15] Altieri, M. A., C. I. Nicholls, A. Henao, and M. A. Lana. 2015. Agroecology and the design of climate change-resilient farming systems. *Agronomy for Sustainable Development* 35:869–890.

[ARP+17] Albrecht, E., T. Reinsch, A. Poyda, F. Taube, and C. Henning. 2017. Klimaschutz durch Wiedervernässung von Niedermoorböden: Wohlfahrtseffekte am Beispiel der Eider-Treene-Region in Schleswig-Holstein. *Berichte über Landwirtschaft* 95(3): ISSN 2196–5099.

[BA20] Bruno, Basso, and John Antle. 2020. *Digital agriculture to design sustainable agricultural systems.* ▶ https://doi.org/10.1038/s41893-020-0510-0. Accessed 13 March 2022.

[Bas20] BASF SE. 2020. *BASF, Boortmalt and SAI Platform collaborate for sustainable barley production.* Press Release. ▶ https://www.basf.com/global/en/media/news-releases/2020/11/p-20-369.html. Accessed 13 March 2022.

[BLB+05] Bellamy, Pat, Peter Loveland, R. Bradley, R. Lark, and Guy Kirk. 2005. Carbon losses from all soils across England and Wales, 1978–2003. *Nature* 437:245–258. ▶ https://doi.org/10.1038/nature04038.

[Bok20] Bockholt, K. 2020. *10 Cent mehr für regionales Lerchenbrot.* ▶ https://www.agrarheute.com/pflanze/getreide/10-cent-mehr-fuer-regionales-lerchenbrot-imagegewinn-plus-600-euroha-567818. Accessed 13 March 2022.

[BOV07] Baker, J. M., T. E. Ochsner, R. T. Venterea, and T. J. Griffis. 2007. Tillage and soil carbon sequestration – What do we really know? *Agriculture, Ecosystems and Environment* 118:1–5.

[BRA15] Francesco Braga. 2015. The sustainable agriculture initiative platform: The first 10 years. *Journal on Chain and Network Science.* ▶ http://www.wageningenacademic.com/doi/pdf/10.3920/JCNS2014.x015. Accessed 13 March 2022.

[BTV+20] Biernat, Lars, Friedhelm Taube, Iris Vogeler, Thorsten Reinsch, Christof Kluß, and Ralf Loges. 2020. Is organic agriculture in line with the EU-Nitrate directive? On-farm nitrate leaching from organic and conventional arable crop rotations. *Agriculture, Ecosystems & Environment* 298:106964. ▶ https://doi.org/ggt6s6.

[Bui21] Business Wire. ▶ https://www.businesswire.com/news/home/20200106005370/en/Indigo-Closes-200M-Financing-to-Support-Continued-Growth-of-Its-Platforms-Including-Indigo-Grain-Marketplace-and-Indigo-Carbon.. Accessed 7 July 2021.

[BWP+20] Bottcher, U., W. Weymann, J. W. M. Pullens, J. E. Olesen, and H. Kage. 2020. Development and evaluation of HUME-OSR: A dynamic crop growth model for winter oilseed rape. *Field Crops Research* 246:107679.

[CBB+09] Churkina, G., V. Brovkin, W. von Bloh, K. Trusilova, F. Jung, and F. Dentener. 2009. Synergy of rising nitrogen depositions and atmospheric CO_2 on land carbon uptake moderately offsets global warming. *Global Biogeochemical Cycles* 23. ▶ https://doi.org/10.1029/2008GB003291.

[CDH+08] Cdconant, R. T., R. A. Drijber, M. L. Haddix, W. J. Parton, E. A. Paul, A. F. Plante, J. Six, and J. M. Steinweg. 2008. Sensitivity of organic matter decomposition to warming varies with its quality. *Global Change Biology* 14:868–877. ▶ https://doi.org/10.1111/j.1365-2486.2008.01541.x.

[CFA19] Cool Farm Alliance. 2019. ▶ https://coolfarmtool.org/cool-farm-alliance/. Accessed 13 March 2022.

[Chu13] Churkina, G. 2013. An introduction to carbon cycle science. In *Land use and the carbon cycle: Advances in integrated science, management, and policy*, Eds. D. Brown, D. Robinson, N. French, and B. Reed, pp. 24–51. Cambridge: Cambridge University Press. ▶ https://doi.org/10.1017/CBO9780511894824.004.

[COS17] COSA. 2020. *The committee on sustainability assessment, simpler resilience measurement: Tools to diagnose and improve how households fare in difficult circumstances from conflict to climate change.* ▶ https://thecosa.org/wp-content/uploads/2018/04/COSA-FORD-Simpler-Resilience-Measurement-Full-20180413.pdf. Accessed 13 March 2022.

[DP04] Dobbs, T. L., and J. N. Pretty. 2004. Agri-environmental stewardship schemes and 'multifunctionality.' *Review of Agricultural Economics* 26(2):220–237.

[DVL20] Deutscher Verband für Landschaftspflege (DVL) e. V. Ed. 2020. Public goods bonus. A concept for the effective remuneration of agricultural environmental and climate protection services within the eco-schemes of the EU Common Agricultural Policy (CAP) beyond 2020, *Developed in cooperation with agriculture, science and administration DVL, Ansbach.* ▶ https://www.dvl.org/uploads/tx_ttproducts/datasheet/DVL-Publication-EN_Public_goods_bonus.pdf.

[EC18] European Commission. 2021. ▶ https://ec.europa.eu/clima/policies/ets_en. Accessed 6 July 2021.

[EC19] European Commission. 2021. ▶ https://ec.europa.eu/info/strategy/priorities-2019-2024/european-green-deal_en. Accessed 6 July 2021.

[EGM+07] Esperschütz, J., A. Gattinger, P. Mäder, M. Schloter, and A. Fließbach. 2007. Response of soil microbial biomass and community structures to conventional and organic farming systems under identical crop rotations. *FEMS Microbiology Ecology* 61:26–37. ▶ https://doi.org/10.1111/j.1574-6941.2007.00318.x.

[EGP08] Enfors, E. I., L. J. Gordon, and G. D. Peterson. 2008. Making investments in dryland development work: Participatory scenario planning in the Makanya catchment. *Tanzania. Ecology and Society* 13:42.

[FAO14] Food and Agriculture Organization of the United Nations (FAO). 2014. Sustainability Assessment of Food and Agriculture systems (SAFA) Guidelines. ISBN 978-92-5-108485-4.

[FAO98] Food and Agriculture Organization of the United Nations (FAO). 1998. *The state of food and agriculture.* AGRIS:E16-E80a, ISBN 9251042004.

[FCS+15] Fountas S., G. Carli, C. G. Sørensen, Z. Tsiropoulos, C. Cavalaris, A. Vatsanidou, B. Liakos, M. Canavari, J. Wiebensohn, and B. Tisserye. 2005. Farm management information systems: Current situation and future perspectives. *Computers and Electronics in Agriculture* 115:40–50.

[FFV14] Frank, M., K. Fischer, and D. Voeste. 2014. BASF: Measurability of shared value creation in agriculture. In *CSR und value chain management,* Ed. Michael D'Heur, pp. 217–236. SpringerGabler.

[FG09] Flint, D. J., and S. L. Golicic. 2009. Searching for competitive advantage through sustainability: A qualitative study in the New Zealand wine industry. *International Journal of Physical Distribution & Logistics Management* 39 (10):841–860. ▶ https://doi.org/10.1108/09600030911011441.

[FIE16] FTM. 2016. *Field to market: The alliance for sustainable agriculture, environmental and socioeconomic indicators for measuring outcomes of on farm agricultural production in the United States (FTM 2016 Third Edition).* ▶ https://fieldtomarket.org/national-indicators-report-2016/report-downloads/. Accessed 13 March 2022.

[FIE18] *Field to Market: The Alliance for Sustainable Agriculture. 2018. Fieldprint Platform.* ▶ https://calculator.fieldtomarket.org/. Accessed 13 March 2022.

[FIE19] Field to Market: The Alliance for Sustainable Agriculture. 2019. *Field to market and SAI platform announce first use of joint equivalency module by leading food companies Barry Callebaut and Unilever.* ▶ https://www.duurzaam-ondernemen.nl/field-to-market-and-sai-platform-announce-first-use-of-joint-equivalency-module-by-leading-food-companies-barry-callebaut-and-unilever/. Accessed 13 March 2022.

[FOG+07] Fließbach, Andreas, Hans-Rudolf Oberholzer, Lucie Gunst, and Paul Mäder. 2007. Soil organic matter and biological soil quality indicators after 21 years of organic and conventional farming. *Agriculture, Ecosystems & Environment* 118:1–4.

[FSG+14] Markus Frank, Peter Saling, Martjn. Gipmans, and Jan Schöneboom. 2014. Life cycle assessment towards a sustainable food supply – A review on BASF's strategy. *Proceedings of the 9th international conference on life cycle assessment in the agri-food sector.*

[FST+15] [FST+15] Fountas, G. C., C. G. Sørensen, Z. Tsiropoulos, C. Cavalaris, A. Vatsanidou, B. Liakos, M. Canavari, J. Wiebensohn, and B. Tisserye. 2015. Farm management information systems: Current situation and future perspectives. *Computers and Electronics in Agriculture* 115:40–50.

[GAE+03][GAE+03] James, N., Galloway, John D., Aber, Jan Willem Erisman, Sybil P. Seitzinger, Robert W. Howarth, Ellis B. Cowling, and B. Jack Cosby. 2003. The nitrogen cascade. *BioScience* 53 (4):341–356.

[GBC+10] Godfray H. C. J., Beddington J. R., Crute I. R., Haddad L., Lawrence L., Muir J. F., Pretty J., Robinson S., Thomas S. M., and Toulmin, C. 2010. Food security: The challenge of feeding 9 billion people. *Science* 327:812–818.

[GHA+21] Giller, K. E., Hijbeek, R., Andersson, J. A., and Sumberg, J. 2021. Regenerative agriculture: An agronomic perspective. *Outlook on Agriculture* 50(1):13–25. ▶ https://doi.org/10.1177/0030727021998063.

[GLE+96] Gregorich, E. G., B. C. Liang, B. H. Ellert, and Drury, C. F. 1996. Fertilization effects on soil organic matter turnover and corn residue C storage. *Soil Science Society of America Journal* 60:472–476. ▶ https://doi.org/10.2136/sssaj1996.03615995006000020019x.

[God16] GODAN. 2016. *A global data ecosystem for agriculture and food.* ▶ https://www.godan.info/

sites/default/files/documents/Godan_Global_Data_Ecosystem_Publication_lowres.pdf. Accessed 28 Sep 2020.

[GPG08] Gordon, L. J., G. D. Peterson, and E. M. Bennett. 2008. Agricultural modifications of hydrological flows create ecological surprises. *Trends in Ecology & Evolution* 23:211–219.

[GSN12+] Giovannucci, Daniele, Sara Scherr, Charlotte Hebebrand, Julie Shapiro, Jeffrey Milder, and Keith Wheeler. 2020. *Food and agriculture: The future of sustainability. A strategic input to the sustainable development in the 21st Century (SD21) project.* ▶ https://sustainabledevelopment.un.org/content/documents/1443sd21brief.pdf. Accessed 13 March 2022.

[GVM+07] Gál, Anita, Tony J. Vyn, Erika Michéli, Eileen J. Kladivko, and William W. McFee. 2019. Soil carbon and nitrogen accumulation with long-term no-till versus moldboard plowing overestimated with tilled-zone sampling depths, Soil and Tillage. *Research* 96(1–2).

[HAC07] Huggins, D. R., R. R. Allmaras, C. E. Clapp, J. A. Lamb, and G. W. Randall. 2007. Corn-soybean sequence and tillage effects on soil carbon dynamics and storage. *Soil Science Society of America Journal* 71:145–154. ▶ https://doi.org/10.2136/sssaj2005.0231.

[HCD+14] Harvey, C. A., M. Chacón, C.I. Donatti, E. Garen, L. Hannah, A. Andrade, L. Bede, D. Brown, A. Calle, J. Chará, C. Clement, E. Gray, M. H. Hoang, P. Minang, A. M. Rodríguez, C. Seeberg-Elverfeldt, B. Semroc, S. Shames, S. Smukler, E. Somarriba, E. Torquebiau, J. van Etten, and E. Wollenberg. 2014. Climate-smart landscapes: Opportunities and challenges for integrating adaptation, Climate-smart landscapes: Opportunities and challenges for integrating adaptation and mitigation in tropical agriculture. *Conservation Letters* 7 (2):77–90.

[HM11] Horlings, L. G., and T. K. Marsden. 2011. Towards the real green revolution? Exploring the conceptual dimensions of a new ecological modernisation of agriculture that could 'feed the world'. *Global Environmental Change* 21:441–452.

[Iaa19] IAASTD. 2019. *International assessment of agricultural knowledge, science and technology for development.* Global Report. Washington, US: Island Press.

[IPC14] IPCC. 2014. *Climate change 2014: Mitigation of climate change. Contribution of working group III to the fifth assessment report of the intergovernmental panel on climate change.* Cambridge, United Kingdom and New York, NY, USA: Cambridge University Press.

[JHR+05] Jones, R. J. A., R. Hiederer, E. Rusco, and L. Montanarella. 2005. Estimating organic carbon in the soils of Europe for policy support. *European Journal of Soil Science* 56:655–671. ▶ https://doi.org/10.1111/j.1365-2389.2005.00728.x.

[JMC+05] Jones, C., C. McConnell, K. Coleman, P. Cox, P. Falloon, D. Jenkinson, and D. Powlson. 2005. Global climate change and soil carbon stocks; predictions from two contrasting models for the turnover of organic carbon in soil. *Global Change Biology* 11:154–166. ▶ https://doi.org/10.1111/j.1365-2486.2004.00885.x.

[KEY09] Keystone Center. 2009. *Field to market: Environmental resource indicators report, environmental resource indicators for measuring outcomes of on-farm agricultural production in the United States (NIR- 2009 First Report).* ▶ https://fieldtomarket.org/national-indicators-report-2016/report-downloads/. Accessed 13 March 2022.

[KEY12] Keystone Center. 2012 V2. *Field to market: Environmental and socioeconomic indicators for measuring outcomes of on-farm agricultural production in the United States,* (NIR-2012 Second Report Version 2). ▶ https://fieldtomarket.org/national-indicators-report-2016/report-downloads/. Accessed 15 July 2020.

[KHS+19] Konefal, J., Hatanaka, M., Strube, J., Glenna, L., and Conner, D. 2019. Sustainability assemblages: From metrics development to metrics implementation in United States agriculture. *Journal of Rural Studies.* ▶ https://doi.org/10.1016/j.jrurstud.2019.10.023.

[Lal07] Lal, R. 2007. Soil science and the carbon civilization. *Soil Science Society of America Journal* 71:1425–1437. ▶ https://doi.org/10.2136/sssaj2007.0001

[Lal14] Lal, R. 2014. Abating climate change and feeding the world through soil carbon sequestration. In *Soil as world heritage,* Ed. Dent D. Dordrecht: Springer. ▶ https://doi.org/10.1007/978-94-007-6187-2_47.

[LKJ07] Linton, J. D., R. Klassen, and V. Jayaramanet. 2007. Sustainable supply chains: An introduction. *Journal of Operations Management* 25:1075–1082.

[LMK+20] Loges, Ralf, Sabine Mues, Christof Kluß, Carsten S. Malisch, Cecilia Loza, Arne Poyda, Thorsten Reinsch, and Friedhelm Taube. 2020. Dairy cows back to arable regions? Grazing leys for eco-efficient milk production systems. *Grassland Science in Europe* 25:400–402. ▶ https://www.europeangrassland.org/fileadmin/documents/Infos/Printed_Matter/Proceedings/EGF2020.pdf.

[LRK+20] Lorenz, Heike, Thorsten Reinsch, Christof Kluß, Friedhelm Taube, and Ralf Loges. 2020. Does the admixture of forage herbs affect the yield performance, yield stability and forage quality of a grass clover ley? *Sustainability* 12:5842. ghcqfq.

[Luc16] Luck, J. D. 2016. Precision ag data usage: Current trends and future opportunities. *Resource Magazine* 23(6):18–19.

[LWM+20] Lefebvre, D., A. Williams, and J. Meersmans, et al. 2020. Modelling the potential for soil carbon sequestration using biochar from sugarcane residues in Brazil. *Scientific Reports* 10:19479. ▶ https://doi.org/10.1038/s41598-020-76470-y.

[MAC90] Roderick John MacRae. 1990. Policies, programs, and regulations to support the transition to sustainable agriculture in Canada. *American Journal of Alternative Agriculture* 5(2):76–92. ▶ https://doi.org/10.1017/S0889189300003325.

[MB20] Maixner, Ed, and Philip Brasher. 2020. *Carbon markets lure farmers, but will benefits be enough to hook them?* ▶ https://www.agri-pulse.com/articles/14880-carbon-markets-lure-farmers-but-are-benefits-enough-to-hook-them. Accessed 12 Feb 2021.

[MCS+14] Meier, Toni, Olaf Christen, Edmund Semler, Gerhard Jahreis, Lieske Voget-Kleschin, Alexander Schrode, and Martina Artmann. 2014. Balancing virtual land imports by a shift in the diet. Using a land balance approach to assess the sustainability of food consumption. Germany as an example. *Appetite* 74:20–34. ▶ https://doi.org/f5stw2.

[NDT17] Neumann, Helge, Uwe Dierking, and Friedhelm Taube. 2017. Erprobung und Evaluierung eines neuen Verfahrens für die Bewertung und finanzielle Honorierung der Biodiversitäts-, Klima- und Wasserschutzleistungen landwirtschaftlicher Betriebe („Gemeinwohlprämie"). *Berichte über Landwirtschaft - Zeitschrift für Agrarpolitik und Landwirtschaft* 95. ▶ https://doi.org/gg5swt.

[NHD+11] Nemecek T., O. Huguenin-Elie, D. Dubois, G. Gaillard, B. Schaller, and A. Chervet. 2011. Life cycle assessment of Swiss farming systems: II. *Extensive and Intensive production, Agricultural Systems* 104(3):233–245.

[Ope20a] OpenTEAM. 2020. ▶ https://openteam.community. Accessed 13 March 2022.

[OPE20b] Oppermann, Rainer, Sonja C. Pfister, and Anja Eirich. 2020. *Sicherung der Biodiversität in der Agrarlandschaft Quantifizierung des Maßnahmenbedarfs und Empfehlungen zur Umsetzung*, S. 191. Institut für Agrarökologie und Biodiversität (IFAB). ISBN 978-3-00-066368-0.

[PB18] Pretty, J., and Z. Pervez Bharucha. 2018. *Sustainable intensification of agriculture. Greening the world's food economy*. Routledge, Abingdon, UK: Earthscan.

[PHB12] Poetz, Katharina, Rainer Haas, and Michaela Balzarova. 2012. Emerging strategic corporate social responsibility partnership initiatives in agribusiness: The case of the sustainable agriculture initiative. *Journal on Chain and Network Science*. Wageningen Academic Publishers. ▶ https://doi.org/10.3920/JCNS2012.x010. Accessed 6 June 2020.

[Pet09] Peterson, H. C. 2009. Transformational supply chains and the "wicked problem" of sustainability: Aligning knowledge, innovation, entrepreneurship, and leadership. *Journal on Chain and Network Science* 9:71–82.

[Pie21] Piepenbrock. 2021. ▶ https://f3.de/food/ruckverfolgbarkeit-als-produktstory-1264.html. Accessed 14 June 2021.

[PRK+16] Poyda, Arne, Thorsten Reinsch, Christof Kluß, Ralf Loges, and Friedhelm Taube. 2016. Greenhouse gas emissions from fen soils used for forage production in northern Germany. *Biogeosciences* 13:5221–5244. ▶ https://doi.org/gcc4t7.

[PWG11] Powlson, D. S., A. P. Whitmore, and K. W. T. Goulding. 2011. Soil carbon sequestration to mitigate climate change: A critical re-examination to identify the true and the false. *European Journal of Soil Science* 62:42–55.

[RBS+17] Röös, Elin, Bojana Bajželj, Pete Smith, Mikaela Patel, David Little, and Tara Garnett. 2017. Greedy or needy? Land use and climate impacts of food in 2050 under different livestock futures. *Global Environmental Change* 47:1–12. ▶ https://doi.org/gcr65w.

[RCL+09] Russell, A. E., C. A. Cambardella, D. A. Laird, D. B. Jaynes, and D. W. Meek. 2009. Nitrogen fertilizer effects on soil carbon balances in midwestern U.S. agricultural systems. *Ecological Applications* 19 (5):1102–1113. ▶ https://doi.org/10.1890/07-1919.1. PMID: 19688919.

[RLV+] Reinsch, Thorsten, Cecilia Loza, Iris Vogeler, Christof Kluß, Ralf Loges, and Friedhelm Taube. Ecological intensification in dairy production: Towards specialised or integrated systems in northwest Europe? ▶ https://www.frontiersin.org/articles/10.3389/fsufs.2021.614348/full.

[SAI09] SAI Sustainable Agriculture Initiative Platform. 2009. *Principles & practices for Sustainable Production of Arable & Vegetable Crops Report*. ▶ https://saiplatform.org/our-work/reports-publications/principles-practices-arable-and-vegetable-crops/. Accessed 10 June 2020.

[SAI16] SAI Sustainable Agriculture Initiative Platform. 2016. *SAI Platform Annual Report*. ▶ https://saiplatform.org/wp-content/uploads/2017/05/sai-platform-annual-report-2016-2.pdf. Accessed 6 June 2020.

[SAI18a] SAI - FSA Web-App. 2018a. Video file. ▶ https://fsatool.sustainabilitymap.org/index.html#!/home. Accessed 13 March 2022.

[SAI18b] SAI Sustainable Agriculture Initiative Platform. 2018. *Donana Berry Project. SAI PLATFORM 6 Avenue Jules Crosnier* 1206 Geneva, Switzerland.

[SAI19a] SAI Sustainable Agriculture Initiative Platform. 2019a. *SAIRISI sustainability and collaboration across the value Chain.* ▶ https://saiplatform.org/?q=SAIRISI%3A+across+value+chain. Accessed 14 June 2020.

[SAI19b] SAI Sustainable Agriculture Initiative Platform. 2019. *European Sugar Beet Project. SAI PLATFORM 6 Avenue Jules Crosnier* 1206 Geneva, Switzerland.

[SBK+09] Senthilkumar, S., B. Basso, A. N. Kravchenko, and G. P. Robertson. 2009. Contemporary evidence of soil carbon loss in the U.S. Corn Belt.. *Soil Science Society of America Journal* 73:2078–2086. ▶ https://doi.org/10.2136/sssaj2009.0044.

[SCP+02] Six, J., R. T. Conant, E. A. Paul, et al. 2002. Stabilization mechanisms of soil organic matter: Implications for C-saturation of soils. *Plant and Soil* 241:155–176. ▶ https://doi.org/10.1023/A:1016125726789.

[SFS+15] Schoeneboom, J., M. Frank, J. Spencer, and P. Saling. 2015. *Enabling farmers to reduce their impact and to show it by Life Cycle Management systems. Abstract 481, Life Cycle Management Conference*, Bordeaux, 2015.

[SFV+12] Saling, Peter, Markus Frank, Dirk Voeste, Martijn Gipmans, Jan Schoeneboom, and Richard Gelder. 2012. *AgBalance – holistic sustainability assessment of agricultural Production.*

[SHE+11] Sutton, Mark A., Clare M. Howard, Jan Willem Erisman, William J. Bealey, Gilles Billen, Albert Bleeker, Alexander F. Bouwman, Peringe Grennfelt, Hans van Grinsven, and Bruna Grizzetti. 2011. The challenge to integrate nitrogen science and policies: The European Nitrogen Assessment approach. In *The European nitrogen assessment*, Eds. Mark A. Sutton, Clare M. Howard, Jan Willem Erisman, William J. Bealey, Gilles Billen, Albert Bleeker, Alexander F. Bouwman, Peringe Grennfelt, Hans van Grinsven, and Bruna Grizzetti, pp. 82–96. Cambridge: Cambridge University Press. ▶ https://doi.org/b2kc78.

[SLK+15] Srinivasarao, C., R. Lal, S. Kundu, and P. Thakur. 2015. Conservation agriculture and soil carbon sequestration. In *Conservation agriculture*, Eds. M. Farooq and K. Siddique. Cham: Springer. ▶ https://doi.org/10.1007/978-3-319-11620-4_19.

[SR17] Seufert, Verena, and Navin Ramankutty. 2017. Many shades of gray—The context-dependent performance of organic agriculture. *Science Advances* 3:e1602638. ▶ https://doi.org/ggcvf7.

[Syn15] Syngenta. 2015. *The good growth plan progress report 2015.* ▶ https://www.syngenta.com/sites/syngenta/files/presentation-and-publication/updated/thegoodgrowthplanprogressreport2015/Syngenta-The-Good-Growth-Plan-Progress-Report-2015-online-EN.PDF. Accessed 20 Feb 2021.

[TBH+11] Tilman D., Balzer, C., Hill, J., and Befort, B. I. 2011. Global food demand and the sustainable intensification of agriculture. *PNAS* 108:20260–20264.

[Tea07] Teasdale, J. R. 2007. Strategies for soil conservation in no-tillage and organic farming systems. *Journal of Soil and Water Conservation* 62:144A–147A.

[Tit14] Tittonell, Pablo. 2014. Ecological intensification of agriculture—sustainable by nature. *Current Opinion in Environmental Sustainability* 8:53–61. https://doi.org/ggmkhm.

[Toe16] Toensmeier, E. 2016. *The carbon farming solution*, 1 ed. Chelsea Green Publishing, White River Junction.

[TRS09] The Royal Society. 2009. *Reaping the benefits: Science and the sustainable intensification of global agriculture.* London, UK.

[TVK+20] Taube, Friedhelm, Iris Vogeler, Christof Kluß, Antje Herrmann, Mario Hasler, Jürgen Rath, Ralf Loges, and Carsten S. Malisch. 2020. Yield progress in forage maize in NW Europe—Breeding progress or climate change effects? *Frontiers in Plant Science* 11:16. ▶ https://doi.org/gg79j2.

[Tyl13] Tylianakis, J. M. 2013. The global plight of pollinators. *Science* 339:1532–1533.

[Uni13] Unilever. 2013. *The sustainable source.* Issue 4, February 2013. ▶ https://www.unilever.com/about/suppliers-centre/sustainable-sourcing-suppliers/certification-vs-self-verification/. Accessed 28 Feb 2021.

[UNI15] 2021. Transforming our world: the 2030 Agenda for Sustainable Development. United Nations General Assembly document A/RES/70/1. ▶ https://sustainabledevelopment.un.org/post2015/transformingourworld/publication. Accessed 13 March 2022.

[Uni18] Unilever. 2018. *Press Release.* ▶ https://www.unilever.com/news/press-releases/2018/unilevers-sustainable-living-plan-continues-to-fuel-growth.html.

[Uni19] Unilever. 2019. *Unilever Sustainable Living Plan 3 Year Performance Summary 2017–2019.* ▶ https://www.unilever.com/Images/uslp-3-year-performance-summary-2017-2019_tcm244-549781_en.pdf. Accessed 28 Feb 2021.

[Ver21] Verra. 2021. ▶ https://verra.org/project/vcs-program/. Accessed 7 July 2021.

[WOR87] World Commission on Environment and Development. 1987. Report of the World Commission on Environment and Development: *Our Common Future*, United Nations General Assembly document A/42/427. ▶ http://www.un-documents.net/wced-ocf.htm. Accessed 13 March 2022.

[WBAE20] WBAE, Achim Spiller, Britta Renner, Lieske Voget-Kleschin, Ulrike Arens-Azevedo, Alfons Balmann, Hans Konrad Biesalski, Regina Birner, Wolfgang Bokelmann, Olaf Christen, Matthias Gauly, Harald Grethe, Uwe Latacz-Lohmann, José Martínez, Hiltrud Nieberg, Monika Pischetsrieder, Matin Qaim, Julia C. Schmid, Friedhelm Taube, and Peter Weingarten. 2020. *"Promoting sustainability in food consumption – Developing an integrated food policy and creating fair food environments"*. Executive summary and synthesis report. Berichte über Landwirtschaft. Special Issue 233. ▶ https://doi.org/gh2zrx.

[Woi07] Axel Woitowitz. 2007. *„Auswirkungen einer Einschränkung des Verzehrs von Lebensmitteln tierischer Herkunft auf ausgewählte Nachhaltigkeitsindikatoren – dargestellt am Beispiel konventioneller und ökologischer Wirtschaftsweise."* PhD Thesis, Technical University of Munich. ▶ http://mediatum.ub.tum.de/?id=619300.

[Wra18] Wrangler. 2018. Sustainability data in the agricultural supply chain – Technical Paper No. 2. ▶ https://kontoorbrands.app.box.com/v/burden-to-benefit. Accessed 28 Feb 2021.

[WS09] Wezel, A., and V. Soldat. 2009. A quantitative and qualitative historical analysis of the scientific discipline of agroecology. *International Journal of Agricultural Sustainability* 7(1):3–18.

Summary

Jörg Dörr and Matthias Nachtmann

Contents

7.1 Key Insights from the Experts Views – 386

7.2 Remaining Challenges and Vision for the Future of the Digital Agricultural Ecosystem – 391

© The Author(s), under exclusive license to Springer-Verlag GmbH, DE, part of Springer Nature 2022
J. Dörr and M. Nachtmann (eds.), *Handbook Digital Farming*,
https://doi.org/10.1007/978-3-662-64378-5_7

7.1 Key Insights from the Experts Views

In this book, renowned experts gave us their views on the digital transformation of the agricultural sector from various perspectives. In the following, we summarize the most important insights into the current state of practice and the state of the art that we have gained from each chapter.

In ▶ Chap. 1, we looked at the current challenges of the digital transformation of the agricultural sector. From a farming perspective, we learned that a number of issues, such as environmental concerns, limited resources, climate change, public perception of farming, agricultural policies, labor availability, an overwhelming amount of new technology, and price volatility, are making the work of farmers and the profitability of farming increasingly difficult. Against this background, options to address these challenges were discussed, such as massive growth, differentiation, exiting, and relocating farming operations. The fast development and adoption of new digital technologies and systems can create a short-term advantage. Digital technologies are expected to assist farmers to better manage growing demands for efficiency, precision, quality, sustainability, and bureaucracy, ultimately helping them to stay in business. In addition, digitalization enables new service offerings and business models. From a sustainability perspective, the role of agriculture in the UN Sustainable Development Goals (SDGs) was discussed, such as its role in ending hunger, promoting human health, and also as a vital source of income. In this context, we outlined why data and technology are essential for achieving the SDGs. We then delved deeper into the impact of agriculture on the environment. The EU Commission's "Farm to Fork" strategy builds on existing, politically defined targets for a more environmentally friendly agriculture. It has a high priority and also sets quantitative environmental targets for agriculture, such as for greenhouse gas emissions, biodiversity, air, soil, and water cleanness. This will bring about changes in the instruments used to shape agriculture, such as regulatory law, taxes, and agricultural subsidies. The digital transformation of agriculture is an instrument that can make an important contribution in this regard to reducing the environmental impact per unit of land as well as per unit of product. We took a closer look at the adoption and acceptance of Digital Farming technologies with a focus on Germany. Here, we learned from a survey that many digital technologies are used by only a small percentage of farmers, and that many of them do not plan to use them in the future. We discussed the reasons behind this, such as high investment requirements, concerns about data sovereignty, poor usability, and compatibility issues, and reflected on how these perceptions have changed over time. Beyond this, we also considered the perspective of digital policy makers at the global, European, and German policy levels and found that there is an overarching consensus on the great potential of digitally transformed agriculture, which is supported by a variety of different conceptual approaches. Finally, we learned about the current state of agricultural digital and data law, including aspects such as normative frameworks, regulations for data sovereignty and agricultural data spaces, self-regulation regarding data sets and data exchange, but also regulated use of personal and non-personal data. In this course, the chances and limits of artificial intelligence, e.g., in terms of safety and liability, were also discussed. We learned that the field of law has a significant impact on the future of Digital Farming by enabling and taming novel technologies.

In ▶ Chap. 2, we took a methodological perspective and focused on methods and frameworks that support the digital transformation of the agricultural ecosystem. We opened with a historical overview

Summary

and roadmap for Digital Farming, exploring its roots, disruptive trends in the agricultural value chain, the need to transform the business models of the upstream sector (from traditional provision of agricultural inputs to platform-based holistic recommendations and applications), and finally the role of the circular economy in this regard. From there, we highlighted key macrotrends that will impact the AgFood system in the future, dividing megatrends into sustainability-driven disruption (e.g., in-vitro meat), society-driven disruption (e.g., robots replacing farmers), and digitally driven disruption (e.g., new sensors for pest and disease control and AI). Furthermore, we discussed three key challenges: dealing with the increasing complexity of the knowledge base, dealing with new players (entering the agricultural value chain from the outside), as well as using the right strategy and identifying the necessary skills to thrive in such a complex ecosystem. From this higher-level perspective, we moved to the farm level and addressed a framework for quantifying the economic benefits that digitalization can bring to farmers, including the fundamentals of economic value creation and identifying cost structures for digital solutions. These considerations directly target one of the main barriers to the adoption of digital solutions as outlined in the beginning: the difficulty in assessing their cost–benefit ratio. However, the ability to quantify the economic benefits is not sufficient in itself to steer an effective digitalization, which is why we broadened our view to include additional influencing factors from a French farming perspective. Recommendations were made to the various stakeholders, i.e., digital tool suppliers, distributors, farmers, and influencers, on how they can take action to enable better acceptance and adoption. With new innovation opportunities in mind, we looked at a methodological tool for developing and analyzing new business models: the Business Model Canvas, which we illustrated with examples from the agricultural domain. We closed the chapter with an experience report on the role of accelerators and partnerships in agriculture. By analogy with the laws of physics and financial portfolio theory, we explored why different ways of doing business are more successful than others. We reflected on the state of technology in the food value chain today and how things need to change, e.g., by moving from stand-alone programs in one company to cross-company innovation platforms.

In ▶ Chap. 3, we took a technological perspective. We began with a look at current systems technology for automation and autonomous machines. Mechanization and automation have been and continue to be an important driver of productivity in agriculture. However, we see a paradigm shift from "bigger, faster, wider" to more sustainability, a growing role of robotics, and increasing autonomy. We reflected on the required competence portfolio of the various stakeholders in the agricultural value chain to be successful in the market. Furthermore, we learned about the technological challenges in the development of autonomous systems, e.g., in terms of sensors/actuators or image processing and AI. It was concluded that there will be a clear trend toward autonomous agricultural systems in the long term, but the rate at which farmers will adopt autonomous systems will vary widely in different markets around the world. In the area of precision farming, we reflected on a wide range of enabling technologies (including GNSS/RTK, various sensor and actuator technologies, GIS, and FMIS) and discussed various applications (such as steering aids, autonomous vehicles, task documentation, and implement control). With regard to autonomous systems, special attention was paid to one particular topic: safe object detection. We identified challenges in safe sensing of the surroundings, mapped them to the system development life cycle, and discussed initial approaches to overcoming these issues. One solution explained is the use of simulation and AI technologies to address the inherent challenge of seasonality in agriculture, where new systems can only be tested and validated during certain parts of the year. Furthermore, we looked at technologies that show promise for overcoming the lack of interoperability in the agricultural ecosystem. In this context, we highlighted reference architectures and platforms that can provide important key components for achieving interoperability. In addition to the architectural perspective, we also raised the question of how such interoperable platforms can be set up in the first place. To this end, we discussed a lean multi-stakeholder approach to ecosystem development. One important aspect in Digital Farming already mentioned is artificial intelligence, which is increasingly being promoted. We discussed where AI is being used, such as in environmental sensing, the use of semantic technologies for data exchange and shared understanding, and in the interpretation, analysis, and decision support for farmers. We learned about the different approaches to machine learning and application examples of artificially intelligent robots such as weed removal robots or robots used to collect field data. Special attention was placed on agricultural data and terminologies, as they are an important prerequisite for effective data exchange between different stakeholders. For the data landscape in the agricultural sector, several initiatives for the description and standardization of data were described, such as ISOBUS, ISOagriNet, or INSPIRE. We explored the key components of a global agricultural data space, including the FAIR principles for data exchange and the role of terminologies, vocabularies, and ontologies. When looking at geo-based data, we elaborated on its importance to the agricultural sector and what a resilient infrastructure that can provide such data might look like, using the example of the Geobox infrastructure in Rhineland-Palatinate, Germany. The Geobox Viewer was discussed as an example of a frontend that can display geo-based data from many different public information sources. We concluded the chapter with a discussion of the technological outlook. In doing so, we addressed the critical enablers for Digital Farming such as advances in sensor technology, AI trends, and communication infrastructures such as 5G, edge computing, and satellites. In addition, we discussed trends in autonomous systems, including UAVs and robots, innovative concepts such as the digital twin in conjunction with blockchain technologies, and the role of digital technologies that can catapult agricultural production into a new era of accessibility, openness, and traceability.

In ► Chap. 4, we set our attention on the agricultural perspective, including supporting services and the field crop production process. In the support services segment, we observed the need for a major shift in mindset: moving from individual product sales to holistic solution offerings. This promises to help increase adoption of Digital Farming by reducing the complexity and economic risk for farmers by bundling Digital Farming products into an integrated solution. The example "field-zone-specific inorganic fertilization" illustrated that many different products are required to provide a digital solution. In this light, the role of the "Key Account Manager Smart Farming" was described as a potential nucleus for connecting all internal and external data and competencies to develop, test, and scale such new solutions. We furthermore looked at the data exchange between different machinery brands. These data need to be seamlessly available for all fields, zones, and crops to enable specific planning, execution, and documentation. In contrast to 5–10 years ago, there are commercial offerings for this purpose today, e.g., agrirouter, DataConnect, Nevonex and JoinData. An overview of the relevant associations (including AEF, Ag-

Gateway) and research projects (including ATLAS, GAIA-X) completed the picture. Moving on to online purchasing options for farmers, we covered three types, including digital marketplaces, reverse digital marketplaces, and e-commerce, which were complemented by a deep dive into ag. supply's offerings and strategy. Pre-seasonal support measures of farmers were complemented by a discussion on agricultural insurance. Here, the increasing availability of farm- and field-specific data opens up new possibilities for additional indices and parameters, i.e., risk profiles and insurance products. This assumption was verified for the area of climate-based production risks. In considering the crop production process, we concentrated first on soil and seed management, as these initial activities in the season lay the foundation for yield levels, as well as water and input management. Digital Farming provides benefits by making soil and seed management much more field-zone-specific. On the other hand, however, this requires much more field data and effective interlinking of soil biology, chemistry, and physics. As an important aspect in this regard, we looked at nutrient supply, which is important not only for yield levels, but also for the CO_2 footprint and the reduction of ground water leaching. In the last few decades, the accuracy of nutrition has evolved from field to field-zone resolution. The main digital components highlighted are metering, spreading, determining machine settings, and GPS-based automation. The dependencies of field-zone-specific nutrient determination and supply, for both mineral and organic fertilizers, completed the picture. At the beginning of the crop production season, weed control and protection against disease and pest infestation constitute important measures. We reflected on both mechanical and chemical weed control, including the innovations of the last 20 years in sensor technology for more precision and convenience (e.g., ultrasound, GPS, and nozzle control).

The topic was closed with a short- and mid-term outlook on the use of camera systems and automation, concluding that the increasing specificity of field sensor technology and application technologies will lead to a more specific and diverse crop protection solution mix. Subsequently, we directed our attention to weather observation with public data, but also with recordings from in-field weather stations. Historically, weather stations were the earliest field sensors that provided temperature, precipitation, wind, and other important data relevant to crop management and/or irrigation. We learned about the different sensor types, data transmission technologies, and future trends such as AI-based virtual weather stations. Weather data is also used for the final process step in crop production: harvest. During harvest, the yield level is the most important measure of the success of the agronomic production season and drives key agronomic decisions. Further related topics included vehicle-based sensing, remote sensing support, and automation, as well as future developments in these areas. We also devoted attention to post-season services and bridged the gap to food chain requirements. In this context, we reflected on direct marketing as an alternative distribution channel for farmers that allows them to address end consumers directly, bypassing trade stages and achieving higher trade margins. On the downside, we learned about increasing administrative overhead (e.g., picking, delivery of bills and invoices) as well as the need for strategic decision-making (e.g., channel management). Within this scope, we highlighted software solutions and platforms and learned about case studies for end products such as bread, cheese, and wool. This included a discussion on the challenges and success factors such as location, generation, and business agility. In addition, the impact of COVID-19 on direct marketing was addressed. We then looked at the food industry and examined how digitalization is changing food sales. We observed that food processing and retail companies, as well as start-ups, are also leveraging new digital technologies, channels, and products to adapt to customer needs. Trends known from the past such as "transparency" and "direct to consumer" are being transferred to new offerings such as crowd farming, crowd butching, marketplaces, or subscription models. The topic area ended with reflections on the B2B2C model, where established players could join forces with the e-food start-up scene to create new business opportunities. We ended the chapter with an outlook on a possible future scenario where AI could be used to better plan crop supply to reduce food waste through digital optimization.

In ► Chap. 5, we took the perspective of agriculture and livestock. The chapter started with the key question of scalability of Digital Farming depending on various farming systems. Based on long years of first-hand experience, the first topic discussed was the "how" of scaling valuable solutions. To this end, two different perspectives were assumed: the product perspective and the farmer's perspective. From the product perspective, we opened with a report on the current state of production in a French cereal production system. French digital products and solutions were described along the agronomic decision cycle (observe, analyze, decide, and act). The outlook included the opportunity of compensating farmers for their agroecological performance using Digital Farming solutions. Compared to cereals, field-grown vegetables have different requirements, e.g., due to the large variety of crop segments (e.g., root, hypocotyl, leafy, and fruits vegetables) and due to marketing (e.g., value per hectare, harvest costs, labor). Along the vegetable production process (including seedling cultivation and planting, fertilization, plant protection, and irrigation up to harvesting and processing) the most important requirements were discussed and solutions were described. Emphasis was placed on the biggest challenges of the future, namely the availability of labor for heavy work, on the one hand, and the development of selective, automated solutions for the vegetable hyper-segments, on the other. Subsequently, the spotlight was placed on the digital transformation of fruit production. Particular attention was paid to the potential of Digital Farming for irrigation, crop load management, and post-harvest fruit quality management. Future digital opportunities include improved crop stress management, better shelf-life management, and thus reduced food waste (e.g., by measuring fruit quality and predicting shelf-life). We learned that the digital transformation of the wine industry is a model for authentic and sustainable agricultural products. Historically, at least Central European wineries cover viticulture, enology, and often also sales. Digital technologies support all three steps and can improve both the process and the product. Future opportunities include data-enriched products, such as using agronomic data to document the high-quality production process. Subsequently, we turned to livestock farming, in particular to the influences of digitalization on animal welfare in dairy farming. The starting point includes a description of the dairy market in Germany (Strategy 2030) and a summary of animal welfare indicators. A conclusion was that digital transformation is directly linked to the long-term success and sustainability of the dairy sector. Finally, a cross-farm comparison was made between different German agricultural regions and a Dutch farm profile. Despite the fact that the five German farmers interviewed come from very different regions (including north, east, south, and west), have different farm sizes (ranging from 100 ha family farm to 1,700 ha joint venture) and have different focus areas in farming (including arable, livestock, and contract farming), they all agreed on two main benefits of Digital Farming: simplified and automated farming processes and decision support. In this context, an overview of publicly available field-specific data in Germany was presented. The chapter ended with a vivid description of a Dutch family farm, including its historical setup, the role of precision farming, as well as sustainability today and in the future. One insight was that a large base of agricultural data (almost 50 years) has a high value for business operations. The additional data is associated with the purchase of new equipment and thus higher costs. This equipment could be financed in the future by compensation payments to farmers for green and blue infrastructure services (green: trees, hedges, etc., blue: water, rivers, ponds).

▶ Chap. 6 addressed the question of how sustainability can be operationalized with the help of Digital Farming. We looked at sustainability associations that play an important role in translating the UN Sustainable Development Goals, among others, into farming practice. This includes visioning, establishing metrics and best practices, developing assessment tools, engaging in multi-stakeholder partnerships, and providing high-quality data and platforms. Three exemplary associations were described in more detail: Field to Market—the Alliance for Sustainable Agriculture, the Sustainable Agriculture Initiative, and the Committee on Sustainability Assessment. Despite the successful establishment and growth of the associations described, significant challenges were identified, including the integration of sustainability systems and platforms. Besides the associations mentioned above, the arguably even more important driver for digitally enabled sustainable agriculture is the AgFood value chain. We looked at case studies on "sustainability assessment" from various industry segments (including AgBalance, Cool Farm Alliance, Syngenta, and Wrangler), including new business models. The conclusion was that FMIS interoperability and sustainability assessment could become an important enabler for sustainable agriculture at scale. The next core topic we looked at was ecological intensification within the context of hybrid agriculture. The core idea involves the re-integration of animal husbandry and arable farming, which combines the benefits of organic and conventional farming in an extended cropping cycle. Also in this context, the conclusion resulted that Digital Farming will play an important role in this transformation process and for the change of the given structures (e.g., spot spray). We then looked at today's highly efficient crop production practices and how they can regain their balance. Digital Farming could support this change by gathering and using knowledge about ecosystems, the impact of agricultural practices, and by helping to adapt to natural processes. This will lead to a greater number of crop types per rotation and thus a greater diversity of crops and farming systems. We concluded this topic area with a discussion on carbon farming, a future business opportunity for Digital Farming. Starting with the carbon and nitrogen cycles, the principles of carbon sequestration through regenerative agriculture were explained and the market opportunities of Digital Farming were described. In conclusion, it became clear that Digital Farming can not only support farms to use their resources more efficiently, but will also serve as a catalyst for new business models.

7.2 Remaining Challenges and Vision for the Future of the Digital Agricultural Ecosystem

In this book, we read about a large variety of mature methods and technologies and also about successful applications of digital technologies in many different farming systems and processes along the complete value chain. Digital Farming is part of the solution towards ecological intensification. All expert views confirm with rationales, insights, and use cases the importance of digital technologies to transform agriculture toward a more sustainable farming. This includes all levels: data, decision support, and automation. The specific application depends on the crop and farm-specific challenge. The second known, but surprisingly consistent conclusion is that experts see that Digital Farming can help farmers to monetize their public goods services, incl. protecting biodiversity, climate, water, soil, etc. Furthermore, there is a momentum for business models, e.g., direct marketing, regional sourcing, alternative proteins, carbon farming.

The reality shows light and shadow. Looking from a distance, especially on the agronomy perspective taken in ▶ Chap. 4 and the farming perspective taken in ▶ Chap. 5, is not it overwhelming? All key farming areas see new products, concepts, and ideas. Farmers could easily run a season with next generation seeder, fertilizer, sprayer, drones, and software. But key questions how to use these new technologies to transform the production system to the next levels are not yet answered. Questions are: How does a better crop rotation look like? How does this create value for the environment, customers, and farmers? How to evaluate this value during, after, and across seasons? How do digital products support these questions best? How can farmers still benefit when taking care for public goods? It's almost like a very big puzzle. All pieces are on the table, the corners of the puzzle are completed, but how to connect the remaining pieces toward sustainable better yields is not completely clear.

From the perspective of the editors, the solution might be an inter-disciplinary process optimization and ecosystem perspective. The rise of software in production and management is very closely linked with standardization of core and support processes to enable interoperability of systems. Translating back to farming, the change from "optimize a product" (e.g., fertilizer big bag, nitrogen sensor, fertilizer spreader, FMIS Nitrogen balance calculator) to "optimize a process output in an ecosystem" perspective (e.g., efficient use of nitrogen, total admin and operation time/costs) might give guidance. This change would automatically need to consider soil type, crop rotation, interim crop, N-balance, nitrification inhibitors, etc. But how can this be done considering that every field and every farm is different, as one of the key farming beliefs says? And the product and technology portfolio is growing faster than ever, but we have to be reminded that digitalization is not an end in itself.

Every process optimization starts with the customer and stakeholder and their needs. We learned about the importance of food safety, affordable quality, choice, and sustainability as major needs of European consumers. That actually means, consumers ask for high product and process quality. As a consequence, we need to enable farmers to plan, pursue, and document this high process and product quality in a way, food and retail brands can differentiate with European farmer products. Therefore, we see the following fields of action:

▪ **Enabling the Farmers**

A key prerequisite for successful adoption of Digital Farming technologies and solutions is the enablement of the farmers to efficiently integrate the right balance of new technologies in his farm operation, crop production processes and routines. This is a major challenge, as the "job description" of a farmer nowadays is already packed with a large variety of needed competencies, ranging from agronomic via technological to business competencies. Now digital technologies are added to this portfolio as well. We want to encourage all farmers to invest time in getting familiar with the new Digital Farming technologies appearing on the markets as they can and will have a strong influence now and in the future and will potentially transform their business. But this challenge does not only affect farmers. We want to encourage researchers, teachers, public and private consulting organizations to consider if their current portfolio in teaching and consultancy reflects the current state of the art in digital technologies. And we have to think if our traditional ways of continuous learning and technology transfer fits to the needs of the future in the agricultural domain. Maybe new ways to convey Digital Farming technologies via small digital units are needed for (continuous) education purposes combined with lowering investment entry hurdles (e.g., funding, tax, or other benefits). Definitely, this portfolio will change and grow in future. And, as usual with digital technologies: with a much shorter expiration data for current digital technology and a much higher release rate for new digital technology. If we master this challenge, we will have farmers that are able to make informed decisions when it comes to Digital Farming solutions, which will help to have an effective digital transformation of the agricultural ecosystem at the right places. As Diana Lenzi, president of the European Council of Young Farmers (CEJA), recently said in a public discussion: Smart farming needs smart farmers.

- **Show Benefits of Digital Technology with Empiricism**

As the digital transformation of the agricultural ecosystem is in the beginning, we currently have not yet achieved a comprehensive body of knowledge about empirically proven benefits and side-effects of Digital Farming solutions, especially given cross-season, multi-crop/harvest, and sustainable benefit calculations. Many experiments took place under controlled conditions and with a small selection of fields. But seen in the plethora of agronomic challenges and Digital Farming solutions around, the empirical evidence is still very limited to certain usage contexts. This means farmers and consultants do not have a solid basis to easily search for new solutions and judging the impact it has on the respective farm. This is especially problematic in the agricultural ecosystem, as farming is highly context-dependent: no two farms are identical. Digital Farming solutions must show their benefit in an environment where all farms are different, climate and weather is changing over the years, soil-types are different, just to name a few relevant parameters that influence yield, among others. In order to improve this situation, researchers and companies in all agricultural segments should aim at having a joint location for storing the empirical evidence and agree on common description scheme to make results comparable. With such a joint effort, the whole agricultural ecosystem will benefit by understanding the dependencies between agronomic challenges and impact of Digital Farming solutions much better, making the digital transformation more efficient.

- **Multi-Dimensional View on Multi-Benefits**

Reflecting on the benefits of Digital Farming solutions brings us to the next big challenge. Many times, the benefits are hard to quantify and reduced by stakeholders to just one dimension: investment in the solution vs. the direct saving, in part because we only focus on existing agricultural policy and given agricultural business models. The benefits of Digital Farming solutions should be seen in a multi-dimensional space including direct savings, impacts on sustainability, impacts on the workforce, cross-season effects, possible prerequisites or enablement of sustainable intensification and of new business models. This will be especially relevant with new CAP and the potential compensation of farmers' activities improving public goods, e.g., via environmental services for biodiversity, climate or water protection. We believe that more research and development toward the multi-dimensional quantification and dependencies of benefit and effects of Digital Farming solutions is needed. This topic is not new. Historically farms produced multiple products, e.g., animals, feed, food, and energy. Making dependencies and economics transparent across seasons and crop rotations is important to manage new on-farm value streams.

- **Interoperability of Digital Farming Products, Services, and Solutions**

A further key challenge in the digital agricultural ecosystem is the lacking interoperability of the thousands of Digital Farming solutions. Farmers have been complaining for a long time about incompatibilities, making their decisions for investing in Digital Farming solutions difficult and once acquired, making their work-processes inefficient or even ineffective. The current landscape of solutions can be characterized as many partly connected islands rather than a continent of Digital Farming solutions. As standardization takes a long time, many projects and initiatives started bottom-up to address this problem, leading potentially to various (compatible or incompatible) solution concepts. Therefore, all stakeholders in this realm should join forces and work toward a common vision and compatible concepts and exchange with the top-down approaches that currently start to have effects on the agricultural domain. One example of such a top-down approach is the initiatives on common data spaces and common service infrastructure. These approaches have the potential to be real game changers in the digital transformation of the agricultural ecosystem. One example is the GAIA-X initiative. Here, various EU member states work together toward a common agricultural data and service space, partly reusing concepts from the bottom-up initiatives and projects. Currently, it is unclear which of the concepts developed by the bottom-up and top-down initiatives will become successful. But a clear tendency toward more interoperability and a better exchange of data and services is evident. This will speed up the digital transformation and will create new business opportunities that farmers as well as all stakeholders in the value chain will benefit from.

- **Scale with Inter-Disciplinary Co-Creation with Food Industry**

The discussed marketing and food trends indicate local sourcing of food supply supports the consumer trend toward local and sustainable production (e.g., replace imported soybean or palm oil with local produce), but also decreases dependencies from world trade issues. Despite the weaker production cost perspective compared to very large farm operations in the USA, Canada, Brazil, or Ukraine, the mid- to large-acreage central European farms can differentiate with high volumes, high product quality, and high process quality. Food industry might benefit in multiple ways from Digital Farming, e.g., via sourcing efficiency, risk management, or market differentiation. Food company purchase departments can benefit from production transparency, yield level forecasts, and a more efficient collaboration with farms, or cooperatives. New crops can be grown with reduced risks, by using digitally supported crop management recommendations. New crops, old varieties, or data-enriched products provide opportunities to differentiate in the market place. The specific scope and size of such opportunities depend on local situation and needs. From a methodical perspective, such rebuilding of historic local value chains (e.g., farmer, mill, baker) can benefit from co-creation methods, connecting agronomy, sustainability, digital and food market experts developing better supply opportunities. The editors see a need for and participate themselves in open discussions to translate needs (e.g., CO_2 footprint reduction, cereal protein content, cultivation of new or old varieties) into next farming practices with support of Digital Farming.

From the viewpoint of the editors, these fields of actions are not new, but challenging nonetheless. Today's opportunity is for all stakeholders to move toward more sustainable farming, and for the need for collaboration and partnerships among stakeholders to be widely accepted, which makes us confident that these challenges will be mastered. The digital transformation of the agricultural sector has the clear potential to bring our society a more sustainable and safe food production, enable a farming that is attractive and rewarding for farmers of all sizes, and enables a whole business ecosystem of small and large companies in the whole agricultural and food value chain. The editors together with all friends of Digital Farming look forward to co-create a more sustainable farming together.

Supplementary Information

Glossary – 398

© The Editor(s) (if applicable) and The Author(s), under exclusive license to Springer-Verlag GmbH,
DE, part of Springer Nature 2022
J. Dörr and M. Nachtmann (eds) *Handbook Digital Farming*,
https://doi.org/10.1007/978-3-662-64378-5

Glossary[1]

Abiotic Non-biotic or non-living (e.g., chemical)

ADAPT Agricultural Data Application Programming Toolkit, defined by AgGateway, support of interoperability in precision and Digital Farming

AgFood Synonym for Agri-Food Sector, Agricultural Food Value Chain

AgGateway AgGateway is a global, non-profit organization with the mission to develop the resources and relationships that drive digital connectivity in global agriculture and related industries.

AgFood Market Synonym for AgFood Sector, Agricultural Food Value Chain

AgFood Sector Synonym for AgFood Market, Agricultural Food Value Chain

Agricultural Food Value Chain Synonym for AgFood Market, Agri-Food Sector

Agricultural Revolution See Agricultural 4.0

Agriculture 4.0 For our book: synonym for digital farming.

Agritechnica Forum for agricultural machinery and future crop production issues

AGROVOC A multilingual controlled vocabulary covering all areas of interest to the Food and Agriculture Organization (FAO) of the United Nations

AgInput Company Company providing production inputs for farmers, incl. seed, fertilizer, crop protection, etc.

AgChemicals Synthetic fertilizers and crop protection products

AgMachinery Company Company providing machineries supporting farmers for soil management, seeding, fertilization, irrigation, crop protection, harvesting and other related agronomic activities

Agro-Ecological Agricultural environment important to consider for more sustainable, resource-efficient, biodiversity protecting, climate saving farming practices

AgTech Company Company providing products for farmers using new technologies, incl. digital, breeding, and other technologies

Allelopathy Biological phenomenon of an organism, that produces biochemicals that influence germination, growth, survival, reproduction of other organisms

Amelioration Improvement of substantial properties of soil

Application Map Geo-referenced record of applied measures in the field

Anthropocene Geological period starting with first significant human impact on Earth's geology and ecosystems

Anthropotechnologies Study and improvements of working and living conditions

Artificial Intelligence Intelligence demonstrated by machines opposed to natural intelligence displayed by humans, examples for artificial intelligence methods are neural networks and machine learning

B2A Business to Administration

B2B Business to Business

B2B2C Business to Business to Consumer

Big Ag Very large agricultural companies

Biodegradation Decomposition of material by living organisms (e.g., microorganisms)

Block Chain One type of digital ledger technology that supports tamper-proof data storage

CAD Computer-aided Design

Carbon Cycle Biogeochemical cycle by which carbon is exchanged with biosphere, geosphere, hydrosphere, atmosphere of the Earth, relevant from a farming perspective, as certain farming measures support carbon dioxide removal from the atmosphere and sequestrate it in soils

Carbon Sequestration Capture of carbon in soil, a potentially key climate protection measure

---

[1] Many terms in this glossary are highly standardized and similar definitions can be found in (agricultural or IT) dictionaries, papers, and platforms.

Glossary

CAN Controller Area Network

Canopy Leaf surface

CF Card Compact Flash Memory Card

CGIAR Consultative Group for International Agricultural Research

Circular Economy Production and consumption philosophy, which involves sharing, leasing, reusing, repairing, refurbishing, and recycling materials and products

CO_2-equivalent Number of metric tons of CO_2 emissions with the same global warming potential as one metric ton of another greenhouse gas, e.g., methane (CH_4), nitrous oxide (N_2O)

Code of Conduct EU Code of conduct on agricultural data sharing by contractual agreement

Colostrum First milk produced by mammals

Commodity High-volume crops such as wheat, soybean, corn, and poultry meat

Contractor Business Outsourcing of single or multiple farming measures from farmer to Contractor

Controlled Traffic Farming Farming system built on permanent wheel tracks where the crop zone and traffic lanes are permanently separated. Minimizes soil compaction

Convolutional Neural Networks Class of deep neural network, most commonly used to analyze images

Crop Management Total decisions and actions to grow and harvest crops, incl. seeding, fertilization, scouting, crop protection, harvest and others

Crop Production System Methods, machinery, inputs and experience used to produce crops, incl. seasonal rotations with other crops (e.g., cereals, potato, oilseed rape)

Crop Registration Observing single or multiple crops as proxy for total fields to derive insights for crop management

Cross Hoeing Mechanical hoeing conducted in two perpendicular directions

Crowdsourcing Get information or input to task or project by engaging large number of people, either paid or unpaid, typically via the internet

CTF Controlled Traffic Farming

CVC Corporate Venture Capital

CRISPR, CRISPR-Cas Gene editing technique

CRM Customer Relationship Management

DSS Decision Support System

Deep Learning Machine learning method based on artificial neural networks

Deep Neural Network One type of artificial neural network (ANN) with multiple layers between the input and output layers

Dendrometer Tool to measure size and volume of trees

Dent Variety Grain corn is field grown corn with high soft starch content

Design Thinking Set of processes through which design concepts are developed

Digital Farming Software-supported optimization and automation of agricultural work and business processes as well as the enablement of innovative business models.

Digital Transformation Synonym for Digitalization, Digitization; Making data available in digital form; supporting work and business processes and new business models by digital technologies reaching a next level of process and product quality

Digitalization Synonym for Digital Transformation, Digitization

Digital Ecosystem A Digital Ecosystem is a socio-technical system in which companies and people cooperate who are independent but expect to gain a mutual advantage from participating. A Digital Ecosystem often has at its center a digital platform that supports this cooperation via ecosystem services.

Digitization Synonym for Digital Transformation, Digitalization

Disruption Change happening in a short time frame in contrast to slower evolutionary developments

Downstream Value chain term, downstream means a player that comes in later phases of the value chain (e.g., end customer)

DTLS Datagram Transport Layer Security

EBITDA Earnings before interest, taxes, depreciation, and amortization

Glossary

Edge System Term often used in the context of edge computing. Computation and data storage is closer to the sources of data and not an IT cloud

Ecological Intensification Strategies that rely on the application of ecological science to the study, design and management of sustainable agriculture in order to achieve higher agricultural input efficiency

Ecophysiological Crop physiology depending on environmental factors

EFDI European Forum of Deposit Insurers

EMC Electronic Mass Flow Control

EOS Earth Observation Satellite

ERP Enterprise-Resource-Planning System

ETSI European Telecommunications Standards Institute

EU Code of conduct Code of conduct on agricultural data sharing by contractual agreement, which was signed by nine organizations and associations

Eutrophication The process by which a body of water becomes enriched in dissolved nutrients (such as phosphates) that stimulate the growth of aquatic plant life usually resulting in the depletion of dissolved oxygen

Evapotranspiration Loss of water from the soil by evaporation or transpiration from plants

Experimental Field Field for testing and comparing alternative products and approaches

Extensification Reduction in inputs and accepting reduced agronomic yield

Farm Accountancy Network Farm accountancy data network (FADN) monitors farms' income and business activities

Fertigation Use of fertilizer-enriched irrigation water

F2F Farm to Fork

FMCG Fast Moving Consumer Goods

FMIS Farm Management Information System

Food Culture Includes all aspects of nutrition incl. food, dishes, table decoration, manners, and rituals

Food Fashion Includes presentation of food, categories and nutrition habits like organic, vegan, etc.

Food Safety State of food in which it does not present a health hazard when consumed

Food Security Sufficient supply of foodstuffs

Forage Food for animals

FDR Frequency Domain Reflectometry

FMCW Frequency-Modulated Continuous Wave

FPGA Field Programmable Gate Array (Programmable Integrated Circuit)

GAIA-X Initiative for the development of an efficient and competitive, secure and trustworthy federation of data infrastructure and service providers for Europe

Gas Chromatography chromatography in which the sample mixture is vaporized and injected into a stream of carrier gas (such as nitrogen or helium) moving through a column containing a stationary phase composed of a liquid or particulate solid and is separated into its component compounds according to their affinity for the stationary phase

Georeferenced Field Boundaries Coordinate system of maps or aerial image, provided, e.g., via shape file format

GIS Geographic Information System

GMO Genetically Modified Organisms

GNSS Global Navigation Satellite System

GODAN Global Open Data for Agriculture and Nutrition

GPU Graphics Processing Unit (processing unit especially design for Graphics operations that make them especially well-suited for high performance, cost-efficient processing in certain applications)

Green Revolution Agriculture 3.0 started 1960s incl. high yielding varieties, synthetic crop protection products, leading increased crop production and food security

Greenhouse Gas Gases that trap heat in the atmosphere, incl. carbon dioxide, methane, nitrous oxide

Growth Stage Describes different development stages of crops, defined in BBCH stages

Harrow Machinery equipment pulled by tractor to break the earth into small pieces ready for planting

HI-Tier Platform IT infrastructure for Program for reporting birth, movement, death, slaughter, etc., according to

Glossary

the Livestock Traffic Ordinance and for displaying animal and herd data.

Hoeing Machinery equipment pulled by tractor with a thin, flat blade set, used esp. in breaking up the soil and in weeding

Holobiont Assemblage of organisms building niches for each other and sharing an economy of nutrients and signals that impact on the functioning of the system as a whole

Homologation Clearance, approval (e.g., of crop protection products)

Hyper-local Weather Station Field- and field zone-specific weather stations

Hyperspectral Remote sensing technology incl. visible and parts of invisible wave length spectrum

Hypocotyl Means "below seed leaf," stem of germinating seedling

IACS Integrated Administration and Control System

ICT Information and communication technologies

Index Insurance Insurance that determines the payout based on a defined such as rainfall or yield index

INSPIRE INfrastructure for SPatial InfoRmation in Europe

Intercropping Multiple cropping practice that involves growing two or more crops in proximity

Interoperability Ability of a system to work with or use parts of other systems

IoT Internet of Things

ISOBUS Standardized communication protocol for the agriculture industry

Kalman Filter Estimation algorithm for compensating inaccurate and uncertain measurements

LiDAR Light Detection and Ranging (optics-based method for determining ranges by targeting objects with a laser)

LoRaWAN Network layer protocol for managing communication between gateways and devices

LPWA Low-Power Wide-Area

M2M Machine-to-machine

Macrofauna Soil organisms which are retained on a 0.5 mm sieve

Master Data Core data of a company, incl. product, customer, production and other data points

MQTT Message Queuing Telemetry Transport

NB-IoT Narrowband IoT

NDVI Normalized difference vegetation index

Newtrition New trends in nutrition

NGSI-LD Information model and API for publishing, querying and subscribing to context information

NIR Near-infrared (optical spectrum in close proximity to the infrared band between approx. 750 and 2500 nm wavelength)

NIRS Near-infrared spectroscopy

NMR Nuclear Magnetic Resonance sensor systems for determination of nutrient content

No-tillage Farming without tillage, including the use of herbicides to suppress weeds

NPK Nitrate, phosphor, potassium (major nutrients in fertilizer)

OAuth2 Authorization method for web applications, desktop applications, mobile phones, and smart devices

Oestrus Recurring period of sexual receptivity and fertility in many female mammals

Off-highway Autonomy Autonomy of vehicles not intended for road traffic

OMA Open Mobile Alliance

Ontology An ontology encompasses a representation, formal naming and definition of the categories, properties and relations between concepts, data and entities in a domain

Paludi Cultures Sustainable alternative to drainage-based agriculture, intended to maintain carbon storage in peatlands

Parallel Swathing System In-field guidance with use of GPS

Parametric Index Insurance Non-traditional insurance product that offers pre-specified payouts based upon a trigger event

Parametric Insurance Non-traditional insurance product that offers pre-specified payouts based upon a trigger event, synonym for parametric index insurance

Particulate Matter Fine dust particles

Patch Cropping Very small area in a field, e.g., grown with weeds, e.g., targeted by spot farming applications

Performance Food Food improving personal performance, incl. creativity, endurance

Phenology Models Growth models for plants

PID Controller Proportional–integral–derivative controller is one type of control loop mechanisms for controllers

PKI Public Key Infrastructure

Plot Part of field or single field trial area

Point Entry Exposure, e.g., of crop protection product on single point into soil and or surface waters

POS Point of Sales

Power harrow Non-rotating tillage for seedbed preparation

Predictive Analytics Statistical analysis of facts to make predictions about the future

Perennial Crops that—unlike annual crops—don't need to be replanted each year

Pneumatic Spreader Fertilizer spreader with pneumatic conveyance of the fertilizer

Production System Synonymous for crop and or livestock production system

PTO Power Take Off

Quantum Computing Computation exploiting the properties of quantum states

Raman Spectroscopy A spectroscopic technique, in which the Raman-spectrum is analyzed to determine the properties (e.g., structure) of the substance

Raster Format A dot matrix data structure which (in contrast to vector format) does not scale without showing artifacts

Raytracing Method for calculating the path of waves or particles through a system

REDD Reducing Emissions from Deforestation and forest Degradation

Regenerative Agriculture Conservation and rehabilitation approach to sustainable agriculture

Retrofit Hardware Technologies to be mounted later on used tractors, e.g., GPS guidance, VRA application, camera systems

RTK Real-time kinematic

SaaS Software as a Service

SAE Society of Automotive Engineers

Section Segment of the total working width of an implement

Sentinel Mission Satellite-based earth observation mission from the Copernicus Program

SC Section Control

SD Card Secure Digital Memory Card

Shape File Exemplary file format used by GIS and FMIS systems, e.g., for "georeferenced field boundaries"

Shear Force Sensor Sensor for detecting non-aligned forces, e.g., pushing a vehicle in an unintended direction

Siamese Network Artificial neural network working in tandem on two different input vectors to compute comparable output vectors

Site-Specific Management Farming practice targeting specific input measures and volumes appropriate for field zones conditions, e.g., seed, fertilizer, crop protection

Slow Food Global Organization/trend promoting authentic food, incl. regional, seasonal

Smart Contract Technology used often together with blockchain technologies to automatically execute, control or document legally relevant events and actions

SoS System of Systems, system can refer to technical or business systems

Spectrometer A scientific instrument/measurement device used to separate and measure spectral components

SPFH Self-propelled Forage Harvester

Glossary

Spot Farming Synonymous for site-specific farming, even patch to crop-specific farming

Steering Assistance System GPS-supported steering of a machinery

Strip Tillage Minimum tillage for soil protection

Stubble Cultivator Non-rotational tillage to loosen the soil and incorporate crop residues into the soil

Synthetic Biology Design and construction of new biological parts, devices, and systems

Task Controller One of the ISOBUS functionalities, e.g., for documenting the performed activities

TDR Time Domain Reflectometry

Tillering Formation of side shoots of a plant (e.g., grass) from the base of the stem

TIM Tractor Implement Management

TLS Transport Layer Security (cryptographic protocol)

TOF Time of Flight (e.g., measured by certain sensor types)

Tractive Power Measure for power of a tractor

Tramline Path tractor uses in the field

Transhumanism Philosophical movement, advocating and predicting the enhancement of the human condition for enhancing longevity, mood and cognitive abilities by sophisticated technologies

Twin Disc Spreader Fertilizer spreader with two rotating centrifugal discs

Unfair Advantage Providing an advantage to partners over the competition that could not be achieved otherwise or would be very hard and resource-intensive to achieve at this point in time by the company on its own

UAV Unmanned Aerial Vehicle (e.g., drone)

Undersown Crops Crop sown with or after the main crop so that it continues to grow after the main crop is harvested

Upstream Value chain term, upstream means focus towards supplier and materials

Vector Format Graphics format which uses connected lines and curves to form polygons and other shapes. Scales infinitely compared to raster format

Vertical Farm Practice of growing crops in vertically stacked layers

Vertical Integration Integration of value adding steps within the supply chain into one controlling organization

Virtual Product Development Practice of developing/prototyping products in a completely digital environment

Volunteer Plant Plant that grows by itself and is not planted intentionally

VRA Variable Rate Application

Printed by Printforce, the Netherlands